Finite Element and Finite Volume Methods for Heat Transfer and Fluid Dynamics

This book introduces the two most common numerical methods for heat transfer and fluid dynamics equations, using clear and accessible language. This unique approach covers all necessary mathematical preliminaries at the beginning of the book for the reader to sail smoothly through the chapters. Students will work step-by-step through the most common benchmark heat transfer and fluid dynamics problems, firmly grounding themselves in how the governing equations are discretized, how boundary conditions are imposed, and how the resulting algebraic equations are solved. Providing a detailed discussion of the discretization steps and time approximations, this graduate textbook has everything an instructor needs to prepare students for their exams and future careers. Each illustrative example shows students how to draw comparisons between the results obtained using the two numerical methods, and at the end of each chapter they can test and extend their understanding by working through the problems provided. A solutions manual is also available for instructors.

Professor J. N. Reddy is a Distinguished Professor, Regents Professor, and holder of the O'Donnell Foundation Chair IV in the Department of Mechanical Engineering at Texas A & M University. As the author of 24 textbooks and several hundred journal papers, and a highly cited researcher, Professor Reddy is internationally recognized for his research and education in applied and computational mechanics. He has won many major awards from professional societies (e.g., the S. P. Timoshenko Medal, the von Karman Medal, the von Neumann Medal, and the Gauss–Newton Medal). He is a member of the US National Academy of Engineering.

Professor N. K. Anand is a Regents Professor and James J. Cain '51 Professor III of Mechanical Engineering at Texas A & M University. He teaches and researches in the broad area of thermal sciences. Professor Anand is a recipient of the Association Former Students Distinguished Achievement Award in Teaching at college level. He is an ASME Fellow and was awarded the 2020 ASME Harry Potter Gold Medal. He developed and continues to teach a course in the application of finite volume techniques to heat transfer and fluid flow for three decades.

Dr. Pratanu Roy is a Staff Scientist at Lawrence Livermore National Laboratory (LLNL), California. Dr. Roy conducts research in computational fluid dynamics (finite volume method, multigrid methods), high-performance computing for CFD, transport in Carbon Capture and Storage (CCS), and turbulence modeling. Dr. Roy is the recipient of the 2015 American Rock Mechanics Association (ARMA) best paper award, the 2020 LLNL Physical and Life Sciences Directorate award, and in 2022 he was awarded National Nuclear Security Administration (NNSA) Defense Programs Awards of Excellence.

"I am delighted to recommend this textbook to beginners and early career researchers wanting to work in computational heat and fluid flow problems. This book is a useful tool for teaching postgraduate and senior undergraduate courses and will be an excellent addition to the bookshelves of senior researchers."

Perumal Nithiarasu, ***Swansea University***

Finite Element and Finite Volume Methods for Heat Transfer and Fluid Dynamics

J. N. REDDY
Texas A & M University

N. K. ANAND
Texas A & M University

P. ROY
Lawrence Livermore National Laboratory, California

CAMBRIDGE
UNIVERSITY PRESS

CAMBRIDGE
UNIVERSITY PRESS

Shaftesbury Road, Cambridge CB2 8EA, United Kingdom

One Liberty Plaza, 20th Floor, New York, NY 10006, USA

477 Williamstown Road, Port Melbourne, VIC 3207, Australia

314–321, 3rd Floor, Plot 3, Splendor Forum, Jasola District Centre, New Delhi – 110025, India

103 Penang Road, #05–06/07, Visioncrest Commercial, Singapore 238467

Cambridge University Press is part of Cambridge University Press & Assessment, a department of the University of Cambridge.

We share the University's mission to contribute to society through the pursuit of education, learning and research at the highest international levels of excellence.

www.cambridge.org
Information on this title: www.cambridge.org/highereducation/isbn/9781009275484
DOI: 10.1017/9781009275453

First published 2023

Printed in the United Kingdom by TJ Books Limited, Padstow Cornwall

A catalogue record for this publication is available from the British Library.

ISBN 978-1-009-27548-4 Hardback

Additional resources for this publication at www.cambridge.org/reddy-anand-roy.

Contents

Preface

Our motivation for writing this book came from the need to introduce the two most popular numerical methods - the finite element method (FEM) and the finite volume method (FVM) - as techniques for solving differential equations arising in heat transfer and fluid dynamics problems. As experienced teachers and users of FEM and FVM courses, we felt well-placed to help students acquire a unified and practical understanding of these methods, which will prepare them for future employment and study.

The book is designed for senior undergraduate and first-year graduate students who have had courses in linear algebra, differential equations, and undergraduate level heat transfer and fluid mechanics, and some programming experience. However, additional courses (or exposure to the topics covered) in heat transfer and fluid mechanics should make the student feel more comfortable with the physical examples discussed in the book.

In both the FEM and FVM, the geometric region of the problem (on which a differential equation is to be solved) is represented as a collection of subregions, called a mesh or a grid. In the FEM (and also in the so-called "control volume finite element method"[1]), these subregions are called finite elements, which have associated interpolation functions defined on each element. Then a second mesh of subregions is overlaid on the first mesh, with the objective of developing the discretized equations. In the FEM, the second mesh of finite elements is used to define the interpolation of the dependent unknown as well as to derive the finite element equations using a method of approximation (e.g., weak-form Galerkin method, least-squares method, and so on) over the element. When the two meshes are exactly the same in the FEM, it is called an isoparametric formulation.

In the FVM, the domain is divided into control volumes and typically a node is deployed at the geometric center of each control volume. Nodes are the locations at which unknowns or dependent variables are calculated. Discretization equations at each node are generated by discretizing flux balance (namely, mass, momentum, or energy) equations over each control volume that surrounds the node. In broad terms, the discretization equations at each node in FVM are

[1] The control volume finite element method (CVFEM) is a misnomer; it is nothing to do with the FEM, other than that CVFEM also uses interpolation functions used in the FEM as the approximation functions. The interpolation theory predates the FEM.

obtained by a combination of numerical integration and differentiation of model differential equations over a control volume. In the FEM, the discretized equations over a finite element involve nodal values of that element only, whereas in the FVM, the discretized equations for each control volume involves the nodal values of control volumes adjacent to the control volume under consideration. Thus, the two methods are different in deriving the discretized equations of the problem: The FEM uses a weighted-integral statement to derive the discretized equations over an element and "assembles" the element equations to obtain the discretized equations of the whole domain. It is fair to characterize the FEM as an integral method, the finite difference method (FDM) as a method in which the derivatives are represented with suitable difference formulas, and the FVM as a combination of the two approaches.

The book is broadly divided into three parts. Part I deals with background needed for the later chapters of the book, Part II is concerned with the finite element method, and Part III is devoted to the finite volume method. This division is useful to the reader in skipping any of the parts, as they are fairly independent of each other.

In the present study of the FEM and FVM, advanced mathematics is intentionally avoided in the interest of simplicity. However, a minimum of mathematical machinery that seemed necessary is included in Chapter 1. In Chapter 2, a review of the governing equations of a continuous medium, with focus on heat transfer and fluid mechanics, is presented. It is here background from a continuum mechanics course or undergraduate-level fluid mechanics and heat transfer courses prove to be helpful. Chapter 3 is devoted to a discussion of various methods for solving linear algebraic equations.

In introducing the FEM in Chapter 4 for steady-state heat transfer, the traditional solid mechanics approach is avoided in favor of the "differential equation" approach, which has broader interpretations than a single special case. Since a large number of physical problems are described by second-order differential equations, they are used as model equations in introducing the method. Considerable attention is devoted to the finite element formulation, the derivation of interpolation functions, and the solution of problems to illustrate the main features of the FEM. The FEM for unsteady heat transfer is introduced in Chapter 5 using the same model equations. Weak forms, semidiscrete finite element models, and time approximations are discussed in detail, and concepts of explicit and implicit formulations are examined. Several numerical examples are presented to illustrate how the numerical stability of the forward difference scheme is connected to the critical time step. Chapter 6 is dedicated to the flows of viscous incompressible flows in two dimensions. The velocity–pressure and penalty function formulations are introduced and associated finite element models are presented for Stokes equations (where the convective terms are neglected) and the Navier–Stokes equations. Several numerical examples of steady as well as unsteady problems of viscous flows are presented.

In Chapter 7, we introduce the FVM as applied to steady as well as transient heat transfer in one and two dimensions. Chapter 8 is devoted to the study of advection–diffusion equations in one and two dimensions. Chapter 9 deals with viscous incompressible flows using the FVM. Finally, Chapter 10 is concerned with some advanced topics, including periodically fully developed flows and heat transfer, natural convection, and multigrid techniques. Both FEM and FVM are used in solving a number of benchmark problems. Emphasis is placed on the modeling issues, such as the selection of mesh, imposition of boundary conditions, solution of equations, and interpretation of the results.

At the end of each chapter, especially in the earlier chapters, a number of exercise problems are included to test and extend the understanding of the concepts discussed. A solutions manual of most problem has been prepared, and it is available through the publisher to teachers who adopt the book in their courses. The computer programs used to numerically solve the problems discussed in the examples of various chapters are not included in this book owing to space limitations and proprietary nature of the programs.

Professor Reddy and Professor Anand thank Texas A&M University for providing them with an academic home and the ecosystem that led to the writing of this book. They also thank Mr. Mitch Wittneben for assisting with the software for the book writing. Contributions of several graduate students and post-docs in proof-reading various parts of the manuscript are gratefully acknowledged. The first two authors thank the post-doctoral fellow Praneeth Nampally and graduate students Buyng-Hee Choi, Daniel Orea, Rey Chavez, and Khoi Ngo for assisting with the proof reading of the manuscript. Dr. Roy thanks his mother and colleagues at Lawrence Livermore National Laboratory for encouraging and supporting his participation in this book project. Dr Roy's contribution was performed under the auspices of the U.S. Department of Energy by Lawrence Livermore National Laboratory under Contract DE-AC52-07NA27344. The authors are also grateful to the staff at Cambridge University Press for their help and support during the production of this book. Finally, authors thank their spouses for their understanding and support while the authors were occupied with the writing of this book.

Symbols

The symbols that are used in the book for various important quantities are defined in the following.

Symbol	Meaning
$\mathbf{a}$	Acceleration vector, $\frac{D\mathbf{v}}{Dt}$
a_{ij}	Coefficients of matrix $[A] = \mathbf{A}$
c_v, c_p	Specific heat at constant volume and pressure, respectively
d	Diameter
ds, dS	Surface elements
dA	Area element ($= dxdy$)
dv	Volume element ($= dxdydz$)
$\mathbf{D}$	Symmetric part of the velocity gradient tensor; that is, $\mathbf{D} = \frac{1}{2}\left[(\boldsymbol{\nabla}\mathbf{v})^{\mathrm{T}} + \boldsymbol{\nabla}\mathbf{v}\right]$
D_{ij}	Rectangular Cartesian components of $\mathbf{D}$
D/Dt	Material time derivative, $\frac{D}{Dt} = \frac{\partial}{\partial t} + \mathbf{v}\cdot\boldsymbol{\nabla}$
$\mathbf{e}_i$	Basis vector in the x_i-direction
$(\hat{\mathbf{e}}_r, \hat{\mathbf{e}}_\theta, \hat{\mathbf{e}}_z)$	Basis vectors in the (r, θ, z) system
$(\hat{\mathbf{e}}_x, \hat{\mathbf{e}}_y, \hat{\mathbf{e}}_z)$	Basis vectors in the (x, y, z) system
$(\hat{\mathbf{e}}_1, \hat{\mathbf{e}}_2, \hat{\mathbf{e}}_3)$	Basis vectors in the (x_1, x_2, x_3) system
$\mathbf{f}$	Body force vector
f_x, f_y, f_z	Body force components in the x-, y-, and z-directions
g	Internal heat generation per unit volume
g_f	Acceleration due to gravity
$\mathbf{I}$	Unit second-order tensor
$\mathbf{g}_f$	Acceleration due to gravity vector
$\mathbf{I}$	Unit second-order tensor
J	Determinant of $\mathbf{J}$ (Jacobian)
$\mathbf{J}$	Jacobian (of transformation) matrix

Symbol	Meaning
k	Thermal conductivity
$\mathbf{k}$	Thermal conductivity tensor
K	Kinetic energy
L	Length
$\hat{\mathbf{n}}$	Unit normal vector in the current configuration
n_i	ith component of the unit normal vector $\hat{\mathbf{n}}$
(n_x, n_y, n_z)	Components of the unit normal vector $\hat{\mathbf{n}}$
P	Hydrostatic pressure; perimeter
q_n	Heat flux normal to the boundary, $q_n = \boldsymbol{\nabla} \cdot \hat{\mathbf{n}}$
$\mathbf{q}$	Heat flux vector; diffusion flux
Q	Heat; mass flow rate; volume rate of flow
r	Radial coordinate in the cylindrical polar system; $r = \lvert\mathbf{r}\rvert$
$\mathbf{r}$	Position vector in cylindrical coordinates, $\mathbf{x}$
(r, θ, z)	Cylindrical coordinate system
R	Residual in the approximation; radius
t	Time
$\mathbf{t}$	Stress vector; traction vector
$\mathbf{t}_i$	Stress vector on x_i-plane, $\mathbf{t}_i = \sigma_{ij}\hat{\mathbf{e}}_j$
T	Temperature
v	Velocity, $v = \lvert\mathbf{v}\rvert$
(v_1, v_2, v_3)	Components of velocity vector $\mathbf{v}$ in (x_1, x_2, x_3) system
(v_r, v_θ, v_z)	Components of velocity vector $\mathbf{v}$ in (r, θ, z) system
$\mathbf{v}$	Velocity vector, $\mathbf{v} = \frac{D\mathbf{x}}{Dt}$
$\mathbf{v}_n$	Velocity vector normal to the plane (whose normal is $\hat{\mathbf{n}}$)
$\mathbf{x}$	Position vector in the current configuration
(x, y, z)	Rectangular Cartesian coordinates
(x_1, x_2, x_3)	Rectangular Cartesian coordinates

Other Symbols

Symbol	Meaning
$\boldsymbol{\nabla}$	Gradient operator with respect to $\mathbf{x}$
∇^2	Laplace operator, $\nabla^2 = \boldsymbol{\nabla} \cdot \boldsymbol{\nabla}$
∇^4	Biharmonic operator, $\nabla^4 = \nabla^2 \nabla^2$
[ ]	Matrix of components of the enclosed tensor
{ }	Column of components of the enclosed vector
$\cdot$	Symbol for the dot product or scalar product
$\times$	Symbol for the cross product or vector product

Greek Symbols

Symbol	**Meaning**
α	Angle; coefficient of thermal expansion; relaxation factor
β	Heat transfer coefficient
γ	Penalty parameter
Γ	Total boundary
δ	Dirac delta; variational symbol
δ_{ij}	Components of the unit tensor, $\mathbf{I}$ (Kronecker delta)
ε_{ijk}	Alternating symbol
ζ	Natural (normalized) coordinate
η	Natural (normalized) coordinate
θ	Angular coordinate in the cylindrical and spherical coordinate systems; angle; absolute temperature
λ	Lagrange multiplier
μ	Viscosity
ξ	Natural (normalized) coordinate
ρ	Mass density
$\boldsymbol{\sigma}$	Stress tensor
σ_{ij}	Components of the stress tensor in the rectangular coordinate system (x_1, x_2, x_3)
$\sigma_{rr}, \sigma_{\theta\theta}, \sigma_{r\theta}, \ldots$	Components of the stress tensor $\boldsymbol{\sigma}$ in the cylindrical coordinate system (r, θ, z)
τ	Shear stress
$\boldsymbol{\tau}$	Viscous stress tensor
ϕ	A typical scalar function; velocity potential; angular coordinate in the spherical coordinate system
ϕ_i	A generic variable
ψ	Stream function
ψ_i	Lagrange interpolation functions
ω	Angular velocity
Ω	Domain of a problem
$\boldsymbol{\Omega}$	Spin tensor or skew symmetric part of the velocity gradient tensor, $(\boldsymbol{\nabla}\mathbf{v})^{\mathrm{T}}$; that is $\boldsymbol{\Omega} = \frac{1}{2}\left[(\boldsymbol{\nabla}\mathbf{v})^{\mathrm{T}} - \boldsymbol{\nabla}\mathbf{v}\right]$
ω_i	Components of vorticity vector $\boldsymbol{\omega}$ in the rectangular coordinate system (x_1, x_2, x_3)
$\omega_x, \omega_y, \omega_z$	Components of vorticity vector $\boldsymbol{\omega}$ in the rectangular coordinate system (x, y, z)

Part I Preliminaries

1
Mathematical Preliminaries

1.1 Introduction

Most engineering systems can be described, with the aid of the laws of physics and observations, in terms of algebraic, differential, and integral equations. In most problems of practical interest, these equations cannot be solved exactly, mostly because of irregular domains on which the equations are posed, variable coefficients in the equations, complicated boundary conditions, and the presence of nonlinearities. Approximate representation of differential and integral equations to obtain algebraic relations among quantities that characterize the system and implementation of the steps to obtain algebraic equations and their solution using computers constitute a *numerical method.*

The process of converting differential or integral equations to a set of algebraic equations is called *discretization* of the equations. The discretization procedure varies from one numerical method to another. The discretization equations are generated at a finite number of points within the domain. The location and logical numbering of these locations at which dependent variables are calculated is referred to as *grid generation.* The locations at which the dependent variables (unknowns) are calculated are referred to as *grid* points or *nodes.* The terms grid generation and mesh generation are used interchangeably. A valid numerical solution must be independent of the grid (or mesh) size used, and such a solution is called a *grid-independent solution* . If a numerical solution procedure is represented by an equilateral triangle (see Fig. 1.1.1) its vertices would represent: (a) discretization of differential equations, (b) grid generation, and (c) the solution to a system of linear algebraic equations.

There are a number of numerical methods to solve differential equations, the most commonly used methods being the finite difference, finite element, boundary element, and finite volume methods. Because of the power of numerical methods and electronic computation, it is possible to include most relevant features of a physical process in the corresponding mathematical model, without worrying about its exact solution by an analytical means. The finite element method (FEM) [1–6] and the finite volume method (FVM) [7–11] are the most popular and powerful numerical methods devised to analyze many heat transfer and fluid flow problems. While there are numerous books dedicated to each of these two methods, there is only one book, by Chung [12], which deals with both methods as well as the finite difference method (FDM). The main focus of the book by Chung [12] is to introduce the readers all three methods.

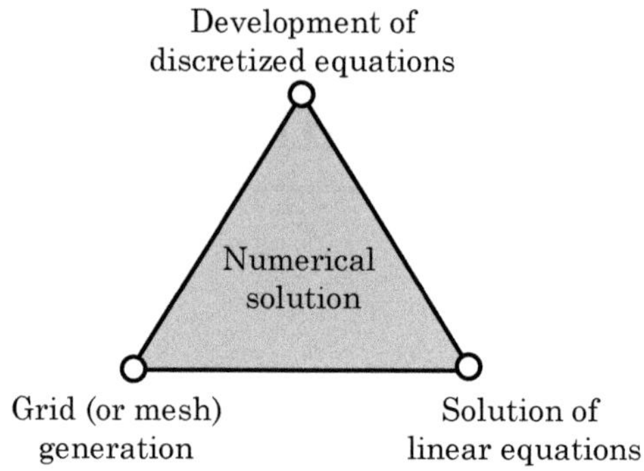

Fig. 1.1.1 Three pillars of numerical solution procedure.

Over the past five decades, computers have made it possible, with the help of suitable mathematical models and numerical methods, to analyze many practical problems of engineering for design and manufacturing. Most of the practicing engineers may end up using a commercial software package or an open source code. But having a background in the numerical methods and having written a computer program based on a numerical method makes them ask the right kind of questions, and be effective users of a commercial software package or an open source code. This book is dedicated to the study of the FEM and FVM as applied to problems of heat transfer and fluid dynamics. The objective is to familiarize the readers with the inner details of these two methods and help them formulate, write their own computer programs, and solve problems of interest to them.

To prepare the readers for an informed walk through various topics that come up in the discussion of differential equations and their solution, we first present some mathematical concepts and definitions. Readers familiar with these topics may skip these and go to Chapter 2.

1.2 Mathematical Models

1.2.1 Preliminary Comments

A set of mathematical relations between variables of a physical system is termed a *mathematical model*. The relationships can be algebraic, differential, and/or integral in nature. Of course, most mathematical models are combinations of algebraic and differential equations. Integral relations usually come through constitutive models (i.e., viscoelastic solids or fluids) and in solving heat transfer problems with radiation. The mathematical models of physical phenomena are often based on fundamental scientific laws of physics such as the principle of conservation of mass, the principle of balance of linear momentum, and the principle of balance of energy (see [13, 14]). The equations resulting from these principles are supplemented by equations that describe the constitutive behavior and boundary and/or initial conditions. A review of the equations of heat transfer and fluid dynamics will be presented in Chapter 2.

1.2.2 Types of Differential Equations

All differential equations can be grouped, based on the number of independent coordinates, into two types: (1) ordinary differential equations (ODEs) are those which contain only one independent coordinate, say x (spatial coordinate) or t (time); and (2) partial differential equations (PDEs) are those which contain two (e.g., x and t or x and y) or more (e.g., x, y, and t; x, y, and z; and x, y, z, and t). Clearly, differential equations describing two- and three-dimensional heat transfer and fluid flow are necessarily PDEs.

Examples of an ODE are $[u = u(t)$ or $u = u(x)]$:

$$a\frac{du}{dt} + bu = f(t), \quad a\frac{d^2u}{dt^2} + c\frac{du}{dt} + pu = f(t) \tag{1.2.1}$$

$$a\frac{du}{dx} + bu = g(x), \quad a\frac{d^2u}{dx^2} + c\frac{du}{dx} + pu = g(x). \tag{1.2.2}$$

Examples of a PDE are:

$$c\frac{\partial u}{\partial t} + \frac{\partial u}{\partial x} = f(x,t), \quad -\frac{\partial}{\partial x}\left(a\frac{\partial u}{\partial x}\right) + c\frac{\partial u}{\partial t} + pu = f(x,t) \tag{1.2.3}$$

$$-\frac{\partial}{\partial x}\left(a_{11}\frac{\partial u}{\partial x}\right) - \frac{\partial}{\partial y}\left(a_{22}\frac{\partial u}{\partial y}\right) = g(x), \quad -\frac{\partial}{\partial x}\left(a\frac{\partial u}{\partial x}\right) + c\frac{\partial^2 u}{\partial t^2} = g(x,t). \tag{1.2.4}$$

In Eqs. (1.2.1)–(1.2.4), a_{11}, a_{22}, a, b, c, p, f, and g are known functions of x, y, and/or t. The variable u is termed a *dependent variable.*

The *order* of a differential equation is the highest-order derivative appearing in the equation. Thus, an ODE or PDE can be first-, second-, or higher-order. The first equation in (1.2.1), the first equation in (1.2.2), and the first equation (1.2.3) are first-order equations and the others are second-order equations. A PDE can have multiple orders, each with respect to independent coordinates, x, y, z, and t.

An nth-order ODE in a dependent variable $u(\xi)$ contains derivatives of order n and less than n with respect to a coordinate ξ. The solution of an nth-order ODE requires n integrations with respect to the independent coordinate ξ. With each integration, one obtains a constant of integration. To determine the constants of integration, one needs n auxiliary conditions on u. When all conditions on u are specified (or known) at one specific value of the independent coordinate, say $\xi = 0$, we say that the ODE describes an *initial value problem* (IVP) and the conditions are called *initial conditions.* When conditions on u are specified at more than one value of the independent coordinate ξ, then the ODE is said to describe a *boundary value problem* (BVP), and the auxiliary conditions are termed as *boundary conditions.* Thus, a BVP has to be second or higher order or contain a set of first order coupled equations, and all first-order ODEs with a single unknown represent IVPs. There are numerical solution techniques that convert a BVP to an IVP (e.g., shooting methods), but the classification of the original equation as a BVP or an IVP follows the above definition.

The distinction between IVPs and BVPs is an important one as it dictates the way the information (direction and speed) propagates within the domain. The understanding of the propagation of information (solution or disturbance) within the domain dictates the kind of numerical scheme we adopt to get a physically realistic solution. In domains described by IVPs, the information propagates at finite speed in one direction; the independent variable could be space or time. Information in such domains could propagate from the present to the future, or left to right, as illustrated in Fig. 1.2.1(a). Thus, in discretizing a differential operator, one should make sure that the discrertization expression is such that no information comes from the future to the present, if it is a time-dependent problem; or no information comes from downstream to upstream, if it is a space-dependent problem. The IVPs are said to exhibit one-way behavior wherein the information propagates in only one direction. These IVPs are also called *marching problems*. On the other hand, in the BVPs, the information propagates from the boundary points into the domain, as shown in Fig. 1.2.1(b).

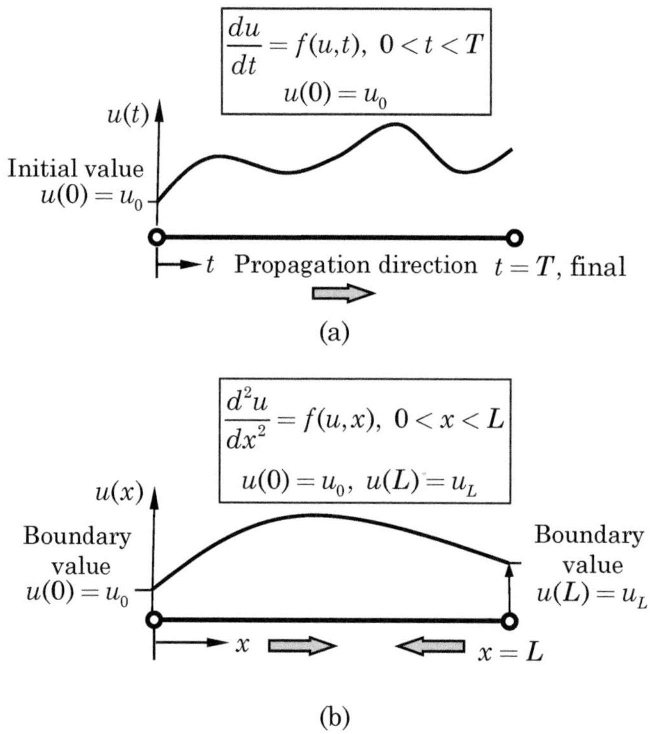

Fig. 1.2.1 (a) Initial value problems (IVPs) and (b) boundary value problems (BVPs) on one-dimensional domains. In IVPs, the initial condition dictates the propagation of the solution into the domain, whereas in the BVPs, the boundary conditions influence the interior solution.

A PDE in the dependent variable $u(\xi, \eta)$, with n derivatives in one independent coordinate ξ and m derivatives in the other independent coordinate η, requires n integrations with respect to ξ and m integrations with respect to η. If the constants that appear due to integration with respect to ξ are determined using known conditions on u and its derivatives at different values of ξ, then the equation describes a BVP with respect to ξ and the conditions are classified as the boundary conditions of the problem. If the constants appearing due to integration with respect to η are determined using known conditions on u and its derivatives at one fixed value (say, at $\eta = 0$), the equation describes an IVP with respect to η and the conditions are classified as initial conditions. When the constants appearing due to integrations with respect to η are also determined using known conditions on u and its derivatives at *different* values of η, the PDE describes a BVP with respect to ξ and η. Thus a PDE can describe a BVP (only) or a boundary value-initial value problem. All time-dependent problems are either pure IVPs or initial value–boundary value problems (IVBPs), when the problems are described by PDEs involving both time and space. However, not all IVPs are time-dependent problems.

In summary, a second-order differential equation in spatial coordinates (ODE or PDE) involves the specification of the dependent variable and its derivatives up to and including order $2n - 1$. On the other hand, an mth-order differential equation in time requires the specification of the dependent variable and its derivatives with respect to time of order $m - 1$ at the initial time (i.e., $t = 0$). A PDE in space and time can be a BVP as well as an IVP (i.e., it requires both boundary conditions and initial conditions). A first-order ODE in a coordinate ξ requires only one condition, which can be at $\xi = 0$ or $\xi = L$, where L is the final value of ξ.

It is important to keep in mind that the differential equation for a system is only valid within a domain (e.g., $0 < x < L$) and not valid at the boundary points (i.e., at $x = 0$ and $x = L$). Of course, the boundary conditions are valid only at the boundary. However, the solution to boundary value or initial value problems are valid both inside the domain and on the boundary.

An ODE or a PDE is called *nonlinear* when the dependent variable(s) or its derivatives appear in a nonlinear form. A system is said to be linear if it satisfies the *Principle of Superposition* and the *Principle of Proportionality.* All of the ODEs and PDEs presented in Eqs. (1.2.1)–(1.2.4) are linear *unless* the coefficients a_{11}, a_{22}, a, b, c, and p are functions of the dependent variable u and/or its derivatives, or the dependent variable manifests as a nonlinear term. One way to determine the linearity of a differential equation is to write it in an operator form

$$Au = f, \tag{1.2.5}$$

where A is an operator acting on the dependent variable u. For example, the eight operators associated with the eight differential equations in Eqs. (1.2.1)–(1.2.4) are

$$\begin{aligned}
&A_1 u = f, \quad A_1 = a\frac{d}{dt} + b; \quad A_2 u = f, \quad A_2 = a\frac{d^2}{dt^2} + c\frac{d}{dt} + p\\
&A_3 u = g, \quad A_3 = a\frac{d}{dx} + b; \quad A_4 u = g, \quad A_4 = a\frac{d^2}{dx^2} + c\frac{d}{dx} + p\\
&\qquad\qquad A_5 u = f, \quad A_5 = c\frac{\partial}{\partial t} + \frac{\partial}{\partial x}
\end{aligned} \tag{1.2.6a}$$

$$
\begin{aligned}
A_6 u = f, \quad & A_6 = -\frac{\partial}{\partial x}\left(a\frac{\partial}{\partial x}\right) + c\frac{\partial}{\partial t} + p \\
A_7 u = g, \quad & A_7 = -\frac{\partial}{\partial x}\left(a_{11}\frac{\partial}{\partial x}\right) - \frac{\partial}{\partial y}\left(a_{22}\frac{\partial}{\partial y}\right) \qquad (1.2.6b) \\
A_8 u = g, \quad & A_8 = -\frac{\partial}{\partial x}\left(a\frac{\partial}{\partial x}\right) + c\frac{\partial^2}{\partial t^2}.
\end{aligned}
$$

An operator A is said to be linear, hence the differential equation (1.2.5) is said to be a linear differential equation, if the following equality [13],

$$
A(\alpha u + \beta v) = \alpha A(u) + \beta A(v) \qquad (1.2.7)
$$

holds for all real numbers α and β. Otherwise, the operator is nonlinear, and the associated differential equation is a nonlinear differential equation. Examples of nonlinear operators are provided by

$$
A_1 = \frac{d^2}{dx^2} + cu\frac{d}{dx} + p, \quad A_2 = -\frac{\partial}{\partial x}\left(\sqrt{u}\frac{\partial}{\partial x}\right) - \frac{\partial}{\partial y}\left(\frac{\partial u}{\partial y}\frac{\partial}{\partial y}\right). \qquad (1.2.8)
$$

All differential equations have two kinds of terms: (1) terms involving the dependent variables and their derivatives of certain order, and (2) terms that contain only the independent variable(s). Of course, the terms containing the dependent variables can also have coefficients that are functions of the independent variables (as well as the dependent variables, when the differential equations are nonlinear). The terms involving only the independent variables (i.e., x, y, and/or t) are typically known as the *source terms.* When the source term in a differential equation is zero, the equation is called a *homogeneous differential equation.* Referring to Eqs. (1.2.1)–(1.2.4), the functions f and g (of x, y, and/or t) are the source terms. Similarly, when a specified initial value or boundary value is nonzero, it is said to be a nonhomogeneous initial or boundary condition. Otherwise, the initial or boundary conditions is called a homogeneous initial or boundary condition. The other way to look at this concept is that, if a dependent variable or its derivative appears in every term of a differential equation or of an initial condition or of a boundary condition then they are referred to as *homogeneous* equations or auxiliary (initial or boundary) conditions respectively. The phrase *general solution* to a differential equation means it is the solution of a homogeneous differential equation. The part of the total solution to a differential equation that satisfies the nonhomogeneous differential equation is known as the *particular solution.*

1.2.3 Examples of Mathematical Models

1.2.3.1 A Boundary Value Problem: the Heat Conduction Problem

As an example of a mathematical model, we consider one-dimensional heat flow in an uninsulated circular cylindrical rod (see Fig. 1.2.2):

$$
-\frac{d}{dx}\left(Ak\frac{dT}{dx}\right) + \beta P(T - T_\infty) = g, \qquad (1.2.9)
$$

where T denotes temperature (°C) above a certain reference temperature, A is the area of cross section (m^2) of the rod, P is the perimeter (m), k is the conductivity [W/(m·°C)], β is the convective heat transfer coefficient [W/(m^2·°C)], and g is the internal heat generation per unit length (W/m). Equation (1.2.9) is a statement of balance of energy [15]. The first term denotes the transfer of energy due to conduction (diffusion) and the second term is the energy transfer due to convection through the surface of the rod.

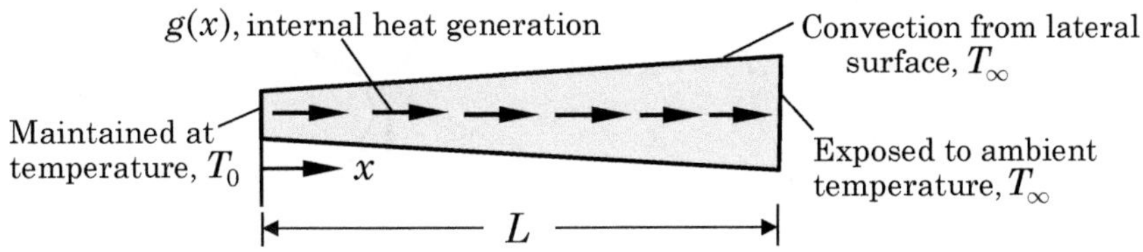

Fig. 1.2.2 One-dimensional heat flow in a rod with internal heat generation $g(x)$.

Equation (1.2.9) is a linear, nonhomogeneous (because the right-hand side is nonzero), second-order differential equation with variable coefficients (because, in general, kA can be a function of x), which can be solved with two known conditions on either temperature T or heat $Q = -kA(dT/dx)$ (but not both) at each of the two boundary points (two ends of the bar).

Suppose that the auxiliary conditions at the boundary points $x = 0$ and $x = L$ are of the form

$$T(0) = T_0, \quad \left[kA\frac{dT}{dx} + \beta A(T - T_\infty)\right]_{x=L} = Q_0, \tag{1.2.10}$$

where T_0 and Q_0 are the specified temperature and specified heat, respectively. These auxiliary conditions are, clearly, boundary conditions. The first condition in Eq. (1.2.10) is called the *Dirichlet* boundary condition. The second condition represents the balance of heat due to conduction $[kA(dT/dx)]$, and convection $[\beta A(T - T_\infty)]$ at $x = L$, and it is known as the *Newton*, *mixed*, or *Robin* boundary condition. In the heat transfer literature, it is also known as the *convective* boundary condition. As a special case, the boundary condition when the end $x = L$ is insulated is given by setting $\beta = 0$ and $Q_0 = 0$ in Eq. (1.2.10). The case where $\beta = 0$ is known as the *Neumann* boundary condition.

Equation (1.2.9) is an ODE because it contains only one independent coordinate, namely, x. Also, Eq. (1.2.9) is a *nonhomogeneous differential equation* because the source term (g) is nonzero. We also note that the boundary conditions in Eq. (1.2.10) are nonhomogeneous, as long as at least one of the three quantities (T_0, T_∞, Q_0) is nonzero.

The problem described by Eqs. (1.2.9) and (1.2.10) is a BVP because its solution requires the specification of the auxiliary conditions at more than one specific value of the independent variable (i.e., $x = 0$ and $x = L$). Quantities k, A, β, P, T_∞, T_0, Q_0, and f are called the *data* or *parameters* of the problem because they are prescribed quantities of the problem.

1.2.3.2 An Initial Value Problem: the Simple Pendulum Problem

Next, consider a simple pendulum, which consists of a bob of constant mass m (kg) attached to one end of a rod of length ℓ (m) with the other end pivoted to a fixed point O (without friction), as shown in Fig. 1.2.3. The bob and the rod are assumed to be rigid (i.e., not deformable) and the rod is massless.

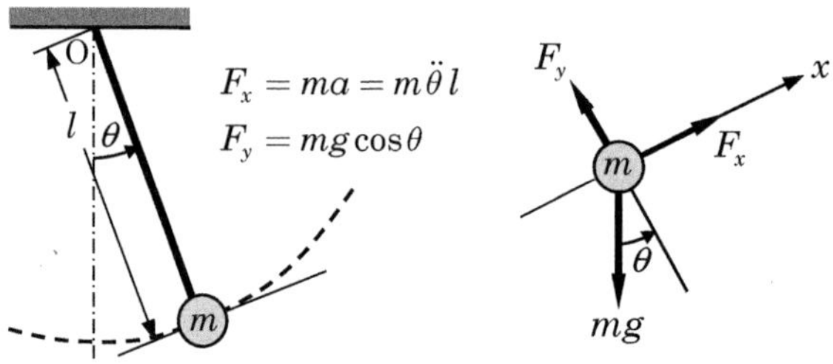

Fig. 1.2.3 Simple pendulum (a bob of mass m is attached to a massless rod of length l), which is connected to point O and free to rotate about the point O.

The equation governing the motion of the simple pendulum can be determined using Newton's Second Law of motion (i.e., the vector sum of externally applied forces on a system is equal to the time rate of change of the linear momentum of the system):

$$m\frac{d^2\theta}{dt^2} + \frac{mg}{\ell}\sin\theta = 0, \tag{1.2.11}$$

where θ is the angular displacement (radians), g is the acceleration due to gravity (m/s^2), and t denotes time (s).

Equation (1.2.11) is nonlinear (in θ) on account of the term $\sin\theta$. As explained before, a check for linearity is to replace the dependent unknown (θ) with a constant multiple of itself and see if the constant can be factored out. For example, replacing θ with $\alpha\theta$ in Eq. (1.2.11), we obtain $m\alpha\frac{d^2\theta}{dt^2} + \frac{mg}{\ell}\sin(\alpha\theta) = 0$. Although α is factored out in the first term (hence the first term is linear), it cannot be factored out in the second term (hence the second term is nonlinear); thus, the whole equation is said to be nonlinear. For small angular motions, $\sin\theta$ can be approximated as $\sin\theta \approx \theta$. Then, Eq. (1.2.3) becomes a homogeneous, second-order, linear ODE (the independent coordinate being t):

$$m\frac{d^2\theta}{dt^2} + \frac{mg}{\ell}\theta = 0. \tag{1.2.12}$$

The solution of Eq. (1.2.11) or Eq. (1.2.12) requires knowledge of conditions on θ and its time derivative $\dot{\theta}$ (angular velocity) at time $t = 0$. These conditions are the initial conditions. Thus, the linear problem involves solving the second-order differential equation (1.2.12) subjected to the initial conditions:

$$\theta(0) = \theta_0, \qquad \left.\frac{d\theta}{dt}\right|_{t=0} = v_0. \tag{1.2.13}$$

Since the problem described by Eqs. (1.2.12) and (1.2.13) requires two known conditions at the same value of the independent coordinate (t), it is an IVP. The initial conditions are said to be homogeneous when both θ_0 and v_0 are zero.

If we wish to solve the nonlinear equation, Eq. (1.2.11), subject to the conditions in Eq. (1.2.13), we may consider using a numerical method because it is not possible to solve Eq. (1.2.11) exactly for large values of θ.

1.2.4 Numerical Solution of First-Order Ordinary Differential Equations

1.2.4.1 The Euler Methods

Let us consider the following first-order ODE in $y(x)$ with an end condition:

$$\frac{dy}{dx} = f(x, y), \quad y(0) = y_0. \tag{1.2.14}$$

Integrating Eq. (1.2.14) over the interval $[x_i, x_{i+1}]$, we obtain

$$\int_{y_i}^{y_{i+1}} dy = \int_{x_i}^{x_{i+1}} f(x, y)dx, \tag{1.2.15}$$

where $y_{i+1} = y(x_{i+1})$ and so on.

One can complete the integration on the right-hand side if one assumes f to remain constant in the interval $[x_i, x_{i+1}]$. But the question is where to evaluate f in this interval when it is a function of x and y (see Fig. 1.2.4). The stability of the method and the complexity of the calculation depend on where $f(x, y)$ or the slope is evaluated in this interval. If the $f(x, y)$ is evaluated at the initial point, then

$$y_{i+1} = y_i + hf(x_i, y_i), \quad h = x_{i+1} - x_i. \tag{1.2.16}$$

In the above equation, an unknown (y_{i+1}) is evaluated in terms of all known quantities. Such a formulation is referred to as an *explicit method*. If $f(x, y)$ is evaluated at the end point x_{i+1}, then we obtain

$$y_{i+1} = y_i + hf(x_{i+1}, y_{i+1}); \quad \text{for } i = 0 \text{ we have } \; y_1 = y_0 + hf(x_0, y_0). \tag{1.2.17}$$

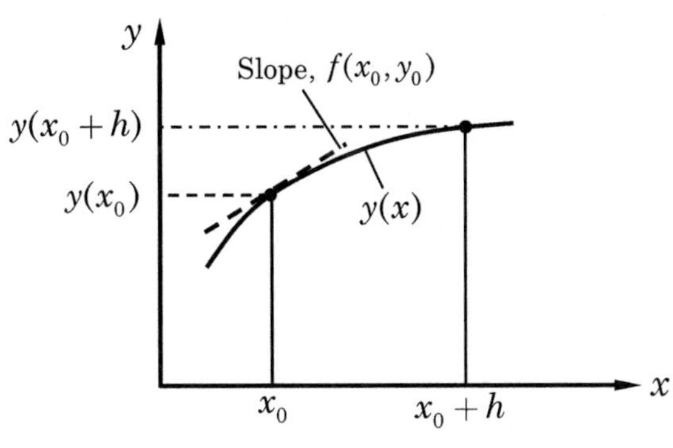

Fig. 1.2.4 Values of a function $y(x)$ at different points, $x = x_0$ and $x = x_0 + h$.

In Eq. (1.2.17), the unknown y_{i+1} is expressed in terms of the unknown y_{i+1}. This type of formulation is known as an *implicit method*. If the slope $f(x, y)$ is represented as the arithmetic mean of slopes at initial and final points, then we obtain

$$y_{i+1} = y_i + \frac{h}{2}\left[f(x_i, y_i) + f(x_{i+1}, y_{i+1})\right]. \tag{1.2.18}$$

In Eq. (1.2.18), the unknown is expressed in terms of both the known (y_i) and the unknown (y_{i+1}) on the right-hand side. This formulation is known as the *semi-implicit method*. It is also known as the *Crank–Nicolson method*. A generalized formulation can be expressed by combining implicit and explicit methods in the following way:

$$y_{i+1} = y_i + h\left[(1-\alpha)f(x_i, y_i) + \alpha f(x_{i+1}, y_{i+1})\right]. \tag{1.2.19}$$

Depending on the value of α, the generalized formulation can reduce to the fully explicit method ($\alpha = 0$), the fully implicit method ($\alpha = 1$), or the (semi-implicit) Crank–Nicolson method ($\alpha = 0.5$). These methods will be discussed later in the context of the finite element method (FEM; see Chapter 5) and the finite volume method (FVM; see Chapter 7) applied to time-dependent problems.

In general, the explicit methods are easy to implement, but are numerically unstable. On the other hand, implicit methods are numerically stable, but are difficult to implement. Implicit methods often warrant solution to a set of linear equations at each integration step. In terms of accuracy, the semi-implicit formulation or Crank–Nicolson method shown above is second-order accurate, whereas the explicit and implicit methods are first-order accurate. However, semi-implicit methods are prone to oscillatory solutions, especially for diffusion equations.

As stated earlier, the slope $f(x, y)$ can be represented as the weighted mean of slopes evaluated at various points within the integration step h. This leads to a family of techniques called *Runge–Kutta methods* [16, 17], which are discussed next. We note that the first-order Runge–Kutta method is the same as the Euler explicit method.

1.2.4.2 Runge–Kutta Family of Methods

Runge–Kutta methods are a family of explicit and implicit methods for numerically integrating ODEs. Runge–Kutta methods can be of nth order, where the value of the slope $f(x, y)$ is evaluated at n locations. In heat transfer and fluid flow problems, second-order (RK-2) or fourth-order Runge–Kutta (RK-4) methods are typically used. In calculating trajectories for deep space exploration, it is common to use eighth order or tenth order Runge–Kutta methods [16]. So there are two questions to answer: (a) what are the coordinates of the intermediary locations where slopes are evaluated, and (b) what are the weights? Details of the derivations are not given here, and readers are referred to [16]. Here, we present the second-order and fourth-order Runge–Kutta methods.

The second-order Runge–Kutta method (RK-2) In the most commonly used second-order Runge–Kutta method [16], the slope is evaluated at two points in the interval $[x_i, x_{i+1}]$, including the initial point:

$$y_{i+1} = y_i + \frac{h}{2}(k_1 + k_2) \tag{1.2.20a}$$
$$k_1 = f(x_i, y_i) \tag{1.2.20b}$$
$$k_2 = f(x_i + h, y_i + hk_1), \tag{1.2.20c}$$

where k_1 and k_2 are slopes, and the corresponding weights are 1/2 and 1/2, respectively.

The fourth-order Runge–Kutta method (RK-4). In the fourth-order Runge–Kutta method, the slopes are evaluated at four locations including the initial point:

$$y_{i+1} = y_i + \frac{h}{6}(k_1 + 2k_2 + 2k_3 + k_4) \tag{1.2.21a}$$
$$k_1 = f(x_i, y_i) \tag{1.2.21b}$$
$$k_2 = f(x_i + \frac{h}{2}, y_i + \frac{hk_1}{2}) \tag{1.2.21c}$$
$$k_3 = f(x_i + \frac{h}{2}, y_i + \frac{hk_2}{2}) \tag{1.2.21d}$$
$$k_4 = f(x_i + h, y_i + hk_3). \tag{1.2.21e}$$

Here k_1, k_2, k_3, and k_4 are slopes, and the corresponding weights are 1/6, 1/3, 1/3, and 1/6, respectively.

Hitherto we discussed only the solution to first-order IVPs. However, often the governing equations are second or higher order. It can be shown that every IVP of an nth order ODE with given initial conditions can be reduced to a system of n first order ODEs by defining $(n-1)$ new variables. We will demonstrate this using the example of the simple pendulum.

1.2.4.3 Implementation of Runge–Kutta Methods for Coupled Systems

In solving coupled system using the Runge–Kutta methods the slope k is represented as k_{IJ}. Here I stands for the constant number (or slope number) and J stands for the dependent variable number.

As an example, consider a coupled system of the form

$$\frac{dy_1}{dx} = f_1(x, y_1, y_2) \tag{1.2.22a}$$
$$\frac{dy_2}{dx} = f_2(x, y_1, y_2) \tag{1.2.22b}$$

subjected to the following initial conditions

$$y_1(x_0) = y_1^0, \quad y_2(x_0) = y_2^0. \tag{1.2.22c}$$

The RK-2 formulation In this case, Eqs. (1.2.20a)–(1.2.20c) applied to the coupled system take the form

$$y_1(x_i + h) = y_1(x_i) + \frac{h}{2}(k_{11} + k_{21}) \tag{1.2.23a}$$

$$y_2(x_i + h) = y_2(x_i) + \frac{h}{2}(k_{12} + k_{22}) \tag{1.2.23b}$$

$$k_{11} = f_1(x_i, y_1^i, y_2^i) \tag{1.2.23c}$$

$$k_{12} = f_2(x_i, y_1^i, y_2^i) \tag{1.2.23d}$$

$$k_{21} = f_1(x_i + h, y_1^i + hk_{11}, y_2^i + hk_{12}) \tag{1.2.23e}$$

$$k_{22} = f_2(x_i + h, y_1^i + hk_{11}, y_2^i + hk_{12}), \tag{1.2.23f}$$

where we have used the notation

$$y_1^i \equiv y_1(x_i), \quad y_2^i \equiv y_2(x_i). \tag{1.2.23g}$$

The RK-4 formulation Equations (1.2.21a)–(1.2.21e) applied to the coupled system take the form

$$y_1(x_i + h) = y_1(x_i) + \frac{h}{6}(k_{11} + 2k_{21} + 2k_{31} + k_{41}) \tag{1.2.24a}$$

$$y_2(x_i + h) = y_2(x_i) + \frac{h}{6}(k_{12} + 2k_{22} + 2k_{32} + k_{42}) \tag{1.2.24b}$$

$$k_{11} = f_1(x_i, y_1^i, y_2^i) \tag{1.2.24c}$$

$$k_{12} = f_2(x_i, y_1^i, y_2^i) \tag{1.2.24d}$$

$$k_{21} = f_1(x_i + \frac{h}{2} y_1^i + \frac{hk_{11}}{2}, y_2^i + \frac{hk_{12}}{2}) \tag{1.2.24e}$$

$$k_{22} = f_2(x_i + \frac{h}{2} y_1^i \frac{hk_{11}}{2}, y_2^i + \frac{hk_{12}}{2}) \tag{1.2.24f}$$

$$k_{31} = f_1(x_i + \frac{h}{2} y_1^i + \frac{hk_{21}}{2}, y_2^i + \frac{hk_{22}}{2}) \tag{1.2.24g}$$

$$k_{32} = f_2(x_i + \frac{h}{2} y_1^i + \frac{hk_{21}}{2}, y_2^i + \frac{hk_{22}}{2}) \tag{1.2.24h}$$

$$k_{41} = f_1(x_i + h, y_1^i + hk_{31}, y_2^i + hk_{32}) \tag{1.2.24i}$$

$$k_{42} = f_2(x_i + h, y_1^i + hk_{31}, y_2^i + hk_{32}). \tag{1.2.24j}$$

1.2.4.4 Numerical Stability

A numerical method must be accurate and stable. If any source of error amplifies or grows as calculations are marched either in time or in an iterative procedure, the method is termed as unstable. There are several techniques reported in the literature to predict stability of an algorithm (e.g., the von Neumann stability analysis). A detailed derivation of the numerical stability criteria is a topic in itself and beyond the scope of this book. However, the numerical stability criterion will be stated as appropriate in the sequel.

Example 1.2.1

Consider a simple pendulum of length ℓ with mass m attached to the end of the rod, oscillating in a frictionless environment. Initially the pendulum is displaced by a small angle θ_0 in the clockwise direction and then the pendulum is allowed to oscillate under the influence of gravity. The governing equations for simple pendulum are [see Eqs. (1.2.11) and (1.2.13)]

$$m\frac{d^2\theta}{dt^2} + \frac{mg}{\ell}\sin\theta = 0 \tag{1.2.25}$$

$$\theta(0) = \theta_0 = \frac{\pi}{12}, \quad \left.\frac{d\theta}{dt}\right|_{t=0} = 0. \tag{1.2.26}$$

Using the data $m = 1$ kg, $\ell = 1$ m, and $g = 9.8$ m/s^2, and a time step size of $\Delta t = 0.05$ s, calculate the displacement $\theta(t)$ and velocity $\frac{d\theta}{dt}$ at $t = 0.05$ s using:

(a) the Euler explicit method,

(b) the fully implicit method,

(c) the second-order Runge–Kutta method, and

(d) the fourth-order Runge–Kutta method.

Solution The first step is to convert the second-order ODE into two first-order ODEs by introducing a new variable. Let

$$w(t) = \frac{d\theta}{dt} \tag{1.2.27a}$$

so that the original second-order equation can be written as a pair of first-order equations:

$$\frac{d\theta}{dt} = w(t) = f_1(t, \theta, w) \tag{1.2.27b}$$

$$\frac{dw}{dt} = -(g/l)\sin\theta = f_2(t, \theta, w) \tag{1.2.27c}$$

with the initial conditions

$$\theta(0) = \frac{\pi}{12}, \quad w(0) = 0. \tag{1.2.27d}$$

In the fully implicit method, we need to linearize the differential equation; otherwise, we will end up solving a nonlinear equation containing a trigonometric function. However, one could solve these equations without linearizing by using explicit methods. In order to have a meaningful consistent comparison of results obtained by all four methods it is imperative we solve the same problem by these four different methods. Accordingly, the linearized differential equation is solved. To linearize the equation we assume that for small angles, $\sin\theta \approx \theta$. Therefore, the equations become,

$$\frac{d\theta}{dt} = w(t) = f_1(t, \theta, w) \tag{1.2.28a}$$

$$\frac{dw}{dt} = -(g/l)\theta = f_2(t, \theta, w) \tag{1.2.28b}$$

$$\theta(0) = \frac{\pi}{12}, \quad w(0) = 0. \tag{1.2.28c}$$

(a) **Euler explicit method**

$$\theta(0.05) = \theta_0 + \Delta t f_1(0, \theta_0, w_0) = \frac{\pi}{12} + 0.05 \times 0 = 0.2617 \text{ rad}$$

$$w(0.05) = w_0 + \Delta t f_2(0, \theta_0, w_0) = 0 - 0.05 \times 9.81 \times \sin\frac{\pi}{12} = -0.1269 \text{ rad/s}.$$

(b) **Fully implicit method** In using the fully implicit method, we use Eqs. (1.2.28a)–(1.2.28c). Using the fully implicit formulation, we obtain

$$\theta(t=0.05)=\theta_1=\theta_0+\Delta t\, f_1(\Delta t,\theta(\Delta t),w(\Delta t))=\frac{\pi}{12}+0.05\times w(t=0.05)$$
$$w(t=0.05)=w_1=w_0+\Delta t\, f_2(\Delta t,\theta(\Delta t),w(\Delta t))=w_0+0.05\times(-9.81\times\theta(t=0.05))$$
$$=0-0.05\times 9.81\times(\frac{\pi}{12}+0.05\times w(t=0.05))=-0.1283-0.0245\,w_1.$$

Solving the above equations, we obtain

$$\theta(0.05)=0.2554\text{ rad},\quad w(0.05)=-0.1252\text{ rad/s}.$$

(c) **RK-2 method**

$$k_{11}=f_1(t_0,\theta_0,w_0)=w_0=0$$
$$k_{12}=f_2(t_0,\theta_0,w_0)=-\frac{g}{l}\sin\theta_0=-9.8\times\sin\frac{\pi}{12}=-2.5364$$
$$k_{21}=f_1(t_0+\Delta t,\theta_0+k_{11}\Delta t,w_0+k_{12}\Delta t)=w_0+k_{12}\Delta t=(0+0.05(-2.5364))=-0.1268$$
$$k_{22}=f_2(t_0+\Delta t,\theta_0+k_{11}\Delta t,w_0+k_{12}\Delta t)=-9.8\times\sin(\frac{\pi}{12}+0.05\times 0)=-2.5364.$$

Therefore, we obtain

$$\theta(0.05)=\theta_0+\frac{\Delta t}{2}(k_{11}+k_{21})=\frac{\pi}{12}+\frac{0.05}{2}(0-0.1268)=0.2586\text{ rad}$$
$$w(0.05)=w_0+\frac{\Delta t}{2}(k_{12}+k_{22})=0+\frac{0.05}{2}(-2.5364-2.5364)=-0.1268\text{ rad/s}.$$

(d) **RK-4 method**

$$k_{11}=f_1(t_0,\theta_0,w_0)=w_0=0$$
$$k_{12}=f_2(t_0,\theta_0,w_0)=-\frac{g}{l}\sin\theta_0=-9.8*\sin\frac{\pi}{12}=-2.5364$$
$$k_{21}=f_1(t_0+\frac{\Delta t}{2},\theta_0+k_{11}\frac{\Delta t}{2}+w_0+k_{12}\frac{\Delta t}{2})=w_0+k_{12}\frac{\Delta t}{2}=-0.0634$$
$$k_{22}=f_2(t_0+\frac{\Delta t}{2},\theta_0+k_{11}\frac{\Delta t}{2}+w_0+k_{12}\frac{\Delta t}{2})=-9.8\times\sin(\theta_0+k_{11}\frac{\Delta t}{2})=-2.5364$$
$$k_{31}=f_1(t_0+\frac{\Delta t}{2},\theta_0+k_{21}\frac{\Delta t}{2}+w_0+k_{22}\frac{\Delta t}{2})=w_0+k_{22}\frac{\Delta t}{2}=-0.0634$$
$$k_{32}=f_2(t_0+\frac{\Delta t}{2},\theta_0+k_{21}\frac{\Delta t}{2}+w_0+k_{22}\frac{\Delta t}{2})=-9.8\times\sin(\theta_0+k_{21}\frac{\Delta t}{2})=-2.5214$$
$$k_{41}=f_1(t_0+\Delta t,\theta_0+k_{31}\Delta t,w_0+k_{32}\Delta t)=w_0+k_{32}\Delta t=-0.1261$$
$$k_{42}=f_2(t_0+\Delta t,\theta_0+k_{31}\Delta t,w_0+k_{32}\Delta t)=-9.8\times\sin[\frac{\pi}{12}+0.05\times(-0.0634)]=-2.5064.$$

Therefore, we obtain

$$\theta(t=0.05)=\theta_0+\frac{\Delta t}{6}(k_{11}+2k_{21}+2k_{31}+k_{41})=0.2586\text{ rad}$$
$$w(t=0.05)=w_0+\frac{\Delta t}{6}(k_{12}+2k_{22}+2k_{32}+k_{42})=-0.1263\text{ rad/s}.$$

Numerical results The results are summarized in Table 1.2.1. The pendulum is displaced clockwise by $(\pi/12)$ or 0.2617 rad and allowed to swing back in the counterclockwise direction. As one can see from Table 1.2.1, all four methods give positive values of displacement. With the exception of the Euler explicit method, the rest gave a value of the displacement that is less than the initial displacement, implying that the pendulum is still to the right of its neutral position ($\theta = 0$) at $t = 0.05$ s. Note that velocities predicted by all four methods at $t = 0.05$ s show negative values, indicating that the pendulum is swinging from the right towards the neutral position. From the accuracy point of view, the RK-4 is fourth-order accurate and the RK-2 is second-order accurate. Both the Euler explicit and fully implicit methods are first-order accurate.

Table 1.2.1 Results of the simple pendulum problem.

Method	$\theta(0.05)$ (rad)	$w(0.05)$ (rad/s)
Euler explicit	0.2617	−0.1269
Fully implicit	0.2554	−0.1252
RK-2	0.2586	−0.1268
RK-4	0.2586	−0.1263
Exact	0.2586	−0.1277

1.2.5 Partial Differential Equations and their Classification

Let us consider the transient (i.e., time-dependent) or evolution problem of finding temperature in an uninsulated cylindrical rod. In this case, we must include *the time rate of change of internal energy* when the balance of energy is considered. The time rate of change of internal energy per unit mass is equal to $c_v(\partial T/\partial t)$, where c_v is the specific heat (or heat capacity) at constant volume (see [15] for details). Since we now have two independent coordinates, namely x and t, ordinary derivatives must be replaced with partial derivatives. Then the balance of energy can be expressed as

$$\rho A c_v \frac{\partial T}{\partial t} - \frac{\partial}{\partial x}\left(kA\frac{\partial T}{\partial x}\right) + \beta P(T - T_\infty) = g(x,t). \tag{1.2.29}$$

Equation (1.2.29) is called a *partial differential equation* (PDE) because it has differentials with respect to two independent coordinates, x and t. The steady-state equation (1.2.9) would become a PDE in two and three dimensions because of additional independent coordinates, y and z. The solution of Eq. (1.2.29) requires two boundary conditions (i.e., conditions on T and/or its derivative at $x = 0$ and $x = L$) and an initial condition (i.e., $T(x,0) = T_0(x)$ at $t = 0$). Hence, the problem associated with Eq. (1.2.29) is both an initial-value and boundary-value problem (IVBP).

The manner in which the information (or disturbance) propagates in a domain impacts the numerical scheme one selects for the solution. In the case of BVPs, information propagates from the boundaries into the domain. In the case of IVPs, information flows from the initial condition in one direction. The direction in which the information propagates is also referred to as the marching direction. In the case of one-dimensional BVPs, the solution at each point in the domain is influenced by the boundary conditions on the left and right ends of the domain. Thus solutions at all points inside the domain are interlinked. In contrast, in an IVP, the solution at any time is influenced by its value at a previous time or upstream. To gain insight into the information propagation in the solution of a PDE (involving both time and space), it is useful to classify PDEs.

Consider a general second-order PDE of the form

$$a\frac{\partial^2 u}{\partial x^2} + b\frac{\partial^2 u}{\partial x \partial y} + c\frac{\partial^2 u}{\partial y^2} + d\frac{\partial u}{\partial x} + e\frac{\partial u}{\partial y} + fu = g(x, y), \tag{1.2.30}$$

where a, b, c, d, e, and f are constants and u is the dependent unknown. Depending on the values of the discriminant $b^2 - 4ac$, we can classify the PDE in Eq. (1.2.30) as

$$\begin{aligned} &b^2 - 4ac > 0, \text{ the PDE is } \textit{hyperbolic} \\ &b^2 - 4ac = 0, \text{ the PDE is } \textit{parabolic} \\ &b^2 - 4ac < 0 \text{ the PDE is } \textit{elliptic.} \end{aligned} \tag{1.2.31}$$

The classification is directly linked to the way information propagates in the computational domain and the pattern of information propagation dictates the type of numerical procedure to be used. The concept of *characteristic curves* is strongly linked to the classification of PDEs (see Table 1.2.2). In a two-dimensional domain, the characteristics are the lines or curves along which information propagates. Discontinuities in the dependent variable, if any, propagate along the characteristic curves. If a PDE possesses real characteristics, then information propagates along these characteristics. If there are no real characteristics (i.e., complex), then there are no preferred paths of information propagation. For an elliptic equation, information propagates at infinite speed on all directions. In a hyperbolic equation, information propagates along characteristic curves at a finite speed. Along these characteristic lines, PDEs can be converted to ODEs. In a parabolic equation, information propagates at a finite speed in one direction. The hyperbolic PDE has two real characteristic curves, the parabolic PDE has one real characteristic curve, and the elliptic PDE has no real characteristic curves. One should note that this classification structure is valid only for *second-order* PDEs. All first-order PDEs are classified as hyperbolic.

Table 1.2.2 Classification of PDEs and characteristics.

$b^2 - 4ac$	Characteristic curves	Classification
< 0	Complex	Elliptic
$= 0$	Real and repeated	Parabolic
> 0	Real and distinct	Hyperbolic

As an example, consider the equation governing the transient temperature distribution $T(x, t)$ in a rod of length L in the absence of internal heat source (i.e., $g = 0$) is given by Eq. (1.2.29). For constant properties, Eq. (1.2.29) takes the form

$$\alpha \frac{\partial T}{\partial t} - \frac{\partial^2 T}{\partial x^2} = 0, \quad \alpha = \frac{\rho c_v}{kA}. \tag{1.2.32}$$

By examining the discriminant of the PDE ($b^2 - 4ac = 0$), we conclude that the given PDE is parabolic ($a = -\alpha$, $e = 1$, and all other parameters are zero).

As another example, consider the two-dimensional steady-state temperature distribution in a homogeneous and isotropic thin rectangular plate of conductivity k with internal heat generation $g(x, y)$. The governing equation is the PDE

$$k\left(\frac{\partial^2 T}{\partial x^2} + \frac{\partial^2 T}{\partial y^2}\right) = g(x, y), \tag{1.2.33}$$

which defines a BVP. For this case, we have $a = c = k$ and all other parameters are zero. Therefore, we have $b^2 - 4ac < 0$, indicating that the equation is elliptic.

The following equations give additional examples of hyperbolic, parabolic, and elliptic PDEs:

$$y\frac{\partial^2 u}{\partial x^2} + \frac{\partial^2 u}{\partial y^2} = 0 \quad (a = y, b = 0, c = 1, d = e = f = 0), \tag{1.2.34}$$

$$b^2 - 4ac = -4y, \quad \text{elliptic for } y > 0, \quad \text{hyperbolic for } y < 0;$$

$$a_{xx}\frac{\partial^2 u}{\partial x^2} + a_{yy}\frac{\partial^2 u}{\partial y^2} = 0 \quad (a = a_{xx}, b = 0, c = a_{yy}, d = e = f = 0), \tag{1.2.35}$$

$$b^2 - 4ac = -4a_{xx}a_{yy}, \quad \text{elliptic for } a_{xx} > 0, a_{yy} > 0,$$
$$\text{hyperbolic for } a_{xx} > 0, a_{yy} < 0 \text{ or } a_{xx} < 0, a_{yy} > 0;$$

$$y\frac{\partial^2 u}{\partial x^2} + x\frac{\partial^2 u}{\partial x \partial y} + y\frac{\partial^2 u}{\partial y^2} = 0 \quad (a = y, b = x, c = y, d = e = f = 0), \tag{1.2.36}$$

$$b^2 - 4ac = x^2 - 4y^2, \quad \text{elliptic for } x < 2y, \quad \text{parabolic for } x = 2y,$$
$$\text{hyperbolic for } x > 2y.$$

1.3 Numerical Methods

1.3.1 Introduction

As discussed before, the exact solution is often difficult when the coefficients of the differential equation(s) are functions of position, the differential equations are nonlinear, or the domain is geometrically complex. In such cases, numerical solution is the only alternative. By a *numerical simulation* of a problem, we mean the solution of the governing equations using a numerical method and a computing tool. Numerical methods typically transform differential equations governing a continuum to a set of discrete algebraic equations, and these algebraic equations are solved using computers.

There are a number of numerical methods. Among them, the finite difference, the finite volume, and the finite element methods are most common, each of them having dominance in certain fields of application. A brief introduction to these methods is presented here with the help of a one-dimensional model BVP described by

$$-\frac{d^2u}{dx^2} + cu = f, \quad 0 < x < L \tag{1.3.1}$$

subjected to the boundary conditions

$$u(0) = u_0, \quad \left[\frac{du}{dx} + bu\right]_{x=L} = Q_0. \tag{1.3.2}$$

Equations (1.3.1) and (1.3.2) are an alternative representation of Eqs. (1.2.9) and (1.2.10) for constant values of kA and $P\beta$, with $u = T - T_\infty$, $c = P\beta/Ak$, $b = \beta/k$, and $u_0 = T(0) - T_\infty$. In the numerical solution, we take $c = 0$, $f = 2$, $b = 0$, $u_0 = 0$, and $Q_0 = 1$, and $L = 1$.

1.3.2 The Finite Difference Method

The focus of this book is the application of FEM and FVM to heat transfer and fluid dynamics problems. However, some introductory material on FDMs are necessary. In both FEM and FVM marching in time is done using finite difference methods to solve transient problems. Furthermore, derivative quantities at walls such as heat flux and shear stress are indeed represented using finite difference expressions. In the FDM, the derivatives of the dependent variable(s) in the differential equation are replaced by difference quotients (or a dependent variable is expanded in a Taylor series) that involve the values of the variable at discrete mesh points of the domain. The resulting algebraic equations are solved for the values of the dependent variables at the mesh points after imposing the boundary conditions.

A Taylor series expansion of a function $f(x)$ that has a continuous $(n+1)$th derivative everywhere in the closed interval $[x_0, x]$ is

$$\begin{aligned} f(x) =& f(x_0) + (x - x_0)f'(x_0) + \frac{(x - x_0)^2}{2!} f''(x_0) + \cdots \\ &+ \frac{(x - x_0)^n}{n!} f^{(n)}(x_0) + R(x), \end{aligned} \tag{1.3.3}$$

where $f' = df/dx$, $f'' = d^2f/dx^2$, and $f^{(n)} = d^n f/dx^n$, and $R(x)$ is the *remainder*

$$R(x) = \frac{(x - x_0)^{n+1}}{(n+1)!} f^{(n+1)}(\xi). \tag{1.3.4}$$

Here $x_0 < \xi < x$ or, if $x < x_0$, $x < \xi < x_0$ [or ξ in (x, x_0)]. For $x = x_0 + h$, for any real number $h > 0$, Eq. (1.3.3) becomes

$$\begin{aligned} f(x_0 + h) =& f(x_0) + hf'(x_0) + \frac{h^2}{2!} f''(x_0) + \frac{h^3}{3!} f'''(x_0) + \cdots \\ &+ \frac{h^n}{n!} f^{(n)}(x_0) + R(x). \end{aligned} \tag{1.3.5}$$

For $x = x_0 - h$, Eq. (1.3.5) becomes

$$f(x_0 - h) = f(x_0) - hf'(x_0) + \frac{h^2}{2!}f''(x_0) - \frac{h^3}{3!}f'''(x_0) + \cdots + (-1)^n\frac{h^n}{n!}f^{(n)}(x_0) + R(x). \tag{1.3.6}$$

If we truncate the Taylor series in Eq. (1.3.5) after the second term on the right-hand side and solve for $f'(x_0)$, we obtain

$$f'(x_0) \approx \frac{f(x_0 + h) - f(x_0)}{h} + O(h). \tag{1.3.7}$$

Here "O" refers to the "order of" the remainder. Equation (1.3.7) is known as the *forward difference formula.* Clearly, the forward difference approximation of a derivative has error (truncation error) of the order h as this is the highest-order term that is dropped for the approximation. Since h is a step size that is measured in relation to the domain length, the value of $h < 1$ and $h > h^2 > h^3, \ldots$. So (1.3.7) is called a first-order accurate forward difference expression or formula. Similarly, truncating the series in Eq. (1.3.6), we obtain

$$f'(x_0) \approx \frac{f(x_0) - f(x_0 - h)}{h} + O(h), \tag{1.3.8}$$

which is known as the *first-order backward difference scheme.* The forward and backward difference schemes can be interpreted as the (discrete) slopes of $f(x)$ at point x_0 using its values at x_0 and $x_0 + h$ or at x_0 and $x_0 - h$, respectively (see Fig. 1.3.1).

Adding Eqs. (1.3.5) and (1.3.6), and solving for the second derivative $f''(x_0)$, we obtain

$$f''(x_0) \approx \frac{f(x_0 - h) - 2f(x_0) + f(x_0 + h)}{2h^2} + O(h^2) \tag{1.3.9}$$

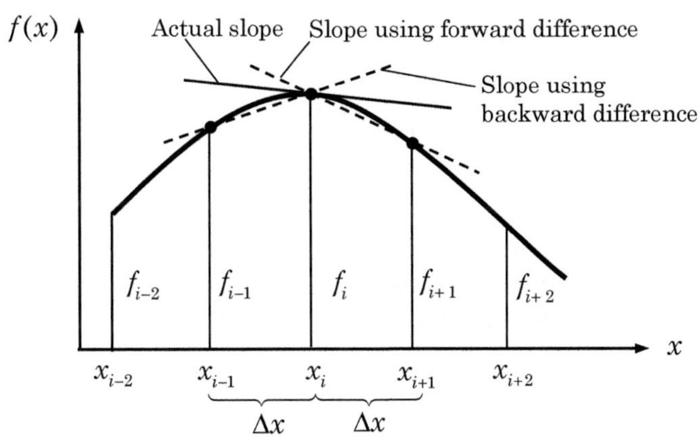

Fig. 1.3.1 Forward and backward schemes to approximate the slope of a function.

which is known as the *second-order central difference formula* for the approximation of the second derivative of the function $f(x)$. Equations (1.3.7), (1.3.8), and (1.3.9) are only a few examples of many possible approximations of first and second derivatives with different orders of errors. For example, a finite difference formula for the first derivative that has second-order accuracy is given by the central difference formula

$$f'(x_0) \approx \frac{f(x_0+h) - f(x_0-h)}{2h} + O(h^2). \tag{1.3.10}$$

In dealing with heat transfer and fluid flow problems, particularly flow of Newtonian fluids, the most commonly occurring differential operators in the model equations are the first and second derivative terms. Accordingly, first order and second order forward, backward, and central difference expressions for first and second derivatives are presented below for the sake of completeness.

Second order forward difference expressions

$$f'(x_0) = \frac{-3f(x_0) + 4f(x_0+h) - f(x_0+2h)}{2h} \tag{1.3.11}$$

$$f''(x_0) = \frac{2f(x_0) - 5f(x_0+h) + 4f(x_0+2h) - f(x_0+3h)}{h^3}. \tag{1.3.12}$$

Second order backward difference expressions

$$f'(x_0) = \frac{3f(x_0) - 4f(x_0-h) + f(x_0-2h)}{2h} \tag{1.3.13}$$

$$f''(x_0) = \frac{2f(x_0) - 5f(x_0-h) + 4f(x_0-2h) - f(x_0-3h)}{h^3}. \tag{1.3.14}$$

As an example of an application of the FDM, we consider a second-order differential equation with mixed boundary conditions (i.e., the Dirichlet boundary condition at $x = 0$ and the Neumann boundary condition at $x = 1$):

$$\begin{aligned} -\frac{d^2u}{dx^2} &= 2, \quad 0 < x < 1 \\ u(0) &= 0, \quad \left[\frac{du}{dx}\right]_{x=1} = 1. \end{aligned} \tag{1.3.15}$$

The exact solution of this simple problem is $u(x) = 3x - x^2$.

In general, the given domain is divided into a set of subintervals, as shown in Fig. 1.3.2(a). The finite difference approximation of the given equation requires the use of the central difference formula in Eq. (1.3.9) with $u_{i-1} \equiv u(x_0 - h)$, $u_i \equiv u(x_0)$ and $u_{i+1} \equiv u(x_0 + h)$:

$$-u_{i-1} + 2u_i - u_{i+1} = 2h^2, \tag{1.3.16}$$

which is valid, strictly speaking, only for an interior mesh point. Suppose that the domain (0, 1) is divided into two equal subintervals of length $\Delta x = h = 0.5$.

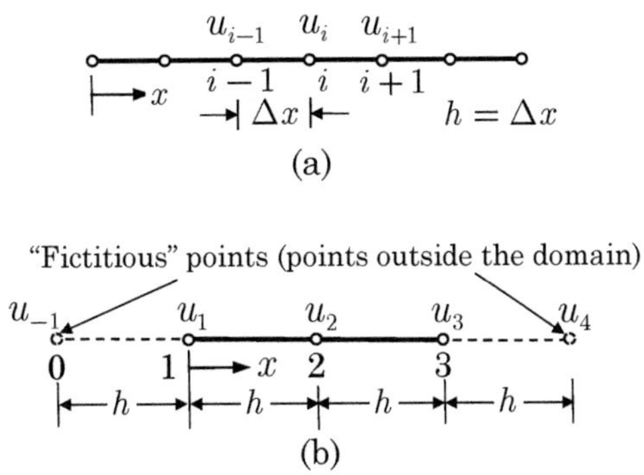

Fig. 1.3.2 (a) Typical stencil associated with the central difference approximation of a second-order differential equation. (b) Discretization of a domain in the FDM (the open circles indicate mesh points).

Then Eq. (1.3.12) for $i = 2$ [see Fig. 1.3.2(b)] gives

$$-u_1 + 2u_2 - u_3 = 2h^2. \tag{1.3.17}$$

Mesh points 1 and 3 are the boundary points, which are subject to boundary conditions. Clearly, the Dirichlet boundary condition at $x = 0$ gives $u_1 = 0$. The Neumann boundary condition at $x = 1$ can be treated in a number of ways. For the same order of accuracy as that used for the approximation of u inside the domain, we may use Eq. (1.3.10) to replace du/dx at $x = 1$ (i.e., $i = 3$) and end up having a *fictitious* mesh point 4:

$$\frac{u_{i+1} - u_{i-1}}{2h} = 1 \quad \Rightarrow \quad u_4 - u_2 = 2h \text{ or } u_4 = 2h + u_2. \tag{1.3.18}$$

Application of the formula from Eq. (1.3.16) at mesh point 3 gives

$$u_2 - 2u_3 + u_4 = -2h^2 \quad \Rightarrow \quad u_3 - u_2 = h^2 + h. \tag{1.3.19}$$

With $u_1 = 0$, Eqs. (1.3.17)–(1.3.19) give $u_2 = 1.25$ and $u_3 = 2$, which coincides with the exact values for this simple problem.

In the interest of avoiding the fictitious mesh point, one may use the second-order backward difference formula to implement the boundary condition at $x = 1$, $(du/dx) = (3u_3 - 4u_2 + u_1)/2h$, resulting in $-4u_2 + 3u_3 = 1$. By solving this equation along with the finite difference expression for u_2, one gets $u_2 = 1.25$ and $u_3 = 2.0$.

1.3.3 The Finite Volume Method

In the FVM a set of *control volumes*[1] that cover the entire domain is selected, including points on the boundary, as shown in Fig. 1.3.3(a). At the center of

[1]The phrase "control domain" in place of the phrase "control volume" is more appropriate because the latter is not meaningful for one- and two-dimensional problems. However, to avoid confusion for readers, we stick with the commonly used terminology.

each control volume, a node (or a mesh point) is located. The discrete values of the dependent variables are calculated at these nodes. The differential equation under consideration is satisfied in an integral sense over each control volume. The discretized form of the equation is obtained through a combination of numerical integration and differentiation in terms of the nodal values. As shown in the following discussion of the method, this leads to a discretized form that closely resembles the finite difference equation valid for a typical mesh point in the discretized domain.

To see the specific steps of obtaining the discretized equations in the FVM, we consider the same example problem as in Eq. (1.3.15). First, we develop the discretized equations for a typical interior node by considering the control volume around the node i [see Fig. 1.3.3(b)].

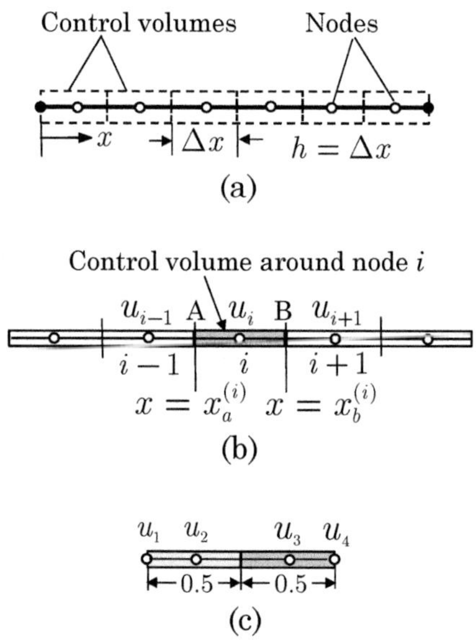

Fig. 1.3.3 (a) Typical node-centered control volume. (b) Discretization of a domain in the FVM. (c) Discretization using two control volumes.

The differential equation is satisfied in the integral sense over the control volume by requiring [see Fig. 1.3.3(b)]

$$0 = \int_{x_a^{(i)}}^{x_b^{(i)}} \left(-\frac{d^2u}{dx^2} - 2 \right) dx. \tag{1.3.20a}$$

By carrying out the integration, we obtain

$$0 = \left[-\frac{du}{dx} \right]_{x_a^{(i)}}^{x_b^{(i)}} - 2\left(x_b^{(i)} - x_a^{(i)} \right). \tag{1.3.20b}$$

Next, we approximate the derivatives $(x_b^{(i)} - x_a^{(i)} = \Delta x = h)$

$$\left[-\frac{du}{dx}\right]_{x_b^{(i)}} \approx \frac{u_i - u_{i+1}}{h}, \quad \left[\frac{du}{dx}\right]_{x_a^{(i)}} \approx \frac{u_i - u_{i-1}}{h}. \tag{1.3.21}$$

Substituting these into Eq. (1.3.20b), we obtain

$$-u_{i-1} + 2u_i - u_{i+1} = 2h^2, \tag{1.3.22}$$

which is identical to the finite difference formula in Eq. (1.3.16) for an interior node which is equidistant from the left- and right-side nodes. Otherwise, the formula must be modified to account for a variable h.

First, we use two control volumes of equal length in the domain (0, 1) (i.e., $h = 0.5$), as shown in Fig. 1.3.3(c). We note that the nodes 1, 2, 3, and 4 are located at $x = 0, 0.25, 0.75$, and 1.0, respectively. Thus, Eq. (1.3.22) is not valid at $x = 0$ and $x = 1.0$ as they are the boundary points, and differential equations are not valid at the boundary points. For control volume 1, which is between $x = 0$ and $x = 0.5$, Eq. (1.3.21) takes the form

$$\left[-\frac{du}{dx}\right]_{x=0.5} \approx \frac{u_2 - u_3}{h}, \quad \left[\frac{du}{dx}\right]_{x=0} \approx \frac{u_2 - u_1}{0.5h}. \tag{1.3.23a}$$

Similarly, for control volume 2, which is between $x = 0.5$ and $x = 1$, we have

$$\left[-\frac{du}{dx}\right]_{x=1} \approx \frac{u_3 - u_4}{0.5h}, \quad \left[\frac{du}{dx}\right]_{x=0.5} \approx \frac{u_3 - u_2}{h}. \tag{1.3.23b}$$

Thus, we have the following two equations from Eq. (1.3.20b) in four nodal values:

$$u_2 - u_3 + 2u_2 - 2u_1 = 2h^2, \quad 2u_3 - 2u_4 + u_3 - u_2 = 2h^2.$$

Since $u_1 = 0$, we have two equations in three unknowns, u_2, u_3, and u_4. The additional equation is given by the flux boundary condition at $x = 1$. Using the backward difference scheme, we can write the boundary condition $du/dx = 1$ as

$$\frac{u_4 - u_3}{0.5h} = 1 \quad \text{or} \quad 2u_4 - 2u_3 = h.$$

In summary, we have the following three equations in three unknowns, u_2, u_3, and u_4: $3u_2 - u_3 = 0.5$, $-u_2 + 3u_3 - 2u_4 = 0.5$, and $-2u_3 + 2u_4 = 0.5$, or, in matrix form,

$$\begin{bmatrix} 3 & -1 & 0 \\ -1 & 3 & -2 \\ 0 & -2 & 2 \end{bmatrix} \begin{Bmatrix} u_2 \\ u_3 \\ u_4 \end{Bmatrix} = \begin{Bmatrix} 0.5 \\ 0.5 \\ 0.5 \end{Bmatrix}. \tag{1.3.24}$$

The solution of these equations is $u_2 = 0.75$, $u_3 = 1.75$, and $u_4 = 2$, whereas the exact solution is $u(0.25) = 0.6875$, $u(0.75) = 1.6875$, and $u(1) = 2$. As the number of control volumes (NCV) is increased the error between the exact and FVM solution decreases, as can be seen from Fig. 1.3.4.

An alternative formulation of the FVM that utilizes half control volumes at the boundary nodes is discussed in **Problem 1.5**.

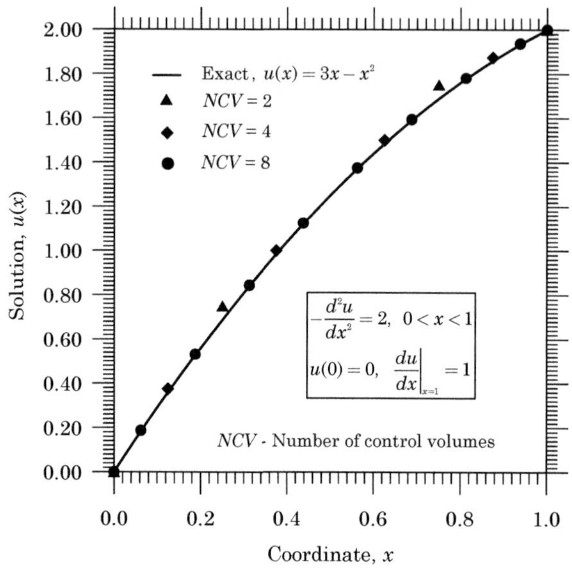

Fig. 1.3.4 Comparison of the FVM solutions with the exact solution for different number of control volumes.

1.3.4 The Finite Element Method

In the FEM, the domain is represented as a collection of nonoverlapping subdomains, called *finite elements* [see Fig. 1.3.5(a)]. However, unlike in the FVM, each element has nodes on its boundary that are used to connect adjacent elements through continuity of the solution (u) and balance of the flux at the nodes common to elements. In addition, the dependent unknown u is explicitly approximated (or interpolated) over the eth element as

$$u(x) \approx u_h^e(x) \equiv \sum_{j=1}^{n} u_j^{(e)} \psi_j^{(e)}(x), \tag{1.3.25}$$

where n is the number of points, called *nodes*, in the element that facilitate the derivation of the interpolation functions $\psi_j^{(e)}$, and $u_j^{(e)}$ are the values of u at the jth node in the element ($j = 1, 2$). For example, a line element with two nodes (i.e., $n = 2$), necessarily the end points, is used to represent the linear approximation of the dependent variable as a linear combination of two nodal values ($u_1^{(e)}, u_2^{(e)}$) of the variable u; then the interpolation functions $\psi_1^{(e)}$ and $\psi_2^{(e)}$ are derived to be:

$$\psi_1^{(e)}(x) = \frac{x_2^e - x}{h_e}, \quad \psi_2^{(e)}(x) = \frac{x - x_1^e}{h_e}. \tag{1.3.26}$$

When $n = 3$, a third node is identified inside the element domain, and the associated approximation becomes quadratic. The approximation in Eq. (1.3.25) provides a means to compute the derivatives and the evaluation of integrals

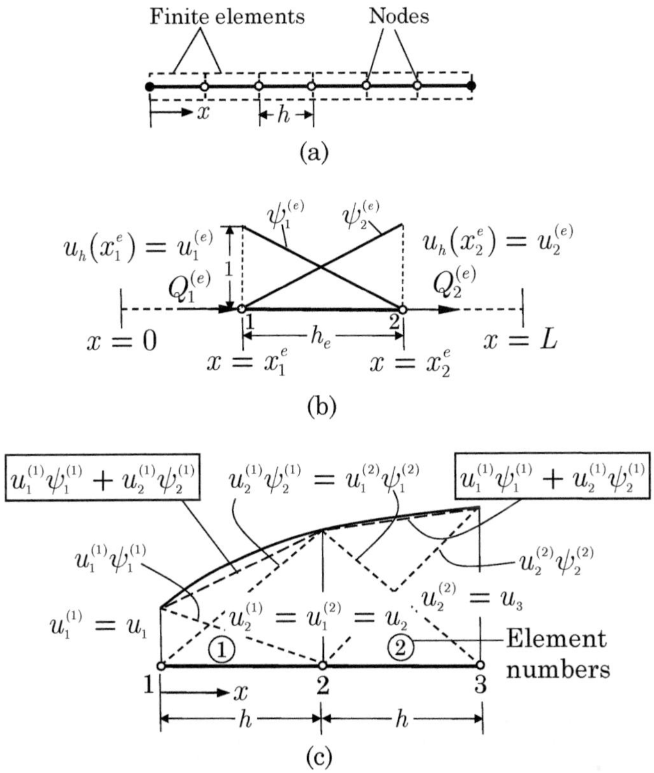

Fig. 1.3.5 (a) Typical (linear) finite element. (b) Discretization of a domain in the FEM. (c) Assembly of two (linear) finite elements with their local and global approximation functions.

involving u and its derivative. In the FEM, the steps involved in the derivation of the discretized equations in algebraic form are independent of the specific value of n; only when the coefficients are to be evaluated as numbers one must select n and associated interpolation functions. In the sequel, the parenthesis around the element label is omitted for brevity.

In the FEM, the governing equation is satisfied in a *weighted-integral sense* (i.e., the differential equation, with all its terms on one side of the equation, is multiplied by a weight function, and the integral of the resulting expression over the element domain is set to zero). In most conventional finite element formulations, one uses the so-called "weak form" obtained by integrating the second-order differential equation once by parts and transferring one of the two derivatives to the weight function. The integration by parts yields a boundary expression containing the weight function and the first derivative of the dependent variable u. The quantity multiplying the weight function in the boundary expression is identified as the *secondary variable* while the dependent variable is called the *primary variable.* Thus, there is a pair of primary and secondary variables: $(u, F(u'))$, where $F(u')$ is a function of the first derivative. In a weak form, the dependent variable can be approximated with first-degree polynomials as opposed to second-degree polynomials required to represent a second-order derivative. The weak form, after substituting the approximation in Eq. (1.3.25), leads to a set of discretized equations among the nodal values of the variable and its dual variable. The element equations are "assembled" using the continuity of the primary variable and balance of the secondary variable at the interelement boundaries to obtain a system of discretized equations among all nodal values of the total domain, which are then solved after imposing the boundary conditions. In general, the FEM has more formulative steps than the FDM or FVM, but FEM is much more modular and easier to implement in a computer for an arbitrary degree of interpolation.

Next, the steps involved in the FEM as applied to the example problem described by Eq. (1.3.15) are discussed. For illustrative purposes, the problem domain (0, 1) is divided into two linear finite elements, as shown in Fig. 1.3.5(b). The approximate solution u_h^e over a typical element is given by Eq. (1.3.25). With the approximation, the differential equation becomes

$$-\frac{d^2u_h^e}{dx^2} - 2 \equiv \mathbb{R}^e(x, u_1^e, u_2^e), \tag{1.3.27}$$

where $\mathbb{R}^e$ denotes the residual (error) in the differential equation. The weighted-integral of the residual and the resulting weak form are set to zero:

$$0 = \int_{x_1^e}^{x_2^e} w_i^e \left(-\frac{d^2u_h^e}{dx^2} - 2\right) dx \tag{1.3.28a}$$

$$= \int_{x_1^e}^{x_2^e} w_i^e \left(\frac{dw_i^e}{dx}\frac{du_h^e}{dx} - 2w_i^e\right) dx - \left[w_i^e(x)\frac{du_h^e}{dx}\right]_{x_1^e}^{x_2^e} \tag{1.3.28b}$$

$$= \int_{x_1^e}^{x_2^e} w_i^e \left(\frac{dw_i^e}{dx}\frac{du_h^e}{dx} - 2w_i^e\right) dx - w_i^e(x_1^e)Q_1^e - w_i^e(x_2^e)Q_2^e, \tag{1.3.28c}$$

where $w_i^e(x)$ is the weight function and

$$Q_1^e \equiv -\left.\frac{du_h^e}{dx}\right|_{x=x_1^e}, \quad Q_2^e \equiv \left.\frac{du_h^e}{dx}\right|_{x=x_2^e}. \tag{1.3.29}$$

We note that in Eq. (1.3.28b) one derivative from u_h^e is transferred to the weight function w_i^e thereby weakening the differentiability of ψ_i^e. That is, the weighted-integral statement in Eq. (1.3.28a) requires at least quadratic approximation of u, whereas the statement in Eq. (1.3.28b) requires only linear approximation of u. Therefore, Eq. (1.3.28b), equivalently Eq. (1.3.28c), is called a *weak form.* The integration by parts used in Eq. (1.3.28b) results in a boundary expression $(w_i^e, du_h^e/dx)$, indicating that du_h^e/dx is the secondary variable and $w_i^e \sim u_h^e$ is the primary variable. The secondary variable evaluated at the left and right nodes of the element are denoted by Q_1^e and Q_2^e, respectively; Q_1^e and Q_2^e are dual to the primary variables u_1^e and u_2^e, respectively. The word *duality* is used here to mean that the pair (u_i^e, Q_i^e) constitutes "cause" and "effect" (i.e., change in one affects the other). When the weight function is taken to be the same as interpolation function $w_i^e = \psi_i^e$ $(i = 1, 2, \ldots, n)$, the finite element model is known as the *weak-form Galerkin finite element model.*

Substitution of Eq. (1.3.25) into the weak form (1.3.28c) and replacing w_i^e with ψ_i^e $(i = 1, 2, \ldots, n)$ yields the discretized equations:

$$\mathbf{K}^e \mathbf{u}^e = \mathbf{f}^e + \mathbf{Q}^e, \tag{1.3.30a}$$

where

$$K_{ij}^e = \int_{x_1^e}^{x_2^e} \frac{d\psi_i^e}{dx}\frac{d\psi_j^e}{dx}\,dx, \quad f_i^e = \int_{x_1^e}^{x_2^e} 2\psi_i^e\,dx. \tag{1.3.30b}$$

For the choice of linear approximation ($n = 2$), Eq. (1.3.30b) takes the form

$$\frac{1}{h_e}\begin{bmatrix} 1 & -1 \\ -1 & 1 \end{bmatrix} = 2 \times \frac{h_e}{2}\begin{Bmatrix} 1 \\ 1 \end{Bmatrix} + \begin{Bmatrix} Q_1^{(e)} \\ Q_2^{(e)} \end{Bmatrix}. \tag{1.3.31}$$

For the uniform mesh of two elements shown in Fig. 1.3.4(b), the assembled equations are ($h_1 = h_2 = h = 0.5$)

$$\frac{1}{h}\begin{bmatrix} 1 & -1 & 0 \\ -1 & 1+1 & -1 \\ 0 & -1 & 1 \end{bmatrix}\begin{Bmatrix} U_1 \\ U_2 \\ U_3 \end{Bmatrix} = h\begin{Bmatrix} 1 \\ 1+1 \\ 1 \end{Bmatrix} + \begin{Bmatrix} Q_1^{(1)} \\ Q_2^{(1)} + Q_1^{(2)} \\ Q_2^{(2)} \end{Bmatrix}, \tag{1.3.32}$$

where $U_1 = u_1^{(1)}$, $U_2 = u_2^{(1)} = u_1^{(2)}$ and $U_3 = u_2^{(2)}$. Note that the equation for an interior global node I (see the second row of the above equation) is the same as in the FDM and FVM:

$$-U_{I-1} + 2U_I - U_{I+1} = 2h^2. \tag{1.3.33}$$

This similarity may hold only for the linear approximation.

The boundary conditions require $U_1 = 0$ and $Q_2^{(2)} = 1.0$ and the balance of the secondary variables at node 2 requires $Q_2^{(1)} + Q_1^{(2)} = 0$. Hence, we obtain ($1/h = 2$)

$$\begin{bmatrix} 2 & -2 & 0 \\ -2 & 4 & -2 \\ 0 & -2 & 2 \end{bmatrix}\begin{Bmatrix} 0 \\ U_2 \\ U_3 \end{Bmatrix} = \begin{Bmatrix} 0.5 \\ 1.0 \\ 0.5 \end{Bmatrix} + \begin{Bmatrix} Q_1^{(1)} \\ 0 \\ 1 \end{Bmatrix}. \tag{1.3.34}$$

Solving the last two equations for U_2 and U_3, we obtain

$$U_2 = 1.25, \quad U_3 = 2.0, \quad Q_1^{(1)} = -3.0. \tag{1.3.35}$$

Using the first equation in Eq. (1.3.34), one can determine $Q_1^{(1)}$ as

$$Q_1^{(1)} = -3.0. \tag{1.3.36}$$

The finite element solution in Eqs. (1.3.35) and (1.3.36) agrees with the exact values of u at $x = 0.5$ and $x = 1.0$ and the exact value of $-du/dx$ at $x = 0$.

1.4 Errors and Convergence

1.4.1 Types of Errors

Numerical solutions involve approximations of various kinds. These approximations lead to errors in the solution. It is important to know the difference between a "mistake" and an "error," although they are often used interchangeably. A mistake is a human attribute. There are many definitions of a mistake in different contexts. In the current context of technical work and scientific computing, a mistake is an unintentional act of writing, deriving, or composing a logical step or an expression. A technical mistake can be either avoided or corrected with effort.

An error is the difference between the actual value and its approximation. The most common types of errors are truncation errors and round-off errors. A *truncation error* is the error that is introduced when an exact mathematical expression is replaced by an approximate one (usually by truncating a series). For example, if we approximate the sine function by the first two nonzero terms of its Taylor series, as in $\sin(x) \approx x - \frac{1}{3!}x^3$ for small x, the resulting error is a truncation error. It is present even with infinite-precision arithmetic, because it is caused by truncation of the infinite Taylor series to form the algorithm. A *round-off error* is a consequence of using finite precision floating-point numbers on computers. Rounding errors are caused by inexactness in the representation of real numbers and the arithmetic operations done with them. Thus, it is the difference between the result produced by a given algorithm using exact arithmetic and the result produced by rounded arithmetic (finite-precision arithmetic). For example, the value of π can be truncated to four decimal points 3.1416 or to eight decimal points 3.141 592 65 by rounding of the actual value (which is an infinite decimal-point number) to a desired decimal point. Similarly, a fractional number such as 1/3 can be approximated as 0.333, 0.3333, or 0.333 333, and so on. When exact numbers (1/3) are represented by approximate numbers (0.333...3, with a finite number of decimal points), we say that the numbers are rounded-off (see [17] and [18] for additional discussion).

The round-off errors and truncation errors have an inverse relationship. The truncation error decreases with a decrease in the time step or grid size. However, small Δx means more mesh points and more calculations, hence more round-off errors [16]. This inverse relation is depicted in Fig. 1.4.1.

For a stable and accurate solution, one must determine optimal values of Δt and Δx so that the total error is the minimum. Also, we should always seek numerical solutions that are independent of the time step and mesh size. For example, if S^k is the solution with Δt_k and S^{k+1} is the solution with Δt_{k+1} then if $|S^{k+1} - S^k| < \varepsilon$ (measured in a suitable norm), where ε is a very small number of the order 10^{-4} or so, then one may declare that the solution obtained with the step size Δt_k is a grid- (or mesh-) independent solution. The goal of a numerical exercise is to find an approximate solution, quantify various types of errors, and develop strategies to minimize them. The word "accuracy" speaks to the proximity of an approximate solution to the true value. For numerical methods involving meshes or grids (which are characterized in terms of the distance between grid points or size of the subdomains), the errors are quantified in terms of the characteristic number, called the mesh parameter, h.

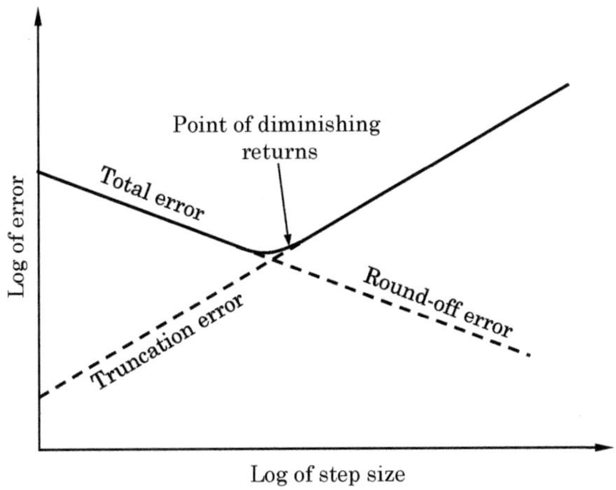

Fig. 1.4.1 Round-off errors versus truncation errors. Total error, which is a summation of round-off errors and truncation errors, decreases with an increase in step size up to some point and then increases with step size due to the increase in the round-off errors.

For a method to be "consistent," the error (e) should decrease and go to zero as the grid is refined; that is, $e \to 0$ as $h \to 0$.

1.4.2 Numerical Convergence

Irrespective of the method used to solve a heat transfer or fluid flow problem by either the FVM or FEM results in a system of linear algebraic equations. These equations are often coupled. Invariably, formulations are implicit for solving incompressible flow and heat transfer problems. Typically, the method adapted for the solution of equations involves a combination of direct and iterative procedures. When an iterative technique is used to solve the equations, the issue of declaring convergence arises (i.e., when to stop iterations).

All numerical methods of solving differential equations result, after the imposition of boundary and initial conditions, in equations of the form

$$\mathbf{KT} = \mathbf{F}. \tag{1.4.1}$$

In Eq. (1.4.1), $\mathbf{T}$ is the $N \times 1$ vector of dependent unknowns, $\mathbf{K}$ is the $N \times N$ coefficient matrix, and $\mathbf{F}$ is the $N \times 1$ source vector. Typically, the coefficient matrix depends on material and geometric parameters of the problem, mesh size, and time increment.

Now suppose that the set of N linear equations can be solved using an iterative method. Let the solution vector at the end of the kth iteration be denoted by $\mathbf{T}^k$. One way to declare convergence is to monitor the change in the solution vector $\mathbf{T}$ during each iteration and declare convergence when the solution vector between two consecutive iterations changes by an a priori specified small value. The measure of change can be the sup-norm or L_2-norm:

$$\max\left|\frac{T_I^{k+1}-T_I^k}{T_I^k}\right| < \epsilon \quad \text{or} \quad \sqrt{\frac{\sum_{I=1}^N |T_I^{k+1}-T_I^k|^2}{\sum_{I=1}^N |T_I^{k+1}|^2}} \le \epsilon, \tag{1.4.2}$$

where ϵ is the error tolerance. If one does not normalize the error, when the true value of the dependent variable is very small, then there may be an appearance of convergence but the solution has really not converged.

Alternatively, one can also monitor residual vector $\mathbf{R}$ at the kth iteration (assuming that $\mathbf{F}$ is independent of $\mathbf{T}$),

$$\mathbf{R}^k \equiv \mathbf{K}^k\mathbf{T}^k - \mathbf{F} \quad \text{or} \quad R_I^k \equiv \sum_{J=1}^N K_{IJ}^k T_J^k - F_I, \tag{1.4.3}$$

by requiring

$$\max\left(|R_I^k|\right) < \epsilon \quad \text{or} \quad \sqrt{\sum_{I=1}^N |R_I^k|^2} \le \epsilon. \tag{1.4.4}$$

In reality, monitoring change in the dependent variable and the residual are equivalent. This can be illustrated through an approximate analysis presented next. Both matrices $\mathbf{K}$ and $\mathbf{F}$ are assumed to be stationary. Let

$$\Delta\mathbf{T}^k = \mathbf{T}^{k+1} - \mathbf{T}^k. \tag{1.4.5}$$

Matrix $\mathbf{K}$ can be decomposed as

$$\mathbf{K} = \mathbf{L} + \mathbf{D} + \mathbf{U}, \tag{1.4.6}$$

where $\mathbf{L}$ is the lower triangular matrix, $\mathbf{U}$ is the upper triangular matrix, and $\mathbf{D}$ is the diagonal matrix.

Invoking the Jacobi (or Picard) iteration, we have

$$\mathbf{D}\mathbf{T}^{k+1} = \mathbf{F} - \mathbf{L}\mathbf{T}^k - \mathbf{U}\mathbf{T}^k. \tag{1.4.7}$$

Subtracting $\mathbf{D}\mathbf{T}^k$ from both sides of Eq. (1.4.7):

$$\begin{aligned}\mathbf{D}\mathbf{T}^{k+1} - \mathbf{D}\mathbf{T}^k &= \mathbf{F} - \mathbf{L}\mathbf{T}^k - \mathbf{U}\mathbf{T}^k - \mathbf{D}\mathbf{T}^k \\ \mathbf{D}\left(\mathbf{T}^{k+1} - \mathbf{T}^k\right) &= \mathbf{F} - (\mathbf{L}+\mathbf{U}+\mathbf{D})\,\mathbf{T}^k = \mathbf{F} - \mathbf{K}\mathbf{T}^k,\end{aligned} \tag{1.4.8}$$

or

$$\mathbf{D}\,\Delta\mathbf{T}^k = -\mathbf{R}^k. \tag{1.4.9}$$

For convergence, as $k \to \infty$, we have $\mathbf{R} \to \mathbf{0}$. If $\mathbf{T}_0$ is the exact solution and error is $\mathbf{E}$ then

$$\mathbf{E}^k = \mathbf{T}_0 - \mathbf{T}^k. \tag{1.4.10}$$

By definition

$$\mathbf{R} = \mathbf{K}\mathbf{T}_0 - \mathbf{F} = \mathbf{0}. \tag{1.4.11}$$

Subtracting Eq. (1.4.3) from Eq. (1.4.11)

$$\begin{aligned} \mathbf{R} - \mathbf{R}^k &= \mathbf{K}\left(\mathbf{T}_0 - \mathbf{T}^K\right) - \mathbf{F} + \mathbf{F} \\ 0 - \mathbf{R}^k &= \mathbf{K}\left(\mathbf{T}_0 - \mathbf{T}^k\right). \end{aligned} \tag{1.4.12}$$

Combining Eqs. (1.4.9) and (1.4.13) we obtain

$$-\mathbf{D}(\Delta \mathbf{T}^k) = \mathbf{K}\mathbf{E}^k \quad \rightarrow \quad -\mathbf{R}^k = \mathbf{K}\mathbf{E}^k. \tag{1.4.13}$$

It is clear that monitoring residuals and changes in dependent variables are equivalent to monitoring error ($\mathbf{E}$). Thus we conclude that monitoring a change in the dependent variable is equivalent to monitoring residuals (see Kim et al. [19]).

1.4.3 Order of Accuracy and Grid Convergence Index

In this section, we will discuss the order of accuracy of a numerical approximation using an example. Let us represent a second derivative $f''(x)$ with the following finite difference approximation:

$$f''(x) = \frac{f(x-h) - 2f(x) + f(x+h)}{h^2}. \tag{1.4.14}$$

We know that the exact differential equation (EDE) = finite difference approximation (FDA) + truncation error (TE)

Using the Taylor series one can show that the above FDA is second-order accurate because the highest-order term dropped in the above approximation is $O(h^2)$ and magnitudes of higher-order terms like $O(h^3)$, $O(h^4)$, ... are less than the $O(h^2)$ term. In grid-based methods, the grid size or grid parameter (h) is far less than the domain characteristic dimension; for example $h/L < 1$ for one-dimensional problems and $h/a < 1$ and $h/b < 1$ for two-dimensional problems with rectangular domains of dimensions $a \times b$. Therefore, it is evident that as $h \to 0$, TE $\to 0$. In other words, by reducing the grid parameter h, we are reducing the truncation error. The goal is to systematically reduce the grid parameter (h) so that the truncation error goes to a minimum. However, the total error is a sum of the truncation error and the round-off error, and further reduction in the grid parameter (h) may cause the round-off error to increase to the point that one cannot indiscriminately decrease the grid parameter (h) (see Fig. 1.4.1). The grid parameter (h) corresponding to the point where further reduction in h results in minimal change in the calculated variable of interest is called the *grid-independent solution*. A process to establish grid independence is often referred to as a *grid sensitivity study* in the literature.

One may monitor a change in variable such as the maximum velocity or maximum temperature or Nusselt number with a change (refinement) in the grid parameter (h) to declare grid independence. For example, consider a situation:

if changes in the Nusselt number are less than 10^{-4}, when the grid parameter (h) is halved in a second-order method and in a fourth-order method, can we conclude both situations are identical? The answer is no. To declare grid independence, one has to consider both the grid parameter (h) and the order of the method (p). To address this, the concept of grid convergence index (GCI) was developed by Roach [20].

For a particular grid, if the calculated value is ϕ and the true value is ϕ_t then the error (e) is given by

$$e = \phi_t - \phi. \tag{1.4.15}$$

The uncertainty estimate, $U_{x\%}$, provides a range $\phi \pm U_{x\%}$ within which the true value ϕ_t probably lies with a probability of $x\%$, that is [21],

$$\phi_t \in [\phi \pm U_{x\%}]. \tag{1.4.16}$$

Now one has to quantify the probability to estimate the uncertainty. The most widely used target is 95% (i.e., there is 95% probability that the true value ϕ_t lies in the range $\phi \pm U_{95\%}$). The 95% confidence level is equivalent to 2σ range of Gaussian distribution. Richardson extrapolation (RE) is based on the assumption [21] that discrete solutions ϕ have a power series representation in grid spacing h. If the formal order of accuracy of an algorithm is known, for example, $O(h^2)$, then reducing the grid size by a factor of two would reduce the error by a factor of four. So based on the assumption made by Richardson,

$$\phi = \phi_t + g_1 h + g_2 h^2 + g_3 h^3 + \cdots, \tag{1.4.17}$$

where $g_1, g_2, \ldots$ are continuous functions of any discretization technique. For a second-order method, $g_1 = 0$. Let us consider ϕ_1 and ϕ_2, two different discrete solutions on two different grids, h_1 and h_2. We obtain

$$\phi_1 = \phi_t + g_2 h_1^2 + \text{H.O.T.} \tag{1.4.18}$$

$$\phi_2 = \phi_t + g_2 h_2^2 + \text{H.O.T.}, \tag{1.4.19}$$

where H.O.T. stands for higher-order terms. Subtracting Eq. (1.4.19) from (1.4.18)

$$\phi_2 - \phi_1 = g_2(h_2^2 - h_1^2), \quad g_2 = \frac{\phi_2 - \phi_1}{h_2^2 - h_1^2}. \tag{1.4.20}$$

Then,

$$\phi_1 = \phi_t + \left(\frac{\phi_2 - \phi_1}{h_2^2 - h_1^2}\right) h_1^2 + \text{H.O.T.} \approx \phi_t + \left(\frac{\phi_2 - \phi_1}{r^2 - 1}\right), \tag{1.4.21}$$

where $r = h_2/h_1$ is the grid ratio

$$\phi_t \approx \phi_1 - \left(\frac{\phi_2 - \phi_1}{r^2 - 1}\right) = \frac{\phi_1 r^2 - \phi_2}{r^2 - 1}. \tag{1.4.22}$$

Therefore, for the most commonly used method (i.e., Newton's half-interval method or grid-doubling method), $r = 2$.

$$\phi_t \approx \frac{4}{3}\phi_1 - \frac{1}{3}\phi_2. \tag{1.4.23}$$

Roache [22] recommends that the GCI should be used to uniformly report grid convergence tests. Consider two grid solutions ϕ_1 and ϕ_2; then from Eq. (1.4.21)

$$\underbrace{\phi_t - \phi_1}_{e_1} = \frac{\phi_2 - \phi_1}{r^2 - 1} = \frac{\overbrace{\phi_1 - \phi_2}^{\varepsilon}}{1 - r^2} \tag{1.4.24}$$

$$e_1 = \frac{\varepsilon}{1 - r^2}. \tag{1.4.25}$$

For a formal pth-order accuracy of the algorithm,

$$e_1 = \frac{\varepsilon}{1 - r^p}, \quad r = \frac{h_2}{h_1} > 1. \tag{1.4.26}$$

If we consider Eq. (1.4.21) with pth-order accuracy,

$$\phi_t = \phi_1 + \frac{\phi_1 - \phi_2}{r^p - 1}, \tag{1.4.27}$$

and subtract ϕ_2 from both sides of the equation, we obtain

$$\underbrace{\phi_t - \phi_2}_{e_2} = \underbrace{\phi_1 - \phi_2}_{-\varepsilon} + \frac{\overbrace{\phi_1 - \phi_2}^{-\varepsilon}}{r^p - 1} \tag{1.4.28}$$

$$e_2 = -\varepsilon - \frac{\varepsilon}{r^p - 1} \tag{1.4.29}$$

$$= \frac{\varepsilon\, r^p}{1 - r^p}. \tag{1.4.30}$$

Roache [22] incorporated a safety factor (F_S) to present error estimates due to various factors and to place all grid convergence studies on the same basis as the Newton half interval method by defining a GCI as

$$GCI_1 = F_S |e_1| \tag{1.4.31}$$

$$GCI_2 = F_S |e_2|. \tag{1.4.32}$$

So when $F_S = 1$, the error band reduces to the best estimate of the error. Roache [22] recommends a value of 3 for F_S (i.e., in a sense GCIs are three times the error estimates).

In order to estimate the discretization error using the GCI method it is necessary to have at least two grid solutions (ϕ_1 and ϕ_2) and the convergence rate p should be known (e.g., $p = 2$ for a second-order discretization scheme). Problems for which the exact solution (ϕ_t) is known allow one to extract p using Eq. (1.4.22). But for cases where the exact solution is not known, we need at least three solutions (ϕ_1, ϕ_2 and ϕ_3) to extract the order p. Using Eq. (1.4.22) we obtain

$$\phi_t \approx \frac{\phi_1 r_{12}^p - \phi_2}{r_{12}^p - 1}. \tag{1.4.33}$$

Then for the most coarse grid ϕ_3,

$$\phi_t \approx \frac{\phi_2 r_{23}^p - \phi_3}{r_{23}^p - 1}. \tag{1.4.34}$$

Equating Eqs. (1.4.33) and (1.4.34), we obtain

$$\frac{\phi_1 r_{12}^p - \phi_2}{r_{12}^p - 1} = \frac{\phi_2 r_{23}^p - \phi_3}{r_{23}^p - 1}. \tag{1.4.35}$$

If $r_{12} = r_{23} = r$ (but need not be), then

$$\frac{\phi_1 r^p - \phi_2}{r^p - 1} = \frac{\phi_2 r^p - \phi_3}{r^p - 1} \tag{1.4.36}$$

$$(\phi_1 - \phi_2) r^p = \phi_2 - \phi_3 \tag{1.4.37}$$

or

$$r^p = \frac{\phi_2 - \phi_3}{\phi_1 - \phi_2}, \tag{1.4.38}$$

$$p \ln(r) = \ln\left(\frac{\phi_2 - \phi_3}{\phi_1 - \phi_2}\right), \tag{1.4.39}$$

$$p = \frac{\ln(\varepsilon_{23}/\varepsilon_{12})}{\ln(r)}, \tag{1.4.40}$$

where $\varepsilon_{23} = \phi_2 - \phi_3$ and $\varepsilon_{12} = \phi_1 - \phi_2$.

Such an analysis is very helpful in evaluating the order of the method. For example, for a finite volume technique, by the nature of its formulation, it is not possible to declare the order of the method using a Taylor series type analysis. But one can use a GCI approach to estimate the order of the method.

1.5 Veracity of Numerical Solutions

1.5.1 Verification and Validation

Numerical solutions are approximate solutions. The question is "how good are the approximate solutions?". In cases where experimental or field data is available one could compare a numerical solution to the experimental or field data to determine the veracity of the numerical solution. However, one must ensure that the computer program written to execute the steps of a numerical method is correct and the mathematical model used in the numerical method has the ability to capture accurately the physics of a physical system or a process being described. Thus, one must ensure that the numerical method is capable of reproducing the exact solution, if one exists, to the problem solved. One must also make sure that the numerical solutions are close to the experimental observations of the response of a system whose mathematical model is numerically simulated. This is known as the "validation" (of the mathematical model).

Roache [22] defines *verification* as solving equations correctly and *validation* as solving the correct mathematical model of the physical process (see also [23]).

The "verification" challenge is to find the exact solution of a mathematical problem under consideration and the "validation" challenge is to find experimental data. Both are not trivial challenges for most real-world problems of complexity. Since a code includes an algorithm (i.e., a systematic procedure to carry out the steps of a numerical method) executed by a computer, it is imperative that a code has to be verified. We note that "verifying" one's numerical solution with someone else's numerical solution does not qualify as a true verification because both numerical solutions could be wrong. When a code is verified by solving a set of problems (each problem differing in the data) which have analytical solutions, we gain the confidence that the series of steps executed to solve a set of PDEs describing the process by a numerical method are correct. In addition, when there is an agreement between a numerical solution and experimental results or field data, the underlying mathematical model is validated. Thus a computer program based on a numerical method and a mathematical model must be both verified and validated in order for us to have the confidence that the results produced are indeed good approximations of the true response of a physical process or a system.

1.5.2 Manufactured Solutions for Verification

The goal of this section is to develop a method of determining the exact solution needed to verify a numerical solution. For more details, the reader is referred to [21] and [24]. It is clear that the verification process depends on the availability of the exact solution of a given PDE for a variety of data (i.e., geometry, boundary conditions, source values, and material parameters). Very few problems, especially in multidimensions, have exact analytical solutions based on conventional mathematical treatments (like direct integration, series method, and separation of variables method). One of the accepted methods is manufactured equation method (MEM) (see [21]). This method is also referred to as manufactured solution method (MSM).

To explain the MSM, suppose that we wish to solve a problem described by a differential operator equation with boundary conditions, posed on a domain Ω with a closed boundary Γ:

$$A(u) = f \quad \text{in} \quad \Omega; \qquad B(u) = q \quad \text{on} \quad \Gamma, \tag{1.5.1}$$

where A and B are linear operators, u is the dependent variable, f is the source term, and q is the boundary value. We assume that the given problem is well-posed and has a solution.

In the MSM, we propose a solution u_m that is continuous and its derivatives of the order up to and including those that appear in the operator A exist and are continuous within the domain, and it satisfies the boundary conditions. The solution must be valid both inside the domain (i.e., satisfy the governing differential equation) and on the boundary (i.e., satisfy the specified boundary conditions). If the boundary conditions are not met by the assumed solution, they must be identified during the process (in a way, we define the problem for which the solution is valid). When the proposed solution is substituted into

$A(u)$ it yields $A(u_m)$, which will not, in most cases, be a zero. In order to satisfy the governing equation $A(u_m) = f$, the residual, $A(u_m)$, must cancel the source term f. Thus, to make the manufactured solution satisfy the governing equation, the source term f is identified as $A(u_m)$, $f = A(u_m)$. Then the proposed solution satisfies the governing equation for the identified source term. To satisfy the boundary conditions, if the assumed solution is not already satisfying them, we evaluate $B(u_m)$ on the boundary Γ and identify it as q. Then the original problem has a manufactured solution u_m that satisfies the operator equation for $f = A(u_m)$ and the boundary conditions, $B(u_m) = q$.

As a specific example of the MSM, we consider steady temperature distribution in a thin, isotropic, rectangular plate of dimensions $a \times b$ (see Fig. 1.5.1). The balance of energy principle, in the absence of internal heat generation, gives the Laplace equation in two dimensions for the temperature field T [the coefficient k multiplying the entire equation is omitted; see Eq. (1.2.33)]:

$$\frac{\partial^2 T}{\partial x^2} + \frac{\partial^2 T}{\partial y^2} = 0. \tag{1.5.2}$$

where k is the thermal conductivity. The prescribed boundary conditions are:

$$T(0, y) = T_1; \quad T(a, y) = T_2; \quad T(x, 0) = T_3; \quad T(x, b) = T_4. \tag{1.5.3}$$

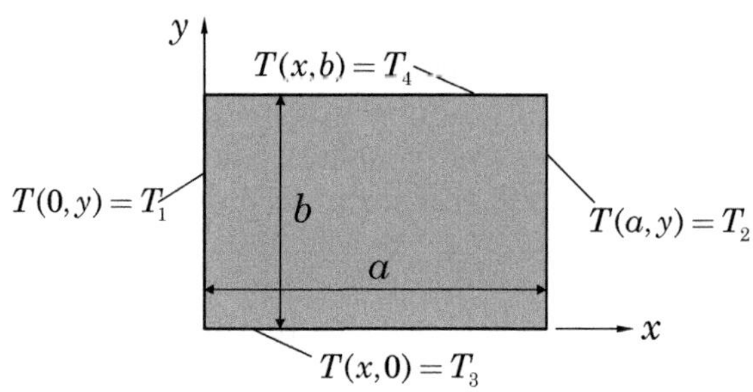

Fig. 1.5.1 Heat diffusion in a thin rectangular plate with specified temperature at all sides (i.e., Dirichlet boundary conditions). Unless $T_1 = T_2 = T_3 = T_4$, the corners have singularities (i.e., nonunique values of the temperature at the corner points).

Next, we propose a manufactured solution of the form

$$T_m(x, y) = T_1 \cos \frac{\pi x}{2a} + T_2 \sin \frac{\pi x}{2a} + T_3 \cos \frac{\pi y}{2b} + T_4 \sin \frac{\pi y}{2b}. \tag{1.5.4}$$

The proposed solution is continuous with all its derivatives in the domain. However, it does not satisfy the specified boundary conditions. By substituting the above proposed solution in Eq. (1.5.4) into the original PDE, that is, Eq.

(1.5.2), we obtain a modified differential equation (MDE),

$$\begin{aligned}
\frac{\partial^2 T_m}{\partial x^2} + \frac{\partial^2 T_m}{\partial y^2} &= \frac{\partial^2}{\partial x^2}\left[T_1 \cos\left(\frac{\pi x}{2a}\right) + T_2 \sin\left(\frac{\pi x}{2a}\right) + T_3 \cos\left(\frac{\pi y}{2b}\right) + T_4 \sin\left(\frac{\pi y}{2b}\right)\right] \\
&+ \frac{\partial^2}{\partial y^2}\left[T_1 \cos\left(\frac{\pi x}{2a}\right) + T_2 \sin\left(\frac{\pi x}{2a}\right) + T_3 \cos\left(\frac{\pi y}{2b}\right) + T_4 \sin\left(\frac{\pi y}{2b}\right)\right] \\
&= \frac{\pi}{2a}\frac{\partial}{\partial x}\left\{ - T_1 \sin\left(\frac{\pi x}{2a}\right) + T_2 \cos\left(\frac{\pi x}{2a}\right)\right\} \\
&+ \frac{\pi}{2b}\frac{\partial}{\partial y}\left\{ - T_3 \sin\left(\frac{\pi y}{2b}\right) + T_4 \cos\left(\frac{\pi y}{2b}\right)\right\} \\
&= -\frac{\pi^2}{4a^2}\left[T_1 \cos\left(\frac{\pi x}{2a}\right) + T_2 \sin\left(\frac{\pi x}{2a}\right)\right] \\
&- \frac{\pi^2}{4b^2}\left[T_3 \cos\left(\frac{\pi y}{2b}\right) + T_4 \sin\left(\frac{\pi y}{2b}\right)\right]. \qquad (1.5.5)
\end{aligned}$$

It is now obvious that

$$S_T = -\frac{\pi^2}{4}\left[\frac{T_1}{a^2}\cos\left(\frac{\pi x}{2a}\right) + \frac{T_2}{a^2}\sin\left(\frac{\pi x}{2a}\right) + \frac{T_3}{b^2}\cos\left(\frac{\pi y}{2b}\right) + \frac{T_4}{b^2}\sin\left(\frac{\pi y}{2b}\right)\right]. \qquad (1.5.6)$$

Thus, the proposed manufactured solution in Eq. (1.5.4) exactly satisfies the MDE in Eq. (1.5.5). The modified boundary conditions can be obtained by evaluating the proposed solution, Eq. (1.5.4), at the boundaries, as explained next.

The new (or modified) boundary conditions of the problem become:

$$\text{On } x = 0 \text{ side:} \quad T(0, y) = T_1 + T_3 \cos\left(\frac{\pi y}{2b}\right) + T_4 \sin\left(\frac{\pi y}{2b}\right) \qquad (1.5.7)$$

$$\text{On } x = a \text{ side:} \quad T(a, y) = T_2 + T_3 \cos\left(\frac{\pi y}{2b}\right) + T_4 \sin\left(\frac{\pi y}{2b}\right) \qquad (1.5.8)$$

$$\text{On } y = 0 \text{ side:} \quad T(x, 0) = T_3 + T_1 \cos\left(\frac{\pi x}{2a}\right) + T_2 \sin\left(\frac{\pi x}{2a}\right) \qquad (1.5.9)$$

$$\text{On } y = b \text{ side:} \quad T(x, b) = T_4 + T_1 \cos\left(\frac{\pi x}{2a}\right) + T_2 \sin\left(\frac{\pi x}{2a}\right). \qquad (1.5.10)$$

The proposed manufactured solution exactly satisfies the differential equation as well as the *modified* boundary conditions given in Eqs. (1.5.7)–(1.5.10). *The manufactured solution (and the problem) have the sole purpose of verifying a numerical solution.* The manufactured problem as well as its solution can be used to verify a numerical technique whose veracity needs to be established. In all grid-based techniques (i.e., numerical methods that have a mesh whose characteristic dimension h decreases as the mesh is refined), the numerical solution should approach the exact solution as $h \to 0$. The above problem with numerical values is solved in Chapter 7 to demonstrate the verification of a solution process using the MSM.

1.6 Present Study

The main focus of the present book is to present two most commonly used numerical methods, namely, the FEM and the FVM, for the solution of problems in heat transfer and fluid mechanics. As discussed previously, the FEM involves integration of the original differential equation (to develop the weak form), and the FDM involves replacing derivatives in the differential equation with discrete values of the unknowns. The FVM shares the features of both the FEM and FDM. Integration has a smoothing effect and smears discontinuities in the solution; as a result the FVM and FEM are diffusive in nature and not preferred to capture strong discontinuities like shock.

In this book, we consider conductive (diffusive) and advective heat transfer and incompressible viscous flows. Both linear and nonlinear problems are considered. The overarching goal of the present study is to understand and predict phenomena described by fluid and thermal transport through the FEM and FVM. Attempt is made to give all basic steps involved in the formulation and analysis of heat transfer and fluid flow problems. The reader will not only have an exposure to the inner workings of the two methods and their applications but will also find their relative merits while gaining insights into the physical phenomena being simulated.

The remainder of the book is organized as follows. After this brief introduction, a summary of the basic equations of heat transfer and fluid flow is presented in Chapter 2 and solution of equations is presented in Chapter 3. Chapter 4 is devoted to an introduction to the FEM as applied to one- and two-dimensional problems of steady-state conduction heat transfer. The derivation of the weak form, interpolations functions, and finite element equations for typical finite elements, assembly of elements, and imposition of boundary conditions are discussed in detail. Numerical examples are presented to illustrate the steps involved in the finite element analysis of one- and two-dimensional problems with a variety of boundary conditions (including convection boundary conditions). Chapter 5 deals with the unsteady counterpart of Chapter 4; semi-discretization and full discretization of the time-dependent heat transfer problems in one and two dimensions are presented. Numerical stability and accuracy, explicit and implicit formulations, and mass lumping are also discussed, and numerical examples are presented to illustrate the ideas. Finite element models of the steady as well as unsteady incompressible viscous flows in two dimensions are discussed in Chapter 6. Two different finite element models, namely, the pressure–velocity (or mixed) and penalty finite element models, are discussed for the case of Stokes flows. Numerical examples of viscous fluid squeezed between parallel plates and wall-driven cavity problems are used to illustrate the selection of meshes, imposition of boundary conditions, and discussion of the numerical results obtained for steady and unsteady cases. Following the finite element models of Stokes flows, the nonlinear finite element models of the Navier–Stokes equations using the penalty function approach are discussed. A number of standard benchmark problems of the Navier–Stokes equations are presented.

In Chapter 7, we introduce the FVM in connection with one- and two-dimensional steady-state and unsteady heat transfer problems. Beginning with the discussion of grid generation using cell-centered control volumes, development of discretized equations using the satisfaction of the integral statement of the governing equations (of heat transfer) over a typical control volume, and inclusion of boundary conditions, and numerical examples, are discussed. Here we use the same examples as those from Chapter 4 to show how the FEM and FVM compare in terms of the accuracy of the numerical results obtained. Chapter 8 contains a study of the advection–diffusion problems of one and two dimensions using the both FEM and FVM. Chapter 8 also introduces a general two-node formulation to address coupled advection–diffusion problems. Chapter 9, is dedicated to the solution of two-dimensional steady and unsteady incompressible viscous flows, including advective (i.e., nonlinear) effects, using the FVM.

Finally, in Chapter 10 we consider some advanced topics: periodically fully developed flows and heat transfer, coupled heat transfer and fluid flow (i.e., natural convection) in two-dimensional domains, and multigrid methods.

The organization of the chapters is based on the thinking that readers and instructors interested in only one method or the other can skip consecutive chapters (e.g., skip Chapters 4–6 if one is not interested in the FEM and omit Chapters 7–10 if the FVM is not of interest. The later chapters are advanced enough for the readers to selectively use the material based on the interest. Of course, the objective of the book is to present both methods for the readers to learn and appreciate their relative merits in the context of their application.

Problems

1.1 Classify the following differential equation as elliptic, parabolic, or hyperbolic on the basis of the value of M:

$$(1 - M^2)\frac{\partial^2 u}{\partial x^2} + \frac{\partial^2 u}{\partial y^2} = 0.$$

1.2 (a) Write Eqs. (1.2.1) and (1.2.2) in the following alternative form using $u = T - T_\infty$:

$$-\frac{d^2 u}{dx^2} + m^2\, u = 0, \quad m = \sqrt{\frac{\beta P}{kA}}, \quad 0 < x < L \tag{1}$$

$$u(0) = u_0, \quad \left[\frac{du}{dx} + \frac{\beta}{k}u\right]_{x=L} = 0. \tag{2}$$

(b) Use the central finite difference approximation to develop the discrete equation. Use the diameter of the rod to be $d = 0.02$ m, length $L = 0.05$ m, and thermal conductivity $k = 50$ W/(m °C); take $T_0 = 320\,°$C, $T_\infty = 20\,°$C, and $\beta = 100$ W/(m^2 · °C), and determine the solution for four and eight intervals.

1.3 Solve **Problem 1.2** for 16 subdivisions and compare the finite difference solution with the analytical solution. *Ans:* The finite difference solutions are

$$\{u\} = \{285.36,\ 271.84,\ 259.37,\ 247.92,\ 237.44,\ 227.89,\ 219.22, 211.42,\ \ldots\ , 176.78\}^{\mathrm{T}}. \tag{1}$$

The analytical solution at the same points is

$$\{u\} = \{285.56,\ 272.25,\ 259.99,\ 248.75,\ 238.48,\ 229.15,\ 220.71, 213.13,\ \ldots\ , 180.66\}^{\mathrm{T}}. \tag{2}$$

1.4 An improvement of Euler's scheme is provided by Heun's scheme, which uses the average of the derivatives at the two ends of the interval to estimate the slope. Applied to the equation

$$\frac{du}{dt} = f(t, u).$$

Heun's scheme has the form

$$u_{i+1} = u_i + \frac{\Delta t}{2}\left[f(t_i, u_i) + f(t_{i+1}, u^0_{i+1})\right], \quad u^0_{i+1} = u_i + \Delta t\, f(t_i, u_i).$$

The second equation is known as the *predictor* equation and the first equation is called the *corrector* equation. Apply Heun's scheme to Eq. (1.3.1) and obtain the numerical solution for $\Delta t = 0.05$.

1.5 Let us revisit the FVM discussed in Section 1.3.3. Instead of dividing the domain into a set of control volumes and then placing a node at the center of each control volume and nodes at the boundary points, first divide the domain into a set of subintervals with each subinterval having nodes at its ends. Then place a control volume around each internal nodes such that the subintervals are bisected, as shown in Fig. P1.5(a). This automatically leaves a half control volume (for a one-dimensional domain) on one side of each boundary node. For the example problem considered in Section 1.3.3, divide the domain (0, 1) into two equal intervals and identify the two end points and the midpoint as the nodes. Then, place a control volume around the midside node and half control volumes around the boundary nodes, as shown in Fig. P1.5(b). Equation (1.3.18) was derived using the full control volume and hence valid for the full control volume around node 2. For the case at hand, it gives (with $i = 2$, $u_1 = 0$, and $h = 0.5$)

$$-u_1 + 2u_2 - u_3 = 2h^2 \quad \rightarrow \quad 2u_2 - u_3 = 0.5.$$

Obtain the additional algebraic equation required using Eq. (1.3.16) in the half control volume at node 3 (note that $du/dx = 1$ at $x = 1$). In particular, show that

$$u_3 - u_2 = h^2 + h,$$

which is the same as Eq. (1.3.15). Also, determine the value of du/dx at $x = 0$ using the half control volume at node 1.

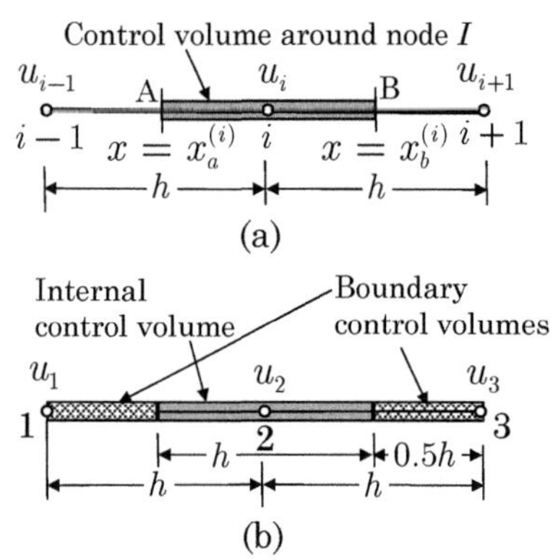

Fig. P1.5

1.6 Show that the approximation used for the first-order derivative at the boundary faces is only first-order accurate.

1.7 Show that a second-order accurate approximation of the derivative du/dx at the west face is given by

$$\left.\frac{du}{dx}\right|_w \approx \frac{9U_2 - U_3 - 8U_1}{3\Delta x}. \tag{1}$$

1.8 Use the second-order approximation derived in **Problem 1.7** to solve the problem described by Eq. (1.3.11). Use two control volumes of equal length, $h = 0.5$.

2

Equations of Heat Transfer and Fluid Mechanics

2.1 Introduction

Because this book is concerned with the numerical solutions to problems of heat transfer and fluid mechanics, it is useful to summarize the governing equations of these two fields, which are closely related. Subject areas as diverse as aerodynamics, biology, combustion, geology and geophysics, manufacturing, and meteorology can be studied using the equations governing heat transfer and fluid mechanics (for detailed discussion of the underlying physics and derivation of the equations, see [14, 15] and [25–35]). In studying these two subjects, we will come across a number of physical quantities that participate in the analytical descriptions of various physical phenomena. First we give a descriptive meaning of quantities introduced in the two subjects before the governing equations are presented.

Flow of heat (i.e., heat transfer) takes place in one or more of four ways: conduction (or diffusion), advection, convection, and radiation. The transfer of heat within a medium due to a diffusion process is called *conduction heat transfer*. *Convection heat transfer* is defined as energy transfer between a body and its surroundings, say, a fluid medium. *Thermal radiation* is a transfer of radiant (electromagnetic) energy emitted by a medium and is due solely to the temperature of the medium. *Advection* is the energy or momentum transported by a bulk motion of a fluid. Readers should note that convection and advection are two different transport phenomenon even though these two are often used interchangeably in the literature.

Fluid mechanics is the subject that deals with the motion of fluids (i.e., gases and liquids) and their interaction with the surroundings. The shear force per unit area in a fluid is proportional to the local velocity gradient. This is known as *Newton's law of viscosity*, and the proportionality parameter is known as the *viscosity*. An *inviscid fluid* is one where the viscosity is assumed to be zero. An *incompressible fluid* is one with constant density. An inviscid and incompressible fluid is termed an *ideal* or a *perfect* fluid. A *real fluid* is one with finite viscosity, and it may or may not be incompressible. Fluids that obey Newton's law of viscosity are termed *Newtonian fluids*; fluids that do not obey this law are called *non-Newtonian* or *generalized* fluids. A non-Newtonian constitutive relation can be of algebraic (e.g., power-law), differential, or integral type.

The analytical and computational study of heat transfer and fluid mechanics involves the solution of equations governing the phenomena. Therefore, it is necessary for us to review the governing equations of these two fields. The equations are derived using the laws of physics, which are independent of the choice of a coordinate system. A coordinate system is useful when a particular problem is solved. Therefore, to present the governing equations in invariant form (i.e., in vector/tensor form that is independent of a coordinate system), we must review certain elements of the algebra and calculus of vectors and tensors. For a detailed discussion of vectors and tensors, the reader may consult the continuum mechanics books by Reddy [14, 15] and references therein.

2.2 Elements of Vectors and Tensors

2.2.1 Introduction

In engineering analysis, we encounter two types of quantities: *scalars* and *nonscalars*. Scalars are just numbers (i.e., have a magnitude) whereas nonscalars have not only a magnitude specified, but also a "direction." Nonscalars that obey certain rules such as "vector addition" and "scalar multiplication" are called *vectors*. In printed material, vectors are commonly denoted by boldface letters, as used in the present study. The magnitude of a vector is a scalar. *Tensors* are more general vectors that are endowed with a magnitude and "direction(s)" and satisfy the rules of vector addition and scalar multiplication.

The magnitude of a vector $\mathbf{A}$ is denoted by $|\mathbf{A}|$, or A. Although a plane area is a scalar, for mathematical convenience we define it as a vector by defining the direction of a plane area by means of a unit vector drawn normal to that plane. The direction of the normal is taken by convention as that in which a right-handed screw advances as it is rotated according to the sense of travel along the boundary curve of the area (see Fig. 2.2.1). If the unit normal vector is denoted by $\hat{\mathbf{n}}$, then a *area vector* can be defined as $\mathbf{A} = A\,\hat{\mathbf{n}}$.

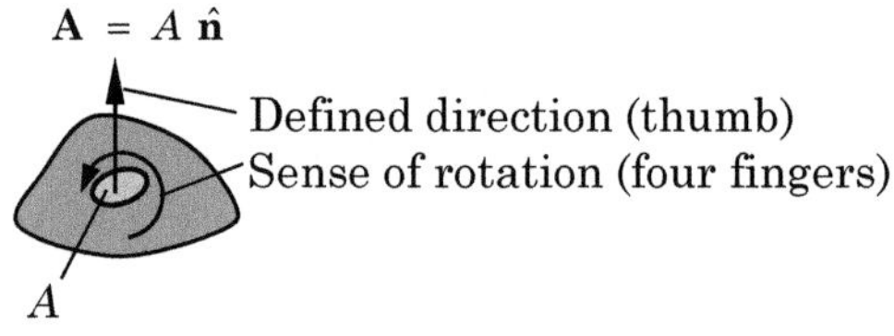

Fig. 2.2.1 Vector representation of a plane area: $\mathbf{A} = A\,\hat{\mathbf{n}}$.

Scalars are also called zeroth-order tensors while vectors are often termed first-order tensors. The higher-order tensors are used to specify a quantity with multiple properties or features. For example, a stress tensor is second order because it is endowed with (in addition to the magnitude) two additional features: the plane on which it acts and the direction of the force vector causing the stress.

2.2.2 Coordinate Systems and Summation Convention

A set of n vectors, $\{\mathbf{A}_1, \mathbf{A}_2, \ldots, \mathbf{A}_n\}$, is said to be *linearly dependent* if a set of n numbers $c_1, c_2, \ldots, c_n$ can be found such that the linear combination $c_1\mathbf{A}_1 + c_2\mathbf{A}_2 + \cdots + c_n\mathbf{A}_n = \mathbf{0}$, where $c_1, c_2, \ldots, c_n$ cannot all be zero. If this expression cannot be satisfied, the set of vectors $\{\mathbf{A}_1, \mathbf{A}_2, \ldots, \mathbf{A}_n\}$ is said to be *linearly independent*. In a three-dimensional space, a set of no more than three linearly independent vectors can be found. Let us choose any set and denote it as $(\mathbf{e}_1,\ \mathbf{e}_2,\ \mathbf{e}_3)$. Each vector $\mathbf{e}_i$ $(i = 1, 2, 3)$ is called a basis vector and the set is called a *basis* (or a base system). A basis is called *orthonormal*, if $\mathbf{e}_i$ are mutually orthogonal and have unit magnitudes. To distinguish the basis $(\mathbf{e}_1, \mathbf{e}_2, \mathbf{e}_3)$ that is not orthonormal from one that is orthonormal, we denote the orthonormal basis by $(\hat{\mathbf{e}}_1, \hat{\mathbf{e}}_2, \hat{\mathbf{e}}_3)$, where

$$
\begin{aligned}
\hat{\mathbf{e}}_1 \cdot \hat{\mathbf{e}}_2 = 0, \quad \hat{\mathbf{e}}_2 \cdot \hat{\mathbf{e}}_3 = 0, \quad \hat{\mathbf{e}}_3 \cdot \hat{\mathbf{e}}_1 = 0 \\
\hat{\mathbf{e}}_1 \cdot \hat{\mathbf{e}}_1 = 1, \quad \hat{\mathbf{e}}_2 \cdot \hat{\mathbf{e}}_2 = 1, \quad \hat{\mathbf{e}}_3 \cdot \hat{\mathbf{e}}_3 = 1.
\end{aligned}
\tag{2.2.1}
$$

Sometimes the notation $(\hat{\mathbf{i}}, \hat{\mathbf{j}}, \hat{\mathbf{k}})$ or $(\hat{\mathbf{e}}_x, \hat{\mathbf{e}}_y, \hat{\mathbf{e}}_z)$ is used in place of $(\hat{\mathbf{e}}_1, \hat{\mathbf{e}}_2, \hat{\mathbf{e}}_3)$; see Fig. 2.2.2. A rectangular Cartesian system is one in which the basis vectors are orthonormal and have constant magnitudes and fixed directions.

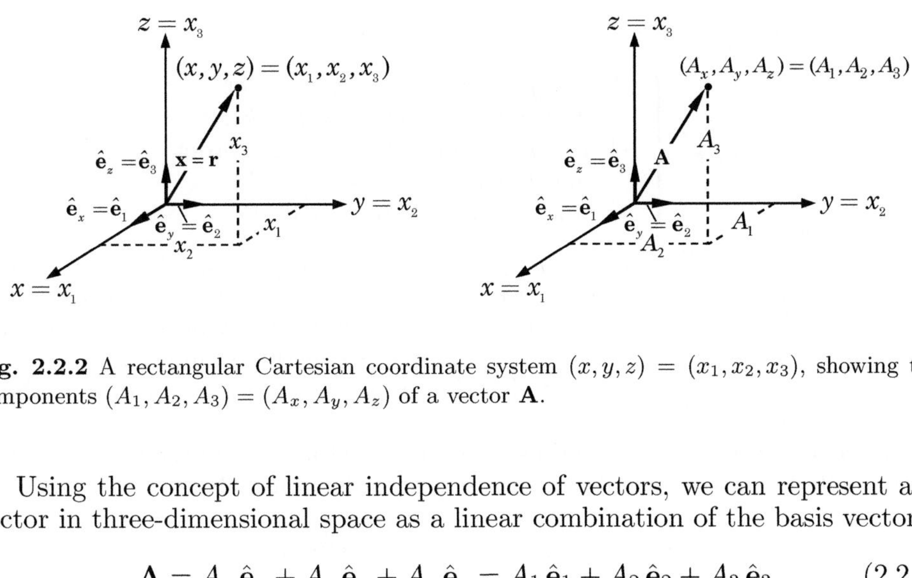

Fig. 2.2.2 A rectangular Cartesian coordinate system $(x, y, z) = (x_1, x_2, x_3)$, showing the components $(A_1, A_2, A_3) = (A_x, A_y, A_z)$ of a vector $\mathbf{A}$.

Using the concept of linear independence of vectors, we can represent any vector in three-dimensional space as a linear combination of the basis vectors:

$$
\mathbf{A} = A_x\,\hat{\mathbf{e}}_x + A_y\,\hat{\mathbf{e}}_y + A_z\,\hat{\mathbf{e}}_z = A_1\,\hat{\mathbf{e}}_1 + A_2\,\hat{\mathbf{e}}_2 + A_3\,\hat{\mathbf{e}}_3. \tag{2.2.2}
$$

The vectors $A_1\hat{\mathbf{e}}_1$, $A_2\hat{\mathbf{e}}_2$, and $A_3\hat{\mathbf{e}}_3$ are called the *vector components* of $\mathbf{A}$, and A_1, A_2, and A_3 are called *scalar components* of $\mathbf{A}$ associated with the basis $(\hat{\mathbf{e}}_1,\ \hat{\mathbf{e}}_2,\ \hat{\mathbf{e}}_3)$. Use of index notation (A_1, A_2, A_3), as opposed to (A_x, A_y, A_z), facilitates writing expressions in succinct form. For example, $\mathbf{A} = A_1\hat{\mathbf{e}}_1 + A_2\hat{\mathbf{e}}_2 + A_3\hat{\mathbf{e}}_3$ can be abbreviated as

$$
\mathbf{A} = \sum_{i=1}^{3} A_i\,\hat{\mathbf{e}}_i. \tag{2.2.3a}
$$

The summation index i is arbitrary as long as the same index is used for both A and $\hat{\mathbf{e}}$. The expression can be further shortened by omitting the summation sign and having the understanding that a repeated index means summation over all values of that index. Thus, the three-term expression $A_1\hat{\mathbf{e}}_1 + A_2\hat{\mathbf{e}}_2 + A_3\hat{\mathbf{e}}_3$ in Eq. (2.2.2) can be simply written as

$$\mathbf{A} = A_i\mathbf{e}_i. \tag{2.2.3b}$$

This notation is called the *summation convention.*

The "dot product" (or scalar product) between two orthonormal base vectors can be expressed as

$$\hat{\mathbf{e}}_i \cdot \hat{\mathbf{e}}_j = \delta_{ij}, \tag{2.2.4a}$$

where δ_{ij} Kronecker delta symbol, defined by

$$\delta_{ij} = \begin{cases} 1, \text{ if } & i = j, \\ 0, \text{ if } & i \neq j. \end{cases} \tag{2.2.4b}$$

Similarly, the "cross product" (or vector product) between two orthonormal base vectors can be defined as

$$\hat{\mathbf{e}}_i \times \hat{\mathbf{e}}_j \equiv e_{ijk}\hat{\mathbf{e}}_k \quad \text{or} \quad e_{ijk} = \hat{\mathbf{e}}_i \times \hat{\mathbf{e}}_j \cdot \hat{\mathbf{e}}_k = \hat{\mathbf{e}}_i \cdot \hat{\mathbf{e}}_j \times \hat{\mathbf{e}}_k, \tag{2.2.5a}$$

where e_{ijk} is the permutation symbol defined by

$$e_{ijk} = \begin{cases} 1, \text{ if } i, j, k \text{ are in a natural order} \\ \quad \text{and not repeated } (i \neq j \neq k), \\ -1, \text{ if } i, j, k \text{ are not in a natural order} \\ \quad \text{and not repeated } (i \neq j \neq k), \\ 0, \text{ if any of } i, j, k \text{ are repeated.} \end{cases} \tag{2.2.5b}$$

In Eq. (2.2.4a), indices i and j appear without repeating. They are called free indices. On the other hand, in Eq. (2.2.5a), indices i and j appear in each expression without repeating but index k is repeated (implying summation over its range) on the right-hand side. The index k is called a dummy index because it can be replaced by any other index (i.e., $e_{ijk}\hat{\mathbf{e}}_k = e_{ijn}\hat{\mathbf{e}}_n = e_{ijm}\hat{\mathbf{e}}_m$), without changing other expressions.

A second-order tensor requires two basis vectors to represent it. We can display all of the components s_{ij} of a second-order tensor $\mathbf{S}$ as

$$\begin{aligned} \mathbf{S} = & s_{11}\,\hat{\mathbf{e}}_1\hat{\mathbf{e}}_1 + s_{12}\,\hat{\mathbf{e}}_1\hat{\mathbf{e}}_2 + s_{13}\,\hat{\mathbf{e}}_1\hat{\mathbf{e}}_3 \\ & + s_{21}\,\hat{\mathbf{e}}_2\hat{\mathbf{e}}_1 + s_{22}\,\hat{\mathbf{e}}_2\hat{\mathbf{e}}_2 + s_{23}\,\hat{\mathbf{e}}_2\hat{\mathbf{e}}_3 \\ & + s_{31}\,\hat{\mathbf{e}}_3\hat{\mathbf{e}}_1 + s_{32}\,\hat{\mathbf{e}}_3\hat{\mathbf{e}}_2 + s_{33}\,\hat{\mathbf{e}}_3\hat{\mathbf{e}}_3 \\ & = s_{ij}\,\hat{\mathbf{e}}_i\hat{\mathbf{e}}_j. \end{aligned} \tag{2.2.6}$$

Equation (2.2.6) shows that a second-order tensor in three-dimensional space has nine independent components in general, each component associated with a certain pair of unit vectors. When the ordering is understood, such as suggested by the form in Eq. (2.2.6), the components of the second-order tensors can be expressed as a matrix and the tensor can be expressed as a product of matrices and column vectors of the basis:

$$[S] = \begin{bmatrix} s_{11} & s_{12} & s_{13} \\ s_{21} & s_{22} & s_{23} \\ s_{31} & s_{32} & s_{33} \end{bmatrix} \quad \text{and} \quad \mathbf{S} = \begin{Bmatrix} \hat{\mathbf{e}}_1 \\ \hat{\mathbf{e}}_2 \\ \hat{\mathbf{e}}_3 \end{Bmatrix}^{\mathrm{T}} [S] \begin{Bmatrix} \hat{\mathbf{e}}_1 \\ \hat{\mathbf{e}}_2 \\ \hat{\mathbf{e}}_3 \end{Bmatrix}, \tag{2.2.7}$$

where s_{ij} is the coefficient in the ith row and the jth column of the matrix.

2.2.3 Calculus of Vectors and Tensors

The symbol grad ϕ is interpreted as an operation on function $\phi(\mathbf{x})$, and it is denoted as $\boldsymbol{\nabla}\phi$, where $\boldsymbol{\nabla}$ has the following meaning:

$$\boldsymbol{\nabla} \equiv \hat{\mathbf{e}}_x \frac{\partial}{\partial x} + \hat{\mathbf{e}}_y \frac{\partial}{\partial y} + \hat{\mathbf{e}}_z \frac{\partial}{\partial z} = \hat{\mathbf{e}}_1 \frac{\partial}{\partial x_1} + \hat{\mathbf{e}}_2 \frac{\partial}{\partial x_2} + \hat{\mathbf{e}}_3 \frac{\partial}{\partial x_3} = \hat{\mathbf{e}}_i \frac{\partial}{\partial x_i}. \tag{2.2.8}$$

The operator $\boldsymbol{\nabla}$ is called the *del operator* . The del operator is a *vector differential* operator. The dot product of the del operator with a vector is called its *divergence* and is denoted by (note that the Cartesian base vectors $\hat{\mathbf{e}}_i$ are constant orthonormal vectors and have vanishing derivatives with respect to x_i)

$$\boldsymbol{\nabla} \cdot \mathbf{A} \equiv \text{div}\mathbf{A} = \hat{\mathbf{e}}_i \frac{\partial}{\partial x_i} \cdot A_j \hat{\mathbf{e}}_j = \frac{\partial A_i}{\partial x_i}. \tag{2.2.9}$$

We note that $\boldsymbol{\nabla} \cdot \mathbf{A}$ is a scalar whereas $\mathbf{A} \cdot \boldsymbol{\nabla}$ is a scalar differential operator. Thus, the del operator does not commute in this sense. The divergence of the gradient is known as the Laplacian and denoted as ∇^2:

$$\text{div}\,(\text{grad}) = \boldsymbol{\nabla} \cdot \boldsymbol{\nabla} \equiv \nabla^2 = \frac{\partial^2}{\partial x_1^2} + \frac{\partial^2}{\partial x_2^2} + \frac{\partial^2}{\partial x_3^2} = \frac{\partial^2}{\partial x_i \partial x_i}. \tag{2.2.10}$$

Note that $\nabla^2 \neq (\partial^2/\partial x_i^2)$ because without the repeated index the meaning is lost and the expression is simply incorrect.

The *curl of a vector* is defined as the del operator operating on a vector by means of the "cross product":

$$\text{curl}\,\mathbf{A} = \boldsymbol{\nabla} \times \mathbf{A} \equiv e_{ijk} \hat{\mathbf{e}}_i \frac{\partial A_k}{\partial x_j}. \tag{2.2.11}$$

The quantity $\hat{\mathbf{n}} \cdot \text{grad}\,\phi$ of a function $\phi(\mathbf{x})$ is called the *normal derivative* of ϕ (i.e., the derivative of ϕ with respect to the coordinate normal to the surface) and it is denoted by

$$\frac{\partial \phi}{\partial n} \equiv \hat{\mathbf{n}} \cdot \nabla \phi = n_i \frac{\partial \phi}{\partial x_i}. \tag{2.2.12}$$

Table 2.2.1 contains a summary of the basic operations and identities for the two commonly used orthogonal curvilinear coordinate systems, namely, the *cylindrical* and the *spherical* coordinate systems (see Fig. 2.2.2).

Table 2.2.1 Base vectors and del and Laplace operators in cylindrical and spherical coordinate systems.

- **Cylindrical coordinate system** (r, θ, z)

$x = r\cos\theta,\ y = r\sin\theta,\ z = z,\ \ \mathbf{R} = r\hat{\mathbf{e}}_r + z\hat{\mathbf{e}}_z,\ \ \mathbf{A} = A_r\hat{\mathbf{e}}_r + A_\theta\hat{\mathbf{e}}_\theta + A_z\hat{\mathbf{e}}_z$

$\hat{\mathbf{e}}_r = \cos\theta\ \hat{\mathbf{e}}_x + \sin\theta\ \hat{\mathbf{e}}_y,\ \ \hat{\mathbf{e}}_\theta = -\sin\theta\ \hat{\mathbf{e}}_x + \cos\theta\ \hat{\mathbf{e}}_y,\ \ \hat{\mathbf{e}}_z = \hat{\mathbf{e}}_z$

$\frac{\partial\hat{\mathbf{e}}_r}{\partial\theta} = -\sin\theta\ \hat{\mathbf{e}}_x + \cos\theta\ \hat{\mathbf{e}}_y = \hat{\mathbf{e}}_\theta,\ \ \frac{\partial\hat{\mathbf{e}}_\theta}{\partial\theta} = -\cos\theta\ \hat{\mathbf{e}}_x - \sin\theta\ \hat{\mathbf{e}}_y = -\hat{\mathbf{e}}_r$

All other derivatives of the base vectors are zero.

$$\boldsymbol{\nabla} = \hat{\mathbf{e}}_r\frac{\partial}{\partial r} + \frac{1}{r}\hat{\mathbf{e}}_\theta\frac{\partial}{\partial\theta} + \hat{\mathbf{e}}_z\frac{\partial}{\partial z}$$

$$\nabla^2 = \frac{1}{r}\left[\frac{\partial}{\partial r}\left(r\frac{\partial}{\partial r}\right) + \frac{1}{r}\frac{\partial^2}{\partial\theta^2} + r\frac{\partial^2}{\partial z^2}\right]$$

$$\boldsymbol{\nabla}\cdot\mathbf{A} = \frac{1}{r}\left[\frac{\partial(rA_r)}{\partial r} + \frac{\partial A_\theta}{\partial\theta} + r\frac{\partial A_z}{\partial z}\right]$$

$$\boldsymbol{\nabla}\times\mathbf{A} = \left(\frac{1}{r}\frac{\partial A_z}{\partial\theta} - \frac{\partial A_\theta}{\partial z}\right)\hat{\mathbf{e}}_r + \left(\frac{\partial A_r}{\partial z} - \frac{\partial A_z}{\partial r}\right)\hat{\mathbf{e}}_\theta + \frac{1}{r}\left[\frac{\partial(rA_\theta)}{\partial r} - \frac{\partial A_r}{\partial\theta}\right]\hat{\mathbf{e}}_z$$

$$\boldsymbol{\nabla}\mathbf{A} = \frac{\partial A_r}{\partial r}\hat{\mathbf{e}}_r\hat{\mathbf{e}}_r + \frac{\partial A_\theta}{\partial r}\hat{\mathbf{e}}_r\hat{\mathbf{e}}_\theta + \frac{1}{r}\left(\frac{\partial A_r}{\partial\theta} - A_\theta\right)\hat{\mathbf{e}}_\theta\hat{\mathbf{e}}_r + \frac{\partial A_z}{\partial r}\hat{\mathbf{e}}_r\hat{\mathbf{e}}_z + \frac{\partial A_r}{\partial z}\hat{\mathbf{e}}_z\hat{\mathbf{e}}_r$$
$$+\frac{1}{r}\left(A_r + \frac{\partial A_\theta}{\partial\theta}\right)\hat{\mathbf{e}}_\theta\hat{\mathbf{e}}_\theta + \frac{1}{r}\frac{\partial A_z}{\partial\theta}\hat{\mathbf{e}}_\theta\hat{\mathbf{e}}_z + \frac{\partial A_\theta}{\partial z}\hat{\mathbf{e}}_z\hat{\mathbf{e}}_\theta + \frac{\partial A_z}{\partial z}\hat{\mathbf{e}}_z\hat{\mathbf{e}}_z$$

- **Spherical coordinate system** (R, ϕ, θ)

$x = R\sin\phi\cos\theta,\ \ y = R\sin\phi\sin\theta,\ \ z = R\cos\phi,\ \ \mathbf{R} = R\hat{\mathbf{e}}_R$

$\mathbf{A} = A_R\hat{\mathbf{e}}_R + A_\phi\hat{\mathbf{e}}_\phi + A_\theta\hat{\mathbf{e}}_\theta$

$\hat{\mathbf{e}}_R = \sin\phi\cos\theta\ \hat{\mathbf{e}}_x + \sin\phi\sin\theta\ \hat{\mathbf{e}}_y + \cos\phi\ \hat{\mathbf{e}}_z$

$\hat{\mathbf{e}}_\phi = \cos\phi\cos\theta\ \hat{\mathbf{e}}_x + \cos\phi\sin\theta\ \hat{\mathbf{e}}_y - \sin\phi\ \hat{\mathbf{e}}_z$

$\hat{\mathbf{e}}_\theta = -\sin\theta\ \hat{\mathbf{e}}_x + \cos\theta\ \hat{\mathbf{e}}_y$

$\hat{\mathbf{e}}_x = \sin\phi\cos\theta\ \hat{\mathbf{e}}_R + \cos\phi\cos\theta\ \hat{\mathbf{e}}_\phi - \sin\theta\ \hat{\mathbf{e}}_\theta$

$\hat{\mathbf{e}}_y = \sin\phi\sin\theta\ \hat{\mathbf{e}}_R + \cos\phi\sin\theta\ \hat{\mathbf{e}}_\phi + \cos\theta\ \hat{\mathbf{e}}_\theta$

$\hat{\mathbf{e}}_z = \cos\phi\ \hat{\mathbf{e}}_R - \sin\phi\ \hat{\mathbf{e}}_\phi$

$\frac{\partial\hat{\mathbf{e}}_R}{\partial\phi} = \hat{\mathbf{e}}_\phi,\ \ \frac{\partial\hat{\mathbf{e}}_R}{\partial\theta} = \sin\phi\ \hat{\mathbf{e}}_\theta,\ \ \frac{\partial\hat{\mathbf{e}}_\phi}{\partial\phi} = -\hat{\mathbf{e}}_R,\ \ \frac{\partial\hat{\mathbf{e}}_\phi}{\partial\theta} = \cos\phi\ \hat{\mathbf{e}}_\theta,$

$\frac{\partial\hat{\mathbf{e}}_\theta}{\partial\theta} = -\sin\phi\ \hat{\mathbf{e}}_R - \cos\phi\ \hat{\mathbf{e}}_\phi$

All other derivatives of the base vectors are zero.

$$\boldsymbol{\nabla} = \hat{\mathbf{e}}_R\frac{\partial}{\partial R} + \frac{1}{R}\hat{\mathbf{e}}_\phi\frac{\partial}{\partial\phi} + \frac{1}{R\sin\phi}\hat{\mathbf{e}}_\theta\frac{\partial}{\partial\theta}$$

$$\nabla^2 = \frac{1}{R^2}\frac{\partial}{\partial R}\left(R^2\frac{\partial}{\partial R}\right) + \frac{1}{R^2\sin\phi}\frac{\partial}{\partial\phi}\left(\sin\phi\frac{\partial}{\partial\phi}\right) + \frac{1}{R^2\sin^2\phi}\frac{\partial^2}{\partial\theta^2}$$

$$\boldsymbol{\nabla}\cdot\mathbf{A} = 2\frac{A_R}{R} + \frac{\partial A_R}{\partial R} + \frac{1}{R\sin\phi}\frac{\partial(A_\phi\sin\phi)}{\partial\phi} + \frac{1}{R\sin\phi}\frac{\partial A_\theta}{\partial\theta}$$

$$\boldsymbol{\nabla}\times\mathbf{A} = \frac{1}{R\sin\phi}\left[\frac{\partial(\sin\phi A_\theta)}{\partial\phi} - \frac{\partial A_\phi}{\partial\theta}\right]\hat{\mathbf{e}}_R + \left[\frac{1}{R\sin\phi}\frac{\partial A_R}{\partial\theta} - \frac{1}{R}\frac{\partial(RA_\theta)}{\partial R}\right]\hat{\mathbf{e}}_\phi$$
$$+\frac{1}{R}\left[\frac{\partial(RA_\phi)}{\partial R} - \frac{\partial A_R}{\partial\phi}\right]\hat{\mathbf{e}}_\theta$$

$$\boldsymbol{\nabla}\mathbf{A} = \frac{\partial A_R}{\partial R}\hat{\mathbf{e}}_R\hat{\mathbf{e}}_R + \frac{\partial A_\phi}{\partial R}\hat{\mathbf{e}}_R\hat{\mathbf{e}}_\phi + \frac{1}{R}\left(\frac{\partial A_R}{\partial\phi} - A_\phi\right)\hat{\mathbf{e}}_\phi\hat{\mathbf{e}}_R + \frac{\partial A_\theta}{\partial R}\hat{\mathbf{e}}_R\hat{\mathbf{e}}_\theta$$
$$+\frac{1}{R\sin\phi}\left(\frac{\partial A_R}{\partial\theta} - A_\theta\sin\phi\right)\hat{\mathbf{e}}_\theta\hat{\mathbf{e}}_R + \frac{1}{R}\left(A_R + \frac{\partial A_\phi}{\partial\phi}\right)\hat{\mathbf{e}}_\phi\hat{\mathbf{e}}_\phi + \frac{1}{R}\frac{\partial A_\theta}{\partial\phi}\hat{\mathbf{e}}_\phi\hat{\mathbf{e}}_\theta$$
$$+\frac{1}{R\sin\phi}\left(\frac{\partial A_\phi}{\partial\theta} - A_\theta\cos\phi\right)\hat{\mathbf{e}}_\theta\hat{\mathbf{e}}_\phi + \frac{1}{R\sin\phi}\left(A_R\sin\phi + A_\phi\cos\phi + \frac{\partial A_\theta}{\partial\theta}\right)\hat{\mathbf{e}}_\theta\hat{\mathbf{e}}_\theta$$

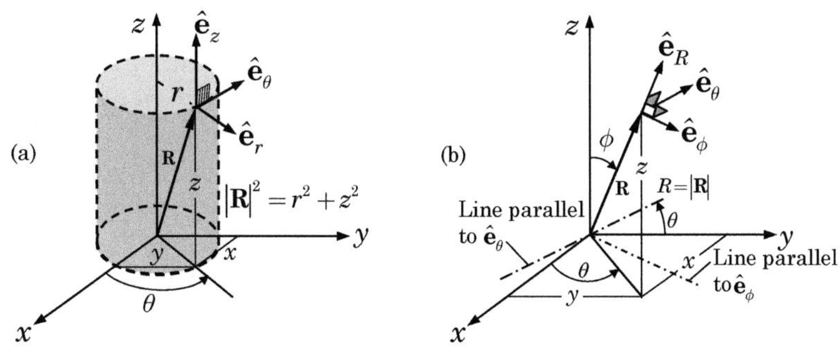

Fig. 2.2.3 (a) Cylindrical coordinate system. (b) Spherical coordinate system.

2.2.3.1 Integral Theorems

Identities between the volume and surface integrals involving the gradient, divergence, or curl of a tensor field (of any order) are useful in the sequel. These identities allow conversion of volume integrals to surface integrals and vice versa. These identities are well known as the gradient (or Green's), divergence, and curl (or Stokes') theorems.

Let Ω denote a region in $\Re^3$ bounded by the closed surface Γ. Let ds be a differential element of the surface and let $\hat{\mathbf{n}}$ be the unit outward normal, and let dv be a differential volume element in Ω. The following relations from advanced calculus hold:

$$\int_\Omega \nabla\phi\, dv = \oint_\Gamma \hat{\mathbf{n}}\,\phi\, ds \quad \text{(gradient theorem)} \tag{2.2.13}$$

$$\int_\Omega \nabla\cdot\mathbf{A}\, dv = \oint_\Gamma \hat{\mathbf{n}}\cdot\mathbf{A}\, ds \quad \text{(divergence theorem)} \tag{2.2.14}$$

$$\int_\Omega \nabla\times\mathbf{A}\, dv = \oint_\Gamma \hat{\mathbf{n}}\times\mathbf{A}\, ds \quad \text{(curl theorem).} \tag{2.2.15}$$

The expression $\hat{\mathbf{n}}\cdot\mathbf{A}\, ds$ denotes the *outflow of* $\mathbf{A}$ *through the differential surface* ds. The integral $\hat{\mathbf{n}}\cdot\mathbf{A}\, ds$ is the total or net outflow through the surrounding surface Δs. This is easier to see if one imagines that $\mathbf{A}$ is a velocity vector $\mathbf{v}$ and the outflow is the amount of fluid flow. In the limit as the region shrinks to a point, the net outflow per unit volume is associated with the *divergence of the vector field.* The integral forms presented in Eqs. (2.2.13)–(2.2.15) are known as the *invariant forms* because they do not depend on a particular coordinate system.

2.2.3.2 The Leibniz Integral Rule

The Leibniz integral rule gives a formula for the differential of a definite integral whose limits are functions of the coordinate with respect to which differentiation is taken. Let $\phi(x, t)$ be a continuous function of position x and time t. Then the Leibniz integral rule is stated as

$$\frac{d}{dt}\int_{A(t)}^{B(t)} \phi(x,t)dx = \int_{A(t)}^{B(t)} \frac{\partial \phi}{\partial t}\, dx + \phi(B(t),t)\frac{dB}{dt} - \phi(A(t),t)\frac{dA}{dt}. \qquad (2.2.16)$$

The three-dimensional version of the Leibniz integral rule for a region occupying the volume $\Gamma(t)$ with the bounding surface $\Gamma(t)$ whose points moving with a velocity $\mathbf{v}_s$ is given by

$$\frac{d}{dt}\int_{\Omega(t)} \phi(\mathbf{x},t)dv = \int_{\Gamma(t)} \frac{\partial \phi}{\partial t}\, dv + \int_{\Gamma(t)} (\mathbf{v}_s \cdot \hat{\mathbf{n}})\phi(s,t)\, ds. \qquad (2.2.17)$$

2.3 Governing Equations of a Continuous Medium

2.3.1 Descriptions of Motion

In this section we review the equations governing flows of viscous fluids and the combined advective and diffusive transport of thermal energy in a material region. The motion of a continuous medium is governed by conservation of mass and balance of momenta and energy.

There are two descriptions of motion of a continuous medium under external stimuli. In the first description, one considers the motion of all matter passing through a *fixed spatial location* and determines various properties (e.g., velocity, pressure, temperature, density, and so on) of the matter that instantly occupies the location. This description is known as the *Eulerian description* or *spatial description*. In the second description, attention is paid to a set of *fixed material particles*, irrespective of the spatial location (i.e., we follow material points). The relative displacements of these particles and the stress caused by external forces are of interest in this case. This description is known as the *Lagrangian description* or *material description*. The Eulerian description is most commonly used to study fluid flows and convective heat transfer, while the Lagrangian description is generally used to study heat transfer in solids and stress and deformation of solid bodies. The velocity of the flow in the Eulerian description, $\mathbf{v}(\mathbf{x}, t)$, is the same as the time rate of change of the position of the fluid particle,

$$\mathbf{v}(\mathbf{x},t) = \frac{d\mathbf{x}}{dt}. \qquad (2.3.1)$$

In this section we present the governing equations of a continuous medium based on the Eulerian description. For additional details, the reader may consult books on continuum mechanics, heat transfer, and fluid mechanics (e.g., see [14, 15] and [33]).

2.3.2 Material Time Derivative

The total time rate of change of a field variable $\phi(\mathbf{x}, t)$ (which can be density, temperature, velocity, or stress) at a fixed position $\mathbf{x}$ is the partial or instantaneous derivative, and the time rate of change of a moving particle $\mathbf{x}(t)$ is called the *Eulerian derivative* or *substantial derivative.* We have

$$\text{Partial derivative:} \quad \left.\frac{d\phi}{dt}\right|_{\mathbf{x}=\text{fixed}} = \frac{\partial \phi}{\partial t} \tag{2.3.2}$$

$$\text{Eulerian derivative:} \quad \left.\frac{d\phi}{dt}\right|_{\mathbf{x}(t)} = \frac{\partial \phi}{\partial t} + \frac{\partial \phi}{\partial x_1}\frac{dx_1}{dt} + \frac{\partial \phi}{\partial x_2}\frac{dx_2}{dt} + \frac{\partial \phi}{\partial x_3}\frac{dx_3}{dt} \tag{2.3.3}$$

$$= \frac{\partial \phi}{\partial t} + \frac{d\mathbf{x}}{dt} \cdot \boldsymbol{\nabla}\phi. \tag{2.3.4}$$

Thus the total time derivative d/dt operator is given by

$$\frac{d}{dt} = \frac{\partial}{\partial t} + \frac{d\mathbf{x}}{dt} \cdot \boldsymbol{\nabla}. \tag{2.3.5}$$

The first term on the right hand side denotes the local time rate of change (i.e., when $\mathbf{x}$ is kept constant) and the second term denotes the advective rate of change, which is due to the change brought about by the new particle occupying the location $\mathbf{x}(t)$. The arbitrary total time derivative operator d/dt corresponds to a change in ϕ following a change in $\mathbf{x}(t)$ with time. If $\mathbf{x}$ corresponds to the position of a fixed material particle, then $d\mathbf{x}/dt = \mathbf{v}$ corresponds to the velocity of the material point. To distinguish the time rates of change following a material particle from other arbitrary changes, we write

$$\frac{D}{Dt} = \frac{\partial}{\partial t} + \mathbf{v} \cdot \boldsymbol{\nabla} \tag{2.3.6}$$

and D/Dt is referred to as the *material time derivative.* An example of the time rate of change is provided by $\phi = \mathbf{v}$

$$\frac{d\mathbf{v}}{dt} = \frac{\partial \mathbf{v}}{\partial t} + (\mathbf{v} \cdot \boldsymbol{\nabla})\,\mathbf{v}. \tag{2.3.7}$$

The material derivative of an integral expression is given by the Leibniz integral rule in Eq. (2.2.17).

Let each element of mass in the continuous medium move with the velocity $\mathbf{v}(\mathbf{x}, t)$ and consider a special region Ω such that the bounding surface Γ is attached to a fixed set of material elements. Then each point of this surface moves with the material velocity, that is, $\mathbf{v}_s = \mathbf{v}$, and the region Ω thus contains a fixed total amount of mass because no mass crosses the boundary surface Γ. To distinguish the time rate of change of an integral over this material region, we replace d/dt by D/Dt and write

$$\frac{D}{Dt}\int_\Omega \phi(\mathbf{x}, t)\, dv \equiv \int_\Omega \frac{\partial \phi}{\partial t}\, dv + \oint_\Gamma \phi\, \hat{\mathbf{n}} \cdot \mathbf{v}\, ds, \tag{2.3.8}$$

which holds for a material region, that is, a region of fixed total mass. The relation between the time derivative following an arbitrary region and the time derivative following a material region (fixed total mass) is obtained by using Eqs. (2.3.8) and (2.2.17):

$$\frac{d}{dt}\int_\Omega \phi(\mathbf{x}, t)\, dv \equiv \frac{D}{Dt}\int_\Omega \phi(\mathbf{x}, t)\, dv + \oint_\Gamma \phi\, \hat{\mathbf{n}} \cdot (\mathbf{v}_s - \mathbf{v})\, ds. \tag{2.3.9}$$

The velocity difference $\mathbf{v} - \mathbf{v}_s$ is the velocity of the material measured relative to the velocity of the surface. The surface integral

$$\oint_\Gamma \phi\, \hat{\mathbf{n}} \cdot (\mathbf{v} - \mathbf{v}_s)\, ds \tag{2.3.10}$$

thus measures the total *outflow* of the property ϕ from the region Ω.

2.3.3 Velocity Gradient Tensor

In fluid mechanics, a measure of deformation is the gradient of velocity $\boldsymbol{\nabla}\mathbf{v}$. The *velocity gradient tensor*, $\mathbf{L}^{\mathrm{T}} = \boldsymbol{\nabla}\mathbf{v}$, where $(\cdot)^T$ denotes the transpose of the enclosed quantity, can be expressed as the sum of the symmetric $\mathbf{D}$ and the skew-symmetric $\mathbf{W}$ tensors:

$$\mathbf{L} \equiv (\boldsymbol{\nabla}\mathbf{v})^{\mathrm{T}} = \tfrac{1}{2}\left[(\boldsymbol{\nabla}\mathbf{v})^{\mathrm{T}} + \boldsymbol{\nabla}\mathbf{v}\right] + \tfrac{1}{2}\left[(\boldsymbol{\nabla}\mathbf{v})^{\mathrm{T}} - \boldsymbol{\nabla}\mathbf{v}\right] \equiv \mathbf{D} + \mathbf{W}, \tag{2.3.11}$$

where $\mathbf{D}$ is called the *rate of deformation tensor* and $\mathbf{W}$ is called the *vorticity tensor* or *spin tensor*:

$$\mathbf{D} = \tfrac{1}{2}\left[(\boldsymbol{\nabla}\mathbf{v})^{\mathrm{T}} + \boldsymbol{\nabla}\mathbf{v}\right], \quad \mathbf{W} = \tfrac{1}{2}\left[(\boldsymbol{\nabla}\mathbf{v})^{\mathrm{T}} - \boldsymbol{\nabla}\mathbf{v}\right]. \tag{2.3.12}$$

When the components of the symmetric tensor $\mathbf{D}$ and the skew-symmetric tensor $\mathbf{W}$ (i.e., $\mathbf{W}^{\mathrm{T}} = -\mathbf{W}$) are displayed as 3×3 matrices, we have

$$[D] = \begin{bmatrix} D_{11} & D_{12} & D_{13} \\ D_{12} & D_{22} & D_{23} \\ D_{13} & D_{23} & D_{33} \end{bmatrix}, \quad [W] = \begin{bmatrix} 0 & W_{12} & W_{13} \\ -W_{12} & 0 & W_{23} \\ -W_{13} & -W_{23} & 0 \end{bmatrix}. \tag{2.3.13}$$

Thus, $\mathbf{D}$ has six independent scalar components while $\mathbf{W}$ has three. The three independent scalar components of $\mathbf{W}$ can be used to define the scalar components of a vector $\mathbf{w}$, called the *axial vector of* $\mathbf{W}$, as follows:

$$W_{ij} = -e_{ijk} w_k, \quad [W] = \begin{bmatrix} 0 & -w_3 & w_2 \\ w_3 & 0 & -w_1 \\ -w_2 & w_1 & 0 \end{bmatrix}. \tag{2.3.14}$$

The scalar components (w_1, w_2, w_3) of $\mathbf{w}$ can be expressed in terms of the scalar components (W_1, W_2, W_3) of $\mathbf{W}$ as

$$w_i = -\tfrac{1}{2} e_{ijk} W_{jk} = \tfrac{1}{2} e_{ijk} \frac{\partial v_k}{\partial x_j} \quad \text{or} \quad \mathbf{w} = \tfrac{1}{2} \boldsymbol{\nabla} \times \mathbf{v}. \tag{2.3.15}$$

Thus, $\mathbf{w}$ is also known as the *vorticity vector*. By virtue of the vector identity, the divergence of the curl of a vector is zero: $\boldsymbol{\nabla} \cdot \boldsymbol{\nabla} \times \mathbf{v}$). Thus, the vorticity vector is divergence-free, $\boldsymbol{\nabla} \cdot \mathbf{w} = 0$. If a velocity vector $\mathbf{v}$ is of the form $\mathbf{v} = \mathbf{W} \cdot \mathbf{x}$ for some skew symmetric tensor $\mathbf{W}$ that is independent of position $\mathbf{x}$, then the motion is a uniform rigid-body rotation about the origin with angular velocity $\mathbf{w}$.

2.3.4 Conservation of Mass

The principle of conservation of mass can be stated as *the time rate of change of mass in a fixed volume is equal to the net rate of flow of mass across the surface*:

$$\frac{D}{Dt}\int_{\Omega}\rho dv = 0, \tag{2.3.15}$$

where ρ is the density (kg/m^3) of the medium. Then from Eqs. (2.2.17) and (2.3.9) (with $\mathbf{v}_s = \mathbf{0}$) we obtain the following equation, known as the *continuity equation*:

$$\frac{\partial \rho}{\partial t} + \boldsymbol{\nabla}\cdot(\rho\mathbf{v}) = 0, \tag{2.3.17}$$

where $\mathbf{v}$ is the velocity vector (m/s), $\boldsymbol{\nabla}$ is the del operator, and the divergence theorem is used to convert the surface integral to the volume integral. The continuity equation (2.3.17) is in conservation (or divergence) form since it can be derived directly from an integral statement of mass conservation. Using the material time derivative D/Dt, the continuity equation (2.3.17) can be expressed in the alternative form

$$\frac{\partial \rho}{\partial t} + \mathbf{v}\cdot\boldsymbol{\nabla}\rho + \rho\boldsymbol{\nabla}\cdot v = \frac{D\rho}{Dt} + \rho\boldsymbol{\nabla}\cdot\mathbf{v} = 0. \tag{2.3.18}$$

For steady-state conditions, the continuity equation (2.3.17) becomes

$$\boldsymbol{\nabla}\cdot(\rho\mathbf{v}) = 0. \tag{2.3.19}$$

When the density changes following a fluid particle are negligible, the continuum is termed *incompressible* and we have $D\rho/Dt = 0$. The continuity equation (2.3.19), known as the *incompressibility condition*, then becomes

$$\boldsymbol{\nabla}\cdot\mathbf{v} = 0. \tag{2.3.20}$$

The continuity equation in Eq. (2.3.17) can be expressed in rectangular Cartesian and orthogonal curvilinear coordinate systems by writing $\boldsymbol{\nabla}$ and $\mathbf{v}$ in the respective coordinate systems. They are summarized here for ready reference and use (see Figs. 2.2.2 and 2.2.3).

Cartesian rectangular coordinate system (x_1, x_2, x_3)

$$\boldsymbol{\nabla} = \hat{\mathbf{e}}_i\frac{\partial}{\partial x_i}, \quad \mathbf{v} = v_i\,\hat{\mathbf{e}}_i \tag{2.3.21a}$$

$$\frac{\partial \rho}{\partial t} + \frac{\partial(\rho v_1)}{\partial x_1} + \frac{\partial(\rho v_2)}{\partial x_2} + \frac{\partial(\rho v_3)}{\partial x_3} = 0. \tag{2.3.21b}$$

Cylindrical coordinate system (r, θ, z)

$$\boldsymbol{\nabla} = \hat{\mathbf{e}}_r\frac{\partial}{\partial r} + \frac{1}{r}\hat{\mathbf{e}}_\theta\frac{\partial}{\partial \theta} + \hat{\mathbf{e}}_z\frac{\partial}{\partial z}, \quad \mathbf{v} = v_r\,\hat{\mathbf{e}}_r + v_\theta\,\hat{\mathbf{e}}_\theta + v_z\,\hat{\mathbf{e}}_z \tag{2.3.22a}$$

$$\frac{\partial \rho}{\partial t} + \frac{1}{r}\left[\frac{\partial(r\rho v_r)}{\partial r} + \frac{\partial(\rho v_\theta)}{\partial \theta} + r\frac{\partial(\rho v_z)}{\partial z}\right] = 0. \tag{2.3.22b}$$

2.3.5 Reynolds Transport Theorem

The material derivative operator D/Dt corresponds to changes with respect to a fixed mass, that is, $\rho\, dv$ is constant with respect to this operator. Therefore, from Eq. (2.3.8) by substituting for $\phi = \rho F(\mathbf{x}, t)$, where F is an arbitrary function, we obtain the result

$$\frac{D}{Dt}\int_\Omega \rho F(\mathbf{x}, t)\, dv = \frac{\partial}{\partial t}\int_\Omega \rho F\, dv + \oint_\Gamma \rho\, \hat{\mathbf{n}}\cdot\mathbf{v}\, F\, ds, \tag{2.3.23}$$

or

$$\begin{aligned}\frac{D}{Dt}\int_\Omega \rho F(\mathbf{x}, t)\, dv &= \int_\Omega \left[\rho\frac{\partial F}{\partial t} + F\frac{\partial \rho}{\partial t} + \boldsymbol{\nabla}\cdot(\rho F\mathbf{v})\right] dv \\ &= \int_\Omega \left[\rho\left(\frac{\partial F}{\partial t} + \mathbf{v}\cdot\boldsymbol{\nabla}F\right) + F\left(\frac{\partial \rho}{\partial t} + \boldsymbol{\nabla}\cdot(\rho\,\mathbf{v})\right)\right] dv.\end{aligned} \tag{2.3.24}$$

Now using the continuity equation, Eq. (2.3.17), and the definition of the material time derivative in Eq. (2.3.5), we arrive at the result

$$\frac{D}{Dt}\int_\Omega \rho F\, dv = \int_\Omega \rho\frac{DF}{Dt}\, dv. \tag{2.3.25}$$

Equation (2.3.25) is known as the *Reynolds transport theorem.* Equation (2.3.25) also holds when F is a vector- or tensor-valued function.

2.3.6 Conservation of Momenta

The principle of balance of linear momentum (or Newton's Second Law of motion) states that *the time rate of change of linear momentum of a given set of particles is equal to the vector sum of all the external forces acting on the particles of the set, provided Newton's Third Law of action and reaction governs the internal forces*:

$$\frac{D}{Dt}\int_\Omega \rho\,\mathbf{v}\, dv = \oint_\Gamma \mathbf{t}\, ds + \int_\Omega \rho\,\mathbf{f}\, dv, \tag{2.3.26}$$

where $\rho\,\mathbf{v}\, dv$ denotes the linear momentum associated with elemental volume dv, $\mathbf{v}$ is the velocity vector, $\mathbf{f}$ is the body force (measured per unit mass), and $\mathbf{t}$ is the surface traction (measured per unit surface area). Using the Reynolds transport theorem, $\mathbf{t} = \hat{\mathbf{n}}\cdot\boldsymbol{\sigma}$, and the divergence theorem, we arrive at

$$\frac{\partial \rho\mathbf{v}}{\partial t} + \boldsymbol{\nabla}\cdot(\rho\mathbf{v}\otimes\mathbf{v}) = \boldsymbol{\nabla}\cdot\boldsymbol{\sigma} + \rho\mathbf{f}, \tag{2.3.27}$$

where $\otimes$ is the tensor (or dyadic) product of two vectors, $\boldsymbol{\sigma}$ is the Cauchy stress tensor (N/m^2), and $\mathbf{f}$ is the body force vector, measured per unit mass and normally taken to be the gravity vector. Equation (2.3.27) describes the motion of a continuous medium, and in fluid mechanics they are also known as the *Navier–Stokes equations* (after invoking the constitutive relations).

The form of the equation of motion in Eq. (2.3.27) is the conservation form that is most often utilized for compressible flows. This equation may be simplified to a form more commonly used with incompressible flows. Expanding the first two derivatives and collecting terms

$$\rho\left(\frac{\partial \mathbf{v}}{\partial t}+\mathbf{v}\cdot\boldsymbol{\nabla}\mathbf{v}\right)+\mathbf{v}\left(\frac{\partial \rho}{\partial t}+\boldsymbol{\nabla}\cdot\rho\mathbf{v}\right)=\boldsymbol{\nabla}\cdot\boldsymbol{\sigma}+\rho\mathbf{f}. \tag{2.3.28}$$

The second term in parentheses is the continuity equation (2.3.17) and Eq. (2.3.28) reduces to

$$\rho\left(\frac{\partial \mathbf{v}}{\partial t}+\mathbf{v}\cdot\boldsymbol{\nabla}\mathbf{v}\right)=\boldsymbol{\nabla}\cdot\boldsymbol{\sigma}+\rho\mathbf{f}, \tag{2.3.29}$$

where the material derivative in Eq. (2.3.7) has been employed.

In the Cartesian rectangular coordinates, Eq. (2.3.28) takes the form ($\boldsymbol{\sigma}=\sigma_{ij}\hat{\mathbf{e}}_i\hat{\mathbf{e}}_j$):

$$\rho\left(\frac{\partial v_i}{\partial t}+v_j\frac{\partial v_i}{\partial x_j}\right)=\frac{\partial \sigma_{ji}}{\partial x_j}+\rho f_i,\quad (i=1,2,3). \tag{2.3.30}$$

In expanded form, we have

$$\begin{aligned}\rho\Big(\frac{\partial v_1}{\partial t}&+v_1\frac{\partial v_1}{\partial x_1}+v_2\frac{\partial v_1}{\partial x_2}+v_3\frac{\partial v_1}{\partial x_3}\Big)\\&=\frac{\partial \sigma_{11}}{\partial x_1}+\frac{\partial \sigma_{21}}{\partial x_2}+\frac{\partial \sigma_{31}}{\partial x_3}+\rho f_1\end{aligned} \tag{2.3.31a}$$

$$\begin{aligned}\rho\Big(\frac{\partial v_2}{\partial t}&+v_1\frac{\partial v_2}{\partial x_1}+v_2\frac{\partial v_2}{\partial x_2}+v_3\frac{\partial v_2}{\partial x_3}\Big)\\&=\frac{\partial \sigma_{12}}{\partial x_1}+\frac{\partial \sigma_{22}}{\partial x_2}+\frac{\partial \sigma_{32}}{\partial x_3}+\rho f_2\end{aligned} \tag{2.3.31b}$$

$$\begin{aligned}\rho\Big(\frac{\partial v_3}{\partial t}&+v_1\frac{\partial v_3}{\partial x_1}+v_2\frac{\partial v_3}{\partial x_2}+v_3\frac{\partial v_3}{\partial x_3}\Big)\\&=\frac{\partial \sigma_{13}}{\partial x_1}+\frac{\partial \sigma_{23}}{\partial x_2}+\frac{\partial \sigma_{33}}{\partial x_3}+\rho f_3.\end{aligned} \tag{2.3.31c}$$

In the cylindrical coordinate system, we have

$$\begin{aligned}\rho\Big(\frac{\partial v_r}{\partial t}&+v_r\frac{\partial v_r}{\partial r}+\frac{v_\theta}{r}\frac{\partial v_r}{\partial \theta}+v_z\frac{\partial v_r}{\partial z}-\frac{v_\theta^2}{r}\Big)\\&=\frac{\partial \sigma_{rr}}{\partial r}+\frac{1}{r}\frac{\partial \sigma_{r\theta}}{\partial \theta}+\frac{\partial \sigma_{rz}}{\partial z}+\frac{1}{r}\left(\sigma_{rr}-\sigma_{\theta\theta}\right)+\rho f_r\end{aligned} \tag{2.3.32a}$$

$$\begin{aligned}\rho\Big(\frac{\partial v_\theta}{\partial t}&+v_r\frac{\partial v_\theta}{\partial r}+\frac{v_\theta v_r}{r}+\frac{v_\theta}{r}\frac{\partial v_\theta}{\partial \theta}+v_z\frac{\partial v_\theta}{\partial z}\Big)\\&=\frac{\partial \sigma_{\theta r}}{\partial r}+\frac{1}{r}\frac{\partial \sigma_{\theta\theta}}{\partial \theta}+\frac{\partial \sigma_{\theta z}}{\partial z}+\frac{\sigma_{\theta r}+\sigma_{r\theta}}{r}+\rho f_\theta\end{aligned} \tag{2.3.32b}$$

$$\begin{aligned}\rho\Big(\frac{\partial v_z}{\partial t}&+v_r\frac{\partial v_z}{\partial r}+\frac{v_\theta}{r}\frac{\partial v_z}{\partial \theta}+v_z\frac{\partial v_z}{\partial z}\Big)\\&=\frac{\partial \sigma_{zr}}{\partial r}+\frac{1}{r}\frac{\partial \sigma_{z\theta}}{\partial \theta}+\frac{\partial \sigma_{zz}}{\partial z}+\frac{\sigma_{zr}}{r}+\rho f_z.\end{aligned} \tag{2.3.32c}$$

The principle of balance of angular momentum can be stated as *the time rate of change of the total moment of momentum of a given set of particles is equal to the vector sum of the moments of the external forces acting on the system.* In the absence of distributed body couples, the principle leads to the symmetry of the stress tensor:

$$\boldsymbol{\sigma} = (\boldsymbol{\sigma})^{\mathrm{T}} \quad \text{or} \quad \sigma_{ji} = \sigma_{ij}, \tag{2.3.33}$$

where the superscript "T" denotes the transpose of the quantity.

2.3.7 Conservation of Energy

The law of balance of energy (or the First Law of Thermodynamics) can be stated as: *the time rate of change of the total energy is equal to the sum of the rate of work done by applied forces and the change of heat content per unit time.* In the general case, the First Law of Thermodynamics can be expressed as (see Reddy [15] for the details)

$$\rho \frac{De}{Dt} = \boldsymbol{\sigma} : \mathbf{D} - \boldsymbol{\nabla} \cdot \mathbf{q} + \rho Q, \tag{2.3.34}$$

where e is the internal energy (measured per unit mass, $\mathrm{J/m^3}$), $\mathbf{D}$ is the symmetric part of the velocity gradient tensor (or rate of deformation tensor), $\mathbf{q}$ is the heat flux vector ($\mathrm{W/m^2}$), and g is the rate of internal heat generation (measured per unit volume, $\mathrm{W/m^3}$). Using the continuity equation (2.3.17), Eq. (2.3.33) can be expressed in the alternative form, known as the conservative form, as

$$\frac{\partial(\rho e)}{\partial t} + \boldsymbol{\nabla} \cdot (\mathbf{v}\rho e) = -\boldsymbol{\nabla} \cdot \mathbf{q} + g + \Phi, \tag{2.3.35}$$

where Φ is a dissipation function defined by (assuming symmetry of $\boldsymbol{\sigma}$)

$$\Phi = \boldsymbol{\sigma} : \boldsymbol{\nabla}\mathbf{v}. \tag{2.3.36}$$

Constitutive relations for e and $\mathbf{q}$ will be defined in the next section, allowing Eq. (2.3.36) to be expressed in terms of the temperature T.

2.3.8 Equation of State

In addition to the conservation laws described above, two thermodynamic relations or equations of state are required. A caloric equation of state relates the internal energy of the material to two thermodynamic variables. Of the many possible forms for this equation (see, e.g., Bejan [25, 26]) one of the most commonly used in fluid dynamics is the functional relation

$$e = e(T, \nu), \tag{2.3.37}$$

where $\nu = 1/\rho$ is the specific volume.

For an incompressible fluid, it can be shown that

$$\rho \frac{De}{Dt} = \rho c_v \frac{DT}{Dt}, \quad c_v = \left.\frac{\partial e}{\partial T}\right|_{\nu}, \tag{2.3.38}$$

where c_v is the specific heat [J/(kg · °C)] at constant volume, which is the same as the specific heat at constant pressure c_p for constant density processes. Thus, for an incompressible fluid, the energy equation (2.3.33) take the form

$$\rho c_v \frac{DT}{Dt} = -\boldsymbol{\nabla} \cdot \mathbf{q} + g + \Phi, \tag{2.3.39}$$

which can also be expressed, by means of the continuity equation (2.3.17), in the alternative form

$$c_v \left[\frac{\partial(\rho T)}{\partial t} + \boldsymbol{\nabla} \cdot (\rho \mathbf{v} T) \right] = -\boldsymbol{\nabla} \cdot \mathbf{q} + g + \Phi. \tag{2.3.40}$$

Under the incompressibility assumption the pressure becomes a part of the total stress and is treated as a mechanical variable. For flows involving buoyancy forces the equation of state is altered to an extended form of the Boussinesq approximation (see Bejan [25, 26]). The extended Boussinesq approximation allows the fluid properties to be expressed as functions of the thermodynamic state (e.g., temperature) and the density ρ to vary with temperature T according to the relation

$$\rho = \rho_0[1 - \beta(T - T_0)], \tag{2.3.42}$$

where β is the heat transfer coefficient (1/°C) and the subscript zero indicates a reference condition. The variation of density as given in Eq. (2.3.42) is permitted only in the description of the body force; the density in all other situations is assumed to be that of the reference state, ρ_0.

2.3.9 Constitutive Equations

The fluids of interest in this book are assumed to be Newtonian (i.e., the constitutive relations are linear). Then, for viscous incompressible flows, the total stress $\boldsymbol{\sigma}$ can be decomposed into hydrostatic and viscous parts:

$$\boldsymbol{\sigma} = -p\mathbf{I} + \boldsymbol{\tau}, \tag{2.3.42}$$

where p is the hydrostatic (mechanical) pressure, $\mathbf{I}$ is the unit tensor and $\boldsymbol{\tau}$ is the viscous stress tensor. For Newtonian fluids, the viscous stress tensor is related to the strain rate tensor $\mathbf{D}$ by

$$\boldsymbol{\tau} = \mathbf{C} : \mathbf{D} \quad (\tau_{ij} = C_{ijkl} D_{kl}), \tag{2.3.43}$$

where $\mathbf{C}$ is the fourth-order tensor of fluid properties and $\mathbf{D}$ is the rate of deformation tensor [see Eq. (2.3.12)]

$$\mathbf{D} = \frac{1}{2}\left[(\boldsymbol{\nabla}\mathbf{v}) + (\boldsymbol{\nabla}\mathbf{v})^T \right]. \tag{2.3.44}$$

For an isotropic fluid (i.e., whose material properties are independent of direction), the fourth-order tensor $\mathbf{C}$ can be expressed in terms of two constants λ and μ, called Lamé constants, and Eq. (2.3.43) takes the form

$$\boldsymbol{\tau} = \lambda(\mathrm{tr}\,\mathbf{D})\mathbf{I} + 2\mu\mathbf{D}, \tag{2.3.45a}$$

where (tr $\mathbf{D}$) denotes the *trace* (or sum of the diagonal elements) of the matrix $\mathbf{D}$. For an incompressible fluid, we have $\operatorname{tr}\mathbf{D} = 0$ and Eq. (2.3.45a) becomes

$$\boldsymbol{\tau} = 2\mu\mathbf{D} \tag{2.3.45b}$$

and the total stress tensor is then

$$\boldsymbol{\sigma} = -p\mathbf{I} + 2\mu\mathbf{D}. \tag{2.3.46}$$

With the stress tensor defined, the dissipation function in the energy equation may be rewritten as

$$\Phi = (-p\mathbf{I} + \boldsymbol{\tau}) : \boldsymbol{\nabla}\mathbf{v}. \tag{2.3.47}$$

Recall from Eq. (2.3.11) that the velocity gradient tensor can be expressed as the sum of the symmetric strain rate tensor $\mathbf{D}$ and the skew-symmetric spin tensor $\mathbf{W}$. When substituted into Eq. (2.3.47) and contracted with the symmetric stress tensor, the dissipation function becomes

$$\Phi = -p\,\boldsymbol{\nabla}\cdot\mathbf{v} + \boldsymbol{\tau} : \mathbf{D}. \tag{2.3.48}$$

For an incompressible fluid, the reversible pressure work term is zero, and using the definition of $\boldsymbol{\tau}$ from Eq. (2.3.45b) the most common form of the dissipation function becomes

$$\Phi = 2\mu\mathbf{D} : \mathbf{D}. \tag{2.3.49}$$

The Fourier heat conduction law states that

$$\mathbf{q} = -\mathbf{k}\cdot\boldsymbol{\nabla}T, \tag{2.3.50a}$$

where $\mathbf{k}$ denotes the conductivity tensor of order two. For an isotropic medium, $\mathbf{k}$ is of the form

$$\mathbf{k} = k\,\mathbf{I}, \tag{2.3.50b}$$

where k denotes the thermal conductivity [W/(m · °C)] of the medium and $\mathbf{I}$ is the unit tensor.

The material coefficients, μ, c, β, and $\mathbf{k}$ are generally functions of the fluid temperature; property variations as a function of a second thermodynamic variable (e.g., pressure) are possible but usually are unimportant for an incompressible fluid. Solid materials may also have specific heat and conductivity that vary with temperature. In addition, the solid conductivity may vary with spatial position and, in general, remains a symmetric second-order tensor (i.e., $\mathbf{k}^{\mathrm{T}} = \mathbf{k}$) for anisotropic materials.

2.4 Summary

In this chapter the fundamental equations of heat transfer in solids and fluids are presented. The principle of conservation of mass, balance of momenta, and balance of energy are stated, and the associated constitutive equations are presented. A summary of the equations governing a continuous medium is presented in vector form. Reference is made to these equations in the subsequent chapters.

Material time derivative:

$$\frac{D}{Dt} = \frac{\partial}{\partial t} + \mathbf{v} \cdot \boldsymbol{\nabla}. \tag{2.4.1}$$

Reynolds transport theorem:

$$\frac{D}{Dt}\int_{\Omega} \rho\, F\, dv = \int_{\Omega} \rho\, \frac{DF}{Dt}\, dv. \tag{2.4.2}$$

Rate of deformation tensor:

$$\mathbf{D} = \frac{1}{2}\Big[\boldsymbol{\nabla}\mathbf{v} + (\boldsymbol{\nabla}\mathbf{v})^{\mathrm{T}}\Big]. \tag{2.4.3}$$

Constitutive equations (for incompressible viscous flow):

$$\begin{aligned} \rho\left(\frac{\partial e}{\partial t} + \mathbf{v}\cdot\boldsymbol{\nabla} e\right) &= \rho c_v\left(\frac{\partial T}{\partial t} + \mathbf{v}\cdot\boldsymbol{\nabla} T\right) \\ \boldsymbol{\sigma} = -p\mathbf{I} + 2\mu\mathbf{D}; &\quad \mathbf{q} = -\mathbf{k}\cdot\boldsymbol{\nabla} T. \end{aligned} \tag{2.4.4}$$

Boussinesq approximation (density variation with temperature):

$$\rho = \rho_0[1 - \beta(T - T_0)]. \tag{2.4.5}$$

Conservation of mass:

$$\frac{\partial \rho}{\partial t} + \boldsymbol{\nabla}\cdot(\rho\mathbf{v}) = 0 \tag{2.4.6}$$

$$\frac{\partial \rho}{\partial t} + \mathbf{v}\cdot\boldsymbol{\nabla}\rho + \rho\boldsymbol{\nabla}\cdot\mathbf{v} = 0. \tag{2.4.7}$$

Balance of momenta:

$$\frac{\partial \rho\mathbf{v}}{\partial t} + \boldsymbol{\nabla}\cdot(\rho\mathbf{v}\otimes\mathbf{v}) = \boldsymbol{\nabla}\cdot\boldsymbol{\sigma} + \rho\mathbf{f} \tag{2.4.8}$$

$$\rho\left(\frac{\partial \mathbf{v}}{\partial t} + \mathbf{v}\cdot\boldsymbol{\nabla}\mathbf{v}\right) = \boldsymbol{\nabla}\cdot\boldsymbol{\sigma} + \rho\mathbf{f}; \quad \boldsymbol{\sigma}^{\mathrm{T}} = \boldsymbol{\sigma}. \tag{2.4.9}$$

For viscous incompressible flows in the presence of temperature change, in view of Eq. (2.4.4), we have $\mathbf{f} = \rho\mathbf{g}_f$, where $\mathbf{g}_f$ is the vector of gravitational force; and omitting dissipation). Then Eq. (2.4.9) becomes

$$\rho\left(\frac{\partial \mathbf{v}}{\partial t} + \mathbf{v}\cdot\boldsymbol{\nabla}\mathbf{v}\right) = -\boldsymbol{\nabla}P + \mu\Big[\nabla^2\mathbf{v} + \boldsymbol{\nabla}\cdot(\boldsymbol{\nabla}\mathbf{v})^{\mathrm{T}}\Big] + \rho_0\mathbf{g}_f[1 - \beta(T - T_0)]. \tag{2.4.10}$$

Balance of energy (neglecting dissipation):

$$\rho c_v\left(\frac{\partial T}{\partial t} + \mathbf{v}\cdot\boldsymbol{\nabla} T\right) = -\boldsymbol{\nabla}\cdot\mathbf{q} + g \tag{2.4.11}$$

$$c_v\left[\frac{\partial(\rho T)}{\partial t}+\boldsymbol{\nabla}\cdot(\rho\mathbf{v}T)\right]=-\boldsymbol{\nabla}\cdot\mathbf{q}+g. \tag{2.4.12}$$

Note that $\mathbf{g}_f$ is the acceleration due to gravity while g is the internal heat generation. Using the Fourier heat conduction equation, Eqs. (2.4.11) and (2.4.12) can be expressed as

$$\rho c_v\left(\frac{\partial T}{\partial t}+\mathbf{v}\cdot\boldsymbol{\nabla}T\right)=\boldsymbol{\nabla}\cdot(k\boldsymbol{\nabla}T)+g \tag{2.4.13}$$

$$c_v\left[\frac{\partial(\rho T)}{\partial t}+\boldsymbol{\nabla}\cdot(\rho\mathbf{v}T)\right]=\boldsymbol{\nabla}\cdot(k\boldsymbol{\nabla}T)+g. \tag{2.4.14}$$

Equations (2.4.13) and (2.4.14) are known as *the convection–diffusion equations.* Note that Eqs. (2.4.12) and (2.4.14) hold only for an incompressible fluid medium. The equation governing heat transfer in *isotropic solids* is given by setting $\mathbf{v}=\mathbf{0}$ in Eq. (2.4.13):

$$\rho c_v\frac{\partial T}{\partial t}=\boldsymbol{\nabla}\cdot(k\boldsymbol{\nabla}T)+g. \tag{2.4.15}$$

It is important to clarify the meaning of the terms "convection" and "advection" used in the literature and in this book. *Convection* is an interface phenomenon on the other hand, *advection* is either energy or momentum transported by the bulk motion of the fluid. So these terms are used appropriately throughout this book. For example, inertial terms involving first derivatives of velocity in Navier–Stokes equations represent *advection.* Buoyancy-driven convective heat transfer over external surfaces is referred to as *free convection*, and buoyancy-driven convective heat transfer within an enclosure is referred to as *natural convection.*

Problems

2.1 Use the index notation to rewrite the following vector expression as a sum or difference of two vector expressions, when $\mathbf{A}$ is a vector-valued function [i.e., $\mathbf{A}=\mathbf{A}(\mathbf{x})$]:

$$\boldsymbol{\nabla}\times(\boldsymbol{\nabla}\times\mathbf{A}).$$

2.2 Let $\mathbf{r}$ denote the position vector $\mathbf{r}=\mathbf{x}=x_i\hat{\mathbf{e}}_i$ $(r^2=x_ix_i)$ and $\mathbf{A}$ be a vector functions of position vector $\mathbf{x}$ with continuous first and second derivatives. Use the index notation to show that:

$$\text{(a)}\quad \nabla^2(r^n)=n(n+1)r^{n-2},\qquad \text{(b)}\quad \boldsymbol{\nabla}\cdot(\boldsymbol{\nabla}\times\mathbf{A})=0.$$

2.3 Show that the volume of the region Ω enclosed by a boundary surface Γ (with elemental distance $d\Gamma$) is

$$\text{volume}=\tfrac{1}{6}\oint_\Gamma \boldsymbol{\nabla}(r^2)\cdot\hat{\mathbf{n}}\,d\Gamma=\tfrac{1}{3}\oint_\Gamma \mathbf{r}\cdot\hat{\mathbf{n}}\,d\Gamma.$$

2.4 Let $\phi(\mathbf{r})$ be a scalar field. Show that

$$\int_\Omega \nabla^2\phi\,d\Omega=\oint_\Gamma\frac{\partial\phi}{\partial n}\,d\Gamma.$$

2.5 Express the components of the *symmetric part* of the gradient of a vector $\mathbf{v}$

$$\mathbf{D} = \frac{1}{2}\left[(\boldsymbol{\nabla}\mathbf{v})^{\mathrm{T}} + (\boldsymbol{\nabla}\mathbf{v})\right]$$

in terms of the components (v_r, v_θ, v_z) of vector $\mathbf{v}$ and definition of the del operator $\boldsymbol{\nabla}$ in the cylindrical coordinate system (see Table 2.2.1). In particular, show that

$$D_{rr} = \frac{\partial v_r}{\partial r}, \quad D_{r\theta} = \frac{1}{2}\left(\frac{1}{r}\frac{\partial v_r}{\partial \theta} + \frac{\partial v_\theta}{\partial r} - \frac{v_\theta}{r}\right), \quad D_{rz} = \frac{1}{2}\left(\frac{\partial v_r}{\partial z} + \frac{\partial v_z}{\partial r}\right)$$

$$D_{\theta\theta} = \frac{v_r}{r} + \frac{1}{r}\frac{\partial v_\theta}{\partial \theta}, \quad D_{\theta z} = \frac{1}{2}\left(\frac{\partial v_\theta}{\partial z} + \frac{1}{r}\frac{\partial v_z}{\partial \theta}\right), \quad D_{zz} = \frac{\partial v_z}{\partial z}.$$

2.6 Express the components of the *skew-symmetric part* of the gradient of a vector $\mathbf{v}$

$$\mathbf{W} = \frac{1}{2}\left[(\boldsymbol{\nabla}\mathbf{v})^{\mathrm{T}} - (\boldsymbol{\nabla}\mathbf{v})\right]$$

in terms of the components of vector $\mathbf{v}$ in the cylindrical coordinate system. In particular, show that

$$W_{r\theta} = \frac{1}{2}\left(\frac{1}{r}\frac{\partial v_r}{\partial \theta} - \frac{v_\theta}{r} - \frac{\partial v_\theta}{\partial r}\right) = -W_{\theta r}$$

$$W_{rz} = \frac{1}{2}\left(\frac{\partial v_r}{\partial z} - \frac{\partial v_z}{\partial r}\right) = -W_{zr}$$

$$W_{z\theta} = \frac{1}{2}\left(\frac{1}{r}\frac{\partial v_z}{\partial \theta} - \frac{\partial v_\theta}{\partial z}\right) = -W_{\theta z}.$$

2.7 For an arbitrary second-order tensor $\mathbf{S}$ show that $\boldsymbol{\nabla}\cdot\mathbf{S}$ in the cylindrical coordinate system is given by

$$\begin{aligned}\boldsymbol{\nabla}\cdot\mathbf{S} = &\left[\frac{\partial S_{rr}}{\partial r} + \frac{1}{r}\frac{\partial S_{\theta r}}{\partial \theta} + \frac{\partial S_{zr}}{\partial z} + \frac{1}{r}\left(S_{rr} - S_{\theta\theta}\right)\right]\hat{\mathbf{e}}_r \\ &+ \left[\frac{\partial S_{r\theta}}{\partial r} + \frac{1}{r}\frac{\partial S_{\theta\theta}}{\partial \theta} + \frac{\partial S_{z\theta}}{\partial z} + \frac{1}{r}\left(S_{r\theta} + S_{\theta r}\right)\right]\hat{\mathbf{e}}_\theta \\ &+ \left[\frac{\partial S_{rz}}{\partial r} + \frac{1}{r}\frac{\partial S_{\theta z}}{\partial \theta} + \frac{\partial S_{zz}}{\partial z} + \frac{1}{r}S_{rz}\right]\hat{\mathbf{e}}_z.\end{aligned}$$

2.8 Suppose that the second-order tensor $\mathbf{D}$ referred to the cylindrical coordinate system $(\hat{\mathbf{e}}_r, \hat{\mathbf{e}}_\theta, \hat{\mathbf{e}}_z)$ is of the form

$$\mathbf{D} = D_{rr}(r,z)\,\hat{\mathbf{e}}_r\,\hat{\mathbf{e}}_r + D_{\theta\theta}(r,z)\,\hat{\mathbf{e}}_\theta\,\hat{\mathbf{e}}_\theta.$$

Determine (a) $\boldsymbol{\nabla}\times\mathbf{D}$ and (b) the gradient of $\boldsymbol{\nabla}\cdot\mathbf{D}$.

2.9 Use the continuity equation and the equation of motion to obtain the *conservation form* of the linear momentum equation

$$\frac{\partial}{\partial t}(\rho\mathbf{v}) + \text{div}\,(\rho\mathbf{v}\mathbf{v} - \boldsymbol{\sigma}) = \rho\mathbf{f}.$$

2.10 Show that

$$\rho\frac{D}{Dt}\left(\frac{v^2}{2}\right) = \mathbf{v}\cdot\boldsymbol{\nabla}\cdot\boldsymbol{\sigma} + \rho\mathbf{v}\cdot\mathbf{f} \qquad (v = |\mathbf{v}|).$$

2.11 Deduce that

$$\boldsymbol{\nabla}\times\left(\frac{D\mathbf{v}}{Dt}\right) \equiv \frac{D\mathbf{w}}{Dt} + \mathbf{w}\boldsymbol{\nabla}\cdot\mathbf{v} - \mathbf{w}\cdot\boldsymbol{\nabla}\mathbf{v}, \tag{1}$$

where $\mathbf{w} \equiv \frac{1}{2}\boldsymbol{\nabla} \times \mathbf{v}$ is the vorticity vector. *Hint:* First establish the identity

$$\boldsymbol{\nabla} \times (\mathbf{A} \times \mathbf{B}) = \mathbf{B} \cdot \boldsymbol{\nabla}\mathbf{A} - \mathbf{A} \cdot \boldsymbol{\nabla}\mathbf{B} + \mathbf{A}\,\boldsymbol{\nabla} \cdot \mathbf{B} - \mathbf{B}\,\boldsymbol{\nabla} \cdot \mathbf{A}, \tag{2}$$

where$\mathbf{A}$ and $\mathbf{B}$ are vector functions of position $\mathbf{x}$.

2.12 Given the following stress field with respect to the cylindrical coordinate system in a continuum that is in equilibrium ($\sigma_{\theta r} = \sigma_{r\theta}$):

$$\sigma_{rr} = 2A\left(r + \frac{B}{r^3} - \frac{C}{r}\right)\sin\theta,$$

$$\sigma_{\theta\theta} = 2A\left(3r - \frac{B}{r^3} - \frac{C}{r}\right)\sin\theta,$$

$$\sigma_{r\theta} = -2A\left(r + \frac{B}{r^3} - \frac{C}{r}\right)\cos\theta,$$

where A, B, and C are constants, determine if the stress field satisfies the equilibrium equations when the body forces are zero. Assume that all other stress components are zero.

2.13 Given the following stress field with respect to the spherical coordinate system in a body that is in equilibrium:

$$\sigma_{RR} = -\left(A + \frac{B}{R^3}\right), \quad \sigma_{\phi\phi} = \sigma_{\theta\theta} = -\left(A + \frac{C}{R^3}\right),$$

where A, B, and C are constants and R is the length of the position vector, determine if the stress field satisfies the equilibrium equations when the body forces are zero and all other stress components are zero.

2.14 Establish the following alternative form of the energy equation (assume the symmetry of the stress tensor, $\boldsymbol{\sigma}^{\mathrm{T}} = \boldsymbol{\sigma}$):

$$\rho\frac{D}{Dt}\left(e + \frac{v^2}{2}\right) = \boldsymbol{\nabla} \cdot (\boldsymbol{\sigma} \cdot \mathbf{v}) + \rho\mathbf{f} \cdot \mathbf{v} + g - \boldsymbol{\nabla} \cdot \mathbf{q}.$$

2.15 Establish the following *thermodynamic form* of the energy equation (assume the symmetry of the stress tensor, $\boldsymbol{\sigma}^{\mathrm{T}} = \boldsymbol{\sigma}$):

$$\rho\frac{De}{Dt} = \boldsymbol{\nabla} \cdot (\boldsymbol{\sigma} \cdot \mathbf{v}) - \mathbf{v} \cdot \boldsymbol{\nabla} \cdot \boldsymbol{\sigma} + g - \boldsymbol{\nabla} \cdot \mathbf{q}.$$

2.16 The total rate of work done by the surface stresses per unit volume is given by $\boldsymbol{\nabla}\cdot(\boldsymbol{\sigma}\cdot\mathbf{v})$. The rate of work done by the resultant of the surface stresses per unit volume is given by $\mathbf{v}\cdot\boldsymbol{\nabla}\cdot\boldsymbol{\sigma}$. The difference between these two terms yields the rate of work done by the surface stresses in deforming the continuum per unit volume. Show that this difference can be written as $\boldsymbol{\sigma} : \mathbf{D}$, where $\mathbf{D}$ is the symmetric part of the velocity gradient tensor.

2.17 Express the energy equation in Eq. (2.4.13) in the following alternative form:

$$\frac{\partial(\rho T)}{\partial t} + \boldsymbol{\nabla} \cdot (\rho\mathbf{v}T) = \boldsymbol{\nabla} \cdot (\alpha\boldsymbol{\nabla}T) + \rho\hat{g},$$

where $\alpha = k/c_v$ and $\hat{g} = g/c_v$.

2.18 Show that the equation governing heat transfer in an isotropic medium, Eq. (2.4.15), takes the following form in the cylindrical coordinate system:

$$\rho c_v\frac{\partial T}{\partial t} - \frac{1}{r}\left[\frac{\partial}{\partial r}\left(rk\frac{\partial T}{\partial r}\right) + \frac{\partial}{\partial\theta}\left(k\frac{\partial T}{\partial\theta}\right) + r\frac{\partial}{\partial z}\left(k\frac{\partial T}{\partial z}\right)\right] = g.$$

3

Solution Methods for Algebraic Equations

3.1 Introduction

All numerical methods, including the FEM and FVM, ultimately result in a set of linear or nonlinear algebraic equations, relating the values of the dependent variables at the nodal points of the mesh. These algebraic equations can be linear or nonlinear in the nodal values of the primary variables, depending on whether the governing differential equations being solved are linear or nonlinear. When the algebraic equations are nonlinear, we linearize them using certain assumptions and techniques, such as the Picard method or Newton's method.

In this chapter, we first discuss the linearization of the nonlinear algebraic equations and two iterative methods of solution. During each iteration, linear equations are being solved. A brief review the solution methods, such as the Gauss elimination, Gauss–Jordan elimination, and lower–upper (LU) decomposition direct methods and Gauss–Seidel iteration methods for the solution of linear equations are discussed (see [16, 17] for the details of these methods; the topic is also briefly visited in Section 3.3).

3.2 Linearization of Nonlinear Equations

3.2.1 Introduction

In the FVM, the final algebraic equations are derived are linear because the linearization procedure is applied along with the approximation of the expressions. This is not the case with the FVM. As a result, the finite element equations of nonlinear differential equations lead to nonlinear for each element of the finite element mesh. The finite element equation is of the form,

$$\mathbf{K}^e(\mathbf{u}^e)\mathbf{u}^e = \mathbf{F}^e, \tag{3.2.1}$$

where

$$\begin{aligned}
&\mathbf{K}^e \text{ is the element coefficient matrix, which depends on the solution vector } \mathbf{u}^e,\\
&\mathbf{u}^e \text{ is the column vector of element nodal values, and} \qquad\qquad (3.2.2)\\
&\mathbf{F}^e \text{ is the column vector of element nodal source values or forcing function.}
\end{aligned}$$

When $\mathbf{K}^e$ is independent of $\mathbf{u}^e$, the matrix coefficients can be evaluated for all elements and the assembled equations can be solved for the global nodal values

u after imposing boundary conditions. When $\mathbf{K}^e$ depends on the unknown solution vector $\mathbf{u}^e$, the matrix coefficients cannot be evaluated. If we can find an approximation to $\mathbf{u}^e$, say $(\mathbf{u}^e)^{(1)}$, then $\mathbf{K}^e((\mathbf{u}^e)^{(1)})$ can be evaluated and assembled. This amounts to linearizing the nonlinear equations, Eq. (3.2.1). Then a next approximation to the solution can be obtained by solving the *assembled* equations,

$$\mathbf{U}^{(2)} = \left[\mathbf{K}(\mathbf{U}^{(1)})\right]^{-1} \mathbf{F}, \tag{3.2.3}$$

where **U** is the external forcing function or global solution vector, **K** is the assembled global coefficient matrix, and **F** is the global source vector. This procedure can be repeated until the approximate solution comes close to the actual solution in some measure. Such a procedure is called an *iterative procedure.*

Here we discuss the following commonly used iterative procedures.

(1) The Picard iteration (or direct iteration) method.
(2) The Newton iteration method.

The details of these methods are discussed next, with the aid of a single nonlinear equation.

Consider the nonlinear equation of the following type for a single unknown:

$$K(u)\, u = F \quad \text{or} \quad R(u) = 0, \tag{3.2.4}$$

where u is the solution to be determined, $K(u)$ is a known function of u, F is the external forcing function, and R is the residual

$$R(u) = K(u)\, u - F. \tag{3.2.5}$$

A plot of the equilibrium path, $R(u, F) = 0$, is shown in Fig. 3.2.1. For any value u_1, $K(u_1)$ denotes the secant and $(\frac{\partial R}{\partial u})|_{u_1}$ denotes the tangent of the curve at $u = u_1$.

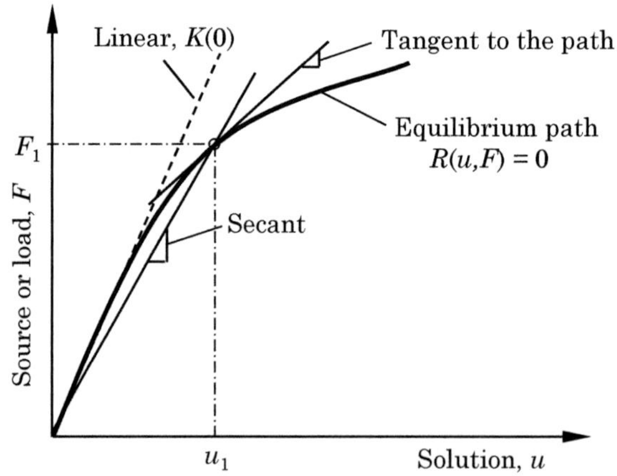

Fig. 3.2.1 A typical force–displacement curve, showing the secant and tangent lines at a point.

3.2.2 The Picard Iteration Method

In the Picard iteration method, also known as the *direct iteration* method or *the method of successive substitutions*, we begin with an initial guess for u, say $u^{(0)}$, $[u^{(0)} = 0$ in Fig. 3.2.2], and determine a first approximation of u by solving the equation

$$u^{(1)} = \left[K\left(u^{(0)}\right)\right]^{-1} F, \tag{3.2.6}$$

$u^{(1)} \neq u$, and a second approximation for u is sought by using the latest known solution $u^{(1)}$ to evaluate K

$$u^{(2)} = \left[K\left(u^{(1)}\right)\right]^{-1} F. \tag{3.2.7}$$

This procedure is continued until the difference between two consecutive approximations of u, namely $u^{(k-1)}$ and $u^{(k)}$, differ by a preselected value. Thus, the algorithm and criterion for convergence may be expressed as

$$\text{algorithm:} \qquad u^{(k)} = \left[K\left(u^{(k-1)}\right)\right]^{-1} F \tag{3.2.8}$$

$$\text{convergence criterion:} \qquad \sqrt{\frac{\left(u^{(k)} - u^{(k-1)}\right)^2}{\left(u^{(k)}\right)^2}} < \epsilon, \tag{3.2.9}$$

where ϵ denotes the *convergence tolerance* and k denotes the iteration number.

A geometric interpretation of the procedure described above is illustrated in Fig. 3.2.2(a) for an initial guess of $u^{(0)} = 0$. At the beginning of iteration k, the secant of the curve $R(u) = 0$ is found at the point $u = u^{(k-1)}$ and the solution $u^{(k)}$ is computed using Eq. (3.2.8). Figure 3.2.2(a) shows the convergence to the true solution u_c, whereas Fig. 3.2.2(b) shows a possible divergence of the algorithm. Thus, the success of the algorithm depends on the nature of the nonlinear curve $R(u) = 0$, the initial guess, and the external forcing function (F).

In the direct iteration method discussed above, the secant is evaluated at each iteration and inverted to obtain the next approximate solution. This can be computationally very expensive when the number of algebraic equations to be solved is large, that is, when $\mathbf{K}$ is a matrix and $\mathbf{K}^{-1}$ is its inverse. When $\mathbf{K}$ has a linear portion (in most problems of interest to us, it does), an *alternative direct iteration* algorithm can be formulated. Let

$$\mathbf{K}(\mathbf{u}) = \mathbf{K}_L + \mathbf{K}_N(\mathbf{u}). \tag{3.2.10}$$

where $\mathbf{K}_L$ and $\mathbf{K}_N(\mathbf{u})$ are the linear and nonlinear parts of K; note the argument $(\mathbf{u})$ of $\mathbf{K}_N$ implies that it depends on the unknown solution $\mathbf{u}$. Note that K_L is the slope at $u = 0$ of the curve $R(u) = 0$. Then we can write

$$\mathbf{u}^{(k)} = [\mathbf{K}_L]^{-1} \left[\mathbf{F} - \mathbf{K}_N^{(k-1)}\, \mathbf{u}^{(k-1)}\right], \tag{3.2.11}$$

where $\mathbf{K}_N^{(k-1)} \equiv \mathbf{K}_N\left(\mathbf{u}^{(k-1)}\right)$. This scheme involves evaluating the nonlinear part $\mathbf{K}_N$ at each iteration, which is computationally less expensive when compared to evaluating $\mathbf{K}_N^{(k-1)}$ and inverting it. The inversion of $\mathbf{K}_L$ is required only once, and it should be saved for subsequent use. The criterion in Eq. (3.2.9) can be used to check for convergence.

(a)

(b)

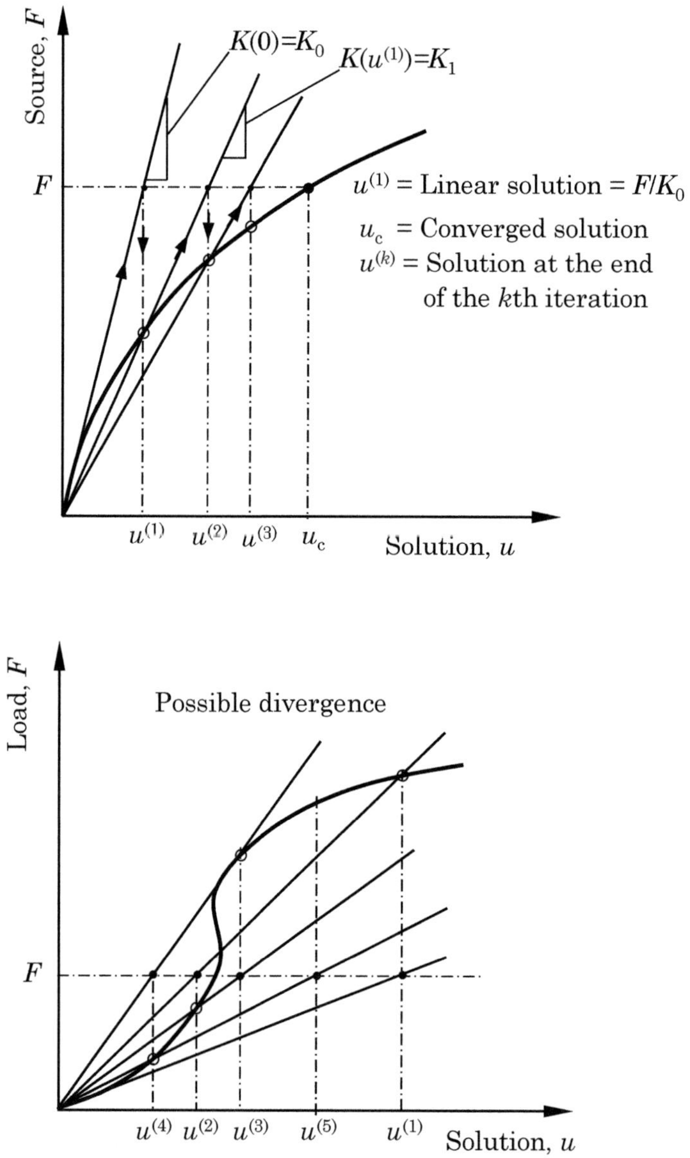

Fig. 3.2.2 A graphical illustration of the direct iteration scheme: $\mathbf{u}^{(k)} = \left[\mathbf{K}\left(\mathbf{u}^{(k-1)}\right)\right]^{-1}\mathbf{F}$, where k denotes the current iteration number. (a) A case of convergence and (b) a case of divergence of the solution.

Geometrically (in a one-dimensional case), the alternative direct iteration algorithm uses the initial slope, that is, the slope at the origin of the curve for *all* iterations, while updating the *effective* force

$$\hat{F} \equiv F - K_N^{(k-1)}\, u^{(k-1)}$$

at each iteration (see Fig. 3.2.3). The rate of convergence of this algorithm, if it converges, is slower than that in Eq. (3.2.8).

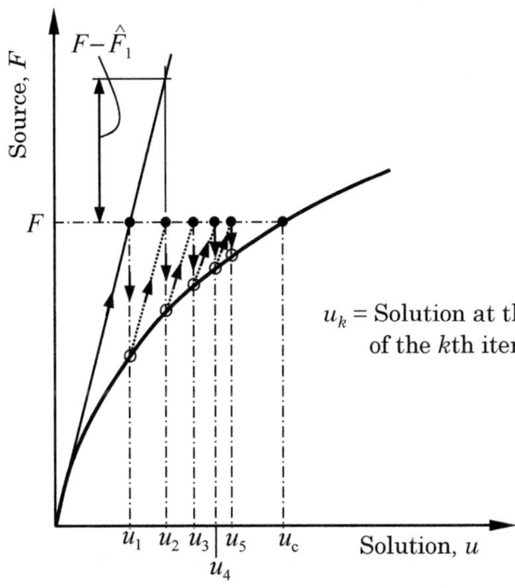

Fig. 3.2.3 A graphical illustration of the modified direct iteration scheme.

As applied to the assembly of finite element equations in Eq. (3.2.1), that is $\mathbf{K}(\mathbf{U})\,\mathbf{U} = \mathbf{F}$, the two algorithms take the following forms:

$$\text{algorithm 1:}\quad \mathbf{U}^k = \left[\mathbf{K}\left(\mathbf{U}^{(k-1)}\right)\right]^{-1}\mathbf{F} \tag{3.2.12}$$

$$\text{algorithm 2:}\quad \mathbf{U}^r = [\mathbf{K}_L]^{-1}\left[\mathbf{F} - \mathbf{K}_N\left(\mathbf{U}^{(k-1)}\right)\mathbf{U}^{(k-1)}\right] \tag{3.2.13}$$

$$\text{convergence criterion:}\quad \sqrt{\frac{\sum_{I=1}^{N}\left(U_I^{(k)} - U_I^{(k-1)}\right)^2}{\sum_{I=1}^{N}\left(U_I^{(k)}\right)^2}} < \epsilon, \tag{3.2.14}$$

where $\mathbf{U}$ denotes the global solution vector. The global matrices $\mathbf{K}$, $\mathbf{K}_L$, and $\mathbf{K}_N$ are assembled from the corresponding element matrices ($\mathbf{u}$ denotes the solution vector at the element level):

$$\begin{aligned} \mathbf{K} \text{ is assembled using } &\ \mathbf{K}^e\left(\mathbf{u}^{(k-1)}\right) \\ \mathbf{K}_L \text{ is assembled using } &\ \mathbf{K}_L^e \\ \mathbf{K}_N \text{ is assembled using } &\ \mathbf{K}_N^e\left(\mathbf{u}^{(k-1)}\right), \end{aligned} \tag{3.2.15}$$

where

$$\mathbf{K}^e = \mathbf{K}_L^e + \mathbf{K}_N^e. \tag{3.2.16}$$

The rate of convergence of the iterative procedure can be accelerated, in certain type of nonlinear behavior, by a *relaxation* procedure in which a weighted average of the last two solutions is used to evaluate $\mathbf{K}^e$ (or $\mathbf{K}_N^e$): $\mathbf{K}^e(\bar{\mathbf{u}}^e)$, where $\bar{\mathbf{u}}^e = \beta \mathbf{u}^{(k-2)} + (1-\beta)\mathbf{u}^{(k-1)}$, and $0 \leq \beta \leq 1$ is called the relaxation or acceleration parameter. In this case Eqs. (3.2.12) and (3.2.13) take the form,

$$\text{algorithm 1} \qquad \mathbf{U}^r = \left[\mathbf{K}(\bar{\mathbf{U}})\right]^{-1}\mathbf{F} \tag{3.2.17}$$

$$\text{algorithm 2} \qquad \mathbf{U}^r = \left[\mathbf{K}_L\right]^{-1}\left(\mathbf{F} - \mathbf{K}_N(\bar{\mathbf{U}})\bar{\mathbf{U}}\right) \tag{3.2.18}$$

$$\bar{\mathbf{U}} = \beta \mathbf{U}^{(k-2)} + (1-\beta)\mathbf{U}^{(r-1)}. \tag{3.2.19}$$

The actual value of β varies from problem to problem.

The computational algorithm of the direct iteration is summarized below. At each load level follow the steps.

(1) Compute element matrices $\mathbf{K}^e$ and $\mathbf{F}^e$ (for transient problems, $\mathbf{K}^e$ and $\mathbf{F}^e$ are to be replaced by $\hat{\mathbf{K}}^e$ and $\hat{\mathbf{F}}^e$) using the solution $\mathbf{U}^{(k-1)}$ from the previous iteration (of current load and/or time). For the first iteration of the subsequent load steps, use $\mathbf{U}_c$, the converged solution of the last load step. For the first load step use $\mathbf{U}_c = \mathbf{0}$, provided $\mathbf{K}$ is not singular, to compute the linear solution.
(2) Assemble the element matrices $\mathbf{K}^e$ and $\mathbf{F}^e$ (or $\hat{\mathbf{K}}^e$ and $\hat{\mathbf{F}}^e$).
(3) Apply the boundary conditions on the assembled set of equations.
(4) Solve the assembled equations.
(5) Check for convergence using Eq. (3.2.14).
(6) If the convergence criterion is satisfied, increase the load to next level, initialize the counter on iterations, and repeat Steps (1)–(5).
(7) If the convergence criterion is not satisfied, check if the maximum number of allowable iterations is exceeded. If yes, terminate the computation printing a message to that effect. If the number of iterations did not exceed the maximum allowed, update $\mathbf{U}^{(k-1)}$ and $\mathbf{U}^{(k)}$ and repeat Steps (1)–(5).

3.2.3 The Newton Iteration Method

Suppose we know the solution $u^{(k-1)}$ of Eq. (3.2.4) at the $(k-1)$th iteration and are interested in seeking solution at the rth iteration. We expand $R(u)$ about the known solution $u^{(k-1)}$ in a Taylor series,

$$R(u) = R\left(u^{(k-1)}\right) + \left(\frac{\partial R}{\partial u}\right)\bigg|_{u^{(k-1)}} \delta u + \frac{1}{2}\left(\frac{\partial^2 R}{\partial u^2}\right)\bigg|_{u^{(k-1)}} (\delta u)^2 + \cdots = 0, \tag{3.2.20}$$

where δu is the increment,

$$\delta u^{(k)} = u^{(k)} - u^{(k-1)}. \tag{3.2.21}$$

Assuming that the second- and higher-order terms in δu are negligible, we can write Eq. (3.2.20) as

$$\begin{aligned}\delta u^{(k)} &= -\left[K_T\left(u^{(k-1)}\right)\right]^{-1} R\left(u^{(k-1)}\right)\\ &= \left[K_T\left(u^{(k-1)}\right)\right]^{-1}\left[F - K\left(u^{(k-1)}\right) u^{(k-1)}\right],\end{aligned} \tag{3.2.22}$$

where K_T is the slope (tangent) of the curve $R(u)$ at $u^{(k-1)}$ (see Fig. 3.2.4):

$$K_T = \left.\frac{\partial R}{\partial u}\right|_{u^{(k-1)}}. \tag{3.2.23}$$

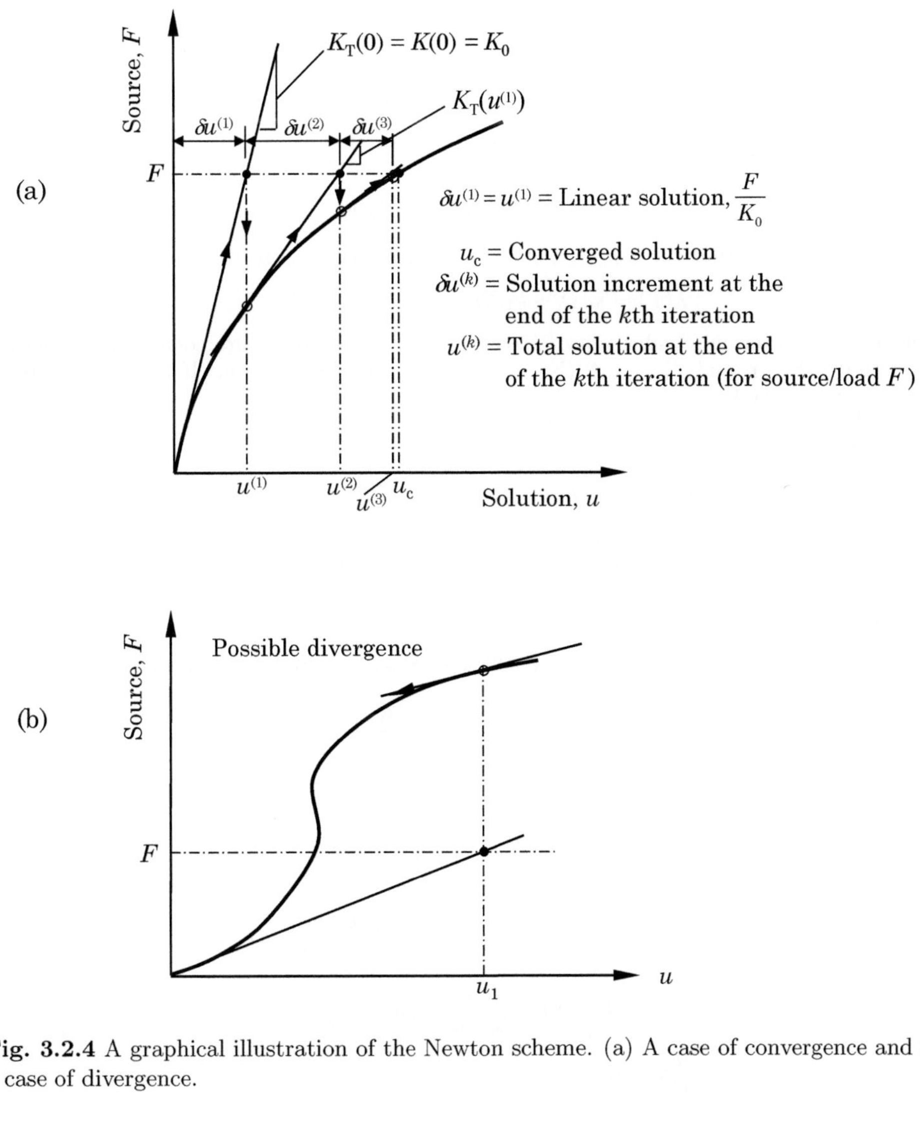

Fig. 3.2.4 A graphical illustration of the Newton scheme. (a) A case of convergence and (b) a case of divergence.

The residual or *imbalance force*, $R\left(u^{(k-1)}\right)$ is gradually reduced to zero if the procedure converges. Equation (3.2.22) gives the increment of u at the kth iteration so that the total solution is

$$u^{(k)} = u^{(k-1)} + \delta u^{(k)}. \tag{3.2.24}$$

The iteration is continued until a convergence criterion, say Eq. (3.2.14), is satisfied. Other convergence criteria include checking the magnitude of the imbalance force. A geometrical interpretation of the Newton procedure is shown in Fig. 3.2.4(a). For most problems, the method has faster convergence characteristics. Figure 3.2.4(b) illustrates possible divergence of the iterative procedure for certain problems.

Example 3.2.1

Consider heat transfer in an isotropic bar of length L m with left end maintained at a temperature of T_0 K and the right end at T_L. Assume that the internal heat generation is $g(x)$ and the surface of the bar is insulated so that there is no convection from the surface. The governing differential equation and boundary conditions of the problem are

$$\begin{aligned} -\frac{d}{dx}\left[k(T)\frac{dT}{dx}\right] &= g, \quad 0 < x < L \\ T(0) = T_0\,\mathrm{K}, \quad T(L) &= T_L\,\mathrm{K}. \end{aligned} \tag{1}$$

In addition, suppose that the conductivity $k(T)$ varies according to the relation

$$k(T) = k_0\left(1 + k_1 \Delta T\right), \quad \Delta T = T - T_{ref}, \tag{2}$$

where k_0 is the thermal conductivity [W/(m K)], k_1 the temperature coefficient of thermal conductivity (K^{-1}), and $T_{ref} = 0$ (K) is the reference temperature. If the finite element equations are of the form

$$\mathbf{K}^e(\mathbf{T}^e)\,\mathbf{T}^e = \mathbf{F}^e, \tag{2}$$

where

$$\begin{aligned} K_{ij}^e &= \int_{x_a}^{x_b} k(T)\frac{d\psi_i^e}{dx}\frac{d\psi_j^e}{dx}\,dx \\ F_i^e &= \int_{x_a}^{x_b} g\psi_i^e\,dx + Q_a\psi_i^e(x_a) + Q_b\psi_i^e(x_b), \end{aligned} \tag{3}$$

determine the tangent matrix coefficients, K_{ij}^T.

Solution The element tangent matrix coefficients are computed from the definition

$$(K^T)_{ij}^e \equiv \frac{\partial R_i^e}{\partial T_j^e} = K_{ij}^e + \sum_{m=1}^{n}\frac{\partial K_{im}^e}{\partial T_j^e}T_m^e - \frac{\partial F_i^e}{\partial T_j^e}, \tag{4}$$

where

$$\begin{aligned} \sum_{m=1}^{n}\frac{\partial K_{im}^e}{\partial T_j^e}T_m^e &= \sum_{m=1}^{n}\frac{\partial}{\partial T_j^e}\left\{\int_{x_a}^{x_b} k_0\left(1 + k_1 T\right)\frac{d\psi_i^e}{dx}\frac{d\psi_m^e}{dx}\,dx\right\}T_m^e \\ &= \sum_{m=1}^{n}\left\{\int_{x_a}^{x_b}\left(0 + k_0 k_1\psi_j^e\right)\frac{d\psi_i^e}{dx}\frac{d\psi_m^e}{dx}\,dx\right\}T_m^e \\ &= \int_{x_a}^{x_b} k_0 k_1\frac{dT}{dx}\frac{d\psi_i^e}{dx}\psi_j^e\,dx \end{aligned} \tag{5}$$

$$\frac{\partial F_i^e}{\partial T_j^e} = 0, \tag{6}$$

and the following identities

$$T = \sum_{m=1}^{n} T_m^e \psi_m^e$$

and

$$\frac{\partial T}{\partial T_j^e} = \delta_{jm}\, \psi_m^e = \psi_j^e$$

are used in arriving at Eq. (6). Thus, the tangent matrix coefficients are given by

$$(K^T)_{ij}^e = K_{ij}^e + \int_{x_a}^{x_b} k_0 k_1 \frac{dT}{dx}\frac{d\psi_i^e}{dx}\psi_j^e \, dx. \tag{7}$$

The Newton method requires that the tangent K_T be computed at each iteration. This can be very expensive when many degrees of freedom are involved. A *modified Newton technique* involves, for a fixed source (or forcing function or "load") step, either keeping K_T fixed while updating the imbalance force at each iteration (see Fig. 3.2.5) or updating K_T only at each preselected number of iterations while updating the imbalance force at each iteration. The latter is known as the *Newton–Raphson method.*

The Newton and modified Newton (i.e, Newton–Raphson) procedures take the following forms when applied to the assembly of element equations:

The Newton procedure

$$\delta \mathbf{U} = -(\mathbf{K}_T)^{-1}\mathbf{R}. \tag{3.2.25}$$

The modified Newton procedure

$$\delta \mathbf{U} = -(\bar{\mathbf{K}}_T)^{-1}\bar{\mathbf{R}}, \tag{3.2.26}$$

where $\mathbf{K}_T$ is the tangent matrix (is calculated at the element level first and then assembled)

$$\mathbf{K}_T = \left.\frac{\partial \mathbf{R}}{\partial \mathbf{U}}\right|_{\mathbf{U}^{k-1}}$$

or

$$(K_T)_{ij} = \frac{\partial R_i}{\partial U_j} = K_{ij} + \sum_{k=1}^{N} \frac{\partial K_{ik}}{\partial U_j} U_k, \tag{3.2.27a}$$

N being the total number of equations in the mesh, and

$$\mathbf{R} = \mathbf{K}\left(\mathbf{U}^{(k-1)}\right)\mathbf{U}^{(k-1)} - \mathbf{F} \tag{3.2.27b}$$

$$\bar{\mathbf{K}}_T = \mathbf{K}_T(\bar{\mathbf{U}}), \quad \bar{\mathbf{K}} = \mathbf{K}(\bar{\mathbf{U}}), \tag{3.2.27c}$$

and $\bar{\mathbf{U}}$ is the solution at the beginning of the current source (or load) increment, and

$$\mathbf{U}^{(k)} = \mathbf{U}^{(k-1)} + \delta \mathbf{U}^{(k)}. \tag{3.2.28}$$

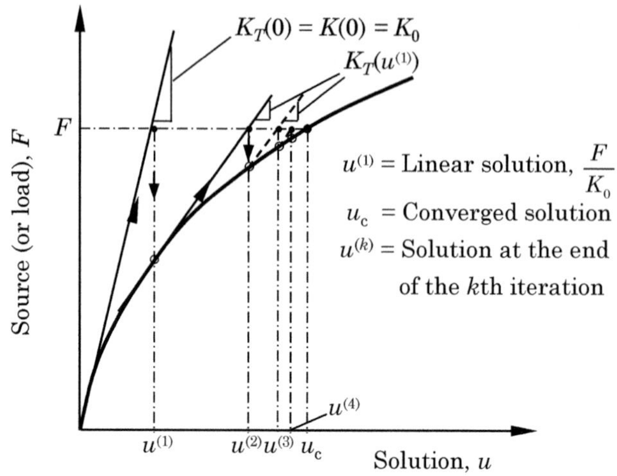

Fig. 3.2.5 The modified Newton scheme.

Since only the increment of the solution is computed in each iteration of the Newton iteration, the incremental equations, Eqs. (3.2.25) and (3.2.26), are subject to the homogeneous form of the specified essential boundary conditions of the problem. Thus after the first iteration of the first source step, any specified nonzero values of $\mathbf{U}$ should be set to zero so that in the subsequent iterations and source increments, $\delta\mathbf{U}$ is subjected to homogeneous boundary conditions.

For each increment of the source value, the following computations are required for the Newton or modified Newton procedure.

(1) Evaluate element matrices $\mathbf{K}^e$ and $\mathbf{F}^e$ (or $\hat{\mathbf{K}}^e$ and $\hat{\mathbf{F}}^e$ for the transient problems), and compute $\mathbf{K}_T^e$ and $\mathbf{R}^e$ using Eqs. (3.2.27a) and (3.2.27b) for an element.
(2) Assemble element matrices $\mathbf{K}_T^e$ and $\mathbf{R}^e$; for the modified Newton iteration procedure, save either assembled $\mathbf{K}_T$ or its inverse for use in subsequent iterations.
(3) Apply the boundary conditions on the assembled set of equations. *Note:* Set the specified boundary conditions on $\mathbf{U}$ to zero after Step (3) in the first iteration of the first load step.
(4) Solve the assembled equations.
(5) Update the solution vector using Eq. (3.2.28).
(6) Check for convergence.
(7) If the convergence criterion is satisfied, increase the source value, initialize the iteration counter, and repeat Steps (1)–(6). For the modified Newton iteration, compute $\mathbf{F}^e - \bar{\mathbf{K}}^e\mathbf{U}^{(k-1)}$ in Step (1) and go to Step (2).
(8) If the convergence criterion is not satisfied, check if the maximum number of iterations allowed is exceeded. If it is, terminate the computation by printing a message. If the maximum allowable number of iterations is not exceeded, go to Step (1).

In order to reduce the number of operations per iteration, in the modified Newton method the same system matrices are used for several iterations. These matrices are updated only at the beginning of each source increment or only when the convergence rate becomes poor. The modified Newton method may require more iterations to reach a new equilibrium point.

3.3 Solution of Linear Equations

3.3.1 Introduction

All numerical methods, including the FEM and FVM, ultimately result in a set of linear algebraic equations among the values of the dependent variable(s) and their dual variables (e.g., displacements and forces or temperatures and heat flux) at the nodal points of the mesh. These linear algebraic equations must be solved after the imposition of boundary (and initial) conditions.

In general, the algebraic equations are of the form

$$\mathbf{A}\mathbf{x} = \mathbf{C}, \tag{3.3.1}$$

where $\mathbf{A}$ is the coefficient matrix resulting from the application of a numerical method such as the FEM or FVM, $\mathbf{x}$ is the column of unknowns, and $\mathbf{C}$ is the known column vector. In the FEM and FVM, $\mathbf{A}$ is a banded sparse matrix that may be either symmetric or asymmetric depending on the characteristics of the governing differential equation(s) describing the physical problem and the discretization method used.

A banded matrix is one in which all elements beyond a diagonal parallel to the main diagonal are zero. The maximum number of nonzero diagonals above or below the main diagonal plus one (to account for the main diagonal) is called *half bandwidth* indexhalf bandwidth (NHBW; the letter N is added to indicate it is an integer) of the matrix [see Fig. 3.3.1(a)]. When the matrix is symmetric, it is sufficient to store elements in upper or lower half bandwidth of the matrix $\mathbf{A}$ [see Fig. 3.3.1(b)] and write solvers to take note of the fact that the coefficients of matrix $\mathbf{A}$, namely a_{ij}, satisfy the condition $a_{ij} = a_{ji}$ for all rows i and columns j of the matrix. When the matrix is not symmetric (i.e., $a_{ij} \neq a_{ji}$ for $i \neq j$), one must store all elements in the full bandwidth (2*NHBW-1) of the matrix.

Returning to the set of n linear algebraic equations, Eq. (3.3.1), the $n \times n$ matrix $\mathbf{A}$, in general, is referred to as the coefficient matrix; but in specific applications it may be called by other names (e.g., it is called the *stiffness matrix* in structural applications). The coefficient matrix $\mathbf{A}$ contains information that is internal to the system being analyzed, such as geometric parameters and thermo-physical properties. It may also depend on the initial guess or previous iteration values of the dependent variable (called the *linearization* of $\mathbf{A}$) in the case of nonlinear problems. The right-hand side vector $\mathbf{C}$ of real numbers represents contributions due to external stimuli such as specified boundary conditions and source or sink terms. For unsteady problems, $\mathbf{C}$ will also include the information from the previous time step. The solution (unknown) vector $\mathbf{x}$ can represent nodal displacements in structural problems and nodal temperatures,

species

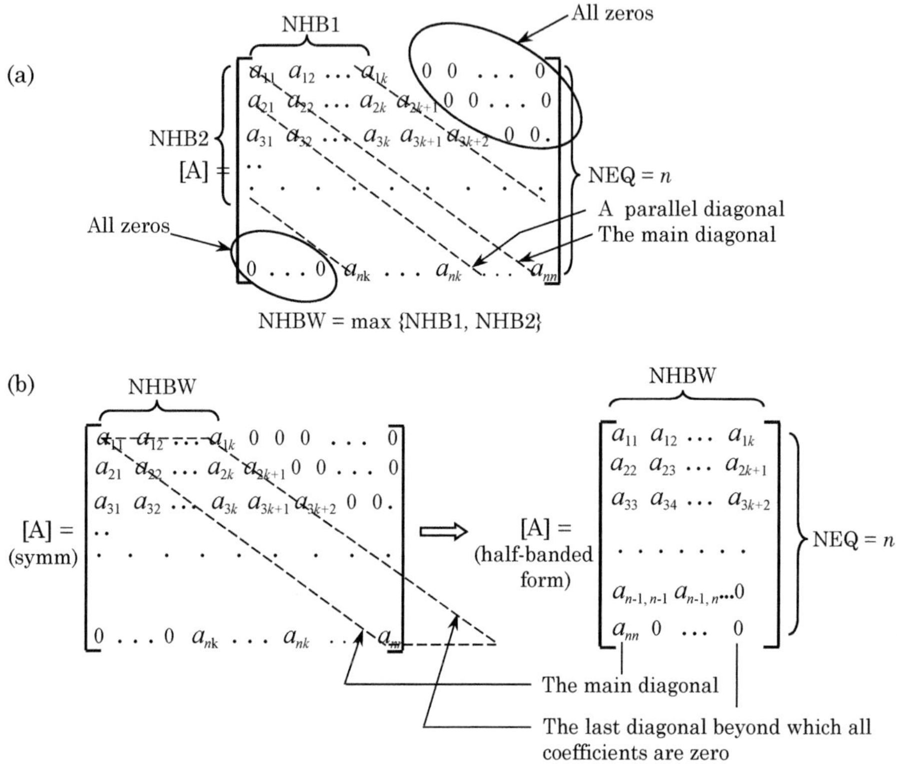

Fig. 3.3.1 Banded symmetric and unsymmetric matrices.

concentrations, velocities, turbulent kinetic energy, and the like, in the case of heat transfer, mass transfer, or fluid flow problems.

If all C_i are zero, then the system in Eq. (3.3.1) is called homogeneous. If any of the right-hand side elements is nonzero (i.e., not all C_i are zero), then the system is called nonhomogeneous. The solution to any linear system can have three possible outcomes:

(1) infinite number of solutions,
(2) no solution, or
(3) unique solution.

For example, consider a system of two linear equations. Each equation can be viewed geometrically as a line. When the lines representing the two equations are identical (e.g., $3x_1 - 6x_2 = 6$ and $x_1 - 2x_2 = 2$), there are an infinite number of solutions because the two equations are the same, leaving one unknown in terms of the other ($x_1 = 2 + 2x_2$). When the two lines are parallel (e.g., $4x_1 - 6x_2 = 6$ and $2x_1 - 3x_2 = 4$), they are inconsistent and no solution exists. If the two lines intersect at a point (e.g., $3x_1 - 6x_2 = 6$ and $x_1 + 2x_2 = 6$), then the solution is unique ($x_1 = 4$ and $x_2 = 1$). We note that the determinant of the coefficient matrix in the first two cases is zero, implying that the linear system does not have a unique solution.

Mathematically, the vector $\mathbf{x}$ is determined by inverting Eq. (3.3.1):

$$\mathbf{x} = \mathbf{A}^{-1}\mathbf{C}, \tag{3.3.2}$$

where $\mathbf{A}^{-1}$ denotes the inverse of the matrix $\mathbf{A}$. When n is very large (say, greater than two digits), computing a formal inverse of $\mathbf{A}$ is computationally expensive (i.e., the CPU time and storage/memory), requiring alternative methods to solve the linear equations.

In general, there are two types of techniques to solve a system of linear algebraic equations:

(1) direct methods,
(2) iterative methods.

3.3.1.1 Direct Methods

Direct methods are those that formally invert the coefficient matrix $\mathbf{A}$, which requires certain fixed number of arithmetic operations (i.e., multiplications and additions), and the end result is an exact solution (within the round-off errors introduced in the numerical evaluations in a computer) of the linear equations. Some of the techniques in this group are:

(1) Cramer's rule,
(2) Gauss elimination method,
(3) Gauss–Jordan method,
(4) LU decomposition.

3.3.1.2 Iterative Methods

The solution of very large systems of equations by direct methods takes large amounts of CPU time and/or memory. In iterative methods of solution, one may start with an initial approximation to the solution and modify the solution in several steps/iterations to bring it closer to the true solution. Once the approximation is sufficiently accurate (using a measure of the error between two consecutive approximate solutions), this is taken to be the solution to the system. Some of the widely used iterative techniques are:

(1) Jacobi iteration,
(2) Gauss–Seidel iteration,
(3) successive over-relaxation (SOR),
(4) alternate direction implicit (ADI),
(5) conjugate gradient,
(6) generalized minimal residuals (GMRES).

In the following sections, we review some of the traditional direct and iterative methods using illustrative examples. Readers may consult linear algebra or numerical methods books [16, 17] for additional details.

3.3.2 Direct Methods

3.3.2.1 Cramer's Rule

A complete derivation of Cramer's rule is not presented here, and only the formula for our use is given. First, we introduce certain necessary terminology. If the row and column containing an element a_{ij} of the square matrix $\mathbf{A}$ are deleted, the determinant of the remaining square matrix is called the *minor* of a_{ij}, and it is denoted here as M_{ij}. The *cofactor* of M_{ij} is defined as

$$A_{ij} = (-1)^{i+j} M_{ij}. \tag{3.3.3}$$

Then the determinant $|\mathbf{A}|$ of a square matrix $\mathbf{A}$ can be defined in terms of its cofactors as

$$|\mathbf{A}| = \sum_{k=1}^{n} a_{ik} A_{ik} \quad \text{or} \quad |\mathbf{A}| = \sum_{k=1}^{n} a_{kj} A_{kj} \tag{3.3.4}$$

for any value of i or j (determinant by expanding along a row or column).

Then Cramer's rule for solving a set of n linear equations, when $|\mathbf{A}| \neq 0$ (assuming that the solution exists) is given by

$$x_i = \frac{1}{|\mathbf{A}|} \sum_{j=1}^{n} a_{ji} c_j, \quad i = 1, 2, \ldots, n. \tag{3.3.5}$$

Now consider the specific case of a 3×3 system of linear equations in three unknowns (x_1, x_2, x_3):

$$\begin{aligned} a_{11}x_1 + a_{12}x_2 + a_{13}x_3 &= c_1 \\ a_{21}x_1 + a_{22}x_2 + a_{23}x_3 &= c_2 \\ a_{31}x_1 + a_{32}x_2 + a_{33}x_3 &= c_3. \end{aligned}$$

For this case, Cramer's rule in Eq. (3.3.5) gives the solution

$$x_1 = \frac{\begin{vmatrix} c_1 & a_{12} & a_{13} \\ c_2 & a_{22} & a_{23} \\ c_3 & a_{32} & a_{33} \end{vmatrix}}{\Delta}, \quad x_2 = \frac{\begin{vmatrix} a_{11} & c_1 & a_{13} \\ a_{21} & c_2 & a_{23} \\ a_{31} & c_3 & a_{33} \end{vmatrix}}{\Delta}, \quad x_3 = \frac{\begin{vmatrix} a_{11} & a_{12} & c_1 \\ a_{12} & a_{22} & c_2 \\ a_{13} & a_{32} & c_3 \end{vmatrix}}{\Delta},$$

where $\Delta = |\mathbf{A}|$,

$$\Delta = \begin{vmatrix} a_{11} & a_{12} & a_{13} \\ a_{21} & a_{21} & a_{23} \\ a_{31} & a_{32} & a_{33} \end{vmatrix}.$$

When $\Delta \neq 0$, we obtain a unique solution. For a homogeneous system the right-hand side of Eq. (3.3.1) is zero (i.e., all c_i are zero). Then the numerator of Eq. (3.3.5) is zero, giving a trivial solution, $x_1 = x_2 = x_3 = 0$.

Example 3.3.1

Consider the following set of linear equations:

$$\begin{aligned} 10x_1 + x_2 + x_3 &= 15 \\ x_1 + 10x_2 + x_3 &= 24 \\ x_1 + x_2 + 10x_3 &= 33. \end{aligned}$$

Determine its solution using Cramer's rule.

Solution First we note that the determinant of the coefficient matrix is

$$\Delta = \begin{vmatrix} 10 & 1 & 1 \\ 1 & 10 & 1 \\ 1 & 1 & 10 \end{vmatrix} = 990 - 9 - 9 = 972.$$

Then the solution is given by

$$x_1 = \frac{1}{972}\begin{vmatrix} 15 & 1 & 1 \\ 24 & 10 & 1 \\ 33 & 1 & 10 \end{vmatrix} = \frac{1}{972}(15 \times 99 - 1 \times 207 - 1 \times 306) = \frac{972}{972} = 1$$

$$x_2 = \frac{1}{972}\begin{vmatrix} 10 & 15 & 1 \\ 1 & 24 & 1 \\ 1 & 33 & 10 \end{vmatrix} = \frac{1}{972}(10 \times 207 - 15 \times 9 + 1 \times 9) = \frac{1944}{972} = 2$$

$$x_3 = \frac{1}{972}\begin{vmatrix} 10 & 1 & 15 \\ 1 & 10 & 24 \\ 1 & 1 & 33 \end{vmatrix} = \frac{1}{972}(10 \times 306 - 1 \times 9 - 15 \times 9) = \frac{2916}{972} = 3.$$

3.3.2.2 Gauss Elimination Method

In the Gauss elimination method, the matrix **A** is algebraically operated until all elements below the main diagonal are zeros of the form (i.e., reduce to the upper-diagonal form):

$$\begin{bmatrix} a'_{11} & a'_{12} & a'_{13} \\ 0 & a'_{21} & a'_{23} \\ 0 & 0 & a'_{33} \end{bmatrix} \begin{Bmatrix} x_1 \\ x_2 \\ x_3 \end{Bmatrix} = \begin{Bmatrix} c'_1 \\ c'_2 \\ c'_3 \end{Bmatrix}. \tag{3.3.6}$$

This step is referred to as *forward elimination*. Once the forward elimination is completed then unknowns are calculated by a process of *backward substitution*. So, starting with the third row in the above set of equations one can write:

$$x_3 = c'_3 / a'_{33}. \tag{3.3.7}$$

Then, using rows 2 and 1 in the above system, we get

$$x_2 = \frac{c'_2 - a'_{23}x_3}{a'_{22}} \tag{3.3.8}$$

$$x_1 = \frac{c'_1 - a'_{12}x_2 - a'_{13}x_3}{a'_{11}}. \tag{3.3.9}$$

This completes the process of *backward substitution.* Now we will use a contrived example problem to illustrate the implementation of direct and iterative methods listed above.

Example 3.3.2

Consider the same system of linear equations as in the case of the previous example, **Example 3.3.1**, and solve them using the Gauss elimination method.

Solution We have

$$\begin{bmatrix} 10 & 1 & 1 \\ 1 & 10 & 1 \\ 1 & 1 & 10 \end{bmatrix} \begin{Bmatrix} x_1 \\ x_2 \\ x_3 \end{Bmatrix} = \begin{Bmatrix} 15 \\ 24 \\ 33 \end{Bmatrix}.$$

Dividing the row 1 (R1) by the coefficient of the diagonal element a_{11} $(= 10)$, $R1' = R1/10$, we obtain

$$\begin{bmatrix} 1 & 0.1 & 0.1 \\ 1 & 10.0 & 1.0 \\ 1 & 1.0 & 10.0 \end{bmatrix} \begin{Bmatrix} x_1 \\ x_2 \\ x_3 \end{Bmatrix} = \begin{Bmatrix} 1.5 \\ 24.0 \\ 33.0 \end{Bmatrix}.$$

Next, we perform the following operations: (1) $R2' = R2 - R1'$ (i.e., subtract the modified row 1, $R1'$, from row 2), and (2) $R3' = R3 - R1'$ (i.e., subtract the modified row 1, $R1'$ from row 2):

$$\begin{bmatrix} 1 & 0.1 & 0.1 \\ 0 & 9.9 & 0.9 \\ 0 & 0.9 & 9.9 \end{bmatrix} \begin{Bmatrix} x_1 \\ x_2 \\ x_3 \end{Bmatrix} = \begin{Bmatrix} 1.5 \\ 22.5 \\ 31.5 \end{Bmatrix}.$$

Now divide $R2'$ by $a_{22}(= 9.9)$ in the above system:

$$\begin{bmatrix} 1 & 0.1 & 0.1000 \\ 0 & 1.0 & 0.0909 \\ 0 & 0.9 & 9.9000 \end{bmatrix} \begin{Bmatrix} x_1 \\ x_2 \\ x_3 \end{Bmatrix} = \begin{Bmatrix} 1.5000 \\ 2.2727 \\ 31.5000 \end{Bmatrix}.$$

Next, we perform $R3'' = R3' - 0.9R2'$:

$$\begin{bmatrix} 1 & 0.1 & 0.1000 \\ 0 & 1.0 & 0.0909 \\ 0 & 0.0 & 9.8181 \end{bmatrix} \begin{Bmatrix} x_1 \\ x_2 \\ x_3 \end{Bmatrix} = \begin{Bmatrix} 1.5000 \\ 22.5000 \\ 29.4555 \end{Bmatrix}.$$

Now starting from $R3''$ we will begin the process of backward substitution and obtain

$$x_3 = 29.4545/9.8181 = 3.0000.$$

Using $R2'$, we obtain

$$x_2 = (2.2727 - 0.0909 * 3) = 2.0000.$$

Finally, using $R1'$:

$$x_1 = (1.5 - 0.1 * 2 - 0.1 * 3) = 1.0000.$$

Thus the solution of the system is

$$\begin{Bmatrix} x_1 \\ x_2 \\ x_3 \end{Bmatrix} = \begin{Bmatrix} 1.0 \\ 2.0 \\ 3.0 \end{Bmatrix}, \tag{3.3.10}$$

which coincides with the solution of **Example 3.3.1**.

3.3.2.3 Gauss–Jordan Method

In the Gauss–Jordan method the coefficient matrix is augmented with the identity matrix, and off-diagonal elements are eliminated in all rows (or equations) as opposed to eliminating off-diagonal elements in subsequent rows (or equations) as in the Gauss elimination method. Thus, through algebraic operations the coefficient matrix is reduced to the identity matrix and the augmented identity matrix becomes the inverse matrix ($\mathbf{A}^{-1}$) and the right-hand side becomes the solution vector. Here we will illustrate these steps using the same example system of equations as before.

Example 3.3.3

Solve the linear equations of **Example 3.3.1** using the Gauss–Jordan method.

Solution Writing the system of equations in matrix form and augmenting the coefficient matrix with the identity matrix, we obtain,

$$\left[\begin{array}{ccc|ccc} 10 & 1 & 1 & 1 & 0 & 0 \\ 1 & 10 & 1 & 0 & 1 & 0 \\ 1 & 1 & 10 & 0 & 0 & 1 \end{array}\right] \begin{Bmatrix} x_1 \\ x_2 \\ x_3 \end{Bmatrix} = \begin{Bmatrix} 15 \\ 24 \\ 33 \end{Bmatrix}.$$

Dividing a row (or equation) by its diagonal element (for example a_{11}) is termed as pivoting or normalizing. So, normalizing $R1$ (i.e., $R1' = R1/a_{11}$),

$$\left[\begin{array}{ccc|ccc} 1 & 0.1 & 0.1 & 0.1 & 0.0 & 0.0 \\ 1 & 10.0 & 1.0 & 0.0 & 1.0 & 0.0 \\ 1 & 1.0 & 10.0 & 0.0 & 0.0 & 1.0 \end{array}\right] \begin{Bmatrix} x_1 \\ x_2 \\ x_3 \end{Bmatrix} = \begin{Bmatrix} 1.5 \\ 24 \\ 33 \end{Bmatrix}.$$

Next, we subtract $R1'$ from $R2$ and $R3$: $R2' = R2 - R1'$ and $R3' = R3 - R1'$ and obtain

$$\left[\begin{array}{ccc|ccc} 1 & 0.1 & 0.1 & 0.1 & 0 & 0 \\ 0 & 9.9 & 0.9 & -0.1 & 1 & 0 \\ 0 & 0.9 & 9.9 & -0.1 & 0 & 1 \end{array}\right] \begin{Bmatrix} x_1 \\ x_2 \\ x_3 \end{Bmatrix} = \begin{Bmatrix} 1.5 \\ 22.5 \\ 31.5 \end{Bmatrix}.$$

Normalizing $R2$, $R2' = R2/a_{22}$,

$$\left[\begin{array}{ccc|ccc} 1 & 0.1 & 0.1000 & 0.100000 & 0.0000 & 0 \\ 0 & 1.0 & 0.0909 & -0.010101 & 0.1010 & 0 \\ 0 & 0.9 & 9.9000 & -0.100000 & 0.0000 & 1 \end{array}\right] \begin{Bmatrix} x_1 \\ x_2 \\ x_3 \end{Bmatrix} = \begin{Bmatrix} 1.5 \\ 2.2727 \\ 31.5 \end{Bmatrix}.$$

Next, $R1'' = R1' - 0.1R2'$ and $R3'' = R3' - 0.9R2'$ gives

$$\left[\begin{array}{ccc|ccc} 1 & 0.0 & 0.09090 & 0.101010 & -0.0101 & 0 \\ 0 & 1.0 & 0.09090 & -0.010101 & 0.1010 & 0 \\ 0 & 0.0 & 9.81819 & -0.090910 & -0.0909 & 1 \end{array}\right] \begin{Bmatrix} x_1 \\ x_2 \\ x_3 \end{Bmatrix} = \begin{Bmatrix} 1.27273 \\ 2.27270 \\ 29.45457 \end{Bmatrix}.$$

Normalizing $R3'$, $R3'' = R3'/a_{33}$, we obtain

$$\left[\begin{array}{ccc|ccc} 1 & 0.0 & 0.0909 & 0.101010 & -0.010100 & 0.00000 \\ 0 & 1.0 & 0.0909 & -0.010101 & 0.101010 & 0.00000 \\ 0 & 0.0 & 1.0000 & -0.009259 & -0.009258 & 0.10185 \end{array}\right] \begin{Bmatrix} x_1 \\ x_2 \\ x_3 \end{Bmatrix} = \begin{Bmatrix} 1.27273 \\ 2.2727 \\ 3.0000 \end{Bmatrix}.$$

Finally, the operations $R1'' = R1' - 0.0909R3''$ and $R2'' = R2' - 0.0909R3''$ give

$$\left[\begin{array}{ccc|ccc} 1 & 0 & 0 & 0.101850 & -0.009259 & -0.009259 \\ 0 & 1 & 0 & -0.009259 & 0.101850 & -0.009259 \\ 0 & 0 & 1 & -0.009259 & -0.009259 & 0.101850 \end{array}\right] \begin{Bmatrix} x_1 \\ x_2 \\ x_3 \end{Bmatrix} = \begin{Bmatrix} 1.0000 \\ 2.0000 \\ 3.0000 \end{Bmatrix}.$$

Thus, the solution of the system is:

$$\begin{Bmatrix} x_1 \\ x_2 \\ x_3 \end{Bmatrix} = \begin{Bmatrix} 1.0 \\ 2.0 \\ 3.0 \end{Bmatrix}.$$

It is evident that the number of arithmetic steps involved in the Gauss–Jordan method is significantly higher than those in the Gauss elimination method. Hence, one could expect significantly more round-off error in the Gauss–Jordan method compared with what would exist in the Gauss elimination method. But the advantage is that along with the solution one obtains the matrix inverse too. Once a matrix inverse for a stiffness matrix is determined ($\mathbf{A}^{-1}$) then the response to different right-hand sides (forcing function) can be obtained by simple matrix manipulation, $\mathbf{x} = \mathbf{A}^{-1}\mathbf{C}$.

3.3.3 Iterative Methods

Consider a sample 3×3 system of the form $\mathbf{Ax} = \mathbf{C}$. In heat transfer and problems we often solve a set of linear equations stemming from a single PDE or a set of nonlinear PDEs. The nonlinearity is buried into the coefficient matrix that is updated at each iteration. So an iterative process lends itself to accommodating updating of the coefficient matrix. Here we will focus on (a) the Jacobi iteration scheme, (b) the Gauss–Seidel iteration scheme, and (c) the successive over-relaxation (SOR) method.

In general, the ith algebraic equation of the system of linear equations can be written as

$$\sum_{j=1}^{N} a_{ij} x_j = c_i. \tag{3.3.11}$$

The above system is said to be diagonally dominant if

$$|a_{ii}| > \sum_{j=1, j\neq i}^{N} |a_{ij}|. \tag{3.3.12}$$

In our example problem, we have $N = 3$. *The diagonal dominance is a sufficient condition for convergence.*

3.3.3.1 Jacobi Iteration

In the Jacobi iteration scheme, the unknowns are calculated by iteration on the following expression:

$$x_i^{(k+1)} = \frac{1}{a_{ii}} \left[c_i - \sum_{j=1, j\neq i}^{N} a_{ij} x_j^{(k)} \right], \tag{3.3.13}$$

where summation on the repeated index i is not assumed.

When the above expression is applied to a 3×3 system,

$$x_1^{(k+1)} = \frac{1}{a_{11}}\left[c_1 - a_{12}x_2^{(k)} - a_{13}x_3^{(k)}\right] \tag{3.3.14}$$

$$x_2^{(k+1)} = \frac{1}{a_{22}}\left[c_2 - a_{21}x_1^{(k)} - a_{23}x_3^{(k)}\right] \tag{3.3.15}$$

$$x_3^{(k+1)} = \frac{1}{a_{33}}\left[c_3 - a_{31}x_1^{(k)} - a_{32}x_2^{(k)}\right]. \tag{3.3.16}$$

The application of the Jacobi iteration to an one-dimensional domain can be depicted as shown in Fig. 3.3.2.

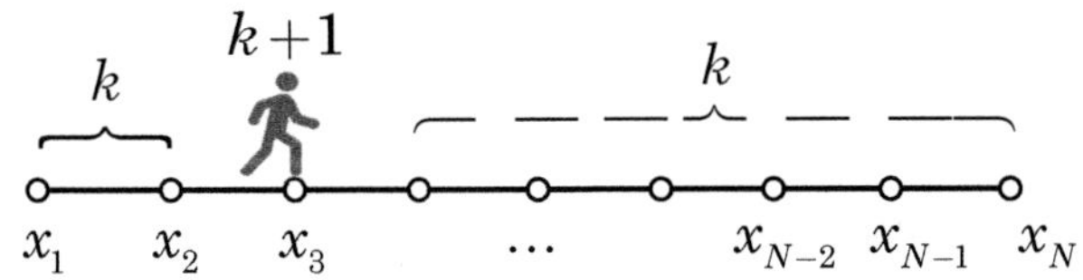

Fig. 3.3.2 Jacobi iteration in a one-dimensional domain.

We will apply the Jacobi iteration scheme to the example problem. In Fig. 3.2.2 an individual walks from left to right visiting each nodal point $i = 1, 2, 3, \ldots, N$ to calculate unknowns at $(k+1)$th iteration. Note that, in this figure, while the unknown at $(k+1)$th iteration is calculated at the ith node, the values of all other unknowns are evaluated from the kth iteration even though the values of $x_1, x_2, \ldots, x_{i-1}$ are available at the $(k+1)$th iteration. We will illustrate the implementation of Jacobi iteration using our example problem.

$$x_1^{(k+1)} = \frac{1}{10}\left[15 - x_2^{(k)} - x_3^{(k)}\right] \tag{3.3.17}$$

$$x_2^{(k+1)} = \frac{1}{10}\left[24 - x_1^{(k)} - x_3^{(k)}\right] \tag{3.3.18}$$

$$x_3^{(k+1)} = \frac{1}{10}\left[33 - x_1^{(k)} - x_2^{(k)}\right]. \tag{3.3.19}$$

In order to start an iterative process one has to begin with an initial guess. To keep the discussion simple for now, we will start with an initial guess of zero, that is

$$x_1^{(0)} = x_2^{(0)} = x_3^{(0)} = 0.$$

The iterative solution is shown in Table 3.3.1.

Table 3.3.1 Results of the Jacobi iteration scheme.

Iteration #	x_1	x_2	x_3
0	0.0000	0.0000	0.0000
1	1.5000	2.4000	3.3000
2	0.9300	1.9200	2.9100
3	1.0170	2.0160	3.0150
4	0.9969	1.9968	2.9967
5	1.0007	2.0006	3.0006
6	0.9999	1.9999	2.9999
7	1.0000	2.0000	3.0000

We know from prior calculations using the Gauss–Jordan method that the exact solution to the system of equations is $x_1 = 1.0, x_2 = 2.0, x_3 = 3.0$. From Table 3.3.1 it is evident that using the Jacobi iteration scheme the solution converges to the exact value up to four decimal places (i.e., the normalized root-mean value of the difference between two consecutive iterations is less than 10^{-4}) in seven iterations.

3.3.3.2 Gauss–Seidel Iteration

In the Gauss–Seidel iteration scheme, the unknowns at the $(k+1)$th iteration are evaluated by using the latest values of the unknowns $(\mathbf{x})$, as shown in Fig. 3.3.3. Here, visitation of nodes starts at $i = 1$ all the way through N at $(k+1)$th iteration. When visiting ith node, information for 1 through $(i-1)$st nodes are used at $(k+1)$th iteration and information for $(i+1)$st node through Nth node is used at kth iteration. In a sense, the Gauss–Seidel iteration scheme is easier to code than the Jacobi iteration scheme as we do not need to store the values of $\mathbf{x}$ from previous iteration and the machine will take the latest available value on the right-hand side of an equation by default. The Gauss–Seidel iteration scheme can be written in a compact form as

$$x_i^{(k+1)} = \frac{1}{a_{ii}}\left[C_i - \sum_{j=1}^{i-1} a_{ij}x_j^{(k+1)} - \sum_{j=i+1}^{N} a_{ij}x_j^{(k)}\right]. \tag{3.3.20}$$

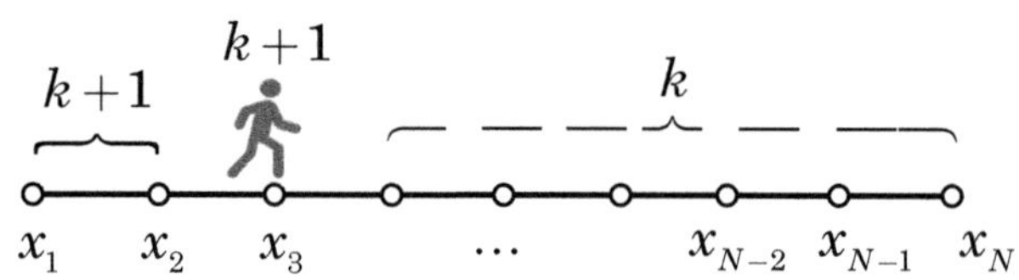

Fig. 3.3.3 The Gauss–Seidel iteration in an one-dimensional domain.

When applied to the example at hand, the three equations are:

$$x_1^{(k+1)} = \frac{1}{10}\left[15 - x_2^{(k)} - x_3^{(k)}\right] \tag{3.3.21}$$

$$x_2^{(k+1)} = \frac{1}{10}\left[24 - x_1^{(k+1)} - x_3^{(k)}\right] \tag{3.3.22}$$

$$x_3^{(k+1)} = \frac{1}{10}\left[33 - x_1^{(k+1)} - x_2^{(k+1)}\right]. \tag{3.3.23}$$

Again, we will start the iterative process with an initial guess of $x_1^{(0)} = x_2^{(0)} = x_3^{(0)} = 0$. The solution by iteration is shown in Table 3.3.2.

Table 3.3.2 Gauss–Seidel iteration.

Iteration #	x_1	x_2	x_3
0	0.0000	0.0000	0.0000
1	1.5000	2.2500	2.9250
2	0.9825	2.0093	3.0008
3	0.9990	2.0000	3.0001
4	1.0000	2.0000	3.0000

From Table 3.3.2, it is clear that the solution has converged to its exact value (up to four decimal places) in four iterations, as opposed to seven iterations in the Jacobi method. This was expected as at each iteration updated values were used in the Gauss–Seidel iteration scheme while in the Jacobi iteration scheme, values from the previous iteration were used. We will skip the further details here as our goal is to present and discuss implementation of a technique to solve two-dimensional problems using a combination of direct and iterative techniques.

3.3.3.3 Successive Over-Relaxation Method (SOR)

In sum, we are always confronted with solving a set of linear equations. These linear system may be the result of discretization of linear or nonlinear PDEs or ODEs. In particular, when a system of linear equations stems from discretization of linear PDEs or ODEs, it may be desirable to accelerate the process of convergence in the Gauss–Seidel iteration scheme without rendering the process unstable. Acceleration of the Gauss–Seidel iteration scheme is accomplished by a technique called *successive over-relaxation* (SOR). Invoke the Gauss–Seidel iteration scheme for an unknown x_i at the $(k+1)$th iteration:

$$x_i^{(k+1)} = \frac{1}{a_{ii}}\left[c_i - \sum_{j=1}^{i-1} a_{ij}x_j^{(k+1)} - \sum_{j=i+1}^{N} a_{ij}x_j^{(k)}\right]. \tag{3.3.24}$$

Add and subtract $x_i^{(k)}$ on the right hand side of the above equation:

$$x_i^{(k+1)} = x_i^{(k)} + \underbrace{\frac{1}{a_{ii}}\left[c_i - \sum_{j=1}^{i-1} a_{ij}x_j^{(k+1)} - \sum_{j=i+1}^{N} a_{ij}x_j^{(k)} - a_{ii}x_i^{(k)}\right]}_{\Delta x_i}. \tag{3.3.25}$$

The second term on the right-hand side of Eq. (3.3.25) is simply the change that is taking place in x_i in going from the kth iteration to the $(k+1)$th iteration

$$\Delta x_i = x_i^{(k+1)} - x_i^{(k)} \tag{3.3.26}$$

$$x_i^{(k+1)} = x_i^{(k)} + \Delta x_i. \tag{3.3.27}$$

One can accelerate the change in x_i by multiplying the change by a factor (ω) that is greater than unity

$$x_i^{(k+1)} = x_i^{(k)} + \omega(\Delta x_i) \tag{3.3.28}$$

$$x_i^{(k+1)} = x_i^{(k)} + \frac{\omega}{a_{ii}}\left[C_i - \sum_{j=1}^{i-1} a_{ij}x_j^{(k+1)} - \sum_{j=i+1}^{N} a_{ij}x_j^{(k)} - a_{ii}x_i^{(k)}\right], \tag{3.3.29}$$

where ω is called the over-relaxation factor. When $\omega > 1$, the method is termed as successive over-relaxation. Typically, the value of ω is between 1 and 2 ($1 \le \omega \le 2$) for stability reasons. Along the same lines, when a system of linear equations is a result of discretization of nonlinear ODEs or PDEs one has to decelerate the process of convergence for stability reasons. Such a technique is called *successive under-relaxation* (SUR) method. Most heat transfer and fluid flow problems are described by nonlinear PDEs, thus warranting under-relaxation ($\omega < 1$). When $\omega = 1$, the method reduces to the Gauss–Seidel iteration technique. In summary, we have

$$\begin{aligned} &\omega = 1 \Rightarrow \text{Gauss–Seidel,} \\ &\omega > 1 \Rightarrow \text{successive over-relaxation (SOR),} \\ &\omega < 1 \Rightarrow \text{successive under-relaxation (SUR).} \end{aligned} \tag{3.3.30}$$

Implementation of SOR for the contrived example is illustrated as

$$x_1^{(k+1)} = x_1^{(k)} + \frac{\omega}{10}\left[15 - x_2^{(k)} - x_3^{(k)} - 10x_1^{(k)}\right] \tag{3.3.31}$$

$$x_2^{(k+1)} = x_2^{(k)} + \frac{\omega}{10}\left[24 - x_1^{(k+1)} - x_3^{(k)} - 10x_2^{(k)}\right] \tag{3.3.32}$$

$$x_3^{(k+1)} = x_3^{(k)} + \frac{\omega}{10}\left[33 - x_1^{(k+1)} - x_2^{(k+1)} - 10x_3^{(k)}\right]. \tag{3.3.33}$$

Starting with an initial guess of $x_1^{(0)} = x_2^{(0)} = x_3^{(0)} = 0$, the solution with each iteration is shown in Table 3.3.3.

Table 3.3.3 Successive over-relaxation (SOR), $\omega = 1.03$.

Iteration #	x_1	x_2	x_3
0	0.0000	0.0000	0.0000
1	1.5450	2.3129	3.0016
2	0.9513	1.9955	3.0054
3	1.0014	1.9994	2.9998
4	1.0000	2.0000	3.0000

From prior calculations we know that the exact solution is $x_1 = 1.0, x_2 = 2.0, x_3 = 3.0$. From Table 3.3.3 the SOR with $\omega = 1.03$ converged in four iterations, which is same as in the case of the Gauss–Seidel iteration scheme. However, SOR converges faster than the Gauss–Seidel iteration scheme provided an optimal value of ω is chosen. Here, no attempt was made to establish the optimal value of ω. The value of $\omega = 1.03$ was chosen arbitrarily.

In closing this section, we note that the solution techniques used in this book are a combination of direct and iterative methods. Specifically, for the FVM we use a combination of the Gauss elimination method and the Gauss–Seidel iteration scheme. A brief review of these methods is presented next.

3.3.4 Iterative Methods for the Finite Volume Method

The algebraic equations derived in the FVM are naturally written in a form suitable for iterative method of solution. For example, consider the algebraic equation

$$a_P U_P^{k+1} - \left[a_W U_W^k + a_E U_E^k + a_S U_S^k + a_N U_N^k + b_P\right] = 0, \tag{3.3.34}$$

which arises in the finite volume discretization of the Poisson equation in two dimensions. The coefficients a_P, a_W, a_E, a_S, a_N, and b_P are known parameters, and U_P^k, U_W^k, U_E^k, U_S^k, and U_N^k are the values of dependent variables at the kth iteration. It is not important to know the details of the discretization method, other than the fact that Eq. (3.3.34) describes a system of linear equations. Then Eq. (3.3.34) can be solved iteratively for U_P^{k+1} by assuming that all the values of U^k with different subscripts are *assumed* to be known from the previous iteration. Let us denote the solution at the kth iteration as U^k; then the solution at the $(k+1)$th iteration can be calculated using

$$U_P^{k+1} = \frac{1}{a_P}\left[a_W U_W^k + a_E U_E^k + a_S U_S^k + a_N U_N^k + b_P\right]. \tag{3.3.35}$$

If U_P^{k+1} values are calculated completely based on the kth iteration values as indicated in Eq. (3.3.35), such a formulation is referred to as the *Jacobi iteration* method (see Fig. 3.3.2).

On the other hand, if U_P^{k+1} is expressed as a combination of values of the kth and $(k+1)$th iterations then such a formulation is referred to as the *Gauss–Seidel iteration scheme* (see Fig. 3.3.3). Then Eq. (3.3.35) becomes

$$U_P^{k+1} = \frac{1}{a_P}\left[a_W U_W^{k+1} + a_E U_E^k + a_S U_S^{k+1} + a_N U_N^k + b_P\right]. \tag{3.3.36}$$

Convergence of these iterative schemes can be declared in a number of different ways. Here we use the normalized relative change in the nodal variables as a measure of the error, and when the error is less than a prescribed tolerance, convergence is declared:

$$\max \left| \frac{U_P^{k+1} - U_P^k}{U_P^k} \right| < \epsilon, \tag{3.3.37a}$$

where ϵ is a small positive number of the order 10^{-3} to 10^{-6}. Another measure of error is

$$\sqrt{\frac{\sum \left(U_P^{k+1} - U_P^k\right)^2}{\sum (U_P^k)^2}} < \epsilon, \tag{3.3.37b}$$

and the sum is taken over the number of nodal values.

The Scarborough criterion [36] is often used to check convergence of the solution when solving linear equations using an iterative method. The Scarborough criterion can be expressed in terms of the values of the coefficients of the discretized equations as

$$\frac{\sum |a_{nb}|}{|a_P|} \begin{cases} \leq 1, & \text{at all nodes,} \\ < 1, & \text{at one node at least,} \end{cases} \tag{3.3.38}$$

where a_P is the coefficient of a typical central node P and the summation in the numerator is taken over all the neighboring nodes of the central node P. For a one-, two-, and three-dimensional problem there will be two (east and west), four (east, west, south, and north), and six (east, west, south, north, top, and bottom) neighbors for each central node, respectively.

If the Scarborough criterion is not satisfied, the Gauss--Seidel iteration scheme is not guaranteed to yield a convergent solution. This criterion is a *sufficient condition*, not a necessary one; that is, one can get convergence, even if the criterion is violated (see Minkowycz [37]).

One may speed up the process of convergence by "over-relaxation" or slow down the process of convergence by "under-relaxation." As we already know, such techniques are referred to as SOR methods and SUR methods, respectively. Under-relaxation may be necessary when the resulting set of discretization are a result of linearization of nonlinear differential equation(s). Under-relaxing makes the iterative process more stable. However, too much under-relaxation can make the convergence very slow. Development of the relaxation procedures for two-dimensional problems are discussed next.

Consider a point-by-point formulation using the Gauss–Seidel iteration scheme for a point P, as shown in Eq. (3.3.36). If we add and subtract U_P^k on the right-hand side of the equation, we obtain

$$U_P^{k+1} = U_P^k + \frac{1}{a_P}\left[a_W U_W^{k+1} + a_E U_E^k + a_S U_S^{k+1} + a_N U_N^k + b_P - a_P U_P^k\right]. \tag{3.3.39}$$

The term inside the square brackets represents the change in U_P in going from the kth to the $(k+1)$th iteration (ΔU_P). During relaxation, the goal is to alter this rate of change so the formulation becomes

$$U_P^{k+1} = U_P^k + \omega(\Delta U_p), \tag{3.3.40}$$

where ω is the relaxation parameter. For $\omega > 1$, SOR; for $\omega < 1$, SUR; and for $\omega = 1$, the Gauss–Seidel iteration method. For under-relaxation ω can be as low as 0.2, and for the over-relaxation it is is usually $1.0 < \omega < 2.0$. If an aggressive over-relaxation value for ω is used, the solution may diverge, that is, the iterative process can become numerically unstable.

Switching to indices notation, the iterative scheme can be set up as a loop over $jt = 2, jtmax - 1$ and a loop over $it = 2, itmax - 1$:

$$\begin{aligned} U_{it,jt}^{k+1} = U_{it,jt}^{k} + \frac{\omega}{a_P}\Big[& a_W\, U_{it-1,jt}^{k+1} + a_E\, U_{it+1,jt}^{k} + a_S\, U_{it,jt-1}^{k+1} \\ & + a_N\, U_{it,jt+1}^{k} - a_P\, U_{it,jt}^{k} + b_P\Big]. \end{aligned} \tag{3.3.41}$$

The above loop of calculations are executed until convergence is reached and is referred to as *point-by-point visitation.* In general, *point-by-point* iterative methods are slow to converge, especially when the number of grid points is large. On the other hand, if we wish to solve Eq. (3.3.11) using a direct method for a very large system ($\sim 10^6$), the computation time will be high. Hence, a *line-by-line procedure* was introduced wherein the linear equations along each line (in the x- or y-coordinate directions) are solved by making an implicit formulation in each direction. This method is a combination of direct and iterative solvers with an objective to accelerate the convergence. The idea is that along each line we will form a tri-diagonal matrix system, where the coefficient matrix is relatively small, and use a direct method (e.g., tri-diagonal matrix algorithm, TDMA) to solve the equation of the form $\mathbf{AU} = \mathbf{C}$. Then we sweep along the x- or y-axis iteratively until the solution converges.

Consider a relaxed form of the discretized equations shown earlier:

$$U_P^{k+1} = U_P^k + \frac{\omega}{a_P}\Big[a_W U_W^{k+1} + a_E U_E^k + a_S U_S^{k+1} + a_N U_N^k - a_P U_P^k + b_P\Big]. \tag{3.3.42}$$

Multiplying the above equation by $\frac{a_P}{\omega}$, and treating the formulation in the x-direction as implicit results in

$$-a_W U_W^{k+1} + \frac{a_P}{\omega} U_P^{k+1} - a_E U_E^{k+1} = a_S U_S^{k+1} + a_N U_N^k + b_P + a_P U_P^k \left(\frac{1}{\omega} - 1\right). \tag{3.3.43}$$

It is evident that the above equation lends itself to be solved using the TDMA or Thomas algorithm along each x-line. The off x-line terms like U_S^{k+1} and U_N^k are treated explicitly and they manifest as nonhomogeneous terms on the right-hand side of the above equation. So, starting from bottom, a tri-diagonal system of equations is solved along each x-line until the top of the domain is reached. Such a sweep is illustrated in Fig. 3.3.4.

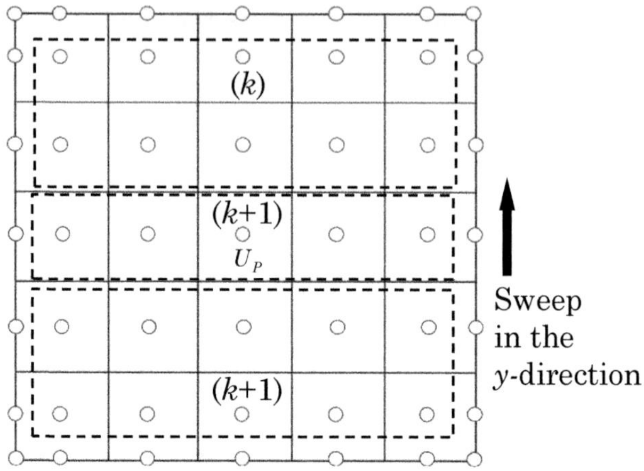

Fig. 3.3.4 Implicit in the x-direction, sweep along the y-direction.

Now treating terms as implicit along the y-direction, Eq. (3.3.42) becomes:

$$-a_S U_S^{k+1} + \frac{a_P}{\omega} U_P^{k+1} - a_N U_N^{k+1} = a_W U_W^{k+1} + a_E U_E^k + b_P + a_P U_P^k \left(\frac{1}{\omega} - 1\right). \tag{3.3.44}$$

Along each y-line, the tri-diagonal system of equations is solved and calculations proceed by sweeping along the x-direction from left to right. Such sweeps are illustrated in Fig. 3.3.5.

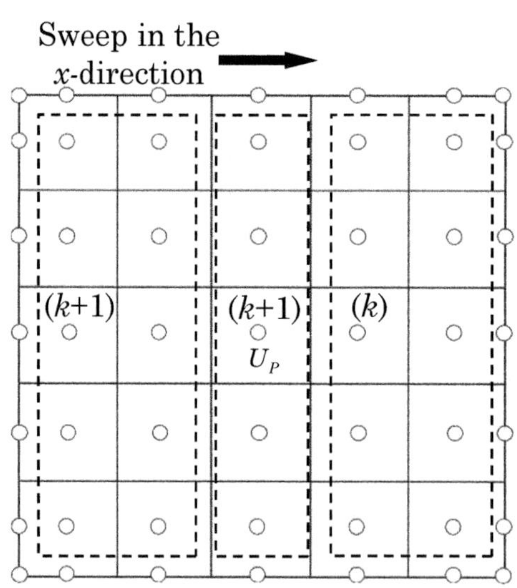

Fig. 3.3.5 Implicit in the y-direction, sweep along the x-direction.

Convergence of the solution can be further accelerated by alternating between the sweeps along the x- and y-directions. This is also known as the *alternating direction implicit method.* Thus, in order to solve the system of equations iteratively, sweeps are first carried out from bottom to top (implicit in x) and from left to right (implicit in y). There may be cases where nonhomogeneous boundary conditions may be on the right (east) boundary or the top (north) boundary. In such cases, if sweeps are restricted to only bottom to top or left to right, the convergence may be delayed. In order to speed up the convergence it is recommended that sweeps are carried out from bottom to top (x-implicit), left to right (y-implicit), top to bottom (x-implicit), and right to left (y-implicit). Such four-directional sweeps constitute one complete iteration. In three-dimensional cases, four-directional sweeps are carried out in each plane. Planes are sliced along all three directions.

There are other mathematically sophisticated methods such as conjugate gradient (CG) method, GMRES method, multigrid methods, and the like [38]. For elliptic equations, multigrid methods usually give the fastest convergence rate among all these methods. For other cases, the line-by-line procedure may or may not converge faster than other methods.

Problems

3.1 Consider the element coefficient matrix in **Example 3.2.1**. Suppose that

$$k(T^e) = a_0^e + a_1^e T^e + a_2^e (T^e)^2 + a_3^e \frac{dT^e}{dx}.$$

Determine the tangent matrix coefficients.

3.2 Consider the coefficient matrix and source vector components

$$K_{ij}^e = \int_{x_a^e}^{x_b^e} \left[a(x,u^e)\frac{d\psi_i^e}{dx}\frac{d\psi_j^e}{dx} + b(x,u^e)\psi_i^e \frac{d\psi_j^e}{dx} + c(x,u^e)\psi_i^e\psi_j^e \right] dx \tag{1}$$

$$F_i^e = \int_{x_a^e}^{x_b^e} f(x)\psi_i^e \, dx + Q_a \psi_i^e(x_a^e) + Q_b \psi_i^e(x_b^e), \tag{2}$$

where u^e is the dependent unknown. Suppose that

$$a(x,u^e) = a_0^e + a_1^e u^e(x) + a_2^e (u^e)^2 + a_3^e \frac{du^e}{dx}, \quad b(x,u^e) = 0, \quad c(x,u^e) = 0. \tag{3}$$

Determine the tangent matrix coefficients.

3.3 Consider the coefficient matrix and source vector components defined in **Problem 3.2** with $a(x,u^e) = \sqrt{2} + u^e$ and $b = c = 0$. Determine the tangent stiffness coefficients.

3.4 Consider the coefficient matrix and source vector components defined in **Problem 3.2** with $a(x,u^e) = 1$ and $b = -2u^e$ and $c = 0$. Determine the tangent stiffness coefficients.

3.5 Consider the following set of three equations in three unknowns:

$$\begin{aligned} x_1 + x_2 + x_3 &= 2 \\ x_1 - x_2 - 3x_3 &= 3 \\ 3x_1 + x_2 - x_3 &= 1. \end{aligned}$$

Does this system of equations have a solution and why?

3.6 Consider the following set of three equations in three unknowns:

$$\begin{aligned} x_1 + 2x_2 + x_3 &= 1 \\ x_1 - x_2 - 2x_3 &= 3 \\ x_1 + x_2 &= 1. \end{aligned}$$

Does the system have a solution?

3.7 Consider the following set of three equations in three unknowns:

$$\begin{aligned} 2x_1 + 3x_2 - x_3 &= 4 \\ x_1 - x_2 + 2x_3 &= 2 \\ x_1 + 2x_2 - x_3 &= 1. \end{aligned}$$

Does the system have a solution and is it unique?

3.8 Consider the following set of three equations in three unknowns:

$$\begin{aligned} x_1 + x_2 + x_3 &= 1 \\ x_1 - x_2 - 3x_3 &= 2 \\ 3x_1 + x_2 - x_3 &= 3. \end{aligned}$$

Determine the solution using Cramer's rule.

3.9 Consider the following set of three equations in three unknowns:

$$\begin{aligned} 3x_1 - x_2 + 2x_3 &= 2 \\ 2x_1 + x_2 + x_3 &= -1 \\ x_1 + 3x_3 &= 2. \end{aligned}$$

Determine the solution using Cramer's rule.

3.10 Consider the following set of three equations in three unknowns:

$$\begin{aligned} -x_1 + 5x_2 + 4x_3 &= 0 \\ x_1 - x_2 + 2x_3 &= 2 \\ x_1 + 2x_2 - x_3 &= 1. \end{aligned}$$

Determine the solution using Cramer's rule.

3.11 Determine the solution of the system of equations in **Problem 3.9** using the Gauss elimination method.

3.12 Determine the solution of the system of equations in **Problem 3.10** using the Gauss elimination method.

3.13 Determine the solution of the system of equations in **Problem 3.9** using the Gauss–Jordan method.

3.14 Determine the solution of the system of equations in **Problem 3.10** using the Gauss–Jordan method.

3.15 Determine the solution of the system of equations in **Problem 3.9** using the Jacobi iteration method (you may write a MatLab program to execute the steps).

3.16 Determine the solution of the system of equations in **Problem 3.10** using the Jacobi iteration method (you may write a MatLab program to execute the steps).

3.17 Determine the solution of the system of equations in **Problem 3.9** using the Gauss–Seidel iteration scheme (you may write a MatLab program to execute the steps).

3.18 Determine the solution of the system of equations in **Problem 3.10** using the Gauss–Seidel iteration scheme (you may write a MatLab program to execute the steps).

Part II The Finite Element Method

4

The Finite Element Method: Steady-State Heat Transfer

4.1 The Basic Idea

In this chapter we will introduce the FEM as a technique of solving differential equations governing a single variable (or dependent variable). Once we understand how the method works, it can be extended to problems governed by coupled PDEs among several unknowns. In particular, equations governing *steady-state* heat transfer in one- and two-dimensional problems are used as the "model" equations to introduce the FEM. A detailed account of the method can be found in the textbooks by Reddy [1] and Surana and Reddy [5, 6].

The FEM has the following basic steps.

(1) The whole domain Ω is discretized as a collection of interconnected and nonoverlapping subdomains, called *finite elements* (see Fig. 4.1.1). The division of the whole into parts is needed to facilitate the use of a low degree of approximation of the field variables; otherwise, one may need a higher degree of approximation if one were to seek the solution on the whole domain, as in the traditional variational methods (e.g., the weighted-residual methods and the Ritz method). Since the domain cannot be divided into an infinite number of parts, the size of the subdomain (element) is finite, hence the phrase *finite element.* Finite elements are geometric shapes that allow the unique derivation of polynomial approximation functions. For the purpose of defining the geometry of the elements, we identify points in the element, called *nodes.* For example, three points in a plane define a triangle uniquely, and the three points are necessarily the vertexes of the triangle. The collection of elements is termed a *finite element mesh.*

(2) Over each representative finite element Ω^e, we develop algebraic relations among the duality pairs (e.g., relations between pairs of dual variables (such as "heats" and "temperatures") using a *weighted* integral statement equivalent to the governing equation. The set of algebraic equations is called a *finite element model.* There are several possible ways, including the subdomain method, least-squares method, and so on, to satisfy the governing equation. Each method results in a different finite element model (e.g., subdomain finite element model, least-squares finite element model, etc.) of the same equation. In this book, we will develop the finite

element using the so-called weak form (equivalent to the Ritz method), which will be discussed in detail in Section 4.2.4. The FVM, in a broad sense, can be viewed as a finite element model in which an integral statement of the governing equation is used (as will be discussed in Chapter 7).

(3) The algebraic relations of all elements are put together, in what is called an *assembly*, using certain physical and mathematical requirements, and one can obtain relations between the duality pairs of the whole system.

(4) Apply the known (boundary) conditions, which have to be converted to known values of the duality pairs at the nodes, and solve the algebraic equations.

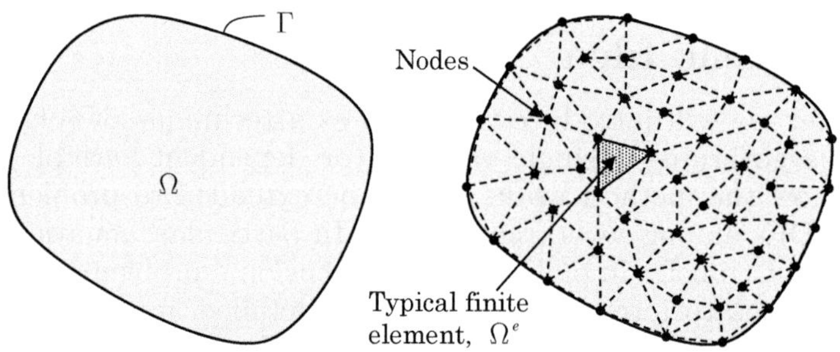

Fig. 4.1.1 Finite element representation of a two-dimensional domain (a set of linear triangles are used, as an example, to discretize the domain).

In general, the approximations used for the geometry and the dependent variable are different. If they are chosen to be the same, we call it an *isoparametric* formulation. In the second step, the conversion of the integral statement to algebraic equations among the nodal values requires an interpolation of the dependent variables of the governing differential equation(s). The functions actually interpolate the unknowns using function values at the nodes. The set of algebraic equations, that is, the finite element model, represents a discrete model of the original differential equation. The third step guarantees that the elements are interconnected and the union (i.e., assembly of elements) satisfies the interelement continuity and balance conditions. The final step accounts for the boundary conditions.

In this chapter, we present the basic steps of deriving the finite element model (i.e., a set of algebraic equations among the duality pairs) as applied to one- and two-dimensional problems described by model (i.e., typical) linear second-order differential equations involving a single dependent unknown. The main objective is to familiarize the reader with the terminology and specific steps involved in the development of the finite element model and its applications in solving some simple problems. For additional derivations and applications of the method, the reader may consult the text books by Reddy [1] and Reddy and Gartling [3].

4.2 One-Dimensional Problems

4.2.1 Model Differential Equation

Consider the differential equation

$$-\frac{d}{dx}\left(a\frac{du}{dx}\right) + b\frac{du}{dx} + cu = g \quad \text{for} \quad 0 < x < L, \tag{4.2.1}$$

where $a = a(x)$, $b = b(x)$, $c = c(x)$, and $g = g(x)$ are the *data* (i.e., known quantities) of the problem and $u(x)$ is the solution to be determined. Equation (4.2.1) is the same as Eq. (1.2.1) when $b = 0$ and $u = T - T_\infty$. When $c = 0$, Eq. (4.2.1) is known as the *advection–diffusion equation*. The data includes the material properties, geometry, and the "source" term. Equation (4.2.1) is to be solved subject to physically admissible end or boundary conditions, such as specifying u or its dual, namely, heat (when a contains the area of cross section; otherwise, it is a heat flux) $Q = -a(du/dx)$ at a boundary point.

4.2.2 Division of the Whole into Parts

The domain $\Omega = (0, L)$ of the uniaxial member (assumed here to be a straight bar) consists of all points between $x = 0$ and $x = L$, and not including the end points. The points $x = 0$ and $x = L$ are the boundary points of the domain. In the FEM, the domain $(0, L)$ is divided into a set of line intervals, $\Omega^e = (x_a^e, x_b^e)$, where x_a^e and x_b^e denote the coordinates of the end points of the line (finite) element Ω^e with respect to the coordinate x, as shown in Fig. 4.2.1. A typical line interval, (x_a^e, x_b^e) is called a *finite element*, $x = x_a^e$ and x_b^e being the end points, and $h_e = x_b^e - x_a^e$ being its length. Since the geometry is simple (a straight line), there is no need to talk about its approximation. However, the solution $u(x)$ can be a constant, linear in x, or a very higher-order function of x; the nature of the solution is dictated by the data and the boundary conditions.

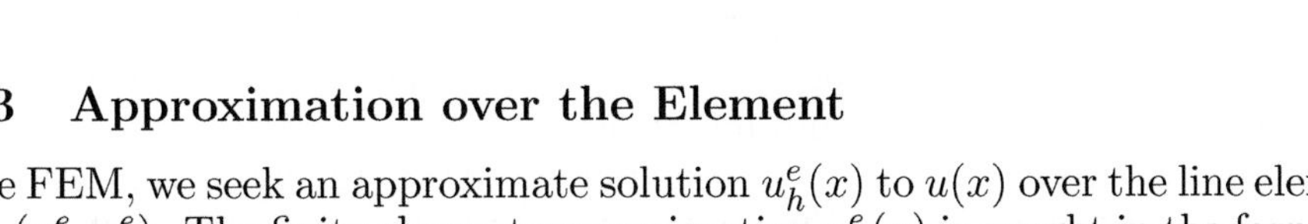

Fig. 4.2.1 A typical (line) finite element in one dimension.

4.2.3 Approximation over the Element

In the FEM, we seek an approximate solution $u_h^e(x)$ to $u(x)$ over the line element $\Omega^e = (x_a^e, x_b^e)$. The finite element approximation $u_h^e(x)$ is sought in the form of a *complete* algebraic polynomial. The word "complete" is used here to mean that all order terms, including the constant term, are included in the polynomial. For example, $u_h^e(x) = c_1 x + c_2 x^2$ and $u_h^e(x) = c_0 + c_2 x^2$ are incomplete because the constant term in the first and linear term in the second are missing. Examples of complete polynomials are provided by $u_h^e(x) = c_0 + c_1 x$ and $u_h(x) = c_0 +$

$c_1x + c_2x^2$. Thus, the $(n-1)$-degree complete polynomial (it contains n terms) is given by

$$u(x) \approx u_h^e(x) = c_0^e + c_2^e x + \cdots + c_{n-1}^e x^{n-1}, \tag{4.2.2}$$

where c_j^e are constants to be determined such that u_h^e satisfies the differential equation (4.2.1) and appropriate end conditions of the element. Since there are n unknown parameters, n linearly independent algebraic relations are needed to determine them. As we will see shortly, it is both necessary and useful to express the parameters c_i^e in terms of the values of u at n selected points, including the end points, of the element. That is, we will express Eq. (4.2.2) as

$$u_h^e(x) = \sum_{j=1}^{n} u_j^e \psi_j^e(x), \tag{4.2.3}$$

where $\psi_j^e(x)$ are interpolation functions systematically derived by equating u_h^e to u_j^e at $x = x_j^e$, the location of the jth node in the element (i.e., one must identify n nodes, including the end points, in the line element). We will return to the construction of ψ_j^e in the sequel.

Substituting the approximate solution in Eq. (4.2.3) into the left-hand side of Eq. (4.2.1), we obtain an expression that, in general, will not be equal to the right-hand side of the equation, $g(x)$, because $g(x)$ is a specified function, independent how the equation is solved. The difference between the two sides of the equation is called the *residual* (i.e., error in the differential equation):

$$-\frac{d}{dx}\left(a\frac{du_h^e}{dx}\right) + b\frac{du_h^e}{dx} + cu_h^e(x) - g(x) \equiv \mathbb{R}^e(x, u_1^e, u_2^e, \ldots, u_n^e) \neq 0.$$

We wish to determine u_j^e $(j = 1, 2, \ldots, n)$ such that the residual $\mathbb{R}^e$ is made zero, in some acceptable sense, over the element.

One way of making the residual zero over the element domain is to set the weighted-integral of the residual to zero:

$$\int_{x_a^e}^{x_b^e} w_i^e(x)\, \mathbb{R}^e(x, u_1^e, u_2^e, \ldots, u_n^e)\, dx = 0, \qquad i = 1, 2, \ldots, n \tag{4.2.4}$$

for each linearly independent choice of the weight functions w_i^e $(i = 1, 2, \ldots, n)$. Equation (4.2.4) provides a set of n algebraic relations among the nodal values u_j^e $(j = 1, 2, \ldots, n)$. The set $\{w_i^e(x)\}_{i=1}^n$ must be linearly independent (not a constant multiple of each other) so that the algebraic equations in Eq. (4.2.4) are also linearly independent and invertible.

There are several choices of w_i^e that may be used. In the present study, we take $w_i^e(x)$ to be the same as the approximation functions $\psi_i^e(x)$. For this particular choice, the weighted-residual method is known as the *Galerkin method.* A different choice of the weight functions will result in a different set of algebraic relations or a different *finite element model* of the same differential equation. In particular, the algebraic relations resulting from the weighted-residual statement with $w_i^e = \psi_i^e$ is called the *Galerkin finite element model.*

We note that the residual $\mathbb{R}^e$ contains the same order derivatives of the dependent unknown $u(x)$ as in the differential equation, Eq. (4.2.1), and hence requires at least a quadratic representation of $u_h^e(x)$. To reduce or weaken the differentiability of ψ_i^e, we mathematically distribute differentiation equally (when the given differential equation is of even order) between w_i^e and u_h^e, and the resulting integral form is known as the *weak form*, because the differentiability on u_h^e (hence on ψ_i^e) is reduced or weakened. The weak-form development is discussed in more detail in Section 4.2.4. The resulting finite element model is termed the *weak-form Galerkin finite element model.*

4.2.4 Derivation of the Weak Form

The first step in the construction of the weak form of Eq. (4.2.1) is the weighted-residual statement in Eq. (4.2.4). The second step is to weaken the differentiability by distributing the differentiation equally (for even-order differential equations) between the solution u_h^e and the weight function w_i^e (which will be replaced with the approximation functions ψ_i^e). The resulting integral statement is termed the *weak form* of Eq. (4.2.1) because it allows the approximation functions ψ_i^e to be of lower order than required in the weighted-integral statement, Eq. (4.2.4). The third and final step involves identification of the secondary variables of the formulation. Typically, the secondary variables are the "causes" that are dual to the primary variables or the "effects." In the present case, the primary variable is the temperature u_h^e and the secondary variable is the heat, $Q_h^e = a\, q_h^e$, q_h^e being the heat flow.

The three-step procedure of constructing the weak form beginning with the governing equation, Eq. (4.2.1) is presented next.

Step 1. *Weighted-integral statement*

$$0 = \int_{x_a^e}^{x_b^e} w_i^e \left[-\frac{d}{dx}\left(a\frac{du_h^e}{dx} \right) + b\frac{du_h^e}{dx} + cu_h^e - g \right] dx. \tag{4.2.5}$$

Step 2. *Trading the differentiation between the dependent unknown u_h^e and the weight function w_i^e*

$$0 = \int_{x_a^e}^{x_b^e} \left(a\frac{dw_i^e}{dx}\frac{du_h^e}{dx} + bw_i^e\frac{du_h^e}{dx} + cw_i^e u_h^e - w_i^e g \right) dx - \left[w_i^e \cdot a\frac{du_h^e}{dx} \right]_{x_a^e}^{x_b^e}. \tag{4.2.6}$$

Step 3. *Identification of the secondary variables and final weak form*

$$0 = \int_{x_a^e}^{x_b^e} \left(a\frac{dw_i^e}{dx}\frac{du_h^e}{dx} + bw_i^e\frac{du_h^e}{dx} + cw_i^e u_h^e - w_i^e g \right) dx - w_i^e(x_a^e)Q_a^e - w_i^e(x_b^e)Q_b^e, \tag{4.2.7}$$

where

$$Q_a^e = Q(x_a^e) = -\left(a\frac{du_h^e}{dx} \right)_{x_a^e}, \qquad Q_b^e = Q(x_b^e) = \left(a\frac{du_h^e}{dx} \right)_{x_b^e}. \tag{4.2.8}$$

The *primary variable* is the variable (u_h^e) appearing in the governing differential equation (for higher-order equations, additional primary variables are introduced; see Reddy [1]). To identify the dual to the primary variable, consider the boundary term appearing in Eq. (4.2.6), namely, the expression

$$\left[w_i^e \cdot a \frac{du_h^e}{dx} \right]_{x_a^e}^{x_b^e}.$$

The coefficient of the weight function w_i^e in the boundary expression, $a(du_h^e/dx)$, is called the *secondary variable*, and its specification constitutes the *natural* or *Neumann boundary condition.* The specification of a primary variable on the boundary constitutes the *essential* or *Dirichlet boundary condition.* Thus, for the model equation at hand, the primary and secondary variables are

$$\text{primary: } u_h^e\,; \quad \text{secondary: } n_x \left(a\frac{du_h^e}{dx} \right) \equiv Q_h^e(x), \tag{4.2.9}$$

where $n_x = -1$ at the left end of the element and $n_x = 1$ at the right.

The primary and secondary variables at the nodes are shown on the typical element in Fig. 4.2.1. It can be viewed as the *free-body diagram* or *control domain* of a typical but arbitrary portion of a bar, with Q_a^e and Q_b^e denoting the heats, as shown in Fig. 4.2.1. For heat conduction problems, Q_a^e denotes the heat input at the left end of the element and Q_b^e denotes the heat output from the right end. Although Q_h^e replaced $a(du_h^e/dx)$, it is *not* considered as a function of u_h^e; it is a variable that is dual to u_h^e. This completes the three-step procedure of constructing the weak form.

Some remarks on the weak-form derivation are in order. The integration-by-parts used in Step 2 has two implications: (1) it reduces the order (or degree) of finite element approximation used to represent u_h^e and (2) it introduces secondary variables Q_a^e and Q_b^e that are physically meaningful in the sense that they can be specified at a point whenever the primary variable u_h^e is not specified there. If a secondary variable Q_h^e resulting from the integration by parts is not a physical quantity, then integration by parts should not be carried out even if it results in the weakening of the differentiability of u_h^e.

4.2.5 Approximation Functions

Recall that the weak form over an element is equivalent to the differential equation, and it contains the end conditions in Eq. (4.2.8) on the heats (or sources) Q_a^e and Q_b^e. Therefore, the approximate solution $u_h^e(x)$ should be selected such that the differentiability (or continuity) conditions implied by the weak form are met and the end conditions on the primary variables $u(x_i^e) = u_i^e$ are satisfied. Since the weak form contains the first-order derivative of u_h^e, any function with a nonzero first derivative would be a candidate for u_h^e. Thus, the finite element approximation u_h^e can be an interpolation, that is, must be equal to u_a^e at x_a^e and u_b^e at x_b^e. The end points that define the geometry and provide a connection to the neighboring elements must be included as interpolation points. Thus, a complete linear polynomial is

$$u_h^e(x) = c_1^e + c_2^e x \tag{4.2.10}$$

is admissible if one can select c_1^e and c_2^e such that

$$u_h^e(x_a^e) \equiv u_1^e = c_1^e + c_2^e\, x_a^e, \qquad u_h^e(x_b^e) \equiv u_2^e = c_1^e + c_2^e\, x_b^e. \tag{4.2.11}$$

Solving for c_1^e and c_2^e in terms of the nodal values u_1^e and u_2^e, we obtain

$$c_1^e = \frac{u_1^e\, x_b^e - u_2^e\, x_a^e}{x_b^e - x_a^e}, \qquad c_2^e = \frac{u_2^e - u_1^e}{x_b^e - x_a^e}. \tag{4.2.12}$$

Substitution of Eq. (4.2.12) for c_i^e into Eq. (4.2.10) yields

$$\begin{aligned} u_h^e(x) = c_1^e &= \frac{u_a^e\, x_b^e - u_b^e\, x_a^e}{x_b^e - x_a^e} + \frac{u_b^e - u_a^e}{x_b^e - x_a^e}\, x \\ &= \psi_1^e(x)\, u_1^e + \psi_2^e(x)\, u_2^e = \sum_{j=1}^{2} \psi_j^e(x)\, u_j^e, \end{aligned} \tag{4.2.13}$$

where (noting that $h_e = x_b^e - x_a^e$)

$$\psi_1^e(x) = \frac{x_b^e - x}{h_e}, \qquad \psi_2^e(x) = \frac{x - x_a^e}{h_e}. \tag{4.2.14}$$

The functions ψ_j^e are known as the *linear Lagrange interpolation functions* (see Fig. 4.2.2), and

$$u_1^e = u_a^e, \qquad u_2^e = u_b^e \tag{4.2.15}$$

are the nodal values of $u_h^e(x)$ at $x = x_a^e$ and $x = x_b^e$, respectively. An element with linear approximation is called a *linear element*. Note that $\psi_j^e(x)$ satisfies the *interpolation property*:

$$\psi_j^e(x_i^e) = \delta_{ij} = \begin{cases} 1, & \text{if } i = j \\ 0, & \text{if } i \neq j, \end{cases} \tag{4.2.16}$$

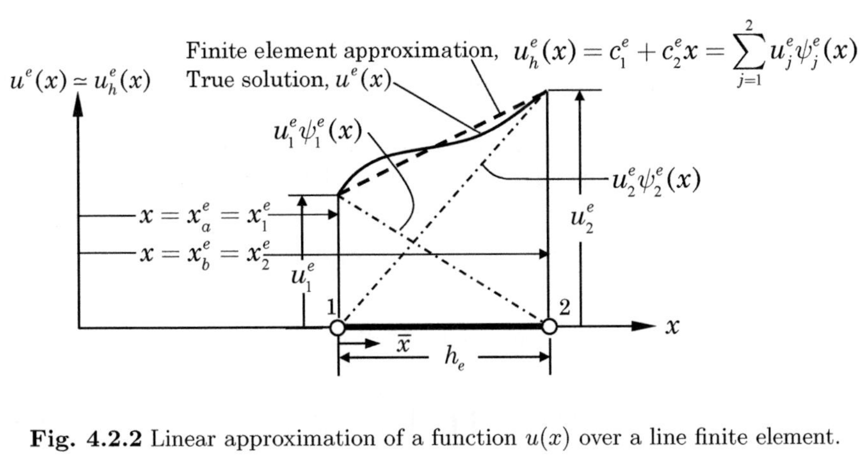

Fig. 4.2.2 Linear approximation of a function $u(x)$ over a line finite element.

where $x_1^e = x_a^e$ and $x_2^e = x_b^e$. In addition, the Lagrange interpolation functions satisfy the property known as the *partition of unity* :

$$1 = \sum_{j=1}^{n} \psi_j^e(x). \tag{4.2.17}$$

If we wish to approximate $u(x)$ with a quadratic polynomial, we write

$$u_h^e(x) = c_1^e + c_2^e x + c_3^e x^2 \tag{4.2.18}$$

and express the three parameters (c_1^e, c_2^e, c_3^e) by evaluating u_h^e at three nodes, including the two end nodes of the element. Identifying the third node at the center of the element, as indicated in Fig. 4.2.3(a), we obtain

$$\begin{aligned} u_1^e &\equiv u_h^e(x_1^e) = c_1^e + c_2^e x_1^e + c_3^e(x_1^e)^2 \\ u_2^e &\equiv u_h^e(x_2^e) = c_1^e + c_2^e x_2^e + c_3^e(x_2^e)^2 \\ u_3^e &\equiv u_h^e(x_3^e) = c_1^e + c_2^e x_3^e + c_3^e(x_3^e)^2, \end{aligned} \tag{4.2.19}$$

where x_i^e $(i = 1, 2, 3)$ are the global coordinates of the nodes of the element (with $x_2^e = x_1^e + 0.5h_e$). Equation (4.2.19) can be solved for (c_1^e, c_2^e, c_3^e) in terms of (u_1^e, u_2^e, u_3^e) and substituted into Eq. (4.2.18) to obtain

$$u_h^e(x) = \psi_1^e(x)u_1^e + \psi_2^e(x)u_2^e + \psi_3^e(x)u_3^e = \sum_{j=1}^{3} \psi_j^e(x)u_j^e, \tag{4.2.20}$$

where $\psi_i^e(x)$ are the quadratic Lagrange interpolation functions shown in Fig. 3.2.3(b), which can be expressed in terms of the element coordinate $\bar{x} = x - x_a^e$ as

$$\begin{aligned} \psi_1^e(\bar{x}) &= \left(1 - \frac{\bar{x}}{h_e}\right)\left(1 - \frac{2\bar{x}}{h_e}\right) \\ \psi_2^e(\bar{x}) &= 4\frac{\bar{x}}{h_e}\left(1 - \frac{\bar{x}}{h_e}\right) \\ \psi_3^e(\bar{x}) &= -\frac{\bar{x}}{h_e}\left(1 - \frac{2\bar{x}}{h_e}\right). \end{aligned} \tag{4.2.21}$$

An element with quadratic approximation of the field variable $u_h^e(x)$ is called a *quadratic element.*

Higher-order Lagrange interpolations of $u(x)$ can be developed along similar lines. Thus, an $(n-1)$-degree Lagrange interpolation of $u(x)$ is given by

$$u_h^e(x) = \psi_1^e(x)u_1^e + \psi_2^e(x)u_2^e + \cdots + \psi_n^e(x)u_n^e = \sum_{j=1}^{n} \psi_j^e(x)u_j^e, \tag{4.2.22}$$

where the Lagrange interpolation functions of degree $(n-1)$ are given by

$$\psi_j^e(x) = \prod_{i=1, i\neq j}^{n} \left(\frac{x - x_i^e}{x_j^e - x_i^e}\right). \tag{4.2.23}$$

The element contains n nodes. In general, the nodes are not equally spaced.

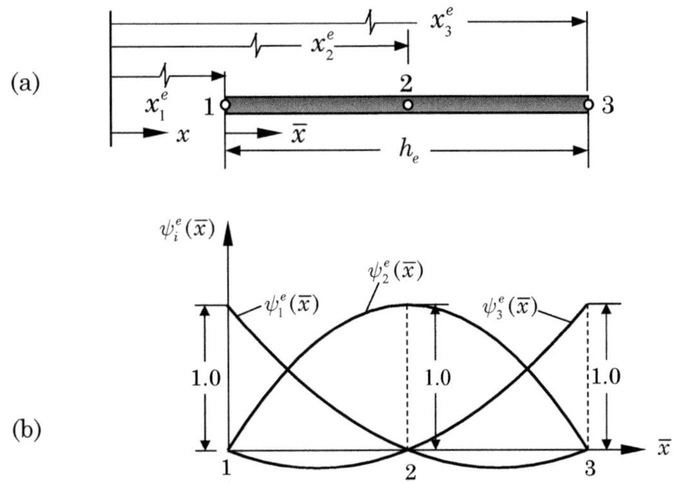

Fig. 4.2.3 (a) Quadratic finite element and (b) associated interpolation functions.

The finite element solution $u_h^e(x)$ must fulfill certain requirements for it be convergent to the actual solution $u(x)$ as the number of elements (h refinement) or the degree of the polynomials (p refinement) is increased. These requirements are listed next.

(1) The approximate solution should be continuous and differentiable as required by the weak form in Eq. (4.2.7).
(2) It should be a complete polynomial, that is, the polynomial includes all lower-order terms, including the constant term, up to the highest-order term used.
(3) It should be an interpolant of the primary variable u_h^e at the nodes of the finite element; it should at least interpolate the solution at the nodes on the boundary of the element.

The first requirement ensures that every term of the weak form has a nonzero contribution. The second requirement is necessary in order to capture all possible states, say, constant, linear, and so on, of the actual solution. For example, if a linear polynomial without the constant term is used to represent the temperature distribution in a one-dimensional system, the approximate solution can never be able to represent a uniform state of temperature field in the element. The third requirement is necessary in order to enforce continuity of the primary variable u_h^e at the nodes where the element is connected to its neighboring elements.

4.2.6 Finite Element Model

Substitution of Eq. (4.2.22) into the weak form in Eq. (4.2.7) gives n *algebraic* equations among the n nodal values u_i^e ($i = 1, 2, \ldots, n$) and Q_i^e of the element, which will be termed the *finite element model.* In order to formulate the finite element model (i.e., derive a set of algebraic relations among u_i^e and Q_i^e) based on the weak form in Eq. (4.2.7), it is *not* necessary to decide the degree of approximation of $u_h^e(x)$ a priori. The equations can be written in generic form in terms of ψ_i^e.

For $n > 2$, the weak form in Eq. (4.2.7) must be modified to include nonzero secondary variables, if any, at the nodes interior to the element (e.g., node 2 in the case of the quadratic element):

$$0 = \int_{x_a}^{x_b} \left(a\frac{dw_i^e}{dx}\frac{du_h^e}{dx} + bw_i^e\frac{du_h^e}{dx} + cw_i^e u_h^e \right) dx - \int_{x_a}^{x_b} w_i^e g\, dx - \sum_{j=1}^{n} w_i^e(x_j^e)\, Q_j^e, \tag{4.2.24}$$

where x_i^e is the global coordinate of the ith node of element $\Omega^e = (x_a^e, x_b^e)$. If nodes 1 and n denote the end points of the element, then $Q_1^e = Q_a^e$ and $Q_n^e = Q_b^e$ represent the *unknown* point sources, and all other Q_i^e $(i = 2, 3, \ldots, n-1)$ are the externally applied (hence, known) point sources at nodes $2, 3, \ldots, n-1$, respectively.

Substituting Eq. (4.2.22) for u_h^e and $w_1^e = \psi_1^e$, $w_2^e = \psi_2^e, \ldots, w_i^e = \psi_i^e, \ldots,$ $w_n^e = \psi_n^e$ into the weak form Eq. (4.2.24), we obtain n algebraic equations. The equations are numbered in such a way that the algebraic equation obtained from the weak form with $w_i^e = \psi_i^e$ is numbered as the ith equation of the set of n equations:

$$0 = \sum_{j=1}^{n} \left[\int_{x_a^e}^{x_b^e} \left(a\frac{d\psi_i^e}{dx}\frac{d\psi_j^e}{dx} + b\psi_i^e\frac{d\psi_j^e}{dx} + c\psi_i^e\psi_j^e \right) dx \right] u_j^e - \int_{x_a^e}^{x_b^e} \psi_i^e g\, dx - Q_i^e$$

$$0 = \sum_{j=1}^{n} K_{ij}^e u_j^e - f_i^e - Q_i^e \tag{4.2.25}$$

for $i = 1, 2, \ldots, n$, where

$$K_{ij}^e = \int_{x_a^e}^{x_b^e} \left(a\frac{d\psi_i^e}{dx}\frac{d\psi_j^e}{dx} + b\psi_i^e\frac{d\psi_j^e}{dx} + c\psi_i^e\psi_j^e \right) dx, \quad f_i^e = \int_{x_a^e}^{x_b^e} g\psi_i^e\, dx. \tag{4.2.26}$$

Note that the interpolation property in Eq. (4.2.16) is used to write

$$\sum_{j=1}^{n} \psi_i^e(x_j^e) Q_j^e = \sum_{j=1}^{n} \delta_{ij}\, Q_j^e = Q_i^e, \tag{4.2.27}$$

where δ_{ij} is the Kronecker delta symbol . In matrix notation, these algebraic equations can be expressed as

$$\mathbf{K}^e \mathbf{u}^e = \mathbf{f}^e + \mathbf{Q}^e \equiv \mathbf{F}^e. \tag{4.2.28}$$

The matrix $\mathbf{K}^e$, which is unsymmetric in the present case (it is symmetric only when $b^e = 0$), is called the *coefficient matrix*, and the column vector $\mathbf{f}^e$ is the *source vector*, and it is a nodal representation of the distributed source $g(x)$.

As already stated, the choice of replacing w_i^e with ψ_i^e amounts to using the Galerkin method of approximation. It is important to note that the classical Galerkin method was based on the weighted-integral statement of the residual – not on the weak form. In fact, utilizing the weak form amounts to using the Ritz method of approximation (see Reddy [1, 13]). Therefore, the finite element model should be called the *Ritz* or the *weak-form Galerkin finite element model.*

Equation (4.2.28) consists of n equations among $n+2$ unknowns, namely, n primary nodal values $(u_1^e, u_2^e, \ldots, u_n^e)$ and two secondary nodal values (Q_1^e, Q_n^e); the values $(u_1^e, u_2^e, \ldots, u_n^e)$ are also called the element *primary nodal degrees of freedom*. Because there are more unknowns than the number of equations, the element equations cannot be solved without assembling all elements of the total domain. The assembly of elements (i.e., putting the elements together) is carried out by imposing the *continuity of the primary variables* and *equilibrium of secondary variables* at nodes common to different elements. Upon assembly and imposition of boundary conditions, we shall obtain exactly the same number of algebraic equations as the total number of unknown primary and secondary nodal degrees of freedom. This is explained next.

The coefficient matrix $\mathbf{K}^e$, and source vector $\mathbf{f}^e$ can be evaluated for a given interpolation of the variable u_h and data (a, c, and g). When a, c, and g are functions of x, it may be necessary to evaluate $\mathbf{K}^e$ and $\mathbf{f}^e$ using numerical integration . When the data (a, c, and g) is element-wise constant, we can evaluate the integrals exactly. Suppose that a_e, c_e, and g_e denote the element-wise constant values of $a(x)$, $c(x)$, and $g(x)$, respectively, over a typical element $\Omega_e = (x_a^e, x_b^e)$. Then the element (matrix) equations are obtained by analytically evaluating the integrals appearing in K_{ij}^e and f_i^e of Eq. (4.2.26). The results are summarized here for a typical line element with linear and quadratic approximations.

Linear element (i.e., element with linear approximation)

$$\psi_1^e(\bar{x}) = 1 - \frac{\bar{x}}{h_e}, \qquad \psi_2^e(\bar{x}) = \frac{\bar{x}}{h_e} \tag{4.2.29a}$$

$$\left(\frac{a_e}{h_e}\begin{bmatrix} 1 & -1 \\ -1 & 1 \end{bmatrix} + \frac{b_e}{2}\begin{bmatrix} -1 & 1 \\ -1 & 1 \end{bmatrix} + \frac{c_e h_e}{6}\begin{bmatrix} 2 & 1 \\ 1 & 2 \end{bmatrix}\right)\begin{Bmatrix} u_1^e \\ u_2^e \end{Bmatrix} = \frac{g_e h_e}{2}\begin{Bmatrix} 1 \\ 1 \end{Bmatrix} + \begin{Bmatrix} Q_1^e \\ Q_2^e \end{Bmatrix}. \tag{4.2.29b}$$

Quadratic element (i.e., element with quadratic approximation)

$$\psi_1^e(\bar{x}) = \left(1 - \frac{2\bar{x}}{h_e}\right)\left(1 - \frac{\bar{x}}{h_e}\right), \; \psi_2^e(\bar{x}) = \frac{4\bar{x}}{h_e}\left(1 - \frac{\bar{x}}{h_e}\right), \; \psi_3^e(\bar{x}) = -\frac{\bar{x}}{h_e}\left(1 - \frac{2\bar{x}}{h_e}\right) \tag{4.2.30a}$$

$$\begin{aligned}\left(\frac{a_e}{3h_e}\begin{bmatrix} 7 & -8 & 1 \\ -8 & 16 & -8 \\ 1 & -8 & 7 \end{bmatrix} + \frac{b_e}{6}\begin{bmatrix} -3 & 4 & -1 \\ -4 & 0 & 4 \\ 1 & -4 & 3 \end{bmatrix} + \frac{c_e h_e}{30}\begin{bmatrix} 4 & 2 & -1 \\ 2 & 16 & 2 \\ -1 & 2 & 4 \end{bmatrix}\right)\begin{Bmatrix} u_1^e \\ u_2^e \\ u_3^e \end{Bmatrix} \\ = \frac{g_e h_e}{6}\begin{Bmatrix} 1 \\ 4 \\ 1 \end{Bmatrix} + \begin{Bmatrix} Q_1^e \\ Q_2^e \\ Q_3^e \end{Bmatrix}.\end{aligned} \tag{4.2.30b}$$

We note that the contribution of a uniform source to the nodes in a quadratic element is nonuniform, that is, $f_i^e \neq g_e h_e/3$; also, the source vector of a quadratic element of length h_e is *not* equivalent to that of two linear elements of length $h_e/2$ each; it is the source vector that is consistent with the coefficient matrix of the quadratic element. We also note that there are more unknowns (because of the heats Q_i^e at the end nodes) than the number of equations, independent of the degree of interpolation functions. Therefore, the element equations cannot be solved without properly assembling finite element equations associated with all elements in the mesh. When only one element is used, there will be only n unknowns because two known values are provided by the boundary conditions.

4.2.7 Axisymmetric Problems

The governing equation for *steady*, axisymmetric one-dimensional heat transfer [i.e., $T = T(r)$ and $g = g(r)$] is obtained from **Problem 2.18** as

$$-\frac{1}{r}\frac{d}{dr}\left(kr\frac{dT}{dr}\right) = g(r), \tag{4.2.31}$$

where g is internal heat generation per unit length. The weak form of this equation over a typical element, $\Omega_e = (r_a^e, r_b^e)$ with area element $d\Omega = r\,dr\,d\theta$ (see Fig. 4.2.4), is obtained using the three-step procedure:

$$0 = \int_0^{2\pi}\int_{r_a^e}^{r_b^e} w_i^e\left[-\frac{1}{r}\frac{d}{dr}\left(kr\frac{dT_h^e}{dr}\right) - g\right] rdrd\theta \tag{4.2.32a}$$

$$= 2\pi\int_0^{2\pi}\int_{r_a^e}^{r_b^e}\left(kr\frac{dw_i^e}{dr}\frac{dT_h^e}{dr} - rg\right)dr - \left[w_i^e\cdot 2\pi k\frac{dT_h^e}{dr}\right]_{r_a^e}^{r_a^e} \tag{4.2.32b}$$

$$= 2\pi\int_0^{2\pi}\int_{r_a^e}^{r_b^e}\left(kr\frac{dw_i^e}{dr}\frac{dT_h^e}{dr} - rg\right)dr - w_i^e(r_a^e)Q_a^e - w_i^e(r_b^e)Q_b^e. \tag{4.2.32c}$$

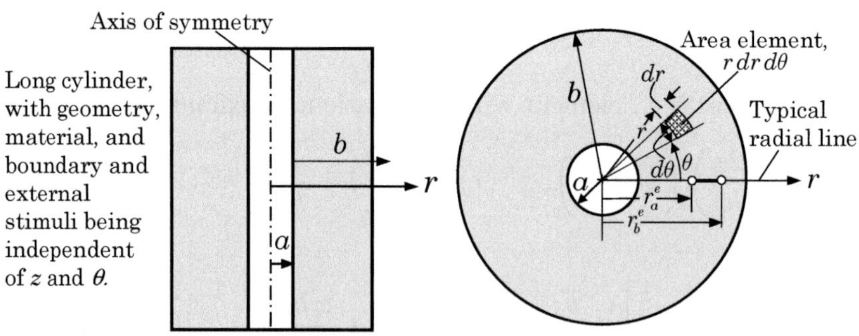

Fig. 4.2.4 Axisymmetric heat transfer in a long cylinder [i.e., the temperature T is independent of the radial coordinate θ and the coordinate z along the length of the cylinder, making $T = T(r)$].

Substituting the finite element approximation in Eq. (4.2.22) with u_h^e replaced by T_h^e and x replaced by r in Eq. (4.2.32c), we obtain

$$\mathbf{K}^e\mathbf{T}^e = \mathbf{f}^e + \mathbf{Q}^e, \tag{4.2.33a}$$

where

$$\begin{aligned} K_{ij}^e &= 2\pi \int_{r_a^e}^{r_b^e} k_e \frac{d\psi_i^e}{dr}\frac{d\psi_j^e}{dr}\, r dr, \quad f_i^e = 2\pi \int_{r_a}^{r_b} \psi_i^e g \; r\, dr \\ Q_1^e &\equiv -2\pi \left(r k_e \frac{dT}{dr} \right)\bigg|_{r_a}, \quad Q_2^e \equiv 2\pi \left(r k_e \frac{dT}{dr} \right)\bigg|_{r_b}. \end{aligned} \tag{4.2.33b}$$

The numerical values of the coefficients K_{ij}^e and f_i^e for element-wise constant values of k_e and g_e are given in Eqs. (4.2.34) and (4.2.35), respectively, for the linear and quadratic approximations of T.

Linear element

$$\begin{aligned} \mathbf{K}^e &= \frac{2\pi k_e}{h_e}(r_a + \tfrac{1}{2}h_e) \begin{bmatrix} 1 & -1 \\ -1 & 1 \end{bmatrix} \\ \mathbf{f}^e &= \frac{2\pi g_e h_e}{6} \begin{Bmatrix} 3r_a + h_e \\ 3r_a + 2h_e \end{Bmatrix}. \end{aligned} \tag{4.2.34}$$

Quadratic element

$$\begin{aligned} \mathbf{K}^e &= \frac{2\pi k_e}{6h_e} \begin{bmatrix} 3h_e + 14r_a & -(4h_e + 16r_a) & h_e + 2r_a \\ -(4h_e + 16r_a) & 16h_e + 32r_a & -(12h_e + 16r_a) \\ h_e + 2r_a & -(12h_e + 16r_a) & 11h_e + 14r_a \end{bmatrix} \\ f^e &= \frac{2\pi g_e h_e}{6} \begin{Bmatrix} r_a \\ 4r_a + 2h_e \\ r_a + h_e \end{Bmatrix}. \end{aligned} \tag{4.2.35}$$

The one-dimensional finite element formulations presented in this section can be implemented into a computer program. Interested readers can consult the textbook by Reddy [1] for the details; the book also contains a description of the finite element program **FEM1D** and its use in the solution of a variety of one-dimensional problems. The program can be downloaded from the website http://mechanics.tamu.edu.

4.2.8 Numerical Examples

In this section we consider two examples to illustrate the ideas presented here to solve one-dimensional heat transfer problems using the FEM (and the program **FEM1D**). The first example is concerned with heat transfer in an uninsulated uniform rod and the second example deals with an axisymmetric heat transfer in a *long* cylinder.

Example 4.2.1

A uniform steel rod of diameter $D = 0.02$ m, length $L = 0.05$ m, and constant thermal conductivity $k = 50$ W/(m·°C) is exposed to ambient air at $T_\infty = 20\,°\mathrm{C}$ with a heat transfer coefficient $\beta = 100$ W/(m^2 · °C). The left end of the rod is maintained at temperature $T(0) = T_0 = 320\,°\mathrm{C}$ and the other end, $x = L$, is either (1) insulated (i.e., no heat flow across the end) or (2) exposed to the ambient air at $T_\infty = 20\,°\mathrm{C}$ with heat transfer coefficient β, as shown in Fig. 4.2.5. Assuming that there is no internal heat generation (i.e., $g = 0$), determine the temperature distribution and the heat input at the left end of the rod using uniform meshes of (a) two linear elements, (b) four linear elements, and (c) two quadratic elements. For the sake of comparison, compute the finite difference solution of the problem with the same number of mesh points in the domain and compare the results with the finite element results.

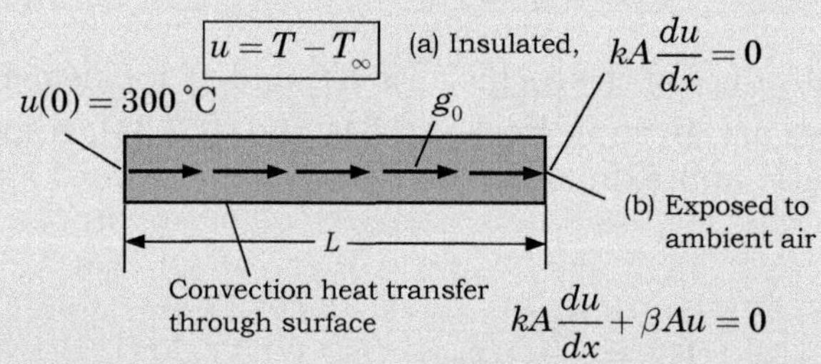

Fig. 4.2.5 Geometry and boundary conditions for heat transfer in an uninsulated rod.

Solution The governing equations of the problem are the same as in Eqs. (1.2.9) and (1.2.10):

$$-\frac{d}{dx}\left(kA\frac{dT}{dx}\right) + \beta P(T - T_\infty) = 0. \tag{1}$$

The boundary conditions are given by

$$T(0) = T_0, \quad \left[kA\frac{dT}{dx} + \beta A(T - T_\infty)\right]_{x=L} = 0. \tag{2}$$

Let $u = T - T_\infty$ so that Eq. (1) reduces to

$$-\frac{d}{dx}\left(kA\frac{du}{dx}\right) + \beta P u = 0 \quad \text{for} \quad 0 < x < L \tag{3}$$

and boundary conditions become

$$u(0) = u_0 = T(0) - T_\infty = 300\,°\mathrm{C}, \quad \left[kA\frac{du}{dx} + \beta A u\right]_{x=L} = 0. \tag{4}$$

The boundary conditions for Case (1) are obtained by setting $\beta = 0$ in Eq. (4). Equation (3) is a special case of Eq. (4.2.1), with $a = kA$, $b = 0$, $c = \beta P$, and $g = 0$.

The exact solution for Case (2) boundary conditions is given by

$$u(x) = u_0\left[\frac{\cosh m(L-x) + (\beta/mk)\sinh m(L-x)}{\cosh mL + (\beta/mk)\sinh mL}\right], \tag{5a}$$

where m is given by ($kA = \pi 50 \times 10^{-4}$ and $\beta P = 2\pi$)

$$m = \sqrt{\frac{\beta P}{Ak}} = \sqrt{\frac{2\pi}{50 \times 10^{-4}\pi}} = 20. \tag{5b}$$

The exact solution for the Case (1) boundary conditions is obtained from Eq. (5a) by setting $\beta = 0$.

Finite Element Solutions

The element equations for typical a linear element are ($kA = 1.570796 \times 10^{-2}$ and $\beta P = 6.283185$)

$$\left(\frac{kA}{h_e}\begin{bmatrix} 1 & -1 \\ -1 & 1 \end{bmatrix} + \frac{\beta P h_e}{6}\begin{bmatrix} 2 & 1 \\ 1 & 2 \end{bmatrix}\right)\begin{Bmatrix} u_1^e \\ u_2^e \end{Bmatrix} = \begin{Bmatrix} Q_1^e \\ Q_2^e \end{Bmatrix}. \tag{6}$$

The element equations for a typical quadratic element are

$$\left(\frac{kA}{3h_e}\begin{bmatrix} 7 & -8 & 1 \\ -8 & 16 & -8 \\ 1 & -8 & 7 \end{bmatrix} + \frac{\beta P h_e}{30}\begin{bmatrix} 4 & 2 & -1 \\ 2 & 16 & 2 \\ -1 & 2 & 4 \end{bmatrix}\right)\begin{Bmatrix} u_1^e \\ u_2^e \\ u_3^e \end{Bmatrix} = \begin{Bmatrix} Q_1^e \\ Q_2^e \\ Q_3^e \end{Bmatrix}. \tag{7}$$

(a) For the choice of <u>*two*</u> *linear elements* ($h = h_1 = h_2 = 0.025$) connected in series [see Fig. 4.2.6(a)], the assembled system of equations is ($kA/h + \beta Ph/3 = 0.68068$ and $-kA/h + \beta Ph/6 = -0.60214$; $u_1^{(1)} = U_1$, $u_2^{(1)} = u_1^{(2)} = U_2$, and $u_2^{(2)} = U_3$)

$$\begin{bmatrix} 0.68068 & -0.60214 & 0.00000 \\ -0.60214 & 1.36136 & -0.60214 \\ 0.00000 & -0,60214 & 0.68068 \end{bmatrix}\begin{Bmatrix} U_1 \\ U_2 \\ U_3 \end{Bmatrix} = \begin{Bmatrix} Q_1^{(1)} \\ Q_2^{(1)} + Q_1^{(2)} \\ Q_2^{(2)} \end{Bmatrix}, \tag{8}$$

where U_I denotes the temperature $u(x) = T - T_\infty$ at the Ith global node. The boundary conditions for Case (1) are

$$U_1 = 300\,^\circ\text{C}, \quad Q_2^{(2)} = 0 \tag{9}$$

and the balance of heats at global node 2 require (because there is no external heat added at the node)

$$Q_2^{(1)} + Q_1^{(2)} = 0. \tag{10}$$

After imposing the conditions in Eqs. (9) and (10), there are three equations in three unknowns, $(Q_1^{(1)}, U_2, U_3)$. The condensed equations for the unknown nodal temperatures (U_2, U_3) are obtained by omitting the first equation (which will be used to calculate $Q_1^{(1)}$

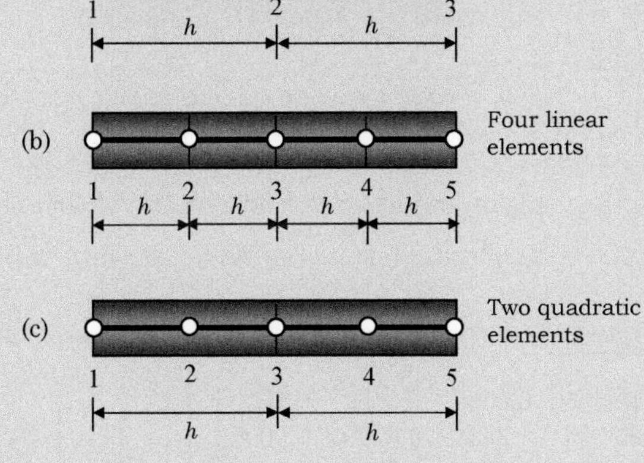

Fig. 4.2.6 Various finite element meshes of heat transfer in a rod (an interface between elements is shown with a vertical line).

later) and using the last two equations of Eq. (8), where the known value $K_{21}U_1 = -0.60214 \times 300 = 180.642$ is moved to the right-hand side of the second equation of Eq. (8):

$$\begin{bmatrix} 1.36136 & -0.60214 \\ -0.60214 & 0.68068 \end{bmatrix} \begin{Bmatrix} U_2 \\ U_3 \end{Bmatrix} = \begin{Bmatrix} 180.642 \\ 0 \end{Bmatrix}. \tag{11}$$

The solution of these two equations is ($T_i = U_i + 20$)

$$U_2 = 217.98\,^\circ\text{C}, \;\; U_3 = 192.83\,^\circ\text{C}; \quad T_2 = 237.98\,^\circ\text{C}, \;\; T_3 = 212.83\,^\circ\text{C}. \tag{12}$$

The heat at $x = 0$, $Q_1^{(1)}$, can be computed using either the first equation in Eq. (8)

$$Q_1^{(1)} = 0.68068\,U_1 - 0.60214\,U_2 = 72.95 \text{ W}. \tag{13}$$

or using the definition (post-computation) in Eq. (4.2.8) ($kA = 157.0796 \times 10^{-4}$),

$$(Q_1^{(1)})_{def} = -kA\frac{du_h}{dx} = kA\left(\frac{U_2 - U_1}{h_1}\right) = kA\left(\frac{300 - 217.98}{0.025}\right) = 51.53 \text{ W}. \tag{14}$$

The *exact solution* for the temperatures at $x = h$ and $x = 2h = L$ and heat at $x = 0$ are

$$u(0.025) = 219.23\,^\circ\text{C}, \;\; u(0.05) = 194.42\,^\circ\text{C}, \;\; Q(0) = 71.78 \text{ W}. \tag{15}$$

The post-computed value of $Q_1^{(1)}$ is not as accurate as that computed from the finite element equations. The difference between the two values will decrease with the mesh refinement.

The boundary conditions for Case (2) are

$$U_1 = 300\,^\circ\text{C}, \quad Q_2^{(2)} = -A\beta U_3 = -0.0314159 U_3, \quad Q_2^{(1)} + Q_1^{(2)} = 0. \tag{16}$$

Hence, the condensed equations [i.e., the second and third equations of the assembled system in Eq. (8)] for the unknown temperatures (U_2, U_3) are obtained by deleting the first row and column of Eq. (8) and taking $K_{21}U_1$ to the right-hand side of the second equation:

$$\begin{bmatrix} 1.36136 & -0.60214 \\ -0.60214 & 0.68068 + 0.03142 \end{bmatrix} \begin{Bmatrix} U_2 \\ U_3 \end{Bmatrix} = \begin{Bmatrix} 180.642 \\ 0.000 \end{Bmatrix}. \tag{17}$$

The solution of these two equations is

$$U_2 = 211.97\,^\circ\text{C}, \quad U_3 = 179.24\,^\circ\text{C}. \tag{18}$$

The heat at global node 1 ($x = 0$) can be computed using the first equation of the assembled equations in Eq. (8)

$$Q_1^{(1)} = 0.68068\,U_1 - 0.60214\,U_2 = 72.95 \text{ W}, \tag{19}$$

whereas from the definition we have

$$(Q_1^{(1)})_{def} = -kA\frac{du_h}{dx} = kA\left(\frac{U_2 - U_1}{h_1}\right) = kA\left(\frac{300 - 211.97}{0.025}\right) = 55.31 \text{ W}. \tag{20}$$

The *exact solution* for the temperatures at $x = h$ and $x = 2h = L$ and heat at $x = 0$ are

$$u(0.025) = 213.13\,^\circ\text{C}, \;\; u(0.05) = 180.66\,^\circ\text{C}, \;\; Q(0) = 75.46 \text{ W}. \tag{21}$$

(b) For the choice of *four linear elements* [see Fig. 4.2.6(b)], the assembled system is

$$\begin{bmatrix} 1.2828 & -1.2435 & 0.0000 & 0.0000 & 0.0000 \\ -1.2435 & 2.5656 & -1.2435 & 0.0000 & 0.0000 \\ 0.0000 & -1.2435 & 2.5656 & -1.2435 & 0.0000 \\ 0.0000 & 0.0000 & -1.2435 & 2.5656 & -1.2435 \\ 0.0000 & 0.0000 & 0.0000 & -1.2435 & 1.2828 \end{bmatrix} \begin{Bmatrix} U_1 \\ U_2 \\ U_3 \\ U_4 \\ U_5 \end{Bmatrix} = \begin{Bmatrix} Q_1^{(1)} \\ Q_2^{(1)} + Q_1^{(2)} \\ Q_2^{(2)} + Q_1^{(3)} \\ Q_2^{(3)} + Q_1^{(4)} \\ Q_2^{(4)} \end{Bmatrix}. \tag{22}$$

The balance conditions require $Q_2^{(1)}+Q_1^{(2)}=0$, $Q_2^{(2)}+Q_1^{(3)}=0$, and $Q_2^{(3)}+Q_1^{(4)}=0$; the boundary conditions for Case (1) are $U_1=300\,^\circ$C and $Q_2^{(4)}=0$. The condensed equations are obtained by deleting the first row and column and modifying the right-hand side to account for the known value of U_1 (i.e., move $-K_{21}U_1=u_0=1.2435\times 300=373.05$ to the right-hand side):

$$\begin{bmatrix} 2.5656 & -1.2435 & 0.0000 & 0.0000 \\ -1.2435 & 2.5656 & -1.2435 & 0.0000 \\ 0.0000 & -1.2435 & 2.5656 & -1.2435 \\ 0.0000 & 0.0000 & -1.2435 & 1.2828 \end{bmatrix} \begin{Bmatrix} U_2 \\ U_3 \\ U_4 \\ U_5 \end{Bmatrix} = \begin{Bmatrix} 373.05 \\ 0.00 \\ 0.00 \\ 0.00 \end{Bmatrix}. \tag{23}$$

The solution of these equations (coincides with the exact solution) is

$$U_2=251.52\,^\circ\text{C},\quad U_3=218.92\,^\circ\text{C},\quad U_4=200.16\,^\circ\text{C},\quad U_5=194.03\,^\circ\text{C}. \tag{24}$$

The heat at node 1 ($x=0$) from the first equation of the assembled system is

$$(Q_1^{(1)})_{equi}=1.2828\,U_1-1.2435\,U_2=72.075\text{ W}, \tag{25a}$$

whereas from the definition of $Q_1^{(1)}$, we have ($kA=157.0796\times 10^{-4}$)

$$(Q_1^{(1)})_{def}=-kA\frac{du_h}{dx}=kA\left(\frac{U_2-U_1}{h_1}\right)=kA\left(\frac{300-251.52}{0.0125}\right)=60.92\text{ W}. \tag{25b}$$

The exact solution is

$$\begin{aligned} &u(0.0125)=251.71\,^\circ\text{C},\quad u(0.025)=219.23\,^\circ\text{C},\quad u(0.0375)=200.52\,^\circ\text{C} \\ &u(0.05)=194.42\,^\circ\text{C},\quad Q(0)=71.78\text{ W}. \end{aligned} \tag{26}$$

The boundary conditions for Case (2) are $U_1=300\,^\circ$C and $Q_2^{(4)}=-kAU_5=0.03142\,U_5$. The condensed equations are obtained by deleting the first row and column and modifying the right-hand side to account for the specified values U_1 and $Q_2^{(4)}$:

$$\begin{bmatrix} 2.5656 & -1.2435 & 0.0000 & 0.0000 \\ -1.2435 & 2.5656 & -1.2435 & 0.0000 \\ 0.0000 & -1.2435 & 2.5656 & -1.2435 \\ 0.0000 & 0.0000 & -1.2435 & \mathbf{1.3142} \end{bmatrix} \begin{Bmatrix} U_2 \\ U_3 \\ U_4 \\ U_5 \end{Bmatrix} = \begin{Bmatrix} 1.2435\times 300 \\ 0 \\ 0 \\ 0 \end{Bmatrix}. \tag{27}$$

The solution of these equations is ($T_i=U_i+20$)

$$U_2=248.57\,^\circ\text{C},\quad U_3=212.85\,^\circ\text{C},\quad U_4=190.56\,^\circ\text{C},\quad U_5=180.31\,^\circ\text{C}. \tag{28}$$

The heat at node 1 ($x=0$) from the first equation of the assembled system is

$$(Q_1^{(1)})_{equi}=1.2828\,U_1-1.2435\,U_2=72.075\text{ W}, \tag{29}$$

whereas from the definition we have

$$(Q_1^{(1)})_{def}=kA\left(\frac{U_2-U_1}{h_1}\right)=157.0796\times 10^{-4}\left(\frac{300-248.57}{0.0125}\right)=64.625\text{ W}. \tag{30}$$

The exact value is $Q(0)=75.46$ W.

(c) For the choice of <u>two</u> *quadratic elements* [see Fig. 4.2.6(c)], the assembled system of equations is

$$\begin{bmatrix} 1.4870 & -1.6650 & 0.2042 & 0.0000 & 0.0000 \\ -1.6650 & 3.4348 & -1.6650 & 0.0000 & 0.0000 \\ 0.2042 & -1.6650 & 2.9740 & -1.6650 & 0.2042 \\ 0.0000 & 0.0000 & -1.6650 & 3.4348 & -1.6650 \\ 0.0000 & 0.0000 & 0.2042 & -1.6650 & 2.9740 \end{bmatrix} \begin{Bmatrix} U_1 \\ U_2 \\ U_3 \\ U_4 \\ U_5 \end{Bmatrix} = \begin{Bmatrix} Q_1^{(1)} \\ Q_2^{(1)} \\ Q_3^{(1)}+Q_1^{(2)} \\ Q_2^{(2)} \\ Q_3^{(2)} \end{Bmatrix}.$$

The balance condition at global node 3 is $Q_3^{(1)} + Q_1^{(2)} = 0$; the boundary conditions for Case (1) are $U_1 = 300\,^\circ\text{C}$, $Q_2^{(1)} = 0$, $Q_2^{(2)} = 0$, and $Q_3^{(2)} = 0$. The condensed equations are obtained by deleting the first row and column and modifying the right-hand side to account for the specified value $U_1 = 300\,^\circ\text{C}$ by taking $K_{21}U_1 = -1.6650 \times 300$ to the right-hand side of the second equation of the above assembled system:

$$\begin{bmatrix} 3.4348 & -1.6650 & 0.000 & 0.000 \\ -1.6650 & 2.9740 & -1.6650 & 0.2042 \\ 0.0000 & -1.6650 & 3.4348 & -1.6650 \\ 0.0000 & 0.2042 & -1.6650 & 2.9740 \end{bmatrix} \begin{Bmatrix} U_2 \\ U_3 \\ U_4 \\ U_5 \end{Bmatrix} = \begin{Bmatrix} 499.5 \\ 0.0 \\ 0.0 \\ 0.0 \end{Bmatrix}. \tag{31}$$

The solution of these equations is ($T_i = U_i + 20$)

$$U_2 = 251.70\,^\circ\text{C}, \quad U_3 = 219.23\,^\circ\text{C}, \quad U_4 = 200.52\,^\circ\text{C}, \quad U_5 = 194.42\,^\circ\text{C}. \tag{32}$$

The heat at node 1 ($x = 0$) from the first equation of the assembled system is

$$(Q_1^{(1)})_{equi} = 1.487\,U_1 - 1.665\,U_2 + 0.2042\,U_3 = 71.79 \text{ W}, \tag{33}$$

whereas from the definition we have

$$\begin{aligned}(Q_1^{(1)})_{def} &= -kA\left(\frac{du_h}{dx}\right)_{x=0} = -kA\left(U_1\frac{d\psi_1}{dx} + U_2\frac{d\psi_2}{dx} + U_3\frac{d\psi_3}{dx}\right)_{x=0} \\ &= -kA\left[U_1\left(-\frac{3}{h_1}\right) + U_2\left(\frac{4}{h_1}\right) + U_3\left(-\frac{1}{h_1}\right)\right] = 70.64 \text{ W}.\end{aligned} \tag{34}$$

The boundary conditions for Case (2) are $U_1 = 300\,^\circ\text{C}$ and $Q_3^{(2)} = -kAU_5 = -0.03142\,U_5$. The condensed equations are obtained by deleting the first row and column and modifying the right-hand side to account for the specified values of U_1 and $Q_3^{(2)}$:

$$\begin{bmatrix} 3.4348 & -1.6650 & 0.0000 & 0.0000 \\ -1.6650 & 2.9740 & -1.6650 & 0.2042 \\ 0.0000 & -1.6650 & 3.4348 & -1.6650 \\ 0.0000 & 0.2042 & -1.6650 & \mathbf{3.0054} \end{bmatrix} \begin{Bmatrix} U_2 \\ U_3 \\ U_4 \\ U_5 \end{Bmatrix} = \begin{Bmatrix} 499.5 \\ 0.0 \\ 0.0 \\ 0.0 \end{Bmatrix}. \tag{35}$$

The solution of these equations is ($T_i = U_i + 20$)

$$U_2 = 248.74\,^\circ\text{C}, \quad U_3 = 213.13\,^\circ\text{C}, \quad U_4 = 190.90\,^\circ\text{C}, \quad U_5 = 180.66\,^\circ\text{C}. \tag{36}$$

The heat at node 1 ($x = 0$) from the first equation of the assembled system is

$$(Q_1^{(1)})_{equi} = 1.487\,U_1 - 1.665\,U_2 + 0.2042\,U_3 = 75.47 \text{ W}, \tag{37}$$

whereas from the definition we have

$$(Q_1^{(1)})_{def} = -kA\left[U_1\left(-\frac{3}{h_1}\right) + U_2\left(\frac{4}{h_1}\right) + U_3\left(-\frac{1}{h_1}\right)\right] = 74.24 \text{ W}. \tag{38}$$

Clearly, the finite element solution obtained with two quadratic elements is very close to the exact solution.

Finite Difference Solutions

In the finite difference scheme, one may think of identifying a number of mesh points (instead of elements), including the boundary points, as shown in Fig. 4.2.7(a). Then derivatives of a function u at an arbitrary point inside the domain are approximated in terms the values of u at several points, as dictated by the truncated Taylor series. For example, the second derivative of u at the ith mesh point, whose coordinate is x_i, may be approximated in terms of the values of u at mesh points $i-1$, i, and $i+1$ using the *central finite difference formula*

$$\left(\frac{d^2u}{dx^2}\right)_i \approx \frac{1}{h^2}\left(u_{i-1} - 2u_i + u_{i+1}\right), \tag{39}$$

where h is the distance between two mesh points. Equation (39) can be viewed as the "element equation," as shown in Fig. 4.2.7(b). We note that the coefficients in the finite difference equation (39) are real numbers (i.e., no evaluations of integrals is required).

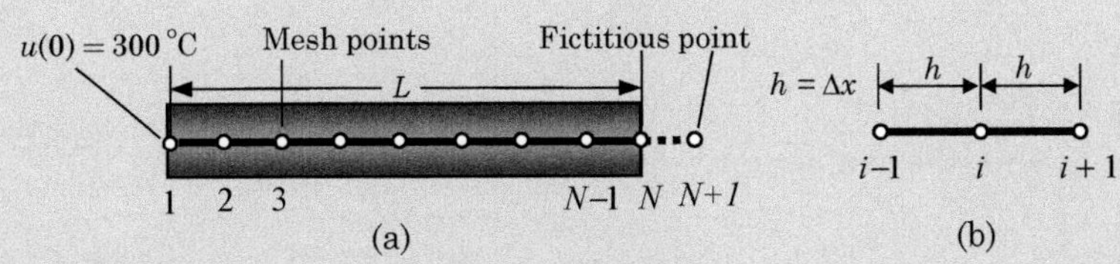

Fig. 4.2.7 Finite difference analysis of heat transfer in a straight rod. (a) Mesh of (base) points and (b) a typical mesh point i with neighboring mesh points $i-1$ and $i+1$.

Since kA is constant in this problem, Eq. (3) can be expressed (by dividing throughout with kA) as

$$-\frac{d^2u}{dx^2} + m^2 u = 0, \quad m^2 = \frac{P\beta}{kA} = 400. \tag{40}$$

Substituting the formula (39) for the second derivative into Eq. (40), we arrive at

$$-u_{i-1} + (2 + m^2 h_i^2)u_i - u_{i+1} = 0, \tag{41}$$

which is valid for any mesh point where u is not specified. By applying the formula in Eq. (41) to nodes $2, 3, \ldots, N$, we obtain the relations among the values of u at the mesh points. The application of Eq. (41) to the last mesh point (N) brings in u_{N+1}, which is for a (fictitious) point outside the domain; u_{N+1} can be expressed in terms of the values of the mesh points inside the domain using the boundary condition at mesh point N.

For Case (1), the boundary condition at mesh point N (i.e., $x = L$) is $(du/dx) = 0$, which can be *approximated* using the central difference formula [error is of the order $O(h^2)$],

$$\left[\frac{du}{dx}\right]_{x=x_i} \approx \frac{u_{i+1} - u_{i-1}}{2h} \tag{42}$$

and we obtain (for $i = N$) the result $u_{N+1} = u_{N-1}$. For Case (2), the boundary condition at $x = L$ is $(du/dx) + (\beta/k)u = 0$. Using, again, the the central difference formula, we obtain

$$\frac{u_{N+1} - u_{N-1}}{2h} + \frac{\beta}{k}u_N = 0 \;\rightarrow\; u_{N+1} = u_{N-1} - 2h\frac{\beta}{k}u_N. \tag{43}$$

(a) First, we choose a mesh of three points ($h = 0.025$), two end points and one in the middle. Applying Eq. (41) at nodes 2 and 3, we obtain ($m^2 = 400$ and $u_4 = u_2$)

$$-u_1 + \left(2 + 400h^2\right)u_2 - u_3 = 0, \quad -u_2 + \left(2 + 400h^2\right)u_3 - u_4 = 0. \tag{44a}$$

Note that u_4 is the value of u at the fictitious node 4.

The boundary conditions for Case (1) are: $u_1 = u(0) = u_0$ and $u_4 = u_2$. Then Eq. (44a) takes the form

$$-u_1 + \left(2 + 400h^2\right) u_2 - u_3 = 0, \quad -2u_2 + \left(2 + 400h^2\right) u_3 = 0. \tag{44b}$$

With the boundary conditions imposed, Eq. (44b) can be written in matrix form as $(2 + 400h^2 = 2.25)$

$$\begin{bmatrix} 2.25 & -1.00 \\ -2.00 & 2.25 \end{bmatrix} \begin{Bmatrix} u_2 \\ u_3 \end{Bmatrix} = \begin{Bmatrix} 300 \\ 0 \end{Bmatrix}. \tag{45}$$

The solution of these equations is

$$u_2 = 220.41\,^\circ\text{C}, \quad u_3 = 195.92\,^\circ\text{C}. \tag{46a}$$

The heat at mesh point 1 ($x = 0$) can be computed using the definition and the forward difference formula for the first derivative [see Eq. (1.3.3)]

$$Q(0) = -kA\left(\frac{du}{dx}\right)_{x=0} = 157.0796 \times 10^{-4}\left(\frac{u_2 - u_1}{h}\right) = 50.01 \text{ W}. \tag{46b}$$

The boundary conditions for Case (2) are ($\beta/k = 2$):

$$u_1 = 300\,^\circ\text{C}, \quad u_4 = u_2 - 4hu_3 = 0 \;\rightarrow\; u_4 = u_2 - 0.1\,u_3. \tag{47}$$

With the boundary conditions imposed, Eq. (44a) can be written in matrix form as

$$\begin{bmatrix} 2.25 & -1.00 \\ -2.00 & 2.25 + \mathbf{0.1} \end{bmatrix} \begin{Bmatrix} u_2 \\ u_3 \end{Bmatrix} = \begin{Bmatrix} 300 \\ 0 \end{Bmatrix}. \tag{48}$$

The solution of these equations is

$$u_2 = 214.45\,^\circ\text{C}, \quad u_3 = 182.51\,^\circ\text{C}. \tag{49}$$

The heat at mesh point 1 ($x = 0$) is

$$Q(0) = -kA\left(\frac{du}{dx}\right)_{x=0} = 157.0796 \times 10^{-4}\left(\frac{u_2 - u_1}{h}\right) = 53.75 \text{ W}. \tag{50}$$

Once we have the nodal values u_{i-1}, u_i and u_{i+1}, values at other mesh points (intermediate to the nodes) can be computed using the "element" approximation

$$u_h^e(\bar{x}) = \sum_{j=1}^{2} u_j^e \psi_j^e(\bar{x}), \quad 0 \le \bar{x} \le h_e, \tag{51}$$

where $u_1^1 = u_{i-1}$, $u_2^1 = u_i = u_1^2$, and $u_2^2 = u_{i+1}$ and ψ_j^e are the linear finite element interpolation functions. Such a formula is not readily available in the FDM but one may use mesh point data to construct an interpolation consistent with the finite difference formula used.

(b) Next, we use a mesh of five points or four equal intervals ($h = 0.0125$). Applying Eq. (41) at mesh points 2, 3, 4, and 5, and using $u_6 = u_4$ for Case (1), we obtain

$$\begin{bmatrix} 2.0625 & -1 & 0 & 0 \\ -1 & 2.0625 & -1 & 0 \\ 0 & -1 & 2.0625 & -1 \\ 0 & 0 & -2 & 2.0625 \end{bmatrix} \begin{Bmatrix} u_2 \\ u_3 \\ u_4 \\ u_5 \end{Bmatrix} = \begin{Bmatrix} 300 \\ 0 \\ 0 \\ 0 \end{Bmatrix}. \tag{52}$$

The solution of these equations is

$$u_2 = 251.89\,^\circ\text{C}, \quad u_3 = 219.53\,^\circ\text{C}, \quad u_4 = 200.89\,^\circ\text{C}, \quad u_5 = 194.80\,^\circ\text{C}. \tag{53}$$

The heat at mesh point 1 (i.e., at $x = 0$) is

$$Q(0) = -kA\frac{du}{dx}\Big|_{x=0} = 157.0796 \times 10^{-4}\,\frac{u_1 - u_2}{h} = 60.46 \text{ W}. \tag{54}$$

For Case (2), using $u_6 = u_4 - 4h\,u_5 = u_4 - 0.05u_5$, the finite difference equations become

$$\begin{bmatrix} 2.0625 & -1 & 0 & 0 \\ -1 & 2.0625 & -1 & 0 \\ 0 & -1 & 2.0625 & -1 \\ 0 & 0 & -2 & 2.0625 + \mathbf{0.05} \end{bmatrix} \begin{Bmatrix} u_2 \\ u_3 \\ u_4 \\ u_5 \end{Bmatrix} = \begin{Bmatrix} 300 \\ 0 \\ 0 \\ 0 \end{Bmatrix}. \tag{55}$$

The solution of these equations is

$$u_2 = 248.95\,^\circ\text{C}, \quad u_3 = 213.47\,^\circ\text{C}, \quad u_4 = 191.32\,^\circ\text{C}, \quad u_5 = 181.13\,^\circ\text{C}. \tag{56}$$

The heat at mesh point 1 ($x = 0$) is

$$Q(0) = kA\Big(-\frac{du}{dx}\Big)_{x=0} = 157.0796 \times 10^{-4}\Big(\frac{u_1 - u_2}{h}\Big) = 64.15 \text{ W}. \tag{57}$$

A comparison of the nodal values of u obtained using the two methods with the exact values for Case (1) and Case (2) is presented in Tables 4.2.1 and 4.2.2, respectively. The tables also contain the value $Q^{def}(0)$ as computed in the two methods using the definition. Both the FDM and FEM solutions are in good agreement with the exact solution (5a) (for Case (1), set $\beta = 0$). We note that the FEM adjusts the nodal values to minimize the error in the approximation over the domain in the weighted-integral sense , whereas the FDM has no such minimization of error but based on truncated a Taylor series approximation. The finite element solution for u at the nodes matches with the exact solution when a uniform mesh of four quadratic elements is used.

Table 4.2.1 Comparison of finite difference and finite element solutions with the exact solution of Case (1): $-\frac{d^2u}{dx^2} + 400\,u = 0,\ 0 < x < L;\ \ u(0) = 300,\ \ [kA\frac{du}{dx}]\Big|_{x=L} = 0.$

	Exact	FEM solution				FDM solution		
	Solution	Linear		Quadratic		Central difference		
x		$N=2$	$N=4$	$N=2$	$N=4$	$N=2$	$N=4$	$N=8$
0.00000	300.00	300.00	300.00	300.00	300.00	300.00	300.00	300.00
0.00625	273.71	—	—	—	273.71	—	—	273.74
0.01250	251.71	—	251.52	251.70	251.71	—	251.89	251.75
0.01875	233.64	—	—	—	233.64	—	—	233.70
0.02500	219.23	217.98	218.92	219.23	219.23	220.41	219.53	219.30
0.03125	208.25	—	—	—	208.25	—	—	208.33
0.03750	200.52	—	200.16	200.52	200.52	—	200.89	200.61
0.04375	195.94	—	—	—	195.94	—	—	196.03
0.05000	194.42	192.83	194.03	194.42	194.42	195.92	194.80	194.51
$Q^{def}(0)$	71.780	51.533	60.923	70.640	71.451	50.000	60.460	65.999

Table 4.2.2 Comparison of finite difference and finite element solutions with the exact solution of Case (2):

$$-\frac{d^2u}{dx^2} + 400\,u = 0,\ 0 < x < L;\ \ u(0) = 300,\ \ \left[kA\frac{du}{dx} + \beta Au\right]\Big|_{x=L} = 0.$$

	Exact	FEM solution				FDM solution		
	Solution	Linear		Quadratic		Central difference		
x		$N=2$	$N=4$	$N=2$	$N=4$	$N=2$	$N=4$	$N=8$
0.00000	300.00	300.00	300.00	300.00	300.00	300.00	300.00	300.00
0.00625	272.25	—	—	—	272.24	—	—	272.27
0.01250	248.75	—	248.57	248.74	248.75	—	248.95	248.80
0.01875	229.15	—	—	—	229.15	—	—	229.22
0.02500	213.13	211.97	212.85	213.13	213.13	214.45	213.47	213.21
0.03125	200.44	—	—	—	200.44	—	—	200.54
0.03750	190.90	—	190.56	190.90	190.90	—	191.32	191.00
0.04375	184.33	—	—	—	184.33	—	—	184.45
0.05000	180.66	179.24	180.31	180.66	180.66	182.51	181.13	180.78
$Q^{def}(0)$	75.457	55.310	64.625	74.239	75.110	53.753	64.151	69.693

The next example deals with heat transfer in a long cylinder with radially symmetric conditions (so that the problem becomes one dimensional).

Example 4.2.2

Consider a *long*, homogeneous, isotropic solid cylinder of outside radius $R_0 = 0.01$ m, conductivity $k = 20$ W/(m·°C), and a constant rate of internal heat generation $g_0 = 2 \times 10^8$ W/m^3. Consider the case where the boundary surface at $r = R_0$ is maintained at a constant temperature $T_0 = 100\,°$C. Calculate the temperature distribution $T(r)$ and heat flux $q(r) = -kdT/dr$ in the cylinder using various uniform meshes of linear finite elements and quadratic elements.

Solution The governing equation for this problem is given by Eq. (4.2.31) with $g = g_0$. The boundary conditions are

$$\left[2\pi kr\frac{dT}{dr}\right]_{r=0} = 0, \quad T(R_0) = T_0. \tag{1}$$

The zero-flux boundary condition, $Q_1^{(1)} = 0$, at $r = 0$ is a result of the radial symmetry at $r = 0$.

The finite element model of the governing equation is given by Eqs. (4.2.33a) and (4.2.33b). For linear and quadratic interpolations of $T(r)$, the element matrices for a typical element are given in Eq. (4.2.34) and (4.2.35), respectively.

For the mesh of one linear element, we have ($r_a = 0$ and $r_2 = h_1 = R_0$)

$$2\pi k\begin{bmatrix} 1 & -1 \\ -1 & 1 \end{bmatrix}\begin{Bmatrix} U_1 \\ U_2 \end{Bmatrix} = \frac{2\pi g_0 R_0^2}{3}\begin{Bmatrix} 1 \\ 2 \end{Bmatrix} + \begin{Bmatrix} Q_1^1 \\ Q_2^1 \end{Bmatrix}. \tag{2}$$

The boundary conditions are $Q_1^1 = 0$ and $U_2 = T_0$. Hence the temperature at node 1 is

$$U_1 = \frac{g_0 R_0^2}{3k} + T_0 \tag{3}$$

and the heat at $r = R_0$ is

$$(Q_2^1)_{equi} = \pi k(U_2 - U_1) - \tfrac{2}{3}\pi g_0 R_0^2 = -\pi g_0 R_0^2. \tag{4}$$

The negative sign indicates that heat is removed from the body. The one-element solution as a function of the radial coordinate r is

$$T_h(r) = U_1\psi_1^1(r) + U_2\psi_2^1(r) = \frac{g_0 R_0^2}{3k}\left(1 - \frac{r}{R_0}\right) + T_0 \tag{5}$$

and the heat flux is

$$q(r) \equiv -k\frac{dT_h}{dr} = \tfrac{1}{3} g_0 R_0. \tag{6}$$

For a mesh of two linear elements ($h_1 = h_2 = \frac{1}{2}R_0$), the assembled equations are

$$\pi k \begin{bmatrix} 1 & -1 & 0 \\ -1 & 1+3 & -3 \\ 0 & -3 & 3 \end{bmatrix} \begin{Bmatrix} U_1 \\ U_2 \\ U_3 \end{Bmatrix} = \frac{\pi g_0 R_0^2}{6} \begin{Bmatrix} \frac{1}{2} \\ 1+2 \\ \frac{1}{2}+2 \end{Bmatrix} + \begin{Bmatrix} Q_1^1 \\ Q_2^1 + Q_1^2 = 0 \\ Q_2^2 \end{Bmatrix}. \tag{7}$$

Imposing the boundary conditions $U_3 = T_0$ and $Q_1^1 = 0$, the condensed equations for the unknown temperatures are

$$\pi k \begin{bmatrix} 1 & -1 \\ -1 & 4 \end{bmatrix} \begin{Bmatrix} U_1 \\ U_2 \end{Bmatrix} = \frac{\pi g_0 R_0^2}{12} \begin{Bmatrix} 1 \\ 6 \end{Bmatrix} + \pi k \begin{Bmatrix} 0 \\ 3T_0 \end{Bmatrix}. \tag{8}$$

The nodal values are

$$U_1 = \tfrac{5}{18}\frac{g_0 R_0^2}{k} + T_0, \quad U_2 = \tfrac{7}{36}\frac{g_0 R_0^2}{k} + T_0. \tag{9}$$

From equilibrium, Q_2^2 is computed as

$$Q_2^2 = -\tfrac{5}{12}\pi g_0 R_0^2 + 3\pi k(U_3 - U_2) = -\pi g_0 R_0^2. \tag{10}$$

The finite element solution becomes

$$\begin{aligned} T_h(r) &= \begin{cases} U_1\psi_1^1(r) + U_2\psi_2^1(r) = \left(\frac{5}{18}\frac{g_0 R_0^2}{k} + T_0\right)\left(1 - \frac{2r}{R_0}\right) + \frac{2r}{R_0}\left(\frac{7}{36}\frac{g_0 R_0^2}{k} + T_0\right) \\ U_2\psi_1^2 + U_3\psi_2^2 = 2\left(\frac{7}{36}\frac{g_0 R_0^2}{k} + T_0\right)\left(1 - \frac{r}{R_0}\right) + T_0\left(\frac{2r}{R_0} - 1\right) \end{cases} \\ &= \begin{cases} \frac{1}{18}\frac{g_0 R_0^2}{k}\left(5 - \frac{3r}{R_0}\right) + T_0, & \text{for } 0 \le r \le \frac{1}{2}R_0 \\ \frac{7}{18}\frac{g_0 R_0^2}{k}\left(1 - \frac{r}{R_0}\right) + T_0, & \text{for } \frac{1}{2}R_0 \le r \le R_0. \end{cases} \end{aligned} \tag{11}$$

The exact solution of the problem is

$$T(r) = \frac{g_0 R_0^2}{4k}\left[1 - \left(\frac{r}{R_0}\right)^2\right] + T_0 \ (^\circ\text{C}) \tag{12a}$$

$$q(r) = \frac{1}{2} g_0 r \ (\text{W/m}^2), \quad Q(R_0) = -\left(2\pi k r \frac{dT}{dr}\right)\bigg|_{R_0} = \pi g_0 R_0^2 \ (\text{W}). \tag{12b}$$

The temperature at the center of the cylinder according to the exact solution is $T(0) = g_0 R_0^2/4k + T_0$, whereas it is $g_0 R_0^2/3k + T_0$ and $5g_0 R_0^2/18k + T_0$ according to the one- and two-element models, respectively.

The finite element solutions obtained using one-, two-, four-, and eight-element meshes of linear elements and four-element mesh of quadratic elements are compared with the exact solution in Table 4.2.3. Convergence of the finite element solutions, $\bar{T} = (T - T_0)k/g_0 R_0^2$, to the exact solution with an increasing number of elements is clear (see Fig. 4.2.8). Figure 4.2.9 shows plots of $\bar{Q}(r) = Q(r)/2\pi R_0 g_0$, where $Q(r) = 2\pi k r dT/dr$, versus $\bar{r} = r/R_0$. The finite element solutions for $Q(r)$ computed at the element centers, not the nodes, are in excellent agreement with the analytical solution.

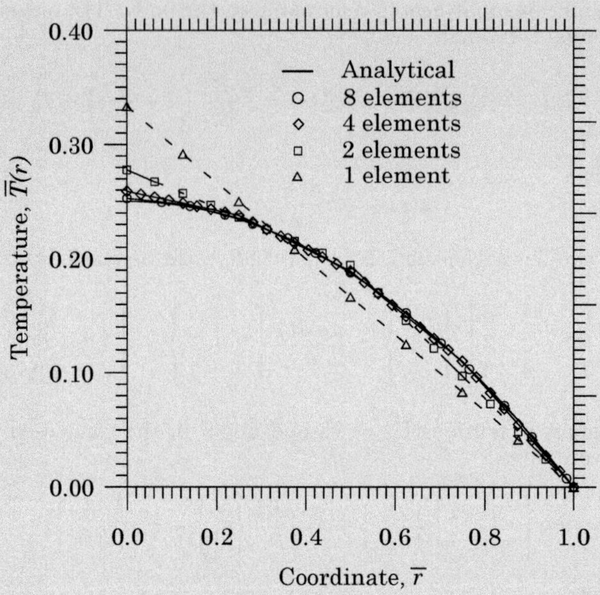

Fig. 4.2.8 Comparison of the finite element solutions with the exact solution, $\bar{T} = (T - T_0)k/g_0R_0^2$, for heat transfer in a radially symmetric problem with cylindrical geometry. The finite element solutions are obtained with one-, two-, four-, and eight-element meshes of linear elements and four-element mesh of quadratic elements.

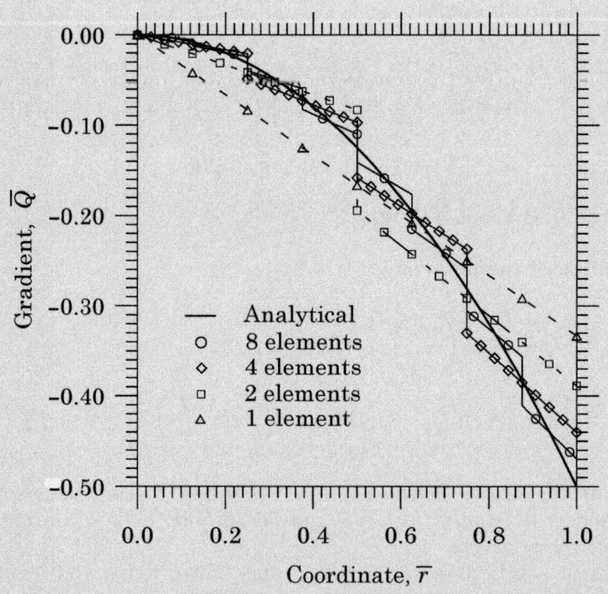

Fig. 4.2.9 Comparison of the finite element solution with the exact solution for the temperature gradient $\bar{Q}(r) = Q(r)/2\pi R_0 g_0$ in a radially symmetric problem with cylindrical geometry. Note that the heat flux is discontinuous at the element interfaces.

Table 4.2.3 Comparison of the finite element and exact solutions for temperature in a radially symmetric circular disc (L – linear and Q – quadratic elements).

	Finite element solution					
$\frac{r}{R_0}$	1L	2L	4L	8L	4Q	Exact
0.000	433.33	377.78	358.73	352.63	350.00	350.00
0.125	391.67	356.94	348.31	347.42	346.09	346.09
0.250	350.00	336.11	337.90	335.27	334.37	334.38
0.375	308.33	315.28	313.59	315.48	314.84	314.84
0.500	266.67	294.44	289.29	287.95	287.50	287.50
0.625	225.00	245.83	249.70	252.65	252.34	252.34
0.750	183.33	197.22	210.12	209.56	209.37	209.38
0.875	141.67	148.61	155.06	158.68	158.59	158.59
1.000	100.00	100.00	100.00	100.00	100.00	100.00

The underlined terms are nodal values and others are the interpolated values.

The steady-state advection–diffusion equation in one dimension is given by

$$\rho u \frac{d\phi}{dx} - \frac{d}{dx}\left(k_x \frac{d\phi}{dx}\right) - g = 0, \quad 0 < x < L, \tag{4.2.36}$$

where u is the x-component of the velocity and k_x is the conductivity in the x direction. The diffusion flux is defined by

$$q_n = -k_x \frac{d\phi}{dx}. \tag{4.2.37}$$

When ρu is known (from the solution of the associated flow problem), Eq. (4.2.36) is a special case of Eq. (4.2.1) with $a = k_x$, $b = \rho u$, and $c = 0$. Hence, the finite element equations presented in Eqs. (4.2.29b) and (4.2.30b) are valid with $c = 0$.

The last example of this section on one-dimensional problems is concerned with the advection–diffusion equation (4.2.36).

Example 4.2.3

We can express Eq. (4.2.36) in dimensionless form by assuming that k_x and ρu are constants. Let us introduce the following quantities

$$\bar{x} = \frac{x}{L}, \quad Pe = \frac{\rho u L}{k_x}, \quad \bar{g} = \frac{L}{\rho u} g, \tag{4.2.38}$$

where Pe is known as the *Péclet number*. Then Eq. (4.2.36) takes the alternative form

$$\frac{d\phi}{d\bar{x}} - \frac{1}{Pe}\frac{d^2\phi}{d\bar{x}^2} - \bar{g} = 0, \quad 0 < \bar{x} < 1 \tag{4.2.39}$$

and the dimensionless diffusion flux is

$$\bar{q}_n = -\frac{1}{Pe}\frac{d\phi}{d\bar{x}}. \tag{4.2.40}$$

Formulate the finite element equations and obtain the assembled equations for uniform meshes of (a) four linear finite elements and (b) two quadratic elements. Then investigate the characteristics of the numerical solutions for $Pe = 10$ and $Pe = 30$ using various meshes, beginning with four linear elements and the following boundary conditions:

$$\phi(0) = 1, \quad \phi(1) = 0. \tag{1}$$

Assume that there is no source or sink, $g = 0$.

Solution The weak-form Galerkin finite element model of Eq. (4.2.39) is (the bar over the quantities is omitted)

$$\mathbf{K}^e \phi^e = \mathbf{Q}^e, \tag{2a}$$

where

$$K_{ij}^e = \int_{x_a^e}^{x_b^e} \left(\psi_i^e \frac{d\psi_j^e}{dx} + \frac{1}{Pe} \frac{d\psi_i^e}{dx} \frac{d\psi_j^e}{dx} \right) dx \tag{2b}$$

and

$$Q_1^e \equiv \left[-\frac{1}{Pe} \frac{d\phi}{dx} \right]_{x_a^e}, \quad Q_2^e \equiv \left[\frac{1}{Pe} \frac{d\phi}{dx} \right]_{x_b^e}. \tag{2c}$$

(a) For a linear finite element, the finite element equations are [see Eq. (4.2.29b); set $a = 1$, $b = 1/Pe$, $c = 0$ and $g = 0$]

$$\left(\frac{1}{2} \begin{bmatrix} -1 & 1 \\ -1 & 1 \end{bmatrix} + \frac{1}{h\,Pe} \begin{bmatrix} 1 & -1 \\ -1 & 1 \end{bmatrix} \right) \begin{Bmatrix} u_1^e \\ u_2^e \end{Bmatrix} = \begin{Bmatrix} Q_1^e \\ Q_2^e \end{Bmatrix}. \tag{3}$$

The assembled system of equations for a uniform mesh of four linear elements is

$$\left(\frac{1}{2} \begin{bmatrix} -1 & 1 & 0 & 0 & 0 \\ -1 & 0 & 1 & 0 & 0 \\ 0 & -1 & 0 & 1 & 0 \\ 0 & 0 & -1 & 0 & 1 \\ 0 & 0 & 0 & -1 & 1 \end{bmatrix} + \frac{1}{h\,Pe} \begin{bmatrix} 1 & -1 & 0 & 0 & 0 \\ -1 & 2 & -1 & 0 & 0 \\ 0 & -1 & 2 & -1 & 0 \\ 0 & 0 & -1 & 2 & -1 \\ 0 & 0 & 0 & -1 & 1 \end{bmatrix} \right) \begin{Bmatrix} U_1 \\ U_2 \\ U_3 \\ U_4 \\ U_5 \end{Bmatrix} = \begin{Bmatrix} Q_1^{(1)} \\ 0 \\ 0 \\ 0 \\ Q_2^{(4)} \end{Bmatrix}. \tag{4}$$

The structure of the equations for an interior global node I is of the form

$$\left(-\frac{1}{2} - \frac{1}{h\,Pe} \right) U_{I-1} + \frac{2}{h\,Pe} U_I + \left(\frac{1}{2} - \frac{1}{h\,Pe} \right) U_{I+1} = 0. \tag{5}$$

(b) For a quadratic finite element, the finite element equations are [see Eq. (4.2.30b)]

$$\left(\frac{1}{6} \begin{bmatrix} -3 & 4 & 1 \\ -4 & 0 & 4 \\ 1 & -4 & 3 \end{bmatrix} + \frac{1}{3h\,Pe} \begin{bmatrix} 7 & -8 & 1 \\ -8 & 16 & -8 \\ 1 & -8 & 7 \end{bmatrix} \right) \begin{Bmatrix} u_1^e \\ u_2^e \\ u_3^e \end{Bmatrix} = \begin{Bmatrix} Q_1^e \\ 0 \\ Q_3^e \end{Bmatrix}. \tag{6}$$

The assembled set of finite element equations for a uniform mesh of two quadratic elements is

$$\left(\frac{1}{6} \begin{bmatrix} -3 & 4 & 1 & 0 & 0 \\ -4 & 0 & 4 & 0 & 0 \\ 1 & -4 & 0 & 4 & 1 \\ 0 & 0 & -4 & 0 & 4 \\ 0 & 0 & 1 & -4 & 3 \end{bmatrix} + \frac{1}{3h\,Pe} \begin{bmatrix} 7 & -8 & 1 & 0 & 0 \\ -8 & 16 & -8 & 0 & 0 \\ 1 & -8 & 14 & -8 & 1 \\ 0 & 0 & -8 & 16 & -8 \\ 0 & 0 & 1 & -8 & 7 \end{bmatrix} \right) \begin{Bmatrix} U_1 \\ U_2 \\ U_3 \\ U_4 \\ U_5 \end{Bmatrix} = \begin{Bmatrix} Q_1^{(1)} \\ 0 \\ 0 \\ Q_3^{(2)} \end{Bmatrix}. \tag{7}$$

When the central difference approximation is used to approximate the derivatives in Eq. (4.2.39), we obtain the same equation as in Eq. (5):

$$\left(\frac{U_{I+1} - U_{I-1}}{2h} \right) - \frac{1}{Pe} \left(\frac{U_{I-1} - 2U_I + U_{I+1}}{h^2} \right) = 0 \tag{8}$$

or

$$\left(-\frac{1}{2} - \frac{1}{h\,Pe}\right)U_{I-1} + \frac{2}{h\,Pe}U_I + \left(\frac{1}{2} - \frac{1}{h\,Pe}\right)U_{I+1} = 0. \tag{9}$$

The exact solution of the problem at hand is given by

$$\phi(x) = \frac{1 - e^{Pe\,(x-1)}}{1 - e^{-Pe}}, \quad \frac{d\phi}{dx} = -\frac{Pe\; e^{Pe\,(x-1)}}{1 - e^{-Pe}}, \quad q_n(x) = \frac{e^{Pe\,(x-1)}}{1 - e^{-Pe}}. \tag{10}$$

Figure 4.2.10 shows a comparison of the exact solution (solid line) with the finite element solutions for meshes of four and ten elements for $Pe = 10$. Clearly, for four linear elements, the numerical solutions exhibit oscillations. The oscillations disappear when the number of elements is greater than $Pe/2$, as shown by Surana and Reddy [5] (see pages 371–378) through a stability analysis. Similar observations can be made for $Pe = 30$. Results using quadratic element are not shown but they too exhibit similar characteristics (see **Problem 3.9**).

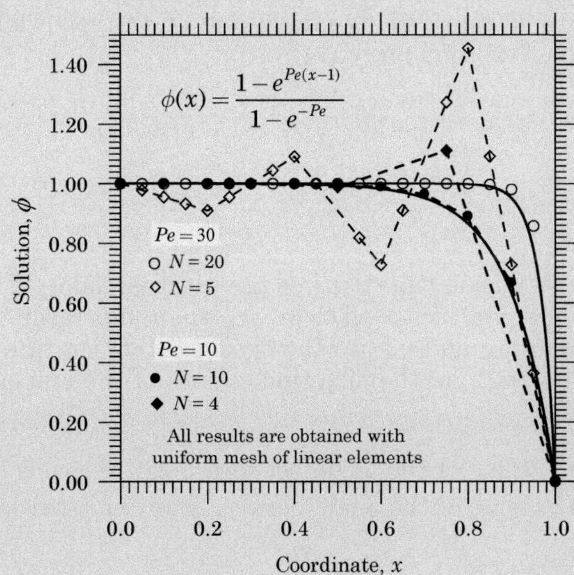

Fig. 4.2.10 A comparison of the numerical solutions against the exact solution for four and ten linear elements for $Pe = 10$ and 30.

4.3 Two-Dimensional Problems

4.3.1 Model Differential Equation

Here we consider a model (i.e., typical) differential equation involving a single unknown in two dimensions. Although there are multitude of engineering problems described by the model equation, we use the equation governing heat transfer in an *orthotropic solid medium* as the model equation. Heat transfer in a fluid medium (including the advection–diffusion problem) will be considered in the sequel. By setting $\mathbf{v} = \mathbf{0}$ in Eq. (2.4.13) for the steady-state case (we shall consider unsteady case in the sequel), we arrive at the following partial differential equation governing the temperature $T = u$:

$$-\left[\frac{\partial}{\partial x}\left(a_{xx}\frac{\partial u}{\partial x}\right)+\frac{\partial}{\partial y}\left(a_{yy}\frac{\partial u}{\partial y}\right)\right]=g(x,y)\quad \text{in } \Omega, \tag{4.3.1}$$

where Ω is a two-dimensional solid domain with boundary Γ, as shown in Fig. 4.3.1(a). Here $a_{xx} = k_{xx}$ and $a_{yy} = k_{yy}$ are the conductivities of the medium (the material coordinate axes are assumed to coincide with the problem coordinates, x and y) in the x- and y-directions, respectively, and $g(x, y)$ is the known internal heat generation (if any) measured per unit area. For an isotropic medium, we set $k_{xx} = k_{yy} = k$. Equation (4.3.1) is also valid for a ground-water flow problem, where u denotes the water head (i.e. velocity potential), a_{xx} and a_{yy} are the permeabilities in the x- and y-directions, respectively, and $g(x, y)$ is the distributed water source.

The function u is required to satisfy, in addition to Eq. (4.3.1) in Ω, certain conditions on the boundary Γ of the bounded closed domain Ω. In a heat transfer problem, u is required to satisfy one of the following two types of boundary conditions at a boundary point:

$$u = \hat{u}(s) \quad \text{on} \quad \Gamma_u \tag{4.3.2a}$$

$$\left(a_{xx}\frac{\partial u}{\partial x}n_x + a_{yy}\frac{\partial u}{\partial y}n_y\right) + \beta(u - u_\infty) = \hat{q}_n(s) \quad \text{on} \quad \Gamma_q, \tag{4.3.2b}$$

where Γ_u and Γ_q are disjoint portions of the total boundary Γ such that $\Gamma = \Gamma_u \cup \Gamma_q$, β is the heat transfer coefficient, u_∞ the temperature of the medium surrounding the body, $\hat{u}$ and $\hat{q}_n$ are the specified temperature and heat flux, respectively, and (n_x, n_y) are the direction cosines of the unit normal vector $\hat{\mathbf{n}}$ on the boundary Γ:

$$\begin{aligned} n_x &= \cos(x, \hat{\mathbf{n}}), \text{cosine of the angle between positive } x\text{-axis and } \hat{\mathbf{n}} \\ n_y &= \cos(y, \hat{\mathbf{n}}), \text{cosine of the angle between positive } y\text{-axis and } \hat{\mathbf{n}}. \end{aligned} \tag{4.3.3}$$

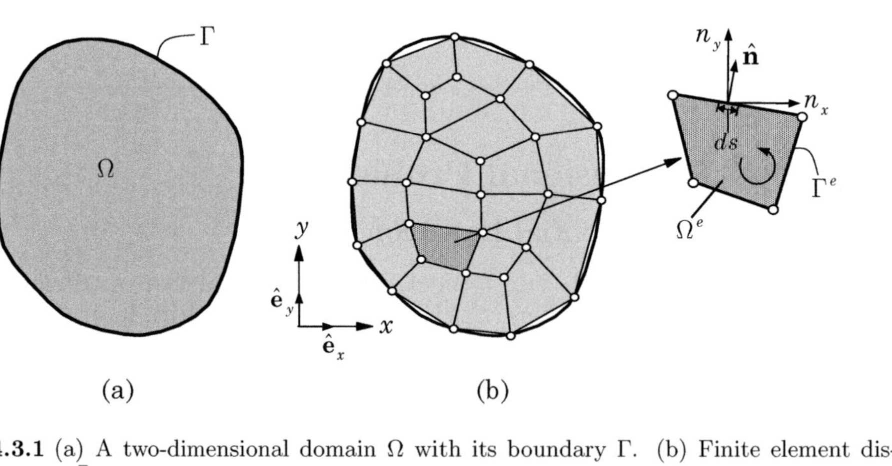

Fig. 4.3.1 (a) A two-dimensional domain Ω with its boundary Γ. (b) Finite element discretization of $\bar{\Omega} = \Omega \cup \Gamma$ and a typical finite element Ω^e with its boundary Γ^e.

4.3.2 Finite Element Approximation

In the FEM, the domain $\bar{\Omega} = \Omega \cup \Gamma$ is divided into a set of closed bounded sub-domains $\bar{\Omega}^e = \Omega^e \cup \Gamma^e$, called finite elements [see Fig. 4.3.1(b)]. Only geometric shapes that qualify as finite elements are those for which approximation functions ψ_i^e can be derived uniquely. It turns out that only triangles and rectangles meet these requirements, as discussed shortly. To keep the formulation steps very general (i.e., not confine the finite element formulation to a specific geometric shape), we denote the domain of a typical element as Ω^e and its boundary by Γ^e. The element $\bar{\Omega}^e$ can be a triangle or quadrilateral in shape, and the degree of interpolation over it can be linear, quadratic, and higher. The union of all elements $\bar{\Omega}^e$ (which may share some boundary segments) is denoted by $\bar{\Omega}^h$, and it is called the *finite element mesh* of the domain Ω. In general, $\bar{\Omega}^h$ may not equal $\bar{\Omega}$ when the boundary Γ is curved. Of course, for polygonal domains, the finite element mesh exactly represents the actual domain, (i.e., $\bar{\Omega}^h = \bar{\Omega}$).

Suppose that the dependent unknown u is approximated over a typical finite element $\bar{\Omega}^e$ by $u_h^e(\mathbf{x})$ in the same way as in the one-dimensional case:

$$u(\mathbf{x}) \approx u_h^e(\mathbf{x}) = \sum_{j=1}^{n} u_j^e \psi_j^e(\mathbf{x}), \quad \mathbf{x} = (x, y) \in \bar{\Omega}^e, \tag{4.3.4}$$

where u_j^e denote the values of the function $u_h^e(\mathbf{x})$ at a selected number of points (i.e., element nodes) in the element $\bar{\Omega}^e$, and ψ_j^e are the Lagrange interpolation functions associated with the element.

A triangle is the simplest two-dimensional geometric shape in two dimensions because it is uniquely defined by three points ($n = 3$) in a plane and the approximated function varies between any two points according to $u_h^e(\mathbf{x}) = c_1^e + c_2^e x + c_3^e y$. A triangle with three nodes per side (a total of six nodes in the element) will uniquely define the geometry while representing the quadratic variation $u_h^e(\mathbf{x}) = c_1^e + c_2^e x + c_3^e y + c_4^e xy + c_5^e x^2 + c_6^e y^2$ uniquely along any of the three sides and inside the element. On the other hand, a pentagon, for example, requires five points to define its geometry uniquely, but there is no five-parameter polynomial in (x, y) that uniquely defines the linear variation along any segment of the pentagon. Therefore, a pentagon is not a classical finite element.

4.3.3 Weak Form

The n nodal values u_j^e in Eq. (4.3.4) must be determined such that the approximate solution $u_h^e(\mathbf{x})$ satisfies Eq. (4.3.1) and interpolates the nodal values of $u_h^e(\mathbf{x})$ over the element Ω^e. As described in Section 4.2.4, we require the governing differential equation to be satisfied only in a weak-form (weighted-integral) sense, with the weight functions being the same as the approximation functions (i.e., use the Galerkin idea). The resulting finite element model (i.e., set of algebraic equations) will be called the *weak-form Galerkin finite element model* or the *Ritz finite element model*.

We use the three-step procedure described in Section 4.2.4 to develop the weak form of Eq. (4.3.1) that accounts for the form of the natural boundary condition in Eq. (4.3.2b) over the typical finite element Ω^e.

Step 1. The first step is to take all nonzero expressions in Eq. (4.3.1) to one side of the equality, multiply the resulting equation with a weight function w_i^e from a set of linearly independent functions $\{w_i^e\}_{i=1}^n$, and integrate the equation over the element domain Ω^e:

$$0 = \int_{\Omega^e} w_i^e \left[-\frac{\partial}{\partial x}\left(a_{xx}\frac{\partial u_h^e}{\partial x}\right) - \frac{\partial}{\partial y}\left(a_{yy}\frac{\partial u_h^e}{\partial y}\right) - g(x,y)\right] dxdy. \tag{4.3.5}$$

The expression in the square brackets of Eq. (4.3.5) represents the residual of the approximation of the differential equation (4.3.1) because $u_h^e(\mathbf{x})$ is only an approximation of $u(\mathbf{x})$. For n independent choices of w_i^e, the weighted-integral statement in Eq. (4.3.5) yields a set of n linearly independent algebraic equations, called the *weighted-residual finite element model.*

Step 2. In the second step, we distribute the differentiation among u_h^e and w_i^e equally so that both u_h^e and w_h^e are required to be differentiable only once with respect to x and y. To achieve this we use the component form of the gradient (or divergence) theorem:

$$\int_{\Omega^e} \frac{\partial}{\partial x}(w_i^e F_x)\, dxdy = \oint_{\Gamma^e} (w_i^e F_x) n_x\, ds \tag{4.3.6}$$

$$\int_{\Omega^e} \frac{\partial}{\partial y}(w_i^e F_y)\, dxdy = \oint_{\Gamma^e} (w_i^e F_y) n_y\, ds, \tag{4.3.7}$$

where n_x and n_y are the components of the unit normal vector $\hat{\mathbf{n}}$

$$\hat{\mathbf{n}} = n_x\hat{\mathbf{e}}_x + n_y\hat{\mathbf{e}}_y = \cos\alpha\ \hat{\mathbf{e}}_x + \sin\alpha\ \hat{\mathbf{e}}_y \tag{4.3.8}$$

on the boundary Γ^e, ds is the arc length of an infinitesimal line element along the boundary, and F_x and F_y are the components of a vector function $\mathbf{F}$. With

$$F_x = a_{xx}\frac{\partial u_h^e}{\partial x}, \quad F_y = a_{yy}\frac{\partial u_h^e}{\partial y} \tag{4.3.9}$$

and identities

$$-w_i^e \frac{\partial F_x}{\partial x} = -\frac{\partial}{\partial x}(w_i^e F_x) + F_x \frac{\partial w_i^e}{\partial x} \tag{4.3.10}$$

$$-w_i^e \frac{\partial F_y}{\partial y} = -\frac{\partial}{\partial y}(w_i^e F_y) + F_y \frac{\partial w_i^e}{\partial y}, \tag{4.3.11}$$

we obtain

$$\begin{aligned} 0 = &\int_{\Omega^e} \left(a_{xx}\frac{\partial w_i^e}{\partial x}\frac{\partial u_h^e}{\partial x} + a_{yy}\frac{\partial w_i^e}{\partial y}\frac{\partial u_h^e}{\partial y} - w_i^e g\right) dxdy \\ &- \oint_{\Gamma^e} w_i^e \left(a_{xx}\frac{\partial u_h^e}{\partial x} n_x + a_{yy}\frac{\partial u_h^e}{\partial y} n_y\right) ds. \end{aligned} \tag{4.3.12}$$

From an inspection of the boundary term in Eq. (4.3.12), we note that u_h^e is the *primary variable*, and specification of u_h^e constitutes the essential boundary condition. The coefficient of the weight function in the boundary expression

$$q_n \equiv a_{xx}\frac{\partial u_h^e}{\partial x}n_x + a_{yy}\frac{\partial u_h^e}{\partial y}n_y \tag{4.3.13}$$

is the *secondary variable*; its specification constitutes the natural boundary condition. By definition, q_n is the positive outward flux normal to the surface as one travels counterclockwise along the boundary Γ^e. This is the reason why the element node numbers are counted in the counterclockwise direction and the evaluation of boundary integrals is carried out in the counterclockwise sense. Physically, the secondary variable q_n denotes the *inward* heat flux normal to the boundary of the element because the heat flux vector $\mathbf{q}$ is given by

$$\mathbf{q} = q_x\hat{\mathbf{e}}_x + q_y\hat{\mathbf{e}}_y, \quad q_x = -a_{xx}\frac{\partial u_h^e}{\partial x}, \quad q_y = -a_{yy}\frac{\partial u_h^e}{\partial y}, \tag{4.3.14}$$

and we have

$$\hat{\mathbf{n}}\cdot\mathbf{q} = n_x q_x + n_y q_y = -\left(a_{xx}\frac{\partial u_h^e}{\partial x}n_x + a_{yy}\frac{\partial u_h^e}{\partial y}n_y\right) = -q_n. \tag{4.3.15}$$

Step 3. The third and last step involves the use the boundary condition from Eqs. (4.3.2b) in Eq. (4.3.12), $q_n = \hat{q}_n - \beta(u - u_\infty)$, and write

$$\begin{aligned} 0 = &\int_{\Omega^e}\left(a_{xx}\frac{\partial w_i^e}{\partial x}\frac{\partial u_h^e}{\partial x} + a_{yy}\frac{\partial w_i^e}{\partial y}\frac{\partial u_h^e}{\partial y} - w_i^e g\right)dA + \oint_{\Gamma^e}\beta w_i^e u_h^e\, ds \\ &- \oint_{\Gamma^e}\beta w_i^e u_\infty\, ds - \oint_{\Gamma^e} w_i^e \hat{q}_n\, ds. \end{aligned} \tag{4.3.16}$$

Since we have started with the model equation (4.3.1) and made use of Eq. (4.3.2b) in arriving at the weak form (4.3.16), it is obvious that the governing differential equation (4.3.1) and the natural boundary condition (4.3.2a) are equivalent to the weak form (4.3.16). The converse of the statement also holds [i.e., the weak form is equivalent to Eqs. (4.3.1) and (4.3.2b)]; for details, see Reddy [1], Reddy and Gartling [3], Gresho and Sani [4], and Surana and Reddy [5, 6].

4.3.4 Finite Element Model

The weak form in Eq. (4.3.16) requires that u_h^e be at least linear in both x and y so that there are no terms in Eq. (4.3.16) that become identically zero. Suppose that u_h^e is represented over a typical finite element Ω^e by a complete polynomial of the form

$$\begin{aligned} u_h^e(x,y) = &\; c_1^e + c_2^e x + c_3^e y + c_4^e xy + c_5^e x^2 + c_6^e y^2 \\ &+ c_7^e x^2 y + c_8^e xy^2 + c_9^e x^2y^2 + \cdots. \end{aligned} \tag{4.3.17}$$

In addition to the property of completeness, we also require *equi-presence* of the x- and y-terms in the polynomial. Thus, one must include the first three terms, the first four terms, or the first six terms, and so on, to derive the interpolation functions. More on this topic will be presented in Section 4.3.6.

For now, we assume that finite element approximation of u_h^e is of the form in Eq. (4.3.4) [i.e., the approximation (4.3.17), for any admissible choice of terms in the polynomial, will reduce to (4.3.4)]. Substituting Eq. (4.3.4) into the weak form (4.3.16), we obtain

$$0 = \sum_{j=1}^{n} \left[\int_{\Omega^e} \left(a_{xx} \frac{\partial w_i^e}{\partial x} \frac{\partial \psi_j^e}{\partial x} + a_{yy} \frac{\partial w_i^e}{\partial y} \frac{\partial \psi_j^e}{\partial y} \right) dxdy + \oint_{\Gamma^e} \beta w_i^e \psi_j^e \, ds \right] u_j^e$$
$$- \int_{\Omega^e} w_i^e g \, dxdy - \oint_{\Gamma^e} \beta w_i^e u_\infty \, ds - \oint_{\Gamma^e} w_i^e \hat{q}_n \, ds. \tag{4.3.18}$$

This equation must hold for any weight function w_i^e from the set $\{w_i^e\}_{i=1}^n$. In particular, it must hold for the case $w_i^e = \psi_i^e$ (i.e., the Galerkin idea).

For each choice of w_i^e we obtain an algebraic relation among $(u_1^e, u_2^e, \ldots, u_n^e)$. The ith algebraic equation is obtained by substituting $w_i^e = \psi_i^e$ into Eq. (4.3.18):

$$\sum_{j=1}^{n} \left(K_{ij}^e + H_{ij}^e \right) u_j^e = f_i^e + P_i^e + q_i^e$$

or

$$(\mathbf{K}^e + \mathbf{H}^e)\, \mathbf{u}^e = \mathbf{f}^e + \mathbf{P}^e + \mathbf{Q}^e \equiv \mathbf{F}^e, \tag{4.3.19}$$

where the coefficients K_{ij}^e, H_{ij}^e ($\mathbf{K}^e$ and $\mathbf{H}^e$ are symmetric matrices), f_i^e, P_i^e, and Q_i^e are defined by

$$K_{ij}^e = \int_{\Omega^e} \left(a_{xx} \frac{\partial \psi_i^e}{\partial x} \frac{\partial \psi_j^e}{\partial x} + a_{yy} \frac{\partial \psi_i^e}{\partial y} \frac{\partial \psi_j^e}{\partial y} \right) dxdy, \quad H_{ij}^e = \oint_{\Gamma^e} \beta \psi_i^e \psi_j^e \, ds$$
$$f_i^e = \int_{\Omega^e} g\, \psi_i^e \, dxdy, \quad P_i^e = \oint_{\Gamma^e} \beta u_\infty \psi_i^e \, ds, \quad Q_i^e = \oint_{\Gamma^e} \hat{q}_n \psi_i^e \, ds. \tag{4.3.20}$$

We note that $\mathbf{H}^e$ and $\mathbf{P}^e$ are nonzero only for elements with advection boundary segments. The evaluations of these coefficients will be discussed in Sections 4.3.6 and 4.3.9.

4.3.5 Axisymmetric Problems

4.3.5.1 Governing Equation

Consider the second-order differential equation in vector form

$$-\boldsymbol{\nabla} \cdot (\mathbf{k} \cdot \boldsymbol{\nabla} u) = g(\mathbf{x}), \tag{4.3.21}$$

where $\boldsymbol{\nabla}$ is the gradient operator, $\mathbf{k}$ is the second-order conductivity tensor, u is the temperature, and g is the internal heat generation. In terms of cylindrical coordinates (r, θ, z), heat transfer in a cylindrically orthotropic material (i.e., when the material axes of an orthotropic material coincide with the r, θ, and z axes), Eq. (4.3.21) takes the form

$$-\frac{1}{r}\frac{\partial}{\partial r}\left(rk_{rr}\frac{\partial u}{\partial r}\right) - \frac{1}{r^2}\frac{\partial}{\partial \theta}\left(k_{\theta\theta}\frac{\partial u}{\partial \theta}\right) - \frac{\partial}{\partial z}\left(k_{zz}\frac{\partial u}{\partial z}\right) = g(r, \theta, z). \tag{4.3.22}$$

When the geometry, loading, and boundary conditions are independent of the circumferential direction (i.e., θ-coordinate direction), the problem is said to be axisymmetric and the governing equation becomes two-dimensional in terms of r and z:

$$-\frac{1}{r}\frac{\partial}{\partial r}\left(rk_{rr}\frac{\partial u}{\partial r}\right) - \frac{\partial}{\partial z}\left(k_{zz}\frac{\partial u}{\partial z}\right) = g(r, z). \tag{4.3.23}$$

In addition, if the problem geometry and data are independent of z, for example when the cylinder is very long, heat transfer is one-dimensional in the radial coordinate r:

$$-\frac{1}{r}\frac{d}{dr}\left(rk_{rr}\frac{du}{dr}\right) = g(r) \quad \text{for} \quad a < r < b, \tag{4.3.24}$$

which is the same as that considered in Section 4.2.7.

4.3.5.2 Finite Element Model

In this section we develop the finite element models of two-dimensional axisymmetric problems described by Eq. (4.3.23). We begin with the development of the weak form, where the volume element dv is replaced by $dv = rdrd\theta dz$.

Following the three-step procedure, we obtain the weak form (the constant multiplier 2π and element label e are omitted in the following)

$$0 = \int_{\Omega_e}\left(k_{rr}\frac{\partial w_i}{\partial r}\frac{\partial u_h}{\partial r} + k_{zz}\frac{\partial w_i}{\partial z}\frac{\partial u_h}{\partial z} - w_i g\right)rdrdz - \oint_{\Gamma_e} w_i q_n \, ds, \tag{4.3.25a}$$

where w_i is the ith weight function and q_n is the flux normal to the boundary

$$q_n = r\left(k_{rr}\frac{\partial u_h}{\partial r}n_r + k_{zz}\frac{\partial u_h}{\partial z}n_z\right). \tag{4.3.25b}$$

Let us assume that $u(r, z)$ is approximated by the finite element interpolation $u_h^e(r, z)$ over a typical two-dimensional element Ω_e (which can be a triangle or quadrilateral in shape and the degree of interpolation is dictated by the number of nodes n in the element)

$$u \approx u_h^e(r, z) = \sum_{j=1}^{n} u_j^e \psi_j^e(r, z), \tag{4.3.26}$$

where interpolation functions $\psi_j^e(r, z)$ are the same as those discussed previously but with $x = r$ and $y = z$. Substitution of Eq. (4.3.26) for u_h^e and ψ_i^e for w_i^e

into the weak form gives the ith algebraic equation of the finite element model

$$0 = \sum_{j=1}^{n} \left[\int_{\Omega_e} \left(k_{rr} \frac{\partial \psi_i^e}{\partial r} \frac{\partial \psi_j^e}{\partial r} + k_{zz} \frac{\partial \psi_i^e}{\partial z} \frac{\partial \psi_j^e}{\partial z} \right) r dr dz \right] u_j^e$$
$$- \int_{\Omega_e} \psi_i^e g \, r \, dr \, dz - \oint_{\Gamma_e} \psi_i^e q_n \, ds, \tag{4.3.27}$$

or

$$0 = \sum_{j=1}^{n} K_{ij}^e u_j^e - f_i^e - Q_i^e \quad \text{or} \quad \mathbf{K}^e \mathbf{u}^e = \mathbf{f}^e + \mathbf{Q}^e, \tag{4.3.28}$$

where

$$K_{ij}^e = \int_{\Omega_e} \left(k_{rr} \frac{\partial \psi_i^e}{\partial r} \frac{\partial \psi_j^e}{\partial r} + k_{zz} \frac{\partial \psi_i^e}{\partial z} \frac{\partial \psi_j^e}{\partial z} \right) r \, dr \, dz \tag{4.3.29a}$$

$$f_i^e = \int_{\Omega_e} \psi_i^e g \, r \, dr \, dz, \quad Q_i^e = \oint_{\Gamma_e} \psi_i^e q_n \, ds. \tag{4.3.29b}$$

When there is advection on the boundary, the heat flux normal to the boundary, q_n, in Eq. (4.3.27) is replaced by $\hat{q}_n - \beta(u - u_\infty)$ [see Eq. (4.3.2b)]; the finite element model becomes

$$\left(\mathbf{K}^e + \mathbf{H}^e \right) \mathbf{u}^e = \mathbf{f}^e + \mathbf{P}^e + \mathbf{Q}^e, \tag{4.3.30a}$$

where

$$K_{ij}^e = \int_{\Omega_e} \left(k_{rr} \frac{\partial \psi_i^e}{\partial r} \frac{\partial \psi_j^e}{\partial r} + k_{zz} \frac{\partial \psi_i^e}{\partial z} \frac{\partial \psi_j^e}{\partial z} \right) r \, dr \, dz, \quad H_{ij}^e = \oint_{\Gamma_e} \beta \psi_i^e \psi_j^e \, ds$$
$$f_i^e = \int_{\Omega_e} \psi_i^e g \, r \, dr \, dz, \quad P_i^e = \oint_{\Gamma_e} \beta u_\infty \psi_i^e \, ds, \quad Q_i^e = \oint_{\Gamma_e} \hat{q}_n \psi_i^e \, ds. \tag{4.3.30b}$$

4.3.6 Approximation Functions and Evaluation of Coefficients for Linear Elements

4.3.6.1 Properties of Interpolation Functions

The finite element approximation $u_h^e(x, y)$ of $u(x, y)$ over an element Ω^e must satisfy the following conditions in order for the approximate solution to converge to the true solution.

(1) $u_h^e(x, y)$ must be continuous as required in the weak form of the problem; that is, all terms in the weak form are represented as nonzero values.

(2) The polynomials used to represent $u_h^e(x, y)$ must be complete and contain both x and y of the same order (i.e., equipresence of x and y). This means all terms, beginning with a constant term up to the highest order desired, in both x and y, should be included in the polynomial.

(3) All terms in the polynomial should be linearly independent.

The number of linearly independent terms in the representation of u_h^e dictates the shape and the number of nodes in the element. It turns out that only triangular and quadrilateral shapes meet the requirements stated above. Here we discuss the interpolation functions of triangular and rectangular elements with linear and quadratic approximations.

4.3.6.2 Linear Triangular Element

An examination of the statement (4.3.18) and the finite element matrices in Eq. (4.3.20) shows that ψ_i^e $(i = 1, 2, \ldots, n)$ should be, at least, a linear function of x and y. The lowest-order linear polynomial that meets this requirements is

$$u_h^e(x, y) = c_1^e + c_2^e\, x + c_3^e\, y. \tag{4.3.31}$$

The polynomial is complete because the constant, linear, and bilinear terms in x and y are included. To write the three parameters (c_1^e, c_2^e, c_3^e) in terms of three values of u_h^e, we must identify three points, (x_i^e, y_i^e), $i = 1, 2, 3$, that uniquely define the geometry of the element Ω^e and allow the imposition of inter-element continuity of the primary variable $u_h^e(x, y)$. Obviously, the geometric shape defined by three points in a two-dimensional domain is a triangle. Thus, the polynomial in Eq. (4.3.31) is associated with a triangular element, with the vertices of the triangle being the three nodes [see Fig. 4.3.2(a)].

The linear interpolation functions for an arbitrary three-node triangular element are derived by evaluating Eq. (4.3.31) at (x_i^e, y_i^e):

$$\begin{aligned}
u_1^e \equiv u_h^e(x_1^e, y_1^e) &= c_1^e + c_2^e x_1^e + c_3^e y_1^e \\
u_2^e \equiv u_h^e(x_2^e, y_2^e) &= c_1^e + c_2^e x_2^e + c_3^e y_2^e \\
u_3^e \equiv u_h^e(x_3^e, y_3^e) &= c_1^e + c_2^e x_3^e + c_3^e y_3^e,
\end{aligned}$$

which can be expressed in matrix form as

$$\begin{Bmatrix} u_1^e \\ u_2^e \\ u_3^e \end{Bmatrix} = \begin{bmatrix} 1 & x_1^e & y_1^e \\ 1 & x_2^e & y_2^e \\ 1 & x_3^e & y_3^e \end{bmatrix} \begin{Bmatrix} c_1^e \\ c_2^e \\ c_3^e \end{Bmatrix}. \tag{4.3.32}$$

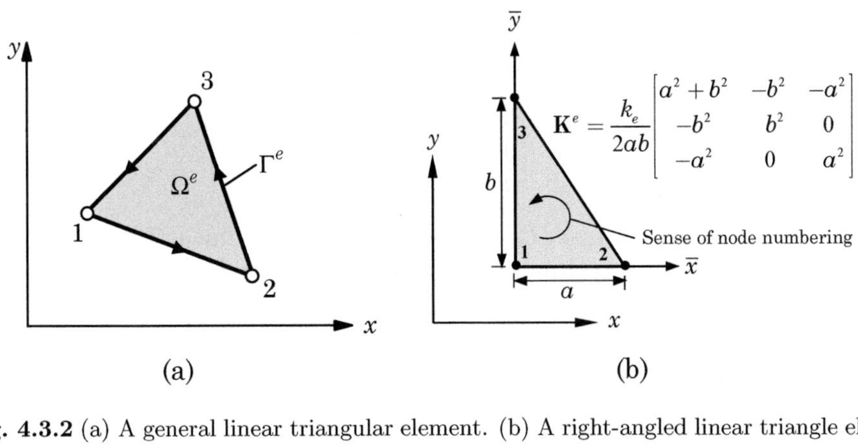

Fig. 4.3.2 (a) A general linear triangular element. (b) A right-angled linear triangle element.

Solving Eq. (4.3.32) for (c_1^e, c_2^e, c_3^e) and substituting into Eq. (4.3.31), we obtain

$$u_h^e(x,y) = \psi_1^e(x,y)u_1^e + \psi_2^e(x,y)u_2^e + \psi_3^e(x,y)u_3^e = \sum_{j=1}^{n=3} \psi_j^e(x,y)u_j^e, \quad (4.3.33a)$$

where

$$\psi_i^e(x,y) = \frac{1}{2A_e}\left(\alpha_i^e + \beta_i^e x + \gamma_i^e y\right), \qquad (i = 1,2,3). \quad (4.3.33b)$$

Here A_e is the area of the triangle [or $2A_e$ is the determinant of the coefficient matrix in Eq. (4.3.32)], and α_i^e, β_i^e, and γ_i^e are the constants which depend only the nodal global coordinates (x_i^e, y_i^e) of the triangular element Ω^e:

$$\alpha_i^e = x_j y_k - x_k y_j; \qquad \beta_i^e = y_j - y_k; \qquad \gamma_i^e = -(x_j - x_k) \quad (4.3.33c)$$

for $i \neq j \neq k$, and i, j, and k permute in a natural order. The interpolation functions ψ_i^e $(i = 1, 2, \ldots, n)$ satisfy the following interpolation properties:

$$(1) \quad \psi_i^e(x_j^e, y_j^e) = \delta_{ij}, \quad (i,j = 1,2,3); \qquad (2) \quad \sum_{i=1}^{3} \psi_i^e(x,y) = 1. \quad (4.3.34)$$

The use of linear approximation of the actual function $u(x,y)$, which is possibly a surface, results in a planar shape. For linear triangular element, the derivatives of u_h^e are constant:

$$\frac{\partial \psi_i^e}{\partial x} = \frac{\beta_i^e}{2A_e}, \quad \frac{\partial \psi_i^e}{\partial y} = \frac{\gamma_i^e}{2A_e}. \quad (4.3.35)$$

The integrals in the definition of the coefficients K_{ij}^e and f_i^e can be evaluated for given data: a_{xx}^e, a_{yy}^e, and g_e (all of which can be functions of x and y). For example, for element-wise constant values of the data, the element coefficient matrix and source vector from Eq. (4.3.20) for linear triangular element become

$$K_{ij}^e = \frac{1}{4A_e}\left(a_{xx}^e \beta_i^e \beta_j^e + a_{yy}^e \gamma_i^e \gamma_j^e\right), \quad f_i^e = \frac{g_e A_e}{3}. \quad (4.3.36)$$

In particular, for a right-angled triangular element with base a and height b, and with node 1 at the right angle [see Fig. 4.3.2(b)], $\mathbf{K}^e$ takes the form

$$\mathbf{K}^e = \frac{a_{xx}^e}{2}\begin{bmatrix} \lambda^e & -\lambda^e & 0 \\ -\lambda^e & \lambda^e & 0 \\ 0 & 0 & 0 \end{bmatrix} + \frac{a_{yy}^e}{2}\begin{bmatrix} \mu^e & 0 & -\mu^e \\ 0 & 0 & 0 \\ -\mu^e & 0 & \mu^e \end{bmatrix}, \quad (4.3.37)$$

where $\lambda^e = b/a$ and $\mu^e = a/b$. Of course, for cases in which the conductivity coefficients a_{xx}^e and a_{yy}^e are functions of (x,y), numerical integration can be used to evaluate the coefficients. We note that the first part of the coefficient matrix contains an aspect ratio b/a and the second part contains the reciprocal of the aspect ratio. When a is very large compared with b, the second part contains large numbers compared with the first part, for the same orders of magnitudes of a_{xx} and a_{yy}; hence, one must maintain an aspect ratio that does not unduly affect the physics.

4.3.6.3 Linear Rectangular Element

The next polynomial that meets the requirements is

$$u_h^e(x, y) = c_1^e + c_2^e x + c_3^e y + c_4^e xy, \tag{4.3.38}$$

which contains four linearly independent terms, $\{1\, x\, y\, xy\}$. In order to represent Eq. (4.3.38) in terms of the values of $u_h^e(x, y)$, an element with four points, with linear variation along any side of the element, must be identified. In general, such an element is a quadrilateral, with four corner points being the nodes. It is simpler to derive the approximation functions for a square or a rectangle (see Fig. 4.3.3). When the element is a quadrilateral in shape, we use coordinate transformations to represent the integrals posed on the quadrilateral element to those over a square geometry and then use numerical integration to evaluate the integrals.

For a linear rectangular element (some times called a *bilinear element* because it contains a term that is linear in both x and y), we have

$$u_h^e(\mathbf{x}) = \sum_{i=1}^{4} u_i^e \psi_i^e(\bar{x}, \bar{y}), \tag{4.3.39}$$

where ψ_i^e are the Lagrange interpolation functions expressed in terms of the element coordinates $(\bar{x}, \bar{y})$,

$$\begin{aligned} \psi_1^e &= \left(1 - \frac{\bar{x}}{a}\right)\left(1 - \frac{\bar{y}}{b}\right), \qquad & \psi_2^e &= \frac{\bar{x}}{a}\left(1 - \frac{\bar{y}}{b}\right), \\ \psi_3^e &= \frac{\bar{x}}{a}\frac{\bar{y}}{b}, & \psi_4^e &= \left(1 - \frac{\bar{x}}{a}\right)\frac{\bar{y}}{b}, \end{aligned} \tag{4.3.40}$$

and $(\bar{x}, \bar{y})$ denote the local coordinates with the origin located at node 1 of the element, and (a, b) denote the horizontal and vertical dimensions of the rectangle, as shown in Fig. 4.3.3.

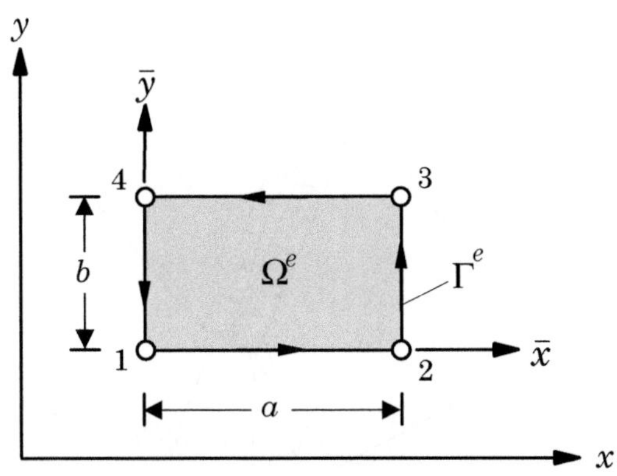

Fig. 4.3.3 The linear rectangular finite element.

The integrals in the definition of K_{ij}^e and f_i^e can be easily evaluated over a rectangular element of sides a and b. For example, for element-wise constant values of the data, we have (see Reddy [1]) the following results:

$$\mathbf{K}^e = a_{xx}^e \mathbf{S}^{11} + a_{yy}^e \mathbf{S}^{22}, \qquad f_i^e = \frac{g_e ab}{4}, \tag{4.3.41a}$$

where the matrices $\mathbf{S}^{11}$ and $\mathbf{S}^{22}$ are defined in Eq. (4.3.41b)($\lambda = b/a$ and $\mu = a/b$):

$$\mathbf{S}^{11} = \frac{1}{6}\begin{bmatrix} 2\lambda & -2\lambda & -\lambda & \lambda \\ -2\lambda & 2\lambda & \lambda & -\lambda \\ -\lambda & \lambda & 2\lambda & -2\lambda \\ \lambda & -\lambda & -2\lambda & 2\lambda \end{bmatrix}, \quad \mathbf{S}^{22} = \frac{1}{6}\begin{bmatrix} 2\mu & \mu & -\mu & -2\mu \\ \mu & 2\mu & -2\mu & -\mu \\ -\mu & -2\mu & 2\mu & \mu \\ -2\mu & -\mu & \mu & 2\mu \end{bmatrix}. \tag{4.3.41b}$$

Because the coefficient matrix $\mathbf{K}^e$ depends on the aspect ratios a/b and b/a, their values should not be too large or too small (compared with unity). Depending on the problem, values of the order $a/b = 10$ are considered to be reasonable (but not of order 100 or greater).

We note that the heat flux contributions to a global node come from the line integrals of the distributed fluxes on the sides of all elements connected at the node. In reality, the Q_i^e $(i = 1, 2, \ldots, n)$ of an element which shares all its sides with other elements are not known because q_n^e on its sides is unknown. However, the sum of Q_i^e from all elements at a global node is equal to heat applied (externally) at the node because the contributions of q_n^e from the neighboring elements at the element interfaces inside the domain is canceled. In other words, q_n^e on side (i, j) of element Ω^e cancels with q_n^f on side (p, q) of element Ω^f when sides (i, j) of element Ω^e and (p, q) of element Ω^f are the same (i.e., at the interface of elements Ω^e and Ω^f; see Fig. 4.3.4). This can be viewed as the balance of the internal flux. When an element boundary coincides with the problem boundary, the nodal contribution Q_i^e due to the distributed flux q_n^e on that boundary can be evaluated with the help of Eq. (4.3.30b).

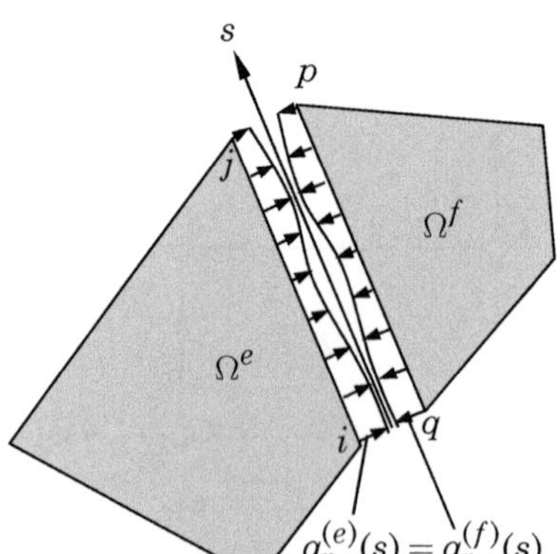

Fig. 4.3.4 Balance of fluxes at the inter-element boundary of elements Ω^e and Ω^f.

4.3.7 Higher-Order Finite Elements

4.3.7.1 Triangular Elements

Higher-order triangular elements (i.e., triangular elements with interpolation functions of higher degree) can be systematically developed with the help of the so-called *area coordinates*. For triangular elements, it is possible to construct three natural coordinates L_i $(i = 1, 2, 3)$, as in the case of one-dimensional elements, which vary in a direction normal to the sides directly opposite each node, as shown in Fig. 4.3.5(a). The coordinates are defined as

$$\psi_i = L_i = \frac{A_i}{A}\,, \qquad A = \text{ area of the triangle,} \tag{4.3.42}$$

where A_i is the area of the triangle formed by nodes j and k and an arbitrary point P in the element, and A is the total area of the element. For example, A_1 is the area of the shaded triangle that is formed by nodes 2 and 3 and point P. Therefore, A_1 is given by [see Fig. 4.3.5(a)]

$$A_1 = \frac{1}{2}\left(\alpha_1 + \beta_1 x + \gamma_1 y\right); \quad \psi_1 = L_1 = \frac{A_1}{A} = \frac{1}{2A}\left(\alpha_1 + \beta_1 x + \gamma_1 y\right). \tag{4.3.43}$$

Similarly, $2A_2 = \alpha_2 + \beta_2 x + \gamma_2 y$ and $2A_3 = \alpha_3 + \beta_3 x + \gamma_3 y$. Thus, for the linear triangular element, we have

$$\psi_i(x, y) = L_i(x, y). \tag{4.3.44}$$

The area coordinates (L_1, L_2, L_3) can be used to construct interpolation functions for higher-order triangular elements. For example, in the case of quadratic triangular element ψ_1 is given by [see Fig. 4.3.5(b)] $\psi_1 = c_1(L_1 - 0)(L_1 - 0.5)$, where c_1 is determined such that $\psi_1 = 1$ when $L_1 = 1$, giving $c_1 = 2$. Hence,

$$\psi_1 = L_1(2L_1 - 1).$$

The explicit forms of the interpolation functions for the linear and quadratic elements are

$$\boldsymbol{\Psi}^e = \begin{Bmatrix} L_1 \\ L_2 \\ L_3 \end{Bmatrix}; \qquad \boldsymbol{\Psi}^e = \begin{Bmatrix} L_1(2L_1 - 1) \\ L_2(2L_2 - 1) \\ L_3(2L_3 - 1) \\ 4L_1L_2 \\ 4L_2L_3 \\ 4L_3L_1 \end{Bmatrix}. \tag{4.3.45}$$

The ordering of the interpolation functions in the above arrays corresponds to the node numbers shown in Figs. 4.3.6(a) and (b). Integrals of the products of the area coordinates L_i over an element can be evaluated using the following formula:

$$\int_A L_1^p\, L_2^q\, L_3^r\; dA = \frac{p!\,q!\,r!}{(p+q+r+1)!}\, A. \tag{4.3.46}$$

(a)

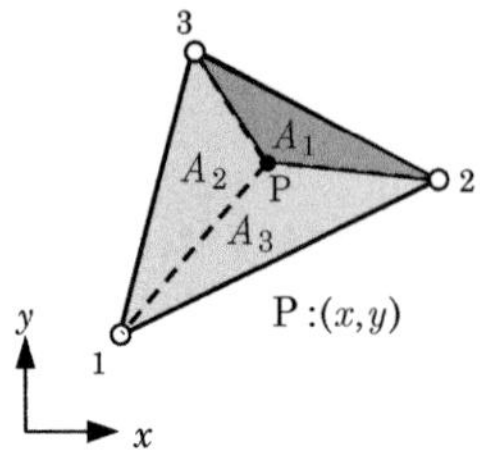

$$2A_1 = \begin{vmatrix} 1 & x & y \\ 1 & x_2 & y_2 \\ 1 & x_3 & y_3 \end{vmatrix} = (x_2 y_3 - x_3 y_2) + (y_2 - y_3)x + (x_3 - x_2)y$$

$$= \alpha_1 + \beta_1 x + \gamma_1 y$$

$$2A_i = \alpha_i + \beta_i x + \gamma_i y; \;\; \psi_i = \frac{A_i}{A} = L_i = \frac{1}{2A}(\alpha_1 + \beta_1 x + \gamma_1 y)$$

(b)

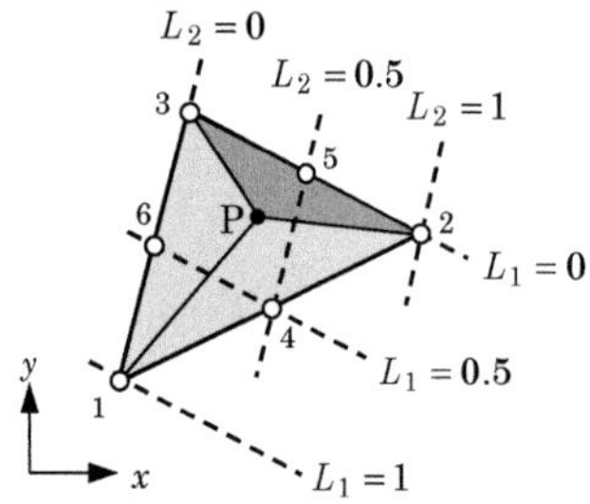

$$\psi_1 = c_1(L_1 - 0)(L_1 - 0.5);$$
$$\psi_1 = 1 \text{ at } L_1 = 1 \text{ gives } c_1 = 2$$
$$\psi_1 = L_1(2L_1 - 1)$$

$$\psi_4 = c_4(L_2 - 0)(L_1 - 0);$$
$$\psi_4 = 1 \text{ at } L_1 = L_2 = 0.5 \text{ gives } c_4 = 4$$
$$\psi_4 = 4L_1 L_2$$

Fig. 4.3.5 (a) Graphical and analytical representations of area coordinates L_i for triangular elements. (b) Determination of ψ_i for the quadratic triangular element.

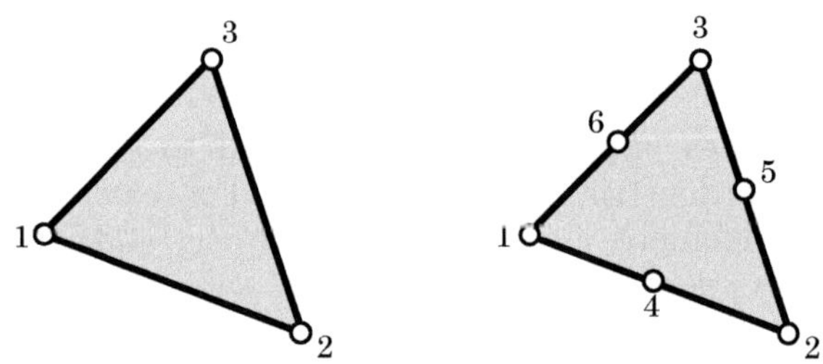

(a) (b)

Fig. 4.3.6 (a) Linear and (b) quadratic triangular elements.

4.3.7.2 Rectangular Elements

The linear and quadratic Lagrange interpolation functions associated with rectangular family of master finite elements can be obtained from the tensor product of the corresponding one-dimensional Lagrange interpolation functions. The one-dimensional linear and quadratic interpolation functions [see, e.g., Eq. (4.2.29a)] can be expressed in terms of the local coordinates ξ, $-1 \leq \xi \leq 1$, as

$$\begin{aligned} &\text{linear:} && \psi_1 = \tfrac{1}{2}(1-\xi), \quad \psi_2 = \tfrac{1}{2}(1+\xi), \\ &\text{quadratic:} && \psi_1 = -\tfrac{1}{2}\xi(1-\xi), \quad \psi_2 = (1-\xi^2), \quad \psi_3 = \tfrac{1}{2}\xi(1+\xi). \end{aligned} \tag{4.3.47}$$

The approximation functions for the liner rectangular element are obtained by taking the product of the column vector with the row vector:

$$\tfrac{1}{2}\begin{Bmatrix}(1-\xi)\\(1+\xi)\end{Bmatrix}\;\tfrac{1}{2}\{(1-\eta)\quad(1+\eta)\}$$

gives four functions, which are placed in the same order as the node numbers shown in Fig. 4.3.7(a):

$$\mathbf{\Psi}^e = \frac{1}{4}\begin{Bmatrix}(1-\xi)(1-\eta)\\(1+\xi)(1-\eta)\\(1+\xi)(1+\eta)\\(1-\xi)(1+\eta)\end{Bmatrix}. \tag{4.3.48}$$

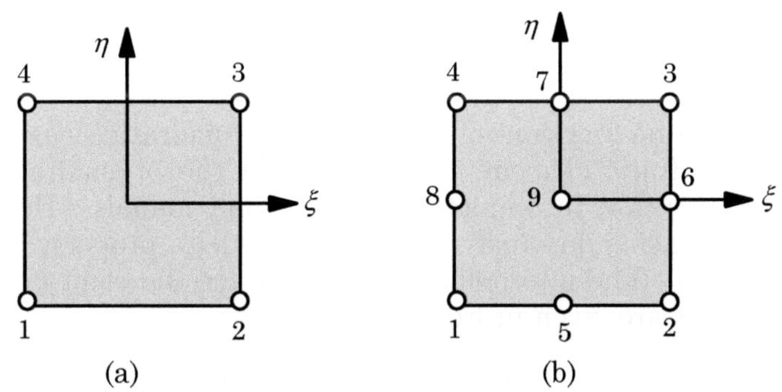

Fig. 4.3.7 (a) Linear and (b) quadratic rectangular elements.

Similarly, the product of the quadratic functions for the two coordinate directions yields the interpolation functions for the quadratic rectangular element [see Fig. 4.3.7(b) for the node numbers]:

$$\mathbf{\Psi}^e = \frac{1}{4}\begin{Bmatrix}(1-\xi)(1-\eta)(-\xi-\eta-1)+(1-\xi^2)(1-\eta^2)\\(1+\xi)(1-\eta)(\xi-\eta-1)+(1-\xi^2)(1-\eta^2)\\(1+\xi)(1+\eta)(\xi+\eta-1)+(1-\xi^2)(1-\eta^2)\\(1-\xi)(1+\eta)(-\xi+\eta-1)+(1-\xi^2)(1-\eta^2)\\2[(1-\xi^2)(1-\eta)-(1-\xi^2)(1-\eta^2)]\\2[(1+\xi)(1-\eta^2)-(1-\xi^2)(1-\eta^2)]\\2[(1-\xi^2)(1+\eta)-(1-\xi^2)(1-\eta^2)]\\2[(1-\xi)(1-\eta^2)-(1-\xi^2)(1-\eta^2)]\\4(1-\xi^2)(1-\eta^2)\end{Bmatrix}. \tag{4.3.49}$$

The *serendipity elements* are those elements that have no interior nodes. These elements have fewer nodes compared with the same-order complete Lagrange elements. The approximation functions ψ_i^e of the serendipity elements are not complete (in the sense that they are not complete polynomials), and they cannot be obtained using tensor products of one-dimensional Lagrange interpolation functions. Instead, an alternative procedure must be employed. The lowest-order serendipity element in two dimensions is the eight-node quadratic element shown in Fig. 4.3.8.

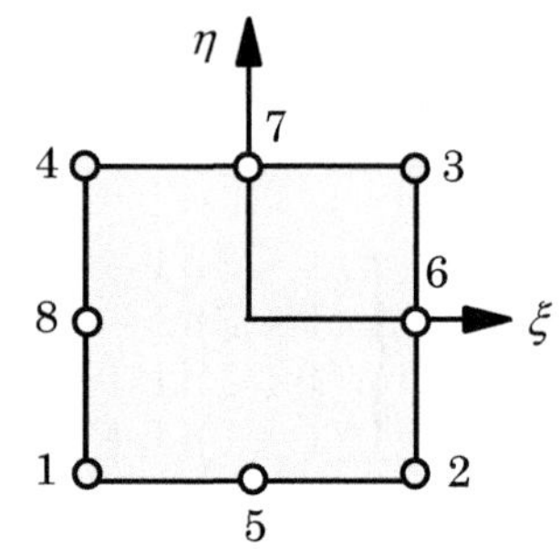

Fig. 4.3.8 Quadratic rectangular serendipity element.

The approximation functions of the eight-node quadratic element, compared with the nine-quadratic element, do not contain the bi-quadratic term $\xi^2\eta^2$ and, therefore, they are incomplete quadratic polynomials. However, ψ_i^e of the serendipity elements do satisfy the Kronecker-delta property [i.e., property (i) in Eq. (4.3.34)]. The interpolation functions for the eight-node quadratic serendipity element are given in Eq. (4.3.50).

$$\mathbf{\Psi}^e = \frac{1}{4}\left\{\begin{array}{c} (1-\xi)(1-\eta)(-\xi-\eta-1) \\ (1+\xi)(1-\eta)(\xi-\eta-1) \\ (1+\xi)(1+\eta)(\xi+\eta-1) \\ (1-\xi)(1+\eta)(-\xi+\eta-1) \\ 2(1-\xi^2)(1-\eta) \\ 2(1+\xi)(1-\eta^2) \\ 2(1-\xi^2)(1+\eta) \\ 2(1-\xi)(1-\eta^2) \end{array}\right\}. \tag{4.3.50}$$

4.3.8 Assembly of Elements

The assembly of finite elements to obtain the equations of the entire domain is based on the following two rules.

(1) Continuity of the primary variable $u_h^e(\mathbf{x})$ (e.g., displacements and temperature).
(2) Equilibrium of secondary variables Q_i^e (e.g., forces and heats).

We illustrate the assembly procedure by considering a finite element mesh consisting of a triangular element and a quadrilateral element, as shown in Fig. 4.3.9.

The nodes of the finite element mesh are called *global nodes*. From the mesh shown in Fig. 4.3.9, it is clear that the following correspondence between global and element nodes exists: nodes 1, 2, and 3 of element 1 correspond to global nodes 1, 2, and 3, respectively. Nodes 1, 2, 3, and 4 of element 2 correspond to global nodes 2, 4, 5, and 3, respectively. Hence, the correspondence between the local and global nodal values of the primary variable u is as follows:

$$u_1^1 = u_1, \quad u_2^1 = u_1^2 = u_2, \quad u_3^1 = u_4^2 = u_3, \quad u_2^2 = u_4, \quad u_3^2 = u_5, \qquad (4.3.51)$$

where the superscripts refer to the element numbers and the subscripts refer to the element node numbers. This amounts to imposing the continuity of the primary variables at the nodes common to elements 1 and 2. Note that the continuity of the primary variables at the interelement nodes guarantees the continuity of the primary variable along the entire interelement boundary, that is $u^1(s) = u^2(s)$, where s denotes a coordinate along the interface of the two elements.

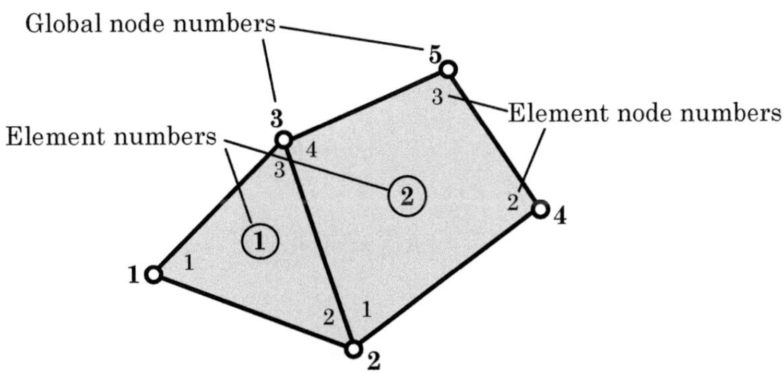

Fig. 4.3.9 Global-local correspondence of nodes for element assembly of two elements.

Next, we consider the equilibrium of secondary variables at the interelement boundaries. At the interface between any two elements, the flux from the two elements should be equal in magnitude and opposite in sign. For the two elements shown in Fig. 4.3.9, the interface is along the side connecting global nodes 2 and 3. Hence, the heat flux q_n^1 on side 2–3 of element 1 should balance, in integral sense, the flux q_n^2 on side 4–1 of element 2 (recall the sign convention on q_n^e):

$$\int_{h_{23}^1} q_n^1 \psi_2^1 \, ds = -\int_{h_{14}^2} q_n^2 \psi_1^2 \, ds \quad \text{or} \quad \int_{h_{23}^1} q_n^1 \psi_3^1 \, ds = -\int_{h_{14}^2} q_n^2 \psi_4^2 \, ds, \qquad (4.3.52)$$

where h_{pq}^e denotes length of the side connecting nodes p and q of element Ω^e.

Now we are ready to assemble the element equations for the two-element mesh. Let K_{ij}^1 $(i, j = 1, 2, 3)$ denote the coefficient matrix corresponding to the triangular element and let K_{ij}^2 $(i, j = 1, 2, 3, 4)$ denote the coefficient matrix corresponding to the quadrilateral element. The element equations of the two elements are written separately first. The element equations of the triangular element are of the form

$$
\begin{aligned}
K_{11}^1 u_1^1 + K_{12}^1 u_2^1 + K_{13}^1 u_3^1 &= f_1^1 + Q_1^1 \\
K_{21}^1 u_1^1 + K_{22}^1 u_2^1 + K_{23}^1 u_3^1 &= f_2^1 + Q_2^1 \\
K_{31}^1 u_1^1 + K_{32}^1 u_2^1 + K_{33}^1 u_3^1 &= f_3^1 + Q_3^1 .
\end{aligned} \tag{4.3.53}
$$

Similarly, the finite element equations for the rectangular element are

$$
\begin{aligned}
K_{11}^2 u_1^2 + K_{12}^2 u_2^2 + K_{13}^2 u_3^2 + K_{14}^2 u_4^2 &= f_1^2 + Q_1^2 \\
K_{21}^2 u_1^2 + K_{22}^2 u_2^2 + K_{23}^2 u_3^2 + K_{24}^2 u_4^2 &= f_2^2 + Q_2^2 \\
K_{31}^2 u_1^2 + K_{32}^2 u_2^2 + K_{33}^2 u_3^2 + K_{34}^2 u_4^2 &= f_3^2 + Q_3^2 \\
K_{41}^2 u_1^2 + K_{42}^2 u_2^2 + K_{43}^2 u_3^2 + K_{44}^2 u_4^2 &= f_4^2 + Q_4^2 .
\end{aligned} \tag{4.3.54}
$$

In order to impose the balance condition in Eq. (4.3.52), it is necessary to add the second equation of element 1 to the first equation of element 2, and also add the third equation of element 1 to the fourth equation of element 2:

$$
\begin{aligned}
&(K_{21}^1 u_1^1 + K_{22}^1 u_2^1 + K_{23}^1 u_3^1) + (K_{11}^2 u_1^2 + K_{12}^2 u_2^2 + K_{13}^2 u_3^2 + K_{14}^2 u_4^2) \\
&\quad = (f_2^1 + Q_2^1) + (f_1^2 + Q_1^2) \\
&(K_{31}^1 u_1^1 + K_{32}^1 u_2^1 + K_{33}^1 u_3^1) + (K_{41}^2 u_1^2 + K_{42}^2 u_2^2 + K_{43}^2 u_3^2 + K_{44}^2 u_4^2) \\
&\quad = (f_3^1 + Q_3^1) + (f_4^2 + Q_4^2).
\end{aligned}
$$

Using the local-global nodal variable correspondence in Eq. (4.3.51), we can rewrite the above equations as

$$
\begin{aligned}
&K_{21}^1 u_1 + (K_{22}^1 + K_{11}^2) u_2 + (K_{23}^1 + K_{14}^2) u_3 + K_{12}^2 u_4 + K_{13}^2 u_5 \\
&\quad = f_2^1 + f_1^2 + (Q_2^1 + Q_1^2) \\
&K_{31}^1 u_1 + (K_{32}^1 + K_{41}^2) u_2 + (K_{33}^1 + K_{44}^2) u_3 + K_{42}^2 u_4 + K_{43}^2 u_5 \\
&\quad = f_3^1 + f_4^2 + (Q_3^1 + Q_4^2).
\end{aligned} \tag{4.3.55}
$$

Now we can impose the conditions in Eq. (4.3.52) by setting appropriate portions of the expressions in parentheses on the right-hand side of Eq. (4.3.55) to zero. This is done by means of the connectivity relations, that is, the correspondence of the local node number to the global node number.

For elements with sides on the boundary of the computational domain where $q_n^e(s)$ is known, we can evaluate the boundary integral

$$
q_i^e = \oint_{\Gamma^e} \hat{q}_n \psi_i^e(s). \tag{4.3.56}
$$

It is necessary to compute such integrals only when Γ^e, *or a portion of it, coincides with the boundary* Γ_q *of the total domain* Ω *on which the flux* q_n *is specified.* When Γ^e falls on the boundary Γ_u of the domain Ω, q_n^e is not known there and can be determined in the post-computation (we note that the primary variable u_h^e is specified on Γ_u).

Another comment is in order on the evaluation of the line integrals. The two-dimensional interpolation functions $\psi_i^e(x, y)$ become one-dimensional function $\psi_i^e(s)$ along the side with coordinate s (see Fig. 4.3.10). Consequently, the evaluation of the line integrals is made simpler, and they are readily available through Eqs. (4.2.29a)–(4.2.30b).

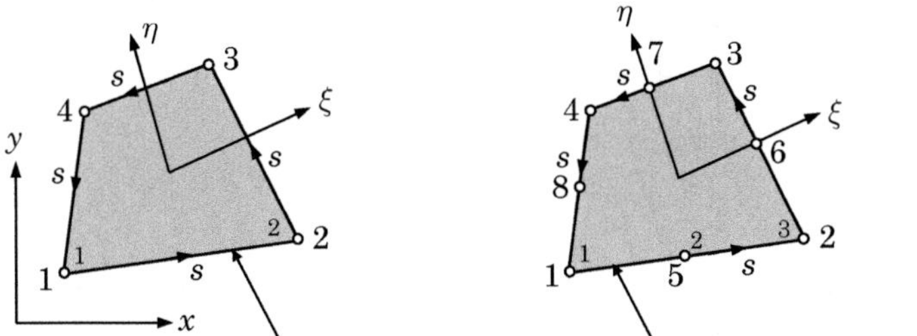

$$\psi_1(s) = \tfrac{1}{2}\left(1 - \frac{s}{h_{12}}\right), \quad \psi_2(s) = \tfrac{1}{2}\frac{s}{h_{12}} \qquad \psi_1(s) = \tfrac{1}{2}\left(1 - \frac{s}{h_{12}}\right)\left(1 - \frac{2s}{h_{12}}\right)$$

Fig. 4.3.10 Evaluation of line integrals on the boundary of two-dimensional elements.

4.3.9 Numerical Examples

Here we consider three numerical examples to illustrate the assembly and imposition of boundary conditions, including advection. The first example deals with the computation of nodal heats due to distributed boundary flux and contributions due to advection on the boundary.

Example 4.3.1

(a) Give the global coefficients K_{55}, $K_{5(10)}$, $K_{(10)(10)}$, $K_{(10)(15)}$, $K_{(15)(15)}$, F_5, F_{10}, and F_{15}; (b) identify the specified boundary conditions; and (c) compute the nodal contributions of various flux distributions and advection shown for a heat transfer problem with the mesh shown in Fig. 4.3.11. Assume that there is no internal heat generation (i.e., $g = 0$).

Solution (a) From Fig. 4.3.11, the required coefficients are

$$\begin{aligned} K_{55} &= K_{22}^4 + H_{22}^4, \quad K_{5(10)} = K_{23}^4 + H_{23}^4 \\ K_{(10)(10)} &= K_{33}^4 + H_{33}^4 + K_{22}^8 + H_{22}^8 \\ K_{(10)(15)} &= K_{23}^4 + H_{23}^4, \quad K_{(15)(15)} = K_{33}^8 + H_{33}^8 + K_{22}^{(12)} \\ F_5 &= P_2^4, \quad F_{10} = P_3^4 + P_2^8, \quad F_{15} = P_3^8 + Q_{15}. \end{aligned} \tag{1}$$

The numerical values of K_{ij}^e can be readily obtained from Eqs. (4.3.37) and (4.3.41a). The numerical values of H_{ij}^e and P_i^e will be computed in part (c).

(b) From Fig. 4.3.11 it is clear that the known primary nodal degrees of freedom (i.e., temperatures) are:

$$U_1 = U_6 = U_{11} = U_{16} = U_{20} = u_0. \tag{2}$$

The known secondary variables (i.e., heats) are

$$Q_2 = Q_2^1 + Q_1^2 = 0; \quad Q_3 = Q_2^2 + Q_1^3 = 0; \quad Q_4 = Q_2^3 + Q_1^4 = 0. \tag{3}$$

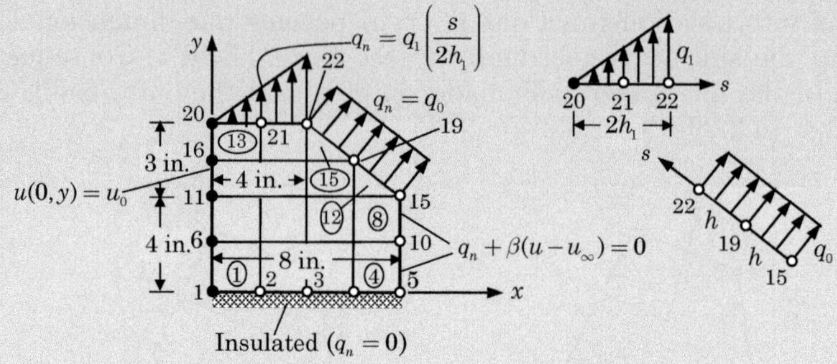

Fig. 4.3.11 A heat transfer problem with various types of boundary conditions.

The computation of the known heats at nodes 15, 19, 20, 21, and 22 is discussed next. The contribution of the uniform flux q_0 to nodes 15, 19, and 22 is computed first using Eq. (4.3.20) or (4.3.56)($h = 2.5$ in.) and utilizing the known result from Eq. (4.2.29b):

$$Q_{15} = \int_0^h q_0 \psi_1^1 \, ds = \frac{q_0 h}{2}, \quad Q_{22} = \int_0^h q_0 \psi_2^2 \, ds = \frac{q_0 h}{2}$$
$$Q_{19} = \int_0^h q_0 \psi_2^1 \, ds + \int_0^h q_0 \psi_1^2 \, ds = q_0 h. \tag{4}$$

The contribution of the linearly varying flux $q_1(s/2h_1)$ to nodes 20, 21, and 22 is computed as follows ($h_1 = 2$ in.):

$$Q_{20} = \int_0^{h_1} q_1 \frac{s}{2h_1} \psi_1^1 \, ds = \int_0^{h_1} q_1 \frac{s}{2h_1}\left(1 - \frac{s}{h_1}\right) ds = \frac{q_1 h_1}{12}$$
$$\begin{aligned} Q_{21} &= \int_0^{h_1} q_1 \frac{s}{2h_1} \psi_2^1 \, ds + \int_{h_1}^{2h_1} q_1 \frac{s}{2h_1} \psi_1^2 \, ds \\ &= \int_0^{h_1} q_1 \frac{s}{2h_1}\left(\frac{s}{h_1}\right) ds + \int_{h_1}^{2h_1} q_1 \frac{s}{2h_1}\left(2 - \frac{s}{h_1}\right) ds \\ &= \frac{q_1 h_1}{6} + \frac{2q_1 h_1}{6} = \frac{q_1 h_1}{2} \end{aligned} \tag{5}$$
$$Q_{22} = \int_{h_1}^{2h_1} q_1 \frac{s}{2h_1} \psi_2^2 \, ds = \int_{h_1}^{2h_1} q_1 \frac{s}{2h_1}\left(\frac{s}{h_1} - 1\right) ds = \frac{5q_1 h_1}{12}.$$

Obviously, Q_{22} just computed should be added to the previously computed value.

(c) Lastly, we compute the convective contributions to nodes 5, 10, and 15. The advection contributes to the element coefficients as well as to the right-hand side. Following the discussion of Section 4.3.4, we have

$$\mathbf{H}^e = \frac{\beta_e h_e}{6}\begin{bmatrix} 2 & 1 \\ 1 & 2 \end{bmatrix}, \quad \mathbf{P}^e = \frac{\beta_e h_e u_\infty}{2}\begin{Bmatrix} 1 \\ 1 \end{Bmatrix}. \tag{6}$$

Since β and $h = 2$ in. is the same for both line elements, we have the following contributions to the global coefficients; note that node 10 has contributions added from the line element below and above:

$$H_{22}^4 = H_{33}^4 = \frac{\beta h}{3}, \quad H_{23}^4 = \frac{\beta h}{6}, \quad P_2^4 = P_3^4 = \frac{\beta u_\infty h}{2}$$
$$H_{22}^8 = H_{33}^8 = \frac{\beta h}{3}, \quad H_{23}^8 = \frac{\beta h}{6}, \quad P_2^8 = P_3^8 = \frac{\beta u_\infty h}{2}. \tag{7}$$

Example 4.3.2

Consider steady-state heat conduction in an isotropic rectangular region of dimensions $3a \times 2a$, as shown in Fig. 4.3.12(a). The origin of the x and y coordinates is taken at the lower left corner such that x is parallel to the side $3a$ and y is parallel to the side $2a$. Boundaries $x = 0$ and $y = 0$ are insulated (i.e., $q_n = 0$), boundary $x = 3a$ is maintained at zero temperature, and boundary $y = 2a$ is maintained at temperature $T = T_0 \cos(\pi x/6)$. Determine the temperature distribution using (1) mesh of linear triangular elements shown in Fig. 4.3.12(b) and (2) mesh of linear rectangular elements shown in Fig. 4.3.12(c).

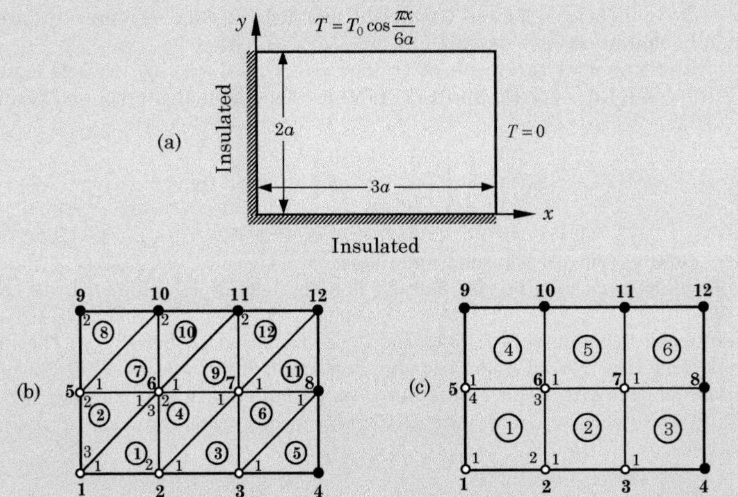

Fig. 4.3.12 Finite element analysis of a heat conduction problem over a rectangular domain: (a) domain; (b) mesh of linear triangular elements; and (c) mesh of linear rectangular elements.

Solution The governing equation is a special case of the model equation (4.3.1) with zero internal heat generation $g = 0$ and coefficients $a_{xx} = a_{yy} = k$. Thus, Eq. (4.3.1) takes the form

$$-k\left(\frac{\partial^2 T}{\partial x^2} + \frac{\partial^2 T}{\partial y^2}\right) = 0. \tag{1}$$

The exact solution of Eq. (1) for the boundary conditions shown in Fig. 4.3.12(a) is

$$T(x,y) = T_0 \frac{\cosh(\pi y/6a)\cos(\pi x/6a)}{\cosh(\pi/3)}. \tag{2}$$

The finite element model of Eq. (1) is given by

$$\mathbf{K}^e \mathbf{T}^e = \mathbf{Q}^e. \tag{3}$$

where T_i^e is the temperature at node i of element Ω_e, and

$$K_{ij}^e = \int_{\Omega_e} k\left(\frac{\partial \psi_i}{\partial x}\frac{\partial \psi_j}{\partial x} + \frac{\partial \psi_i}{\partial y}\frac{\partial \psi_j}{\partial y}\right) dA, \quad Q_i^e = \oint_{\Gamma_e} q_n \psi_i \, ds. \tag{4}$$

Suppose that we use a 3×2 mesh (i.e., three subdivisions along the x-axis and two subdivisions along the y-axis) of linear triangular elements and then a 3×2 mesh of linear rectangular elements, as shown in Figs. 4.3.12(b) and 4.3.12(c). Both meshes have the same number of global nodes, namely 12, but differing numbers of elements. There are six unknown nodal values associated with nodes 1, 2, 3, 5, 6, and 7.

Mesh of Triangular Elements. The global node numbers, element numbers, and element node numbers used are shown in Fig. 4.3.12(b). Of course, the global node numbering and element numbering is arbitrary although the global node numbering dictates the size of the half bandwidth of the assembled equations, which in turn affects the computational time of Gauss elimination methods used in the solution of algebraic equations in a computer. The element node numbering scheme should be the one that is used in the development of element interpolation functions. By a suitable numbering of the element nodes, all similar elements can be made to have identical element coefficient matrices. Such considerations are important only when hand calculations are carried out.

For a typical element of the mesh of triangles in Fig. 4.3.12(b), the element coefficient matrix is given by [see Eq. (4.3.26) and Fig. 4.3.2 for the element matrix and element geometry],

$$\mathbf{K}^e = \frac{k}{2}\begin{bmatrix} 1 & -1 & 0 \\ -1 & 2 & -1 \\ 0 & -1 & 1 \end{bmatrix}, \quad \mathbf{f}^e = \mathbf{0}, \tag{5}$$

where k is the conductivity of the medium. Note that the element matrix is independent of the size of the element, as long as the element is a right-angle triangle with its base equal to its height.

The assembly of the elements follows the logic discussed earlier. From the mesh in Fig. 4.3.12(b), it is clear that global node 1 is connected only to global nodes 2, 5, and 6; hence, the equation associated with node 1, numbered as global equation 1, is of the form

$$K_{11}U_1 + K_{12}U_2 + K_{15}U_5 + K_{16}U_6 = F_6.$$

Similarly, global node 2 is connected only to global nodes 1, 3, 6, and 7, giving the global equation associated with node 2 as

$$K_{21}U_1 + K_{22}U_2 + K_{23}U_3 + K_{26}U_6 + K_{27}U_7 = F_2,$$

and so on. The global coefficients in terms of element coefficients are

$$\begin{aligned} &K_{11} = K_{11}^1 + K_{33}^2 = \tfrac{1}{2}(1+1)k, \quad K_{12} = K_{12}^1 = \tfrac{1}{2}(-1)k \\ &K_{15} = K_{32}^2 = \tfrac{1}{2}(-1)k, \quad K_{16} = K_{13}^1 + K_{31}^2 = 0 + 0, \quad \text{etc.} \\ &K_{21} = K_{21}^1 = \tfrac{1}{2}(-1)k, \quad K_{22} = K_{22}^1 + K_{11}^3 + K_{33}^4 = \tfrac{1}{2}(2+1+1)k, \quad \text{etc.} \\ &F_1 = Q_1^1 + Q_3^2, \quad F_2 = Q_2^1 + Q_1^3 + Q_3^4, \quad \text{etc.} \end{aligned} \tag{6}$$

The boundary conditions on the global nodal temperatures and heats are

$$\begin{aligned} U_4 = U_8 = U_{12} = 0, \quad U_9 = T_0, \quad U_{10} = \tfrac{\sqrt{3}}{2}T_0, \quad U_{11} = \tfrac{1}{2}T_0 \\ F_1 = F_2 = F_3 = F_5 = 0, \end{aligned} \tag{7}$$

and the balance of internal heat flow requires that

$$F_6 = F_7 = 0. \tag{8}$$

Thus, the unknown primary variables and secondary variables are:

$$U_1, \quad U_2, \quad U_3, \quad U_5, \quad U_6, \quad U_7; \quad F_4, \quad F_8, \quad F_9, \quad F_{10}, \quad F_{11}, \quad F_{12}. \tag{9}$$

Condensed Equations for the Unknown Temperatures. We write the six finite element equations for the six unknown primary variables. These equations come from rows 1, 2, 3, 5, 6, and 7 (corresponding to the same global nodes) of the assembled system of matrix equations:

$$\begin{aligned} K_{11}U_1 + K_{12}U_2 + K_{15}U_5 + K_{16}U_6 &= F_1 = 0 \\ K_{21}U_1 + K_{22}U_2 + K_{23}U_3 + K_{26}U_6 + K_{27}U_7 &= F_2 = 0 \\ &\vdots \\ K_{72}U_2 + K_{73}U_3 + K_{76}U_6 + K_{77}U_7 + K_{7(11)}U_{11} &= F_7 = 0. \end{aligned} \tag{10}$$

Using the boundary conditions and the values of K_{IJ}, we obtain

$$\begin{aligned} k(U_1 - \tfrac{1}{2}U_2 - \tfrac{1}{2}U_5) &= 0 \\ k(-\tfrac{1}{2}U_1 + 2U_2 - \tfrac{1}{2}U_3 - U_6) &= 0 \\ k(-\tfrac{1}{2}U_2 + 2U_3 - U_7) &= 0 \\ k(-\tfrac{1}{2}U_1 + 2U_5 - U_6 - \tfrac{1}{2}U_9) &= 0 \quad (U_9 = T_0) \\ k(-U_2 - U_5 + 4U_6 - U_7 - U_{10}) &= 0 \quad (U_{10} = \tfrac{\sqrt{3}}{2}T_0) \\ k(-U_3 - U_6 + 4U_7 - U_{11}) &= 0 \quad (U_{11} = \tfrac{1}{2}T_0), \end{aligned} \tag{11}$$

or, in the matrix form

$$\frac{k}{2}\begin{bmatrix} 2 & -1 & 0 & -1 & 0 & 0 \\ -1 & 4 & -1 & 0 & -2 & 0 \\ 0 & -1 & 4 & 0 & 0 & -2 \\ -1 & 0 & 0 & 4 & -2 & 0 \\ 0 & -2 & 0 & -2 & 8 & -2 \\ 0 & 0 & -2 & 0 & -2 & 8 \end{bmatrix} \begin{Bmatrix} U_1 \\ U_2 \\ U_3 \\ U_5 \\ U_6 \\ U_7 \end{Bmatrix} = \frac{k}{2} \begin{Bmatrix} 0 \\ 0 \\ 0 \\ T_0 \\ \sqrt{3}T_0 \\ T_0 \end{Bmatrix}. \tag{12}$$

The solution of these equations is (in °C)

$$\begin{aligned} U_1 &= 0.6362\,T_0, \quad U_2 = 0.5510\,T_0, \quad U_3 = 0.3181\,T_0 \\ U_5 &= 0.7214\,T_0, \quad U_6 = 0.6248\,T_0, \quad U_7 = 0.3607\,T_0. \end{aligned} \tag{13}$$

Evaluating the exact solution in Eq. (2) at the nodes, we have (in °C)

$$\begin{aligned} T_1 &= 0.6249\,T_0, \quad T_2 = 0.5412\,T_0, \quad T_3 = 0.3124\,T_0 \\ T_5 &= 0.7125\,T_0, \quad T_6 = 0.6171\,T_0, \quad T_7 = 0.3563\,T_0. \end{aligned} \tag{14}$$

Condensed Equations for the Unknown Secondary Variables. The secondary variable at node 8, for example, can be computed from the eighth finite element equation

$$\begin{aligned} F_8 &= K_{83}U_3 + K_{87}U_7 = \left(K_{13}^6 + K_{31}^5\right) U_3 + \left(K_{12}^6 + K_{21}^{11}\right) U_7 \\ &= -0.3607\,kT_0. \end{aligned} \tag{15}$$

Since q_n is the negative of the heat flux normal to the boundary, F_8 is the negative of the heat at node 8. Thus heat at node 8 is

$$Q_8 = -F_8 = 0.3607\,kT_0. \tag{16}$$

Mesh of Rectangular Elements. For a 3 × 2 mesh of linear rectangular elements [see Fig. 4.3.12(c)], the element coefficient matrix is given by Eqs. (4.3.41a) and (4.3.41b):

$$\mathbf{K}^e = \frac{k}{6}\begin{bmatrix} 4 & -1 & -2 & -1 \\ -1 & 4 & -1 & -2 \\ -2 & -1 & 4 & -1 \\ -1 & -2 & -1 & 4 \end{bmatrix}, \quad \mathbf{f}^e = \mathbf{0}. \tag{17}$$

The assembled global coefficients are

$$\begin{aligned}&K_{11}=K_{11}^1,\quad K_{12}=K_{12}^1,\quad K_{15}=K_{14}^1,\quad K_{16}=K_{13}^1,\quad K_{22}=K_{22}^1+K_{11}^2\\&K_{23}=K_{12}^2,\quad K_{25}=K_{24}^1,\quad K_{26}=K_{23}^1+K_{14}^2,\quad K_{27}=K_{13}^2,\quad \text{etc.}\\&F_1=Q_1^1,\quad F_2=Q_2^1+Q_1^2,\quad F_3=Q_2^2+q_1^3,\quad F_4=q_2^3,\quad \text{etc.}\end{aligned}\tag{18}$$

The boundary conditions for the present mesh remain the same as those for the mesh of triangles. The condensed equations for the unknown temperatures are

$$\frac{k}{6}\begin{bmatrix}4&-1&0&-1&-2&0\\-1&8&-1&-2&-2&-2\\0&-1&8&0&-2&-2\\-1&-2&0&8&-2&0\\-2&-2&-2&-2&16&-2\\0&-2&-2&0&-2&16\end{bmatrix}\begin{Bmatrix}U_1\\U_2\\U_3\\U_5\\U_6\\U_7\end{Bmatrix}=\frac{k}{6}\begin{Bmatrix}0\\0\\0\\T_0+\sqrt{3}T_0\\2T_0+\sqrt{3}T_0+T_0\\\sqrt{3}T_0+T_0\end{Bmatrix}.\tag{19}$$

The solution of these equations is

$$\begin{aligned}&U_1=0.6128\,T_0,\quad U_2=0.5307\,T_0,\quad U_3=0.3064\,T_0\\&U_5=0.7030\,T_0,\quad U_6=0.6088\,T_0,\quad U_7=0.3515\,T_0.\end{aligned}\tag{20}$$

The value of the heat at node 8 is given by ($K_{83}=K_{31}^3$, $K_{87}=K_{34}^3+K_{21}^6$, and $K_{8(11)}=K_{24}^6$)

$$\begin{aligned}Q_8=-F_8&=-K_{83}U_3-K_{87}U_7-K_{8(11)}U_{11}\\&=\tfrac{2}{6}k\,(U_3+U_7+U_{11})=0.3860\,kT_0\ \text{W}.\end{aligned}\tag{21}$$

Table 4.3.1 contains a comparison of the finite element solutions with the analytical solution in Eq. (2) for three different meshes of linear triangular and rectangular elements (results are independent of a and k). We note the fact that there are only half as many elements in the mesh of rectangles compared with the mesh of triangles. It is interesting to note that meshes of triangles provide an upper bound while the meshes of rectangles provide a lower bound to the exact solution.

Table 4.3.1 Comparison of the nodal temperatures $T(x,y)/T_0$, obtained using various finite element meshes with the analytical solution in Eq. (2).

		FEM solutions						
		Triangles			Rectangles			Analytical
x	y	3×2	6×4	12×8	3×2	6×4	12×8	solution
0.0	0.0	0.6362	0.6278	0.6256	0.6128	0.6219	0.6242	0.6249
0.5	0.0	—	0.6064	0.6043	—	0.6007	0.6029	0.6036
1.0	0.0	0.5510	0.5437	0.5418	0.5307	0.5386	0.5405	0.5412
1.5	0.0	—	0.4439	0.4424	—	0.4398	0.4413	0.4419
2.0	0.0	0.3181	0.3139	0.3128	0.3064	0.3110	0.3121	0.3124
2.5	0.0	—	0.1625	0.1619	—	0.1610	0.1615	0.1617
0.0	1.0	0.7214	0.7148	0.7131	0.7030	0.7102	0.7119	0.7125
0.5	1.0	—	0.6904	0.6888	—	0.6860	0.6877	0.6882
1.0	1.0	0.6248	0.6190	0.6176	0.6088	0.6150	0.6166	0.6171
1.5	1.0	—	0.5054	0.5042	—	0.5022	0.5034	0.5038
2.0	1.0	0.3607	0.3574	0.3565	0.3515	0.3551	0.3560	0.3563
2.5	1.0	—	0.1850	0.1846	—	0.1838	0.1843	0.1844

The next example deals with advection heat transfer in a cooling fin treated as a two-dimensional domain with no heat transfer through the thickness (into the plane of the paper) of the fin.

Example 4.3.3

Consider a rectangular cooling fin shown in Fig. 4.3.13 has its base, $x = 0$, maintained at 300 °C and exposed to advection on its remaining boundary. Assume that the dimension normal to the xy-plane is very long so that the heat flow in that direction is negligible. Analyze the problem using the finite element mesh shown in Fig. 4.3.13.

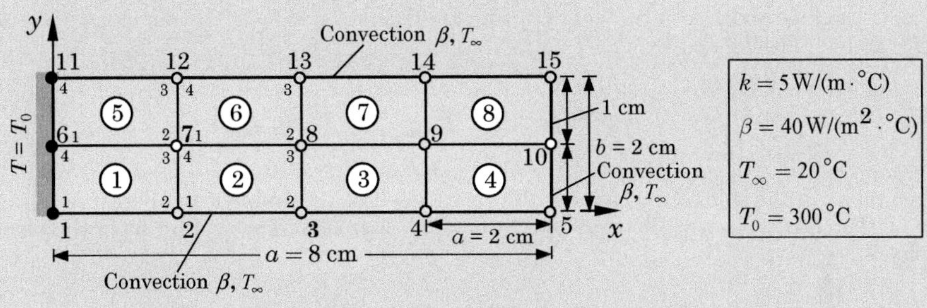

Fig. 4.3.13 Mesh of rectangular elements to discretize a rectangular cooling fin.

Solution The problem does have a symmetry about $y = b/2$. In heat transfer problems, a line of symmetry requires that the flux normal to the line of symmetry is zero. Here we consider both cases: (1) the computational domain is the full body and (2) the computational domain is the symmetric half body.

Considering the full domain, temperatures at nodes 1, 6, and 11 are specified and there are 12 unknown nodal temperatures that need to be determined. Thus, there are 12 equations that constitute the condense equations for the primary unknowns (temperatures). Here we give the global equations associated couple of nodes, one interior and another on the boundary. For example, the finite element equation for node 7 is given by

$$K_{71}U_1 + K_{72}U_2 + K_{73}U_3 + K_{76}U_6 + K_{77}U_7 + K_{78}U_8 + K_{7(11)}U_{11} + K_{7(12)}U_{12} + K_{7(13)}U_{13} = F_7, \tag{1}$$

where K_{IJ} and F_I are the global coefficients, which can be expressed in terms of the element coefficients K^e_{ij} [whose values are readily available from Eqs. (4.3.41a) and (4.3.41b)] and Q^e_I ($f^e_i = 0$ for all elements as there is no internal heat generation specified). Equation (1) becomes

$$(K^1_{32} + K^2_{41})U_2 + K^2_{42}U_3 + (K^2_{43} + K^6_{12})U_8 + (K^5_{23} + K^6_{14})U_{12} + K^6_{13}U_{13} = -K^1_{31}U_1 - (K^1_{34} + K^5_{21})U_6 - K^5_{24}U_{11} + Q^1_3 + Q^2_4 + Q^6_1 + Q^5_2, \tag{2}$$

where $U_1 = U_6 = U_{11} = T_0$ and $Q^1_3 + Q^2_4 + Q^6_1 + Q^5_2 = 0$ by balance of heats at node 7, although individual Q^e_i is not known (and can be determined in the postcomputation either by definition or by element equations).

Similarly, the global finite element equation for node 3 can be written as

$$K_{32}U_2 + K_{33}U_3 + K_{34}U_4 + K_{37}U_7 + K_{38}U_8 + K_{39}U_9 = Q^2_2 + Q^3_1, \tag{3}$$

where $Q_2^2 + Q_1^3$ is equal to (or balanced by) the heat due to advection (verify to yourself)

$$\begin{aligned}
Q_2^2 + Q_1^3 &= \int_{1-2} q_n^2 \psi_2^2(s)\, ds + \int_{1-2} q_n^3 \psi_1^3(s)\, ds \\
&= \int_{1-2} \beta_{12}^2 (U_2\psi_1^2 + U_3\psi_2^2 - T_\infty)\psi_2^2(s)\, ds \\
&\quad + \int_{1-2} \beta_{12}^3 (U_3\psi_1^3 + U_4\psi_2^3 - T_\infty) q_n^3 \psi_1^3(s)\, ds \\
&= H_{12}^2 U_2 + H_{22}^2 U_3 + H_{11}^3 U_3 + H_{21}^2 U_4 + P_2^2 + P_1^3. \qquad (4)
\end{aligned}$$

Here q_n^e is the flux and β_{12}^e is the heat transfer coefficient on side 1-2 of element e ($e = 2, 3$), and H_{ij}^e and P_i^e are defined by

$$H_{ij}^e = \int_{p-q} \beta_{pq}^e \psi_i^e \psi_j^e\, ds, \quad P_i^e = \int_{p-q} T_\infty \beta_{pq}^e \psi_i^e\, ds. \qquad (5)$$

Here p-q refers to the side of the element that is exposed to the ambient temperature T_∞. The values of H_{ij}^e and P_i^e are readily available from Eqs. (4.2.29b). Thus, in terms of the element coefficients, we have

$$\begin{aligned}
&(K_{12}^2 + H_{12}^2)U_3 + (K_{22}^2 + H_{22}^2 + K_{11}^3 + H_{11}^3)U_3 + K_{12}^3 U_4 \\
&+ K_{37}U_7 + K_{38}U_8 + K_{39}U_9 = P_2^2 + P_2^3. \qquad (6)
\end{aligned}$$

It is clear that the element coefficients due to advection on the boundary of element Ω_e, H_{ij}^e, are added to K_{ij}^e; similarly, the contribution P_i^e is added to the right-hand side f_i^e of the element equations.

Numerical results obtained (with computer code **FEM2D**) are presented in Table 4.3.2 for four different meshes of rectangular elements in the upper half domain (using the symmetry). Results for the 4 × 2L mesh in the full domain are also included to show that symmetry does exist (one can also see the symmetry from the results of the full domain analysis, i.e., the nodal values above and below the line $y = 0.01$ m for any fixed x are the same). The convergence of the results with mesh refinement is also clear from the results presented in Table 4.3.2.

Table 4.3.2 Nodal temperatures $T(x, 0.02)$ of the cooling fin obtained using various finite element meshes of rectangles (see Fig. 4.3.13 for the geometry and boundary conditions).

	FEM solutions				
	Half domain				Full domain
x	4 × 1L	8 × 1L	16 × 2L	8 × 1Q	4 × 2
0.00	300.00	300.00	300.00	300.00	300.00
0.01	——	227.61	227.56	227.65	——
0.02	176.31	178.53	178.15	178.07	176.31
0.03	——	141.34	141.28	141.26	——
0.04	112.99	113.86	113.95	113.98	112.99
0.05	——	93.84	94.07	94.07	——
0.06	78.57	79.69	79.90	79.98	78.57
0.07	——	70.27	70.50	70.58	——
0.08	63.97	64.86	65.08	65.15	63.97

4.4 Summary

The present chapter has been devoted to a study of: (1) the weak-form Galerkin (or Ritz) finite element models of one- and two-dimensional problems involving generalized Poisson's equation (governing plane and axisymmetric problems) typical of heat transfer, (2) a derivation of interpolation functions for basic one- and two-dimensional elements, and (3) finite element models of unsteady problems of one- and two-dimensional heat transfer. A number of numerical examples have been presented to illustrate the ideas presented in the chapter.

The advantages of the weak-form finite element models of second-order BVPs are: (a) identification of the primary and secondary variables – a duality that exists in all phenomena of nature, (b) weakening the differentiability required of approximation functions used for the primary variables and leading to C^0-approximations, and hence simpler finite elements, and (c) possible symmetry of the finite element coefficient matrix for most problems of engineering physics. In general, the derivation of the weak forms from the governing differential equations is subject to the requirement that the resulting secondary variables be physically meaningful in that they can be specified on the portions of the boundary where the corresponding primary variables are unknown. A disadvantage of the C^0-approximation is that the derivatives of the variables being approximated are discontinuous across interelement boundaries.

The strong point of methods that use global approximations (i.e., no subdivision of the domain into elements, as in the traditional Ritz and Galerkin methods) is that they do not introduce discontinuities in the derivatives of the solution between elements. At the moment there is no clear consensus on the overall effectiveness and robustness of such computational procedures to be competitive with or displace the current weak-form finite element technology. The FVM to be discussed in Chapter 7 provides a competitive alternative to the FEM.

Readers interested in further details on computer implementation of the ideas presented in this chapter may consult the textbook by Reddy [1]. The textbook also contains detailed discussions of programs **FEM1D** and **FEM2D** and illustrative examples on how to prepare the input to these programs. The fortran source codes **FEM1D** and **FEM2D**, developed by the first author, are available from the website http://mechanics.tamu.edu.

Problems

The readers may find additional examples and problems in the textbook by Reddy [1] to test and extend their understanding of the FEM as applied to linear heat transfer problems.

ONE-DIMENSIONAL PROBLEMS

4.1 Consider the differential equation

$$-\frac{d}{dx}\left(a\frac{du}{dx}\right)+\frac{d^2}{dx^2}\left(b\frac{d^2u}{dx^2}\right)+cu=f,$$

where a, b, c, and f are known functions of position x. (a) Develop the weak form over a typical element $\Omega^e=(x_a^e,x_b^e)$ such that the bilinear form is symmetric, (b) identify the bilinear and linear forms and construct the quadratic functional, (c) discuss the type of finite element approximation of u that may be used, and (d) develop the finite element model of the equation.

4.2 Derive the Lagrange cubic interpolation functions for a four-node (one-dimensional) element with equally spaced nodes (see Fig. P4.2) using the alternative procedure based on interpolation properties listed in Eqs. (4.2.23) and (4.2.24). Use the local coordinate $\bar{x}$ for simplicity.

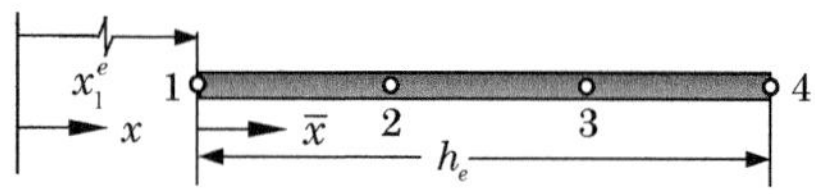

Fig. P4.2

4.3 Derive the finite element model of the differential equation

$$-\frac{d}{dx}\left(a(x)\frac{du}{dx}\right) + cu = f(x) \quad \text{for} \quad 0 < x < L \tag{1}$$

over an element while accounting for the mixed boundary condition of the following form at both ends of the element

$$\left[n_x\left(a(x)\frac{du}{dx}\right) + \beta(u - u_c)\right] = P, \tag{2}$$

where $n_x = -1$ at $x = x_a$ and $n_x = 1$ at $x = x_b$.

4.4 Rewrite Eq. (4.2.1) as a pair of first-order equations

$$-\frac{dv}{dx} + cu - f = 0, \quad \frac{du}{dx} - \frac{v}{a} = 0, \tag{1}$$

and develop the weak-form Galerkin finite element model of the pair of equations keeping both u and v as unknowns and weaken the differentiability on v only.

4.5 The following differential equation arises in connection with heat transfer in a plane wall:

$$-\frac{d}{dx}\left(k\frac{dT}{dx}\right) = g_e \quad \text{for} \quad 0 < x < L$$

$$T(0) = T_0, \quad \left[k\frac{dT}{dx} + \beta(T - T_\infty)\right]\Big|_{x=L} = 0,$$

where T is the temperature, k the thermal conductivity, T_∞ is the ambient temperature at $x = L$, and g_e the internal heat generation. Take the following values for the data: $g_e = 0$, $L = 0.1$ m, $k = 0.01$ W m$^{-1}{}^\circ$C^{-1}, $\beta = 25$ W m$^{-2}{}^\circ$C^{-1}, $T_0 = 50\,^\circ$C, and $T_\infty = 5\,^\circ$C. Solve the problem using two linear finite elements for temperature values at $x = 0.5L$ and $x = L$, and heat at $x = 0$. *Answer:* $T_2 = 27.59\,^\circ$C, $T_3 = 5.179\,^\circ$C, $Q_1^{(1)} = 4.482$ W m^{-2} $= -Q_2^{(2)}$.

4.6 An insulating wall is constructed of three homogeneous layers with conductivities k_1, k_2, and k_3 in intimate contact (see Fig. P4.6). Under steady-state conditions, the temperatures at the boundaries of the layers are characterized by the external surface temperatures T_1 and T_4 and the interface temperatures T_2 and T_3. Formulate the problem to determine the temperatures T_i ($i = 1, 2, 3, 4$) when the ambient temperatures T_0 and T_5 (at the left and right, respectively) and the (surface) film coefficients β_0 and β_5 are known for the following cases. Case (1) $T_0 = T_5 = 20\,^\circ\text{C}$ and $\beta_0 = \beta_5 = 500\ \text{W/m}^2\cdot\text{C}$. Case (2) $T_0 = 120\,^\circ\text{C}$, $T_5 = 20\,^\circ\text{C}$, $\beta_0 = 500\ \text{W/m}^2\cdot\text{C}$, and $\beta_5 = 560\ \text{W/m}^2\cdot\text{C}$. Case (3) $T_1 = 100\,^\circ\text{C}$, $T_5 = 20\,^\circ\text{C}$, and $\beta_5 = 500\ \text{W/m}^2\cdot\text{C}$. Assume that there is no internal heat generation and that the heat flow is one-dimensional.

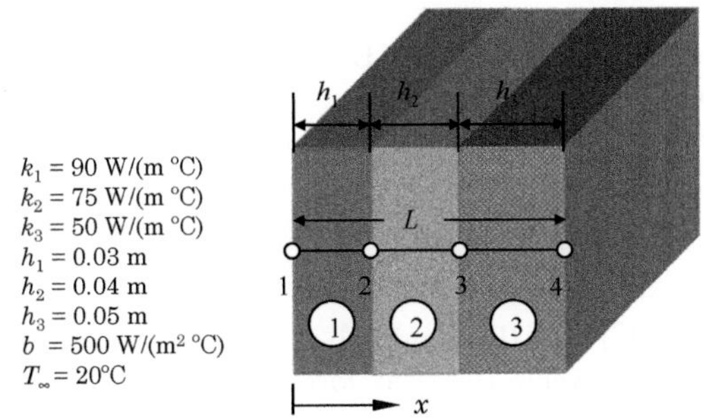

Fig. P4.6

4.7 Consider steady heat conduction in a wire of circular cross-section with an electrical heat source. Suppose that the radius of the wire is R_0, its electrical conductivity is K_e (Ω^{-1}/cm), and it is carrying an electric current density of I (A/cm ^{2}). During the transmission of an electric current, some of the electrical energy is converted into thermal energy. The rate of heat generation per unit volume is given by $g_e = I^2/K_e$. The governing equations of the problem are

$$-\frac{1}{r}\frac{d}{dr}\left(rk\frac{dT}{dr}\right) = g_e \quad \text{for} \quad 0 \le r \le R_0, \qquad \left(rk\frac{dT}{dr}\right)\bigg|_{r=0} = 0, \quad T(R_0) = T_0.$$

Determine the distribution of temperature in the wire using (a) one quadratic element and (b) two quadratic elements, and compare the finite element solution at eight equal intervals with the exact solution (see **Example 4.2.2**). Also, determine the heat flow, $Q = -2\pi R_0 k(dT/dr)|_{R_0}$, at the surface using (i) the temperature field and (ii) the discrete balance equation.

4.8 Consider the steady radial heat flow from the inside to the outside of a thick-walled hollow cylinder of internal radius a and external radius b. The inside wall is maintained at temperature T_a and the outside wall is maintained at T_b, as shown in Fig. P4.8. Determine the temperature distribution in the wall for a uniform mesh of (a) four linear elements and (b) two quadratic elements.

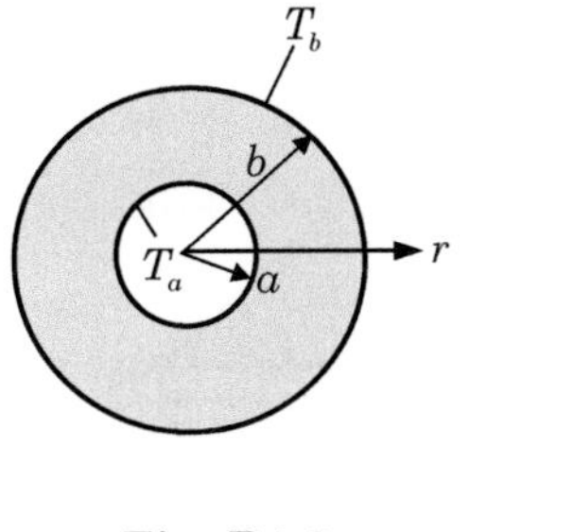

Fig. P4.8

4.9 Solve the advection–diffusion problem of **Example 4.2.3** using (a) 50 linear elements, (b) 100 linear elements (c) 25 quadratic elements, and (d) 50 quadratic elements for $Pe = 100$. Plot and tabulate the results.

Two-Dimensional Problems

4.10 Evaluate the coefficients K_{ij}^e and F_i^e of Eq. (4.3.20) for a linear triangular element when a_{xx}, a_{yy}, f, h_c and u_c are constants.

4.11 Repeat **Problem 4.10** for a linear rectangular element.

4.12 Consider the PDE governing heat transfer in an axisymmetric geometry

$$-\frac{1}{r}\frac{\partial}{\partial r}\left(rk_{rr}\frac{\partial T}{\partial r}\right) - \frac{\partial}{\partial z}\left(k_{zz}\frac{\partial T}{\partial z}\right) = f(r,z), \tag{1}$$

where $(k_{rr},\ k_{zz})$ and f are the conductivities and internal heat generation per unit volume, respectively. Develop the weak form and associated finite element model.

4.13 Consider the steady-state heat transfer governed by the Laplace equation

$$-\left(\frac{\partial^2 u}{\partial x^2} + \frac{\partial^2 u}{\partial y^2}\right) = 0. \tag{1}$$

The boundary conditions for the problem are:

$$u(0,y) = y^2\ ,\ u(x,0) = x^2\ ,\ u(1,y) = 1 - y\ ,\ u(x,1) = 1 - x. \tag{2}$$

Determine the unknown nodal value U_3 using the finite element mesh shown in Fig. P4.13.

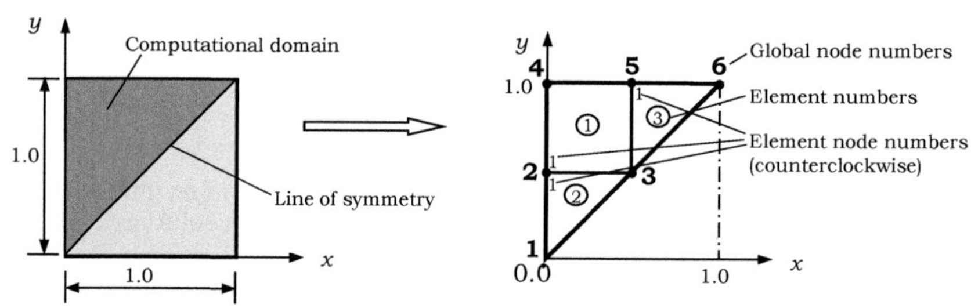

Fig. P4.13

4.14 The nodal temperatures of a linear triangular element (see Fig. P4.14) in the finite-element analysis of the heat transfer problem governed by $-k\nabla^2 u = f_0$ are $u_1^e = U_{10} = 389.79$, $u_2^e = U_{12} = 337.19$, and $u_3^e = U_{11} = 395.08$. Find (a) the gradient of the temperature ∇u in the element, and (b) the places where the $u = 392$ isoline intersects the element boundary.

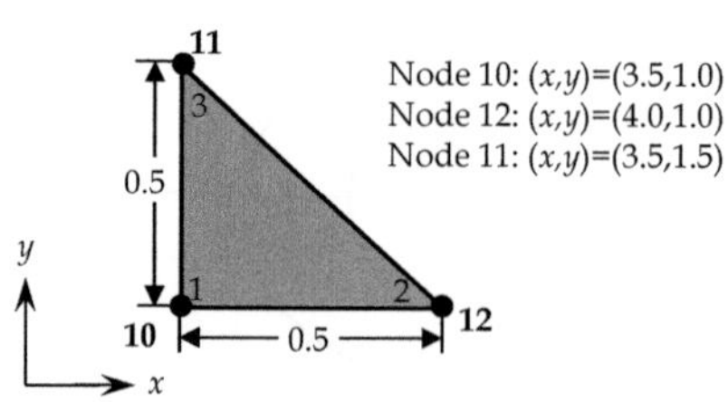

Fig. P4.14

4.15 Consider steady-state heat transfer governed by the differential equation $-k\nabla^2 u = 0$ in the domain shown in Fig. P4.15. Write the global finite element equation associated with global node 8 and identify all of the coefficients (other than unknown U_I) in the equation in terms of the element coefficients K_{ij}^e (algebraic) and problem data only (i.e., you must give the numerical values of other contributions other than K_{ij}^e).

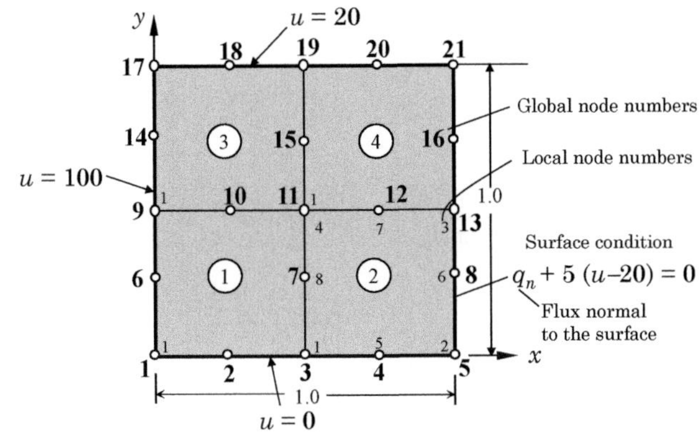

Fig. P4.15

4.16 Consider steady-state heat transfer governed by the differential equation

$$-\frac{\partial}{\partial x}\left(k\frac{\partial u}{\partial x}\right) - \frac{\partial}{\partial y}\left(k\frac{\partial u}{\partial y}\right) = f_0 \tag{1}$$

over the domain shown in Fig. P4.16.

(a) Write the finite element equation associated with global node 1 in terms of element coefficients (include only the nonzero contributions).

(b) Compute the contribution of the flux q_0 to global nodes 1 and 4.

(c) Compute the contribution of the boundary condition $q_n + 5u = 0$ to global node 1 using the coefficients H_{ij}^e and P_i^e.

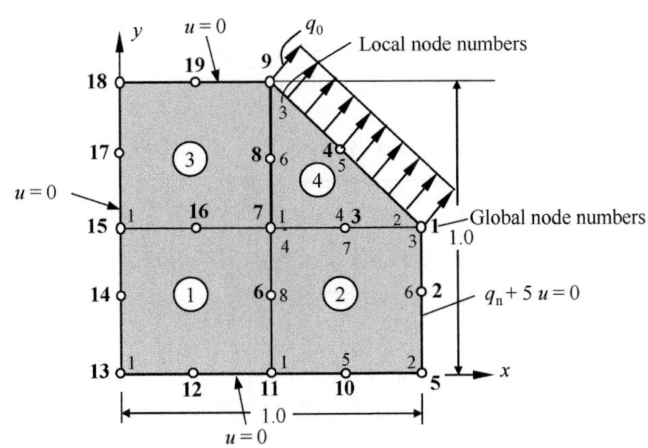

Fig. P4.16

4.17 Consider a *long* circular cylinder whose surface temperature is a function of the angular coordinate θ as sown in Fig. P4.17. The governing equation is given by

$$k\frac{1}{r}\left[\frac{\partial}{\partial r}\left(r\frac{\partial T}{\partial r}\right)+\frac{1}{r}\frac{\partial^2 T}{\partial \theta^2}\right]=0. \tag{1}$$

Determine the temperature field in the interior of the body using the mesh sown. The analytical solution is given by

$$T(r,\theta)=A_0+\sum_{n=1}^{\infty} r^n\left(A_n\sin n\theta+B_n\cos n\theta\right), \tag{2}$$

where

$$A_0=\frac{1}{2\pi}\int_0^{2\pi} f(\theta)d\theta,\quad A_n=\frac{1}{\pi}\int_0^{2\pi} f(\theta)\sin m\theta\, d\theta,\quad B_n=\frac{1}{\pi}\int_0^{2\pi} f(\theta)\cos m\theta\, d\theta. \tag{3}$$

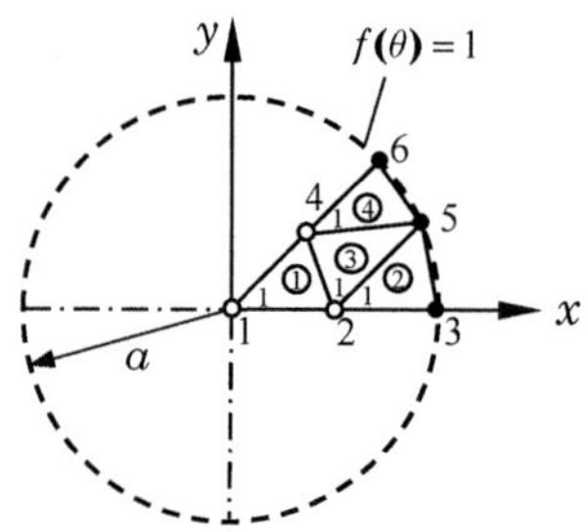

Fig. P4.17

4.18 Consider steady-state heat transfer governed by the differential equation

$$-\frac{\partial}{\partial x}\left(k\frac{\partial T}{\partial x}\right)-\frac{\partial}{\partial y}\left(k\frac{\partial T}{\partial y}\right)=0 \tag{1}$$

over the domain shown in Fig. P4.18. Exploiting the symmetry, analyze the problems using (a) 2×2 mesh of bilinear elements (solid and dashed lines) and (b) 2×2 mesh of linear triangular elements.
The exact solution to the problem is

$$T(x,y)=4k\sum_{n=0}^{\infty}\frac{\sin(\lambda_n x)\,\sinh(\lambda_n y)}{\lambda_n\,\sinh(\lambda_n)},\qquad \lambda_n=(2n+1)\pi. \tag{2}$$

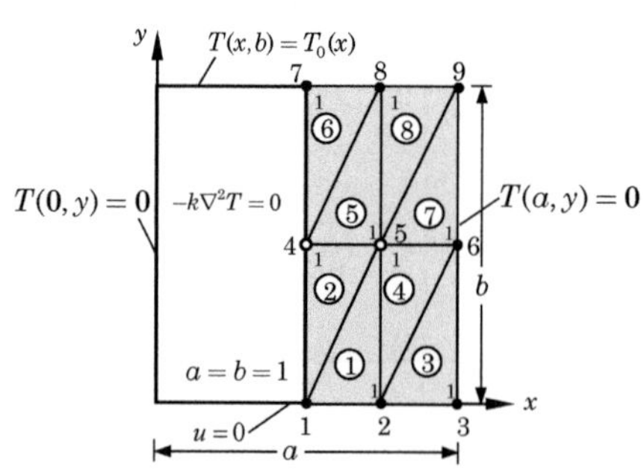

Fig. P4.18

4.19 A series of heating cables has been placed underground, as shown in Fig. P4.19. Assume that the ground (soil) has conductivities of $a_{xx} = k_{xx} = 10$ W/(cm· °C) and $a_{yy} = k_{yy} = 15$ W/(cm· °C), the upper surface is exposed to a temperature of $-5\,^\circ$C [$\beta = 5$ W/(cm^2· K)], and the lower surface has zero heat flow normal to the boundary. Assume that each cable is producing a heat of 250 W/cm. Use a 8×16 mesh of linear triangular (or rectangular) elements in the computational domain (exploit the symmetry available in the problem), and formulate the problem, that is, give (a) element matrices for a typical element, (b) boundary conditions on primary and secondary variables, and (c) advection boundary contributions.

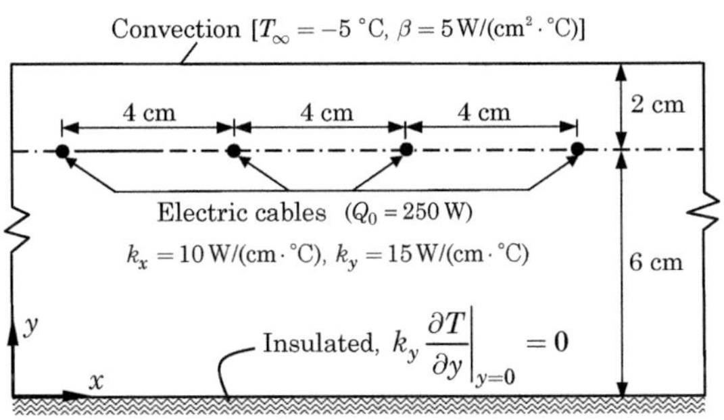

Fig. P4.19

5

The Finite Element Method: Unsteady Heat Transfer

5.1 Introduction

In Chapter 4 we considered finite element analysis of steady state heat transfer. When external stimuli (e.g., boundary conditions and internal heat generation) are independent of time, heat transfer in a medium may attain a steady state; otherwise, the temperature field changes with time (i.e., unsteady state). The governing equations of unsteady heat transfer are obtained using the principle of balance of energy. When unsteady equations are solved the temperature field reaches a steady state if the external stimuli are independent of time (i.e., the time dependence decays with time).

In this chapter, finite element formulation of unsteady heat transfer is considered. The formulation adopted here is decoupled in space and time, where the spatial discretization is carried out first to reduce the PDEs in time and space to ODEs in time only. Then time approximations are used to reduce the ODEs to algebraic equations. We begin with one-dimensional problems, where most concepts of time approximation are introduced (see [1, 6]).

5.2 One-Dimensional Problems

5.2.1 Model Equation

The principle of balance of energy for a one-dimensional heat flow results in an equation of the type

$$c_1\frac{\partial u}{\partial t} - \frac{\partial}{\partial x}\left(a\frac{\partial u}{\partial x}\right) + c_0 u = g(x,t), \quad 0 < x < L, \quad t > 0, \tag{5.2.1}$$

where u denotes the temperature above a reference temperature ($u = T - T_\infty$), $c_1 = \rho c_p A$, $a = kA$, $c_0 = P\beta$, k denotes the thermal conductivity, ρ is the mass density, β is the heat transfer coefficient, A is the cross-sectional area, P is the perimeter, c_p is the specific heat at constant pressure, and g is the internal heat generation per unit length, all of which can, in general, be known functions of position x and time t.

5.2.2 Steps in Finite Element Model Development

The finite element model of transient problems is carried out in two stages:

(1) *Spatial approximation.* In this step the solution u of the equation under consideration is approximated by expressions of the form

$$u(x,t) \approx u_h^e(x,t) = \sum_{j=1}^{n} u_j^e(t)\psi_j^e(x). \tag{5.2.2}$$

This type of approximation is consistent with the concept of "separation of variables" (i.e., time and space dependencies are separated). By taking nodal values to be functions of time, we presume that the nodes have different values of the variable at different times, as illustrated in Fig. 5.2.1). This step involves the weak-form development of the governing equation over the element domain without consideration of time and substitution of approximation (5.2.2) into the weak form. The end result is a set of ODEs, called *semidiscrete finite element model.*

(2) *Time approximation.* Here the system of ODEs in time from Step (1) are further approximated, using suitable finite difference formulae. This step allows conversion of the system of ODEs into a set of algebraic equations among $u_j^e(t_{s+1})$ $(j = 1, 2, \ldots, n)$ at time t_{s+1} $[= (s+1)\Delta t$, when uniform time increment Δt is used]. Here s denotes the time step number, and $s = 0$ corresponds to solution at time $t = 0$.

All time-approximation schemes seek to find the vector of nodal values $\mathbf{u}$ at time t_{s+1} using the known values of $\mathbf{u}$ from the previous times. At the end of the two-stage approximation, one has a spatially continuous solution at discrete intervals of time:

$$u(x,t_s) \approx u_h^e(x,t_s) = \sum_{j=1}^{n} u_j^e(t_s)\psi_j^e(x) \quad (s = 0, 1, \ldots). \tag{5.2.3}$$

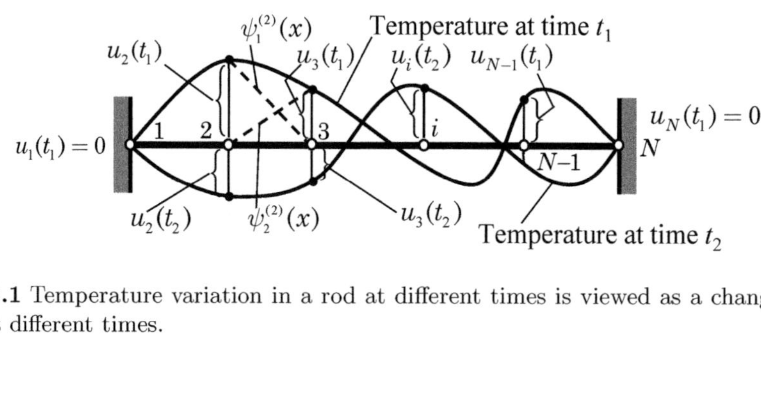

Fig. 5.2.1 Temperature variation in a rod at different times is viewed as a change in nodal values at different times.

5.2.3 Weak Form

We begin with the development of the weak form of Eq. (5.2.1) over an element for any arbitrary time, $t > 0$, followed by the derivation of the semidiscrete finite element model using the approximation of the form in Eq. (5.2.2). Following the three-step procedure of constructing the weak form of a differential equation (all of the terms in the differential equation have already been considered), we obtain

$$\begin{aligned} 0 &= \int_{x_a^e}^{x_b^e} \left[a \frac{dw_i}{dx} \frac{\partial u}{\partial x} + w_i \left(c_0 u + c_1 \frac{\partial u}{\partial t} - g \right) \right] dx + \left[-w_i \cdot a \frac{\partial u}{\partial x} \right]_{x_a^e}^{x_b^e} \\ &= \int_{x_a^e}^{x_b^e} \left[a \frac{dw_i}{dx} \frac{\partial u}{\partial x} + w_i \left(c_0 u + c_1 \frac{\partial u}{\partial t} - g \right) \right] dx \\ &\quad - Q_1^e w_i(x_a^e) - Q_2^e w_i(x_b^e), \end{aligned} \tag{5.2.4a}$$

where $w_i(x)$ is the weight function, and

$$Q_1^e = \left[-a \frac{\partial u}{\partial x} \right]_{x_a^e}, \quad Q_2^e = - \left[-a \frac{\partial u}{\partial x} \right]_{x_b^e} \tag{5.2.4b}$$

are the heats at the ends of the element.

5.2.4 Semidiscrete Finite Element Model

Substituting $w_i = \psi_i^e(x)$ (to obtain the ith equation of the system) and Eq. (5.2.2) into Eq. (5.2.4a), we obtain

$$\begin{aligned} 0 &= \sum_{j=1}^{n} u_j \int_{x_a^e}^{x_b^e} \left(a \frac{d\psi_i^e}{dx} \frac{d\psi_j^e}{dx} + c_0 \psi_i^e \psi_j^e \right) dx + \sum_{j=1}^{n} \frac{du_j^e}{dt} \int_{x_a^e}^{x_b^e} c_1 \psi_i^e \psi_j^e \, dx \\ &\quad - \int_{x_a^e}^{x_b^e} \psi_i^e g \, dx - Q_1 \psi_i^e(x_a^e) - Q_2 \psi_i^e(x_b^e) \\ &= \sum_{j=1}^{n} \left(K_{ij}^e u_j^e + C_{ij}^e \frac{du_j^e}{dt} \right) - F_i^e, \end{aligned} \tag{5.2.5a}$$

where

$$\begin{aligned} K_{ij}^e &= \int_{x_a^e}^{x_b^e} \left(a \, \frac{d\psi_i^e}{dx} \frac{d\psi_j^e}{dx} + c_0 \, \psi_i^e \psi_j^e \right) dx \\ f_i^e &= \int_{x_a^e}^{x_b^e} \psi_i^e g \, dx, \quad C_{ij}^e = \int_{x_a^e}^{x_b^e} c_1 \, \psi_i^e \psi_j^e \, dx \\ Q_i^e &= Q_1^e \psi_i(x_a^e) + Q_2^e \psi_i^e(x_b^e). \end{aligned} \tag{5.2.5b}$$

In matrix form, Eq. (5.2.5a) takes the form

$$\mathbf{K}^e \mathbf{u}^e + \mathbf{C}^e \dot{\mathbf{u}}^e = \mathbf{F}^e, \tag{5.2.6}$$

where the superposed dot indicates differentiation with respect to time, $\mathbf{F}^e = \mathbf{f}^e + \mathbf{Q}^e$.

We note that the matrix coefficients C_{ij}^e are the same as those in the second part of the coefficients, with c_0 replaced by c_1. In other words, for a constant value of c_1, the element matrices for linear and quadratic elements are available from Eq. (4.2.29b) and (4.2.30b), respectively:

$$\mathbf{C}_L^e = \frac{c_1^e h_e}{6} \begin{bmatrix} 2 & 1 \\ 1 & 2 \end{bmatrix}, \quad \mathbf{C}_Q^e = \frac{c_1^e h_e}{30} \begin{bmatrix} 4 & 2 & -1 \\ 2 & 16 & 2 \\ -1 & 2 & 4 \end{bmatrix}. \tag{5.2.7}$$

5.2.5 Time Approximations

5.2.5.1 Alpha-Family of Approximation

To obtain a set of algebraic equations from the semidiscrete finite element model (5.2.6), we use finite difference representation of the time derivatives. We use a first-order differential equation for a representative variable u to discuss the time approximation. Then we generalize the development to matrix equations in (5.2.6) for vector $\mathbf{u}$.

Consider the first-order differential equation

$$a\frac{du}{dt} + cu = g(t), \quad 0 < t < T \quad \text{and} \quad u(0) = u_0, \tag{5.2.8}$$

where $a \neq 0$, c, and u_0 are constants (i.e., not function of t), and g is a function of time t. The exact solution of the problem consists of two parts: the homogeneous and particular solutions. The homogeneous solution is

$$u^h(t) = Ae^{-kt}, \quad k = \frac{c}{a}, \tag{5.2.9a}$$

where A is constant of integration. The particular solution is given by

$$u^p(t) = \frac{1}{a}e^{-kt}\left(\int_0^t e^{k\tau} g_i(\tau)\, d\tau\right). \tag{5.2.9b}$$

The complete solution is given by

$$u(t) = e^{-kt}\left(A + \frac{1}{a}\int_0^t e^{k\tau} g_i(\tau)\, d\tau\right). \tag{5.2.9c}$$

The Taylor series expansions of function $u(t)$ about $t = t_s$ and $t = t_{s+1}$ are given by

$$u(t) = u(t_s) + (t - t_s)\dot{u}(t_s) + \tfrac{1}{2!}(t - t_s)^2\ddot{u}(t_s) + \cdots \tag{5.2.10a}$$
$$u(t) = u(t_{s+1}) + (t - t_{s+1})\dot{u}(t_{s+1}) + \tfrac{1}{2!}(t - t_{s+1})^2\ddot{u}(t_{s+1}) + \cdots. \tag{5.2.10b}$$

In particular, for $t = t_{s+1}$ in Eq. (5.2.10a), we have

$$u(t_{s+1}) = u(t_s) + (t_{s+1} - t_s)\dot{u}(t_s) + \tfrac{1}{2!}(t_{s+1} - t_s)^2\ddot{u}(t_s) + \tfrac{1}{3!}(t_{s+1} - t_s)^3\dddot{u}(t_s) + \cdots.$$

If we truncate the series after the second term and solve for $\dot{u}(t_s)$, we obtain

$$\dot{u}(t_s) = \frac{u(t_{s+1}) - u(t_s)}{t_{s+1} - t_s} + O(\Delta t_{s+1}). \tag{5.2.11a}$$

The accuracy of the approximation is said to be of the order $O(\Delta t_{s+1})$, where $\Delta t_{s+1} = t_{s+1} - t_s$. Thus we have

$$\dot{u}(t_s) \approx \frac{u(t_{s+1}) - u(t_s)}{t_{s+1} - t_s} \quad \text{or} \quad \dot{u}^s \approx \frac{u^{s+1} - u^s}{\Delta t_{s+1}}, \tag{5.2.11b}$$

which is known as the *forward difference* or *Euler's* scheme. The name forward difference indicates that the slope at time $t = t_s$ is obtained using the difference in the value of u at the next (forward) time $t = t_{s+1}$ and the current $t = t_s$ (see Fig. 5.2.2). Next let $t = t_s$ in Eq. (5.2.11b). We arrive at

$$\dot{u}(t_{s+1}) \approx \frac{u(t_{s+1}) - u(t_s)}{t_{s+1} - t_s} \quad \text{or} \quad \dot{u}^{s+1} \approx \frac{u^{s+1} - u^s}{\Delta t_{s+1}}, \tag{5.2.11c}$$

which is known, for the obvious reason, as the *backward difference* scheme (see Fig. 5.2.2).

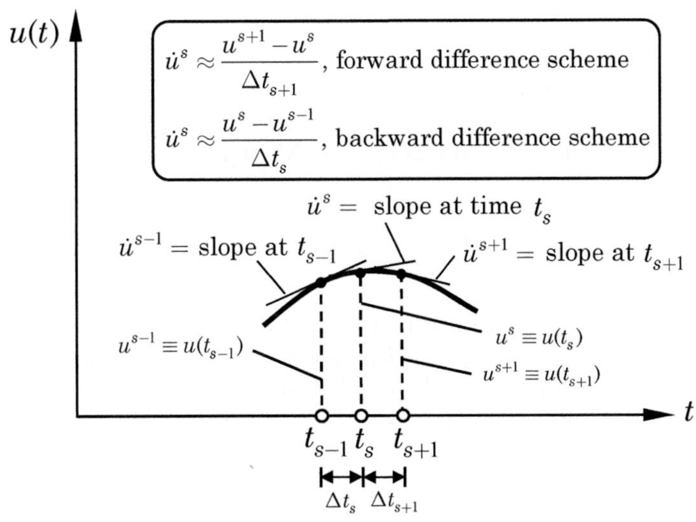

Fig. 5.2.2 Approximation of the time derivative of a function, u, using forward and backward difference schemes.

The forward and backward difference schemes can be put into a single scheme with the help of a parameter, α, $0 \leq \alpha \leq 1$. The two schemes in Eqs. (5.2.11b) and (5.2.11c) can be expressed in a single equation as

$$(1 - \alpha)\dot{u}^s + \alpha\,\dot{u}^{s+1} = \frac{u^{s+1} - u^s}{\Delta t_{s+1}} \quad \text{for } 0 \leq \alpha \leq 1, \tag{5.2.12}$$

where u^s denotes the value of $u(t)$ at time $t = t_s = \sum_{i=1}^{s} \Delta t_i$, and $\Delta t_{s+1} = t_{s+1} - t_s$ is the $(s+1)$st time increment. The scheme in Eq. (5.2.12) is known as the *alpha-family of approximation*, in which a weighted average of the time derivatives of u at two consecutive time steps is approximated by linear interpolation of the values of the variable at the two different times (see Fig. 5.2.2). When we set $\alpha = 0$ in Eq. (5.2.12), we obtain the forward difference scheme, and when $\alpha = 1$, we obtain the backward difference scheme. For $\alpha = 1/2$, the scheme in Eq. (5.2.12) is known as the *Crank–Nicolson scheme* [2]. If the total time $(0, T]$ is divided into equal time steps, say Δt, then $t_s = s\Delta t$. In the rest of the book, we assume that the time steps are of uniform size. Equation (5.2.12) can be expressed as

$$u^{s+1} = u^s + \Delta t\left[(1-\alpha)\dot{u}^s + \alpha\,\dot{u}^{s+1}\right], \quad \text{for } 0 \leq \alpha \leq 1. \tag{5.2.13}$$

Returning to Eq. (5.2.8), we note that it is valid for all times $t > 0$. In particular, it is valid at times $t = t_s$ and $t = t_{s+1}$. Hence, from Eq. (5.2.8) we have

$$\dot{u}^s = \frac{1}{a}\left(g^s - cu^s\right), \quad \dot{u}^{s+1} = \frac{1}{a}\left(g^{s+1} - cu^{s+1}\right). \tag{5.2.14}$$

Substituting the expressions from Eq. (5.2.14) into Eq. (5.2.13), we arrive at

$$(1-\alpha)\left(g^s - cu^s\right) + \alpha\left(g^{s+1} - cu^{s+1}\right) = a\left(\frac{u^{s+1} - u^s}{\Delta t}\right).$$

Solving for u^{s+1}, we obtain

$$\left[a + \alpha\,\Delta t\, c\right] u^{s+1} = \left[a - (1-\alpha)\,\Delta t\, c\right] u^s + \Delta t\left[\alpha\, g^{s+1} + (1-\alpha)g^s\right], \tag{5.2.15a}$$

or

$$u^{s+1} = \frac{a - (1-\alpha)\,\Delta t\, c}{a + \alpha\,\Delta t\, c}\, u^s + \Delta t\frac{\alpha\, g^{s+1} + (1-\alpha)g^s}{a + \alpha\,\Delta t\, c}. \tag{5.2.15b}$$

Thus, Eq. (5.2.15b) can be used repeatedly (i.e., march in time) to obtain the solution at times $t = t_{s+1}, t_{s+2}, \ldots, t_N$, where N is the number of time steps required to reach the final time T (or the solution reaches steady state). At the very beginning, that is, when $s = 0$, the solution u^1 is calculated using the initial value u^0:

$$u^1 = \frac{a - (1-\alpha)\,\Delta t\, c}{a + \alpha\,\Delta t\, c}\, u^0 + \Delta t\,\frac{\alpha\, g^1 + (1-\alpha)g^0}{a + \alpha\,\Delta t\, c}. \tag{5.2.16}$$

5.2.5.2 Numerical Stability

Since the time-marching scheme in Eq. (5.2.15b) uses the previous time step solution u^s, which itself is an approximate solution, there is a possibility that the error introduced in u^s may be amplified when u^{s+1} is computed, and the error may grow unboundedly with time. When the error grows without bounds, the underlying scheme is said to be *unstable.* Here we wish to determine the conditions under which the error remains bounded.

Consider Eq.(5.2.15b) in operator form

$$u^{s+1} = Bu^s + \bar{F}^{s,s+1}, \tag{5.2.17a}$$

where

$$B = \frac{a - (1-\alpha)\,\Delta t\, c}{a + \alpha\,\Delta t\, c}, \quad \bar{F}_{s,s+1} = \Delta t\, \frac{\alpha\, g^{s+1} + (1-\alpha)\, g^s}{a + \alpha\,\Delta t\, c}. \tag{5.2.17b}$$

If the magnitude of the operator B, known as the *amplification operator*, is greater than 1, $|B| > 1$, the error will be amplified during each time step. On the other hand, if the magnitude is equal to or less than unity, the error will not to grow with time. Therefore, in order for the scheme to be stable it is necessary that $|B| \leq 1$:

$$|B| = \left| \frac{a - (1-\alpha)\,\Delta t\, c}{a + \alpha\,\Delta t\, c} \right| \leq 1. \tag{5.2.18}$$

The above equation places a restriction on the magnitude of the time step for certain values of α. When the error remains bounded for any time step [i.e., condition (5.2.18) is trivially satisfied for any value of Δt], it is known as a *stable* scheme. If the error remains bounded only when the time step remains below a certain value [in order to satisfy Eq. (5.2.18)], it is said to be a *conditionally stable* scheme. In the FEM, the coefficients a and c appearing in Eq. (5.2.18) depend on problem material parameters (which cannot be changed) and the element length (which can be selected). Hence, for a given mesh there is a value of Δt that makes the scheme conditionally stable (more details on this topic can be found in the book by Surana and Reddy [6]).

5.2.5.3 Consistency and Convergence

In addition to the truncation error introduced in approximating the time derivative, round-off errors can be introduced because of the finite arithmetic used in the computations. The numerical scheme (5.2.17a) is said to be *consistent* with the continuous problem (5.2.8) if the round-off and truncation errors go to zero as $\Delta t \to 0$. *Accuracy* of a numerical scheme is a measure of the closeness between the approximate solution and the exact solution, whereas the stability of a solution is a measure of the boundedness of the approximate solution with time. As one might expect, the size of the time step as well as the mesh can influence both accuracy and stability. When we construct an approximate solution, we like it to converge to the true solution when the number of elements or the degree of approximation is increased and the time step Δt is decreased. A time-approximation scheme is said to be *convergent* if, for fixed t_s and Δt, the numerical value u^s converges to its true value $u(t_s)$ as $\Delta t \to 0$. Accuracy is measured in terms of the rate at which the approximate solution converges. If a numerical scheme is both stable and consistent, it is also convergent.

5.2.6 Fully Discretized Finite Element Equations

We now have the tools necessary to convert the set of ODEs in Eq. (5.2.6) to a set of algebraic equations, much the same way a single differential equation

(5.2.8) was converted to the algebraic equation (5.2.15b). As applied to a vector of time derivatives of the nodal values, $\dot{\mathbf{u}}$, the alpha-family of approximation in Eq. (5.2.13) takes the form

$$\mathbf{u}^{s+1} = \mathbf{u}^s + a_2\dot{\mathbf{u}}^s + a_1\dot{\mathbf{u}}^{s+1} \quad \text{for } 0 \leq \alpha \leq 1 \tag{5.2.19a}$$

$$a_1 = \alpha\,\Delta t, \quad a_2 = (1-\alpha)\Delta t. \tag{5.2.19b}$$

Evaluating Eq. (5.2.19a) for times $t = t_s$ and $t = t_{s+1}$, we have

$$\mathbf{C}\dot{\mathbf{u}}^s + \mathbf{K}\mathbf{u}^s = \mathbf{F}^s \tag{5.2.20a}$$

$$\mathbf{C}\dot{\mathbf{u}}^{s+1} + \mathbf{K}\mathbf{u}^{s+1} = \mathbf{F}^{s+1}, \tag{5.2.20b}$$

where it is assumed that the matrices $\mathbf{C}$ is independent of time. Premultiplying both sides of Eq. (5.2.19a) with $\mathbf{C}$, we obtain

$$\mathbf{C}(\mathbf{u}^{s+1} - \mathbf{u}^s) = a_1\mathbf{C}\dot{\mathbf{u}}^{s+1} + a_2\mathbf{C}\dot{\mathbf{u}}^s. \tag{5.2.20c}$$

Substituting for $\mathbf{C}\dot{\mathbf{u}}^s$ and $\mathbf{C}\dot{\mathbf{u}}^{s+1}$ from Eqs. (5.2.20a) and (5.2.20b), respectively, we arrive at

$$\mathbf{C}(\mathbf{u}^{s+1} - \mathbf{u}^s) = a_1\left(\mathbf{F}^{s+1} - \mathbf{K}\mathbf{u}^{s+1}\right) + a_2\left(\mathbf{F}^s - \mathbf{K}\mathbf{u}^s\right). \tag{5.2.21}$$

Solving for the vector $\mathbf{u}^{s+1}$, we obtain the recursive relation (i.e., a relation between solutions at two consecutive time steps)

$$\hat{\mathbf{K}}\mathbf{u}^{s+1} = \bar{\mathbf{K}}\mathbf{u}^s + \bar{\mathbf{F}}^{s,s+1} \equiv \hat{\mathbf{F}}^{s,s+1}, \tag{5.2.22a}$$

where

$$\hat{\mathbf{K}} = \mathbf{C} + a_1\mathbf{K}, \quad \bar{\mathbf{K}} = \mathbf{C} - a_2\mathbf{K}$$
$$\bar{\mathbf{F}}^{s,s+1} = \left(a_1\mathbf{F}^{s+1} + a_2\mathbf{F}^s\right). \tag{5.2.22b}$$

Equations (5.2.22a) and (5.2.22b) are valid for a typical finite element, they hold for any problem, independent of the dimension and method of spatial approximation, as long as the end result is Eq. (5.2.6). The assembly, imposition of boundary conditions, and solution of the assembled equations are the same as described before for steady state problems. Calculation of $\hat{\mathbf{K}}$ and $\bar{\mathbf{F}}$ at time $t = 0$ requires knowledge of the initial conditions $\mathbf{u}^0$ and the time variation of $\mathbf{F}$.

All numerical schemes with $\alpha \geq \frac{1}{2}$ (e.g., backward difference and Crank–Nicolson schemes) are stable independent of the mesh. Numerical schemes obtained with $\alpha < \frac{1}{2}$ (e.g., the forward difference method), the alpha-family of approximations is stable only if the time step satisfies the following (stability) condition (see Surana and Reddy [6] for details):

$$\Delta t < \Delta t_{cr} \equiv \frac{2}{(1-2\alpha)\lambda_{max}}, \quad \alpha < \frac{1}{2}, \tag{5.2.23}$$

where λ_{max} is the largest eigenvalue of the associated eigenvalue problem (i.e., assume $\mathbf{u}(t) = \mathbf{u}_0 e^{-\lambda t}$ and $\mathbf{Q}(t) = \mathbf{Q}_0 e^{-\lambda t}$ in Eq. (5.2.6)):

$$(-\lambda \mathbf{C} + \mathbf{K})\, \mathbf{u}_0 = \mathbf{Q}_0, \tag{5.2.24}$$

where $\mathbf{u}_0$ and $\mathbf{Q}_0$ are the amplitudes of the solution and the right-hand side, respectively,

$$\mathbf{u} = \mathbf{u}_0 e^{-\lambda t}, \quad \mathbf{Q} = \mathbf{Q}_0 e^{-\lambda t}, \tag{5.2.25}$$

and the λ is the rate of decay of the solution with time when the system is free of any external source (i.e., after assembly and imposition of homogeneous boundary conditions, the balance of $\mathbf{Q}_0$ results in a zero right-hand side of the eigenvalue problem).

The stability and accuracy characteristics of the following well-known time-approximation schemes should be noted:

$$\alpha = \begin{cases} 0, & \text{the forward difference scheme (conditionally stable),} \\ & \text{order of accuracy } = O(\Delta t); \\ \frac{1}{2}, & \text{the Crank–Nicolson scheme (stable),} \\ & \text{order of accuracy } = O(\Delta t)^2; \\ 1, & \text{the backward difference scheme (stable),} \\ & \text{order of accuracy } = O(\Delta t). \end{cases} \tag{5.2.26}$$

Of these schemes, the Crank–Nicolson scheme ($\alpha = 0.5$) is the most commonly used stable scheme [39].

5.3 Two-Dimensional Problems

5.3.1 Model Equation

For unsteady case, the model equation (4.3.1) in a two-dimensional region Ω with total boundary Γ becomes

$$c_1 \frac{\partial u}{\partial t} - \frac{\partial}{\partial x}\left(a_{xx}\frac{\partial u}{\partial x}\right) - \frac{\partial}{\partial y}\left(a_{yy}\frac{\partial u}{\partial y}\right) = g(x,y,t), \quad (x,y) \text{ in } \Omega \text{ for } t > 0. \tag{5.3.1}$$

The boundary conditions (on different parts of Γ) are of the form given in Eqs. (4.3.2a) and (4.3.2b). The initial condition is of the form

$$u(x,y,0) = u_0(x,y) \quad \text{in } \Omega, \tag{5.3.2}$$

where u_0 is the initial (i.e., at time $t = 0$) value of u, which must be known for all points of the domain. The finite element model development follows the same steps as for the steady state problems, but in two stages as described for one-dimensional problems.

5.3.2 Weak Form

The weak form of Eq. (5.3.1) over an element Ω_e is obtained by the procedure followed in Section 3.3: multiply Eq. (5.3.1) with a weight function $w_i(x,y)$ and integrate over the element Ω_e (not in time), integrate-by-parts (only with respect to x and y) those terms which have the second-order spatial derivatives, and replace the coefficient of the weight function (namely, the secondary variable, q_n) in the boundary integral with $q_n = \hat{q}_n - \beta(u - u_\infty)$. The expressions containing c_1 and f are alone in this process. We obtain

$$0 = \int_{\Omega_e} \left[w_i \left(c_1 \frac{\partial u}{\partial t} + f \right) + a_{xx} \frac{\partial w}{\partial x}\frac{\partial u}{\partial x} + a_{yy} \frac{\partial w}{\partial y}\frac{\partial u}{\partial y} \right] dx\, dy$$
$$+ \oint_{\Gamma_e} \beta u w \, ds - \oint_{\Gamma_e} (\beta u_\infty + \hat{q}_n)\, w\, ds, \tag{5.3.3}$$

where the circle on the line integrals indicates that the integration is to be carried on the entire boundary Γ_e of the element Ω_e, independent of the geometry of the element.

5.3.3 Semidiscrete Finite Element Model

The *semidiscrete* finite element model is obtained from Eq. (5.3.3) by substituting a finite element approximation for the dependent variable, u:

$$u(x,y,t) \approx u_h(x,y,t) = \sum_{j=1}^{n} u_j^e(t)\psi_j^e(x,y), \tag{5.3.4}$$

where u_j^e denotes the value of $u(x,y,t)$ at the nodal location (x_j^e, y_j^e) in the element at time t. The ith differential equation (in time) of the weak-form Galerkin finite element model is obtained by substituting $w_i = \psi_i^e(x,y)$ and substituting for u from Eq. (5.3.4) into Eq. (5.3.3):

$$0 = \sum_{j=1}^{n} \left(C_{ij}^e \frac{du_j^e}{dt} + (K_{ij}^e + H_{ij}^e)\, u_j^e \right) - f_i^e - Q_i^e - P_i^e, \tag{5.3.5a}$$

where

$$C_{ij}^e = \int_{\Omega_e} c_1 \psi_i^e \psi_j^e \, dxdy, \quad M_{ij}^e = \int_{\Omega_e} c\psi_i^e \psi_j^e \, dx\, dy$$
$$K_{ij}^e = \int_{\Omega_e} \left(a_{11} \frac{\partial \psi_i^e}{\partial x}\frac{\partial \psi_j^e}{\partial x} + a_{22} \frac{\partial \psi_i^e}{\partial y}\frac{\partial \psi_j^e}{\partial y} \right) dxdy,$$
$$f_i^e = \int_{\Omega_e} g\psi_i^e \, dxdy, \qquad Q_i^e = \oint_{\Gamma_e} q_n \psi_i^e \, ds$$
$$H_{ij}^e = \oint_{\Gamma_e} \beta \psi_i^e \psi_j^e \, ds, \qquad P_i^e = \oint_{\Gamma_e} \beta u_\infty \psi_i^e \, ds.$$

We note that $\hat{q}_n$ now denotes the external flux normal to the element boundary. In matrix form, Eq. (5.3.5a), which represents a set of ordinary differential equations in time, becomes

$$\mathbf{C}^e\dot{\mathbf{u}} + (\mathbf{K}^e + \mathbf{H}^e)\,\mathbf{u}^e = \mathbf{f}^e + \mathbf{Q}^e + \mathbf{P}^e \equiv \mathbf{F}^e. \tag{5.3.6}$$

5.3.4 Fully Discretized Model

In view of the detailed discussion on how to convert equations of the form (5.3.6) using the alpha-family of approximations in Section 5.2.3, we immediately obtain the following fully discretized finite element model:

$$\hat{\mathbf{K}}\mathbf{u}^{s+1} = \hat{\mathbf{F}}^{s,s+1}, \tag{5.3.7a}$$

where

$$\begin{aligned}\hat{\mathbf{K}} &= \mathbf{C} + a_1\bar{\mathbf{K}},\\ \hat{\mathbf{F}}^{s,s+1} &= \left(\mathbf{C} - a_2\bar{\mathbf{K}}\right)\mathbf{u}^s + \left(a_1\mathbf{F}^{s+1} + a_2\mathbf{F}^s\right),\end{aligned} \tag{5.3.7b}$$

and

$$a_1 = \alpha\Delta t, \qquad a_2 = (1-\alpha)\Delta t. \tag{5.3.7c}$$

Equation (5.3.7a), after assembly and imposition of boundary conditions, is solved at each time step for the nodal values u_j^{s+1} at time $t_{s+1} = (s+1)\Delta t$. At time $t = 0$ (i.e., $s = 0$), $\hat{\mathbf{F}}$ is computed using the initial values $\mathbf{u}^0$ and the known source vectors $\mathbf{f}$ and $\mathbf{P}$, and known flux vector $\mathbf{Q}$.

5.4 Explicit and Implicit Formulations and Mass Lumping

The solution of the fully discretized equations (after assembly and imposition of boundary and initial conditions) requires the inversion of $\hat{\mathbf{K}}$ appearing in Eqs. (5.3.7a) and (5.3.7b) to march forward in time and find the solution at different times. This can be an enormous computational expense, depending on the size of the mesh and the number of time steps. For example, if one needs to solve these equations for 100 time steps, the cost is equivalent to solving 100 steady state problems. Thus, it is of practical interest to find ways to reduce the computational cost. It is clear that if the element coefficient matrix $\hat{\mathbf{K}}^e$ were a diagonal matrix, then the assembled or global coefficient matrix $\hat{\mathbf{K}}$ would be diagonal, and there is no inversion required to solve for the values of the temperature at the Ith global node, U_I^{s+1} (i.e., simply divide each equation with the diagonal element):

$$U_I^{s+1} = \frac{1}{\hat{K}_{(II)}}\left(\sum_{J=1}^{NEQ} \bar{K}_{IJ}^s U_J^s + \hat{F}_I^{s,s+1}\right) \quad \text{(no sum on } I\text{)}. \tag{5.4.1}$$

Formulations that require the inversion of $\hat{\mathbf{K}}$ (because it is not diagonal) are termed *implicit formulations* and those in which no inversion is required are called *explicit formulations*.

In the FEM, *no* time-approximation scheme results in a diagonal matrix $\hat{\mathbf{K}}$ because the matrix $\mathbf{C}$ appearing in $\hat{\mathbf{K}}$ is not diagonal by definition (or consistent with the formulation), unless the interpolation functions ψ_i^e are orthogonal. A matrix $\mathbf{C}$ computed according to the definition is called a *consistent (mass) matrix*. In real-world transient analysis, where hundreds of thousands of degrees of freedom are involved, we must find ways to reduce the cost of solving the equations (which is by far the most time-consuming part of the analysis). We note that even when $\mathbf{C}$ is diagonal, $\hat{\mathbf{K}}$ is not diagonal because of $\mathbf{K}$, unless $\alpha = 0$. Thus, the speed of computation can be increased enormously if (a) we select $\alpha = 0$ (forward difference scheme) and (2) diagonalize $\mathbf{C}$ so that the formulation becomes explicit.

There are several ways of constructing diagonal mass matrices by lumping the mass at the nodes, while preserving the total mass. Two such approaches are used. In *row-sum lumping*, the sum of the elements of each row of the consistent (mass) matrix $\mathbf{C}$ is used as the diagonal element and the off-diagonal elements are set to zero:

$$C_{(ii)}^e = \sum_{j=1}^{n} \int_{x_a^e}^{x_b^e} \rho \psi_i^e \psi_j^e \, dx = \int_{x_a^e}^{x_b^e} \rho \psi_i^e \, dx, \tag{5.4.2}$$

where the property $\sum_{j=1}^{n} \psi_j^e = 1$ of the interpolation functions is used and (ii) implies no summation on i. For example, when c_1 is element-wise constant, the consistent matrices associated with the linear and quadratic one-dimensional elements are

$$\mathbf{C}_c^e = \frac{c_1^e h_e}{6}\begin{bmatrix} 2 & 1 \\ 1 & 2 \end{bmatrix}, \quad \mathbf{C}_c^e = \frac{c_1^e h_e}{30}\begin{bmatrix} 4 & 2 & -1 \\ 2 & 16 & 2 \\ -1 & 2 & 4 \end{bmatrix}. \tag{5.4.3a}$$

From Eq. (5.4.1), the associated diagonal matrices for the linear and quadratic one-dimensional elements are:

$$\mathbf{C}_d^e = \frac{c_1^e h_e}{2}\begin{bmatrix} 1 & 0 \\ 0 & 1 \end{bmatrix}, \quad \mathbf{M}_L^e = \frac{c_1^e h_e}{6}\begin{bmatrix} 1 & 0 & 0 \\ 0 & 4 & 0 \\ 0 & 0 & 1 \end{bmatrix}. \tag{5.4.3b}$$

Here, subscripts d and c refer to diagonalized and consistent mass matrices, respectively.

The second method is known as *proportional lumping*; the diagonal elements of the lumped matrix are computed to be proportional to the diagonal elements of the consistent matrix while conserving the total mass of the element [(ii) means no sum on i]:

$$C_{(ii)}^e = \alpha^e \int_{x_a^e}^{x_b^e} c_1^e \psi_i^e \psi_i^e \, dx, \quad \alpha^e = \frac{\int_{x_a^e}^{x_b^e} m_e \, dx}{\sum_{i=1}^{n} \int_{x_a^e}^{x_b^e} c_1^e \psi_i^e \psi_i^e \, dx}, \tag{5.4.4}$$

where c_1^e is the property $\rho_e c_v^e A_e$ per unit length. For constant c_1^e, in the case of linear element we have

$$\alpha^e = \frac{c_1^e h_e}{\left(\frac{c_1^e h_e}{2} + \frac{c_1^e h_e}{2}\right)} = 1; \quad C_{11}^e = C_{22}^e = \frac{c_1^e h_e}{2}.$$

Thus, the proportional lumping gives the same lumped mass matrices as those obtained in the row-sum technique.

The use of a lumped mass matrix in transient analysis can save computational time in two ways. First, for a forward difference scheme, lumped (mass) matrix results in explicit algebraic equations, not requiring matrix inversions. Second, the critical time step required with the lumped matrix is larger than that required with the consistent matrix because the eigenvalues associated with lumped matrices are smaller than those obtained with the consistent matrix. Although explicit formulations are computationally less expensive compared to implicit formulations, they are less accurate in predicting the transient response because of (mass) lumping of a continuous property. Implicit formulations are more accurate but have smaller critical time steps (for conditionally stable schemes).

5.5 Numerical Examples

5.5.1 One-Dimensional Problems

In this section we consider two examples to illustrate the ideas presented here to solve one-dimensional transient heat transfer problems using the FEM (and program **FEHEAT1D**). The first example is concerned with heat transfer in a plane wall (whose dimensions into the plane of the paper are very large compared with the wall thickness L) and the second example deals with an axisymmetric heat transfer in a *long* cylinder.

Example 5.5.1

Suppose a plane wall, initially at temperature T_0, is suddenly subjected on the left side to temperature T_∞ (see Fig. 5.5.1). Determine the time history of the temperature field in the wall when the right side is: (a) subjected to temperature T_∞, (b) insulated, or (b) exposed to ambient temperature, T_∞ (i.e., solve the problem for three different sets of boundary conditions).

Solution The governing equation of the problem is given by [a special case of Eq. (5.2.1) with $A = 1$ and internal heat generation $g = 0$]:

$$\rho c_p \frac{\partial T}{\partial t} - \frac{\partial}{\partial x}\left(k \frac{\partial u}{\partial x}\right) = 0, \quad 0 < x < L, \quad t > 0. \tag{5.5.1}$$

The boundary conditions of the problem are (the left end is maintained at T_∞ and the right end subjected to one of three boundary conditions)

$$T(0,t) = T_\infty, \quad \text{(a)} \quad T(L,t) = T_\infty, \quad \text{(b)} \quad \left[\frac{\partial T}{\partial x}\right]_{x=L} = 0,$$
$$\text{(c)} \quad \left[\frac{\partial T}{\partial x} + \frac{\beta}{k}(T - T_\infty)\right]_{x=L} = 0. \tag{5.5.2}$$

For convenience, we introduce the following normalization of the dependent variable, time, and spatial coordinate:

$$\xi = \frac{x}{L}, \quad \tau = \frac{k}{\rho c_p L^2} t, \quad \theta = \frac{T - T_\infty}{T_0 - T_\infty}. \tag{5.5.3}$$

Then Eq. (5.5.1) becomes

$$\frac{\partial \theta}{\partial \tau} - \frac{\partial^2 \theta}{\partial \xi^2} = 0, \quad 0 < \xi < 1, \quad \tau > 0. \tag{5.5.4}$$

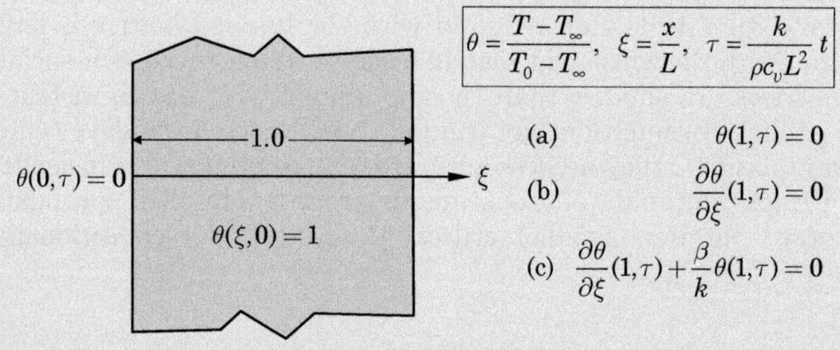

Fig. 5.5.1 A plane wall initially at temperature T_0 is subjected to three different sets of boundary conditions.

and the boundary conditions in Eq. (5.5.2) take the form (note that k is assumed to be a constant)

$$\theta(0,\tau) = 0, \quad \text{(a)} \ \theta(1,\tau) = 0, \quad \text{(b)} \ \left[\frac{\partial \theta}{\partial \xi}\right]_{\xi=1} = 0, \quad \text{(c)} \ \left[\frac{\partial \theta}{\partial \xi} + \frac{\beta}{k}\theta\right]_{\xi=1} = 0. \tag{5.5.5}$$

Fully Discretized Equations for Two-Element Mesh

First we consider a uniform mesh of two linear elements to illustrate the formulation and then present finite element solutions obtained using refined meshes. The finite element model is given by

$$\mathbf{C}^e \dot{\boldsymbol{\theta}}^e + \mathbf{K}^e \boldsymbol{\theta}^e = \mathbf{Q}^e, \tag{5.5.6a}$$

where $\boldsymbol{\theta}^e$ is the vector of nodal values (θ_1^e, θ_2^e) and

$$C_{ij}^e = \int_{\xi_a^e}^{\xi_b^e} \psi_i^e \psi_j^e \, d\xi, \quad K_{ij}^e = \int_{\xi_a^e}^{\xi_b^e} \frac{d\psi_i^e}{d\xi}\frac{d\psi_j^e}{d\xi} \, d\xi. \tag{5.5.6b}$$

For the choice of linear interpolation functions, the semidiscrete equations of a typical element are

$$\frac{h_e}{6}\begin{bmatrix} 2 & 1 \\ 1 & 2 \end{bmatrix}\begin{Bmatrix} \dot{\theta}_1^e \\ \dot{\theta}_2^e \end{Bmatrix} + \frac{1}{h_e}\begin{bmatrix} 1 & -1 \\ -1 & 1 \end{bmatrix}\begin{Bmatrix} \theta_1^e \\ \theta_2^e \end{Bmatrix} = \begin{Bmatrix} Q_1^e \\ Q_2^e \end{Bmatrix}, \tag{5.5.7}$$

where h_e is the length of the element ($h_1 = h_2 = 0.5$). Using of the alpha-family of approximation, with time step $\Delta\tau$, we obtain [see Eqs. (5.2.22a) and (5.2.22b)]

$$(\mathbf{C} + \Delta\tau\, \alpha \mathbf{K})\boldsymbol{\theta}^{s+1} = [\mathbf{M} - \Delta\tau(1-\alpha)\mathbf{K}]\boldsymbol{\theta}^s + \Delta\tau[\alpha \mathbf{Q}^{s+1} + (1-\alpha)\mathbf{Q}^s] \tag{5.5.8}$$

or

$$\begin{bmatrix} \frac{1}{6}+2\alpha\Delta\tau & \frac{1}{12}-2\alpha\Delta t \\ \frac{1}{12}-2\alpha\Delta\tau & \frac{1}{6}+2\alpha\Delta\tau \end{bmatrix} \begin{Bmatrix} \theta_1^e \\ \theta_2^e \end{Bmatrix}^{s+1} = \begin{bmatrix} \frac{1}{6}-2(1-\alpha)\Delta\tau & \frac{1}{12}+2(1-\alpha)\Delta\tau \\ \frac{1}{12}+2(1-\alpha)\Delta\tau & \frac{1}{6}-2(1-\alpha)\Delta\tau \end{bmatrix} \begin{Bmatrix} \theta_1^e \\ \theta_2^e \end{Bmatrix}^{s} + \begin{Bmatrix} \bar{Q}_1^1 \\ \bar{Q}_2^1 \end{Bmatrix}, \tag{5.5.9}$$

where $\bar{Q}_i^1 = \Delta\tau\left[\alpha(Q_i^1)^{s+1} + (1-\alpha)(Q_i^1)^s\right]$.

Assembling the finite element equations of the two elements, we obtain ($\theta_1^1 = U_1$, $\theta_2^1 = \theta_1^2 = U_2$, and $\theta_1^2 = U_3$)

$$\begin{bmatrix} \frac{1}{6}+2\alpha\Delta\tau & \frac{1}{12}-2\alpha\Delta\tau & 0 \\ \frac{1}{12}-2\alpha\Delta\tau & \frac{1}{3}+4\alpha\Delta\tau & \frac{1}{12}-2\alpha\Delta\tau \\ 0 & \frac{1}{12}-2\alpha\Delta t & \frac{1}{6}+2\alpha\Delta\tau \end{bmatrix} \begin{Bmatrix} U_1 \\ U_2 \\ U_3 \end{Bmatrix}^{s+1} = \begin{bmatrix} \frac{1}{6}-2(1-\alpha)\Delta\tau & \frac{1}{12}+2(1-\alpha)\Delta\tau & 0 \\ \frac{1}{12}+2(1-\alpha)\Delta\tau & \frac{1}{3}-4(1-\alpha)\Delta\tau & \frac{1}{12}+2(1-\alpha)\Delta t \\ 0 & \frac{1}{12}+2(1-\alpha)\Delta\tau & \frac{1}{6}-2(1-\alpha)\Delta\tau \end{bmatrix} \begin{Bmatrix} U_1 \\ U_2 \\ U_3 \end{Bmatrix}^{s} + \begin{Bmatrix} \bar{Q}_1^1 \\ \bar{Q}_2^1 + \bar{Q}_1^2 \\ \bar{Q}_2^2 \end{Bmatrix}. \tag{5.5.10}$$

Case (a) The boundary and balance conditions of the problem require

$$U_1^s = 0, \quad U_3^s = 0, \quad \bar{Q}_2^1 + \bar{Q}_1^2 = 0 \quad \text{for all } s > 0 \text{ (i.e., } \tau > 0). \tag{1}$$

The condensed equation for the primary unknown nodal value U_2 at time τ_{s+1} is given by

$$\left[\tfrac{1}{3} + 4\alpha\Delta\tau\right] U_2^{s+1} = \left[\tfrac{1}{3} - 4(1-\alpha)\Delta\tau\right] U_2^s, \tag{2}$$

which is can be solved repeatedly for $s = 0, 1, 2, \ldots$. At time $\tau = 0$ (i.e., $s = 0$) we have $U_2^0 = 0$.

Case (b) The boundary and balance conditions for this case require

$$U_1^s = 0, \quad \bar{Q}_2^2 = 0, \quad \bar{Q}_2^1 + \bar{Q}_1^2 = 0 \quad \text{for all } s > 0, \tag{3}$$

and the condensed equations are

$$\begin{bmatrix} \frac{1}{3}+4\alpha\Delta\tau & \frac{1}{12}-2\alpha\Delta\tau \\ \frac{1}{12}-2\alpha\Delta\tau & \frac{1}{6}+2\alpha\Delta\tau \end{bmatrix} \begin{Bmatrix} U_2 \\ U_3 \end{Bmatrix}^{s+1} = \begin{bmatrix} \frac{1}{3}-4(1-\alpha)\Delta\tau & \frac{1}{12}+2(1-\alpha)\Delta\tau \\ \frac{1}{12}+2(1-\alpha)\Delta\tau & \frac{1}{6}-2(1-\alpha)\Delta\tau \end{bmatrix} \begin{Bmatrix} U_2 \\ U_3 \end{Bmatrix}^{s}, \tag{4}$$

which are to be solved repeatedly for $s = 0, 1, 2, \ldots$ (with $U_2^0 = U_3^0 = 0$).

Case (c) The boundary and balance conditions for this case are

$$U_1^s = 0, \quad \bar{Q}_2^2 = -\Delta\tau\, H\left[\alpha U_3^{s+1} + (1-\alpha)U_3^s\right], \quad \bar{Q}_2^1 + \bar{Q}_1^2 = 0 \quad \text{for all } s > 0, \tag{5}$$

where $H = \beta L/k$. Then the condensed equations are

$$\begin{bmatrix} \frac{1}{3}+4\alpha\Delta\tau & \frac{1}{12}-2\alpha\Delta\tau \\ \frac{1}{12}-2\alpha\Delta\tau & \frac{1}{6}+2\alpha\Delta\tau+H\alpha\Delta\tau \end{bmatrix} \begin{Bmatrix} U_2 \\ U_3 \end{Bmatrix}^{s+1} = \begin{bmatrix} \frac{1}{3}-4(1-\alpha)\Delta\tau & \frac{1}{12}+2(1-\alpha)\Delta\tau \\ \frac{1}{12}+2(1-\alpha)\Delta\tau & \frac{1}{6}-2(1-\alpha)\Delta\tau+H(1-\alpha)\Delta\tau \end{bmatrix} \begin{Bmatrix} U_2 \\ U_3 \end{Bmatrix}^{s}, \tag{6}$$

which can be solved repeatedly for $s = 0, 1, 2, \ldots$ (with $U_2^0 = U_3^0 = 0$).

Determination of Critical Time Increments
Next we must determine the critical time increment required for the forward difference scheme because it is a conditionally stable scheme. This also gives an idea of the time increment to be selected for the Crank–Nicolson and forward difference schemes. For accuracy reasons, we often select the actual time increment to be much smaller than the critical time increment (say, one-tenth of the critical time increment).

To determine the critical time increment ($\Delta\tau_{cr} = 2/\lambda_{max}$), one must solve the eigenvalue problem associated with the chosen mesh for the problem. For the problem at hand, the eigenvalue problem associated with Eq. (5.5.6a) is obtained by replacing (we assume that the solution is of decay type) θ^e with $\theta^e(\xi,\tau) = \theta^0_e(\xi)e^{-\lambda\tau}$ and $Q^e(\tau) = Q^0_e\, e^{-\lambda\tau}$, where λ is called an eigenvalue. We obtain

$$(\mathbf{K}^e - \lambda\mathbf{C}^e)\,\boldsymbol{\theta}^e_0 = \mathbf{Q}^e_0. \tag{5.5.11}$$

This equation, after the assembly and imposition of boundary conditions, yields N eigenvalues $\lambda_1, \lambda_2, \ldots, \lambda_N$, where N is the number of unspecified (i.e., unknown) primary nodal degrees of freedom of the mesh. For example, the eigenvalue problem associated with Case (a) is (after assembly of two element equations)

$$\left(2\begin{bmatrix} 1 & -1 & 0\\ -1 & 2 & -1\\ 0 & -1 & 1\end{bmatrix} - \frac{\lambda}{12}\begin{bmatrix} 2 & 1 & 0\\ 1 & 4 & 1\\ 0 & 1 & 2\end{bmatrix}\right)\begin{Bmatrix} U^0_1\\ U^0_2\\ U^0_3\end{Bmatrix} = \begin{Bmatrix} (Q^0)^1_1\\ (Q^0)^1_2 + (Q^0)^2_1\\ (Q^0)^2_2\end{Bmatrix}. \tag{7}$$

After imposing the boundary and balance conditions $U^0_1 = U^0_3 = 0$ and $(Q^0)^1_2 + (Q^0)^2_1 = 0$, we obtain (by deleting the first and last rows and columns) $(4 - 4\lambda/12) = 0$ giving the single eigenvalue $\lambda = 12$. Hence, we have $\Delta\tau_{cr} = 2/\lambda = 0.1666$.

For Case (b), the boundary and balance conditions are: $U^0_1 = 0$, $(Q^0)^2_2 = 0$, and $(Q^0)^1_2 + (Q^0)^2_1 = 0$. Then by deleting the first row and column in Eq. (7), we obtain

$$\left(2\begin{bmatrix} 2 & -1\\ -1 & 1\end{bmatrix} - \frac{\lambda}{12}\begin{bmatrix} 4 & 1\\ 1 & 2\end{bmatrix}\right)\begin{Bmatrix} U^0_2\\ U^0_3\end{Bmatrix} = \begin{Bmatrix} 0\\ 0\end{Bmatrix}. \tag{8}$$

The solution of these linear equations require setting the determinant of the coefficient matrix to zero:

$$0 = 8(1-\bar{\lambda})^2 - (2+\bar{\lambda})^2, \quad \bar{\lambda} = \frac{\lambda}{12}. \tag{9}$$

The two roots of this quadratic equation yield $\lambda_1 = 31.6891$ and $\lambda_2 = 2.5967$. Hence, the critical time increment is $\Delta\tau_{cr} = 2/\lambda_1 = 0.0631$.

For Case (c), the boundary and balance conditions are: $U^0_1 = 0$, $(Q^0)^2_2 = -HU^0_3$ ($H = \beta L/k$, and $(Q^0)^1_2 + (Q^0)^2_1 = 0$. For $H = 2$ (as an example), the eigenvalue problem becomes

$$\left(2\begin{bmatrix} 2 & -1\\ -1 & 1+\mathbf{1}\end{bmatrix} - \frac{\lambda}{12}\begin{bmatrix} 4 & 1\\ 1 & 2\end{bmatrix}\right)\begin{Bmatrix} U^0_2\\ U^0_3\end{Bmatrix} = \begin{Bmatrix} 0\\ 0\end{Bmatrix} \tag{8}$$

and by setting the determinant of the coefficient matrix to zero gives $\lambda_1 = 42.1423$ and $\lambda_2 = 5.8577$. Hence, the critical time increment is $\Delta\tau_{cr} = 2/\lambda_1 = 0.0475$.

In summary, the critical time increments, using uniform mesh of 2 linear elements, for the three types of boundary conditions are:

$$\text{(a)}\quad \Delta\tau_{cr} = 0.1666, \quad \text{(b)}\quad \Delta\tau_{cr} = 0.0631, \quad \text{(c)}\quad \Delta\tau_{cr} = 0.0475. \tag{10}$$

Stability and Accuracy of Solutions
Next we investigate the stability characteristics of the transient response for Case (a) using the forward difference ($\alpha = 0$) scheme. First we note that the initial conditions and boundary conditions at the boundary nodes are in conflict (i.e., initial conditions require all Us to be unity, where as the boundary conditions require $U_1 = U_3 = 0$). It is important to specify the

initial conditions consistent with the boundary conditions. For example, in Case (a), we take $U_1^0 = 0$, $U_2^0 = 1.0$, and $U_3^0 = 0$, although initially the whole wall is at unit temperature. In general, *when there is a conflict between the initial and boundary conditions, the numerical values for the first few time steps will not be accurate.* When $\Delta\tau = 0.167 > \Delta\tau_{cr} = 0.1666$ is used to carry out the transient analysis, the solution $\theta(0.5, \tau)$ is clearly diverges (i.e., unstable) but slowly and when $\Delta\tau = 0.175$ is used, the divergence is rapid, as can be seen from Fig. 5.5.2. For the same time increment, the Crank–Nicolson scheme ($\alpha = 0.5$) is both stable and nonoscillatory. When $\Delta\tau = 0.16 < \Delta\tau_{cr} = 0.1666$ is used to carry out the analysis, the solution is oscillatory but does not diverge (i.e., stable).

To obtain nonoscillatory and accurate solution for a fixed mesh, one has to take a much smaller time increment than the critical time increment *and* refine the mesh. Typically, one-tenth of the critical time increment is suggested. Figure 5.5.3 contains plots of $\theta(0.5, \tau)$ versus τ for the forward difference scheme ($\alpha = 0.0$) and time increments $\Delta\tau = 0.05, 0.02, 0.01$, and 0.001. The solution obtained with Crank–Nicolson scheme ($\alpha = 0.5$) and $\Delta\tau = 0.01$ and 0.001 is also included for comparison. Clearly, as the time increment is reduced the solution obtained with the forward difference and the Crank–Nicolson schemes give the same solution (they are indistinguishable in the plots for $\Delta\tau = 0.001$). However, they are not close to the analytical solution [for Case (a)]:

$$\theta(\xi, \tau) = 4\sum_{n=0}^{\infty} \frac{1}{\lambda_n} \sin(\lambda_n \xi)\, e^{-\lambda_n^2 \tau}, \quad \lambda_n = (2n+1)\pi. \tag{5.5.12}$$

In order to get close to the analytical solution, one must refine the mesh. The four- and eight-element solutions using the Crank–Nicolson scheme with $\Delta\tau = 0.001$ clearly converge to the analytical solution. The steady state solution (of zero everywhere) is reached approximately between $\tau = 0.6$ and 0.8.

Table 5.5.1 contains numerical values of the normalized temperature $\theta(0.5, \tau)$ as a function of the normalized time τ predicted by various time-approximation schemes ($\Delta\tau = 0.001$) and meshes for Case (a) boundary conditions. The critical time increment for the four-element mesh is $\Delta\tau_{cr} = 0.01578$ and that for the eight-element mesh is $\Delta\tau_{cr} = 0.00291$. Thus, the time increment used for the results in Table 5.5.1 is less than the critical time increment so that the forward difference scheme is stable for both meshes.

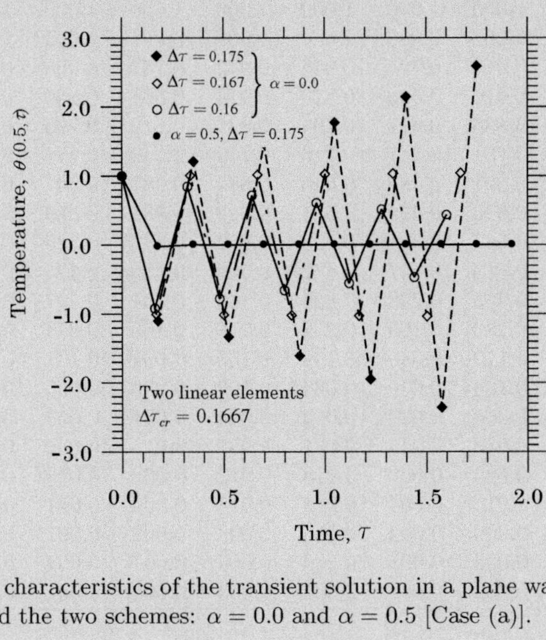

Fig. 5.5.2 Stability characteristics of the transient solution in a plane wall using various time increment values and the two schemes: $\alpha = 0.0$ and $\alpha = 0.5$ [Case (a)].

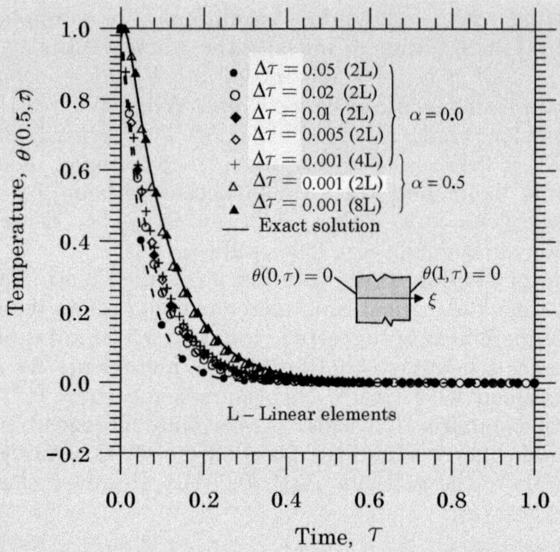

Fig. 5.5.3 Transient solutions in a plane wall as predicted by the Crank–Nicolson scheme ($\alpha = 0.5$) and forward difference scheme ($\alpha = 0.0$) [Case (a)].

Table 5.5.1 A comparison of the finite element solutions obtained using three time-approximation schemes with the analytical solution of transient heat conduction in a plane wall for Case (a) boundary conditions ($\Delta\tau = 0.001$; L stands for linear elements).

	$\alpha = 0.0$		$\alpha = 1.0$			$\alpha = 0.5$			
τ	4L	8L	2L	4L	8L	2L	4L	8L	Exact
0.00	1.000	1.000	1.000	1.000	1.000	1.000	1.000	1.000	1.000
0.01	1.026	1.002	0.886	1.034	1.008	0.887	1.032	1.000	0.999
0.02	0.963	0.974	0.785	0.966	0.982	0.787	0.966	0.980	0.975
0.03	0.879	0.910	0.696	0.879	0.914	0.698	0.880	0.913	0.917
0.04	0.797	0.836	0.617	0.794	0.835	0.619	0.796	0.836	0.846
0.05	0.719	0.761	0.547	0.716	0.759	0.549	0.718	0.760	0.772
0.06	0.649	0.691	0.485	0.645	0.687	0.487	0.647	0.689	0.702
0.07	0.586	0.626	0.429	0.581	0.622	0.431	0.583	0.624	0.637
0.08	0.528	0.567	0.381	0.523	0.562	0.383	0.526	0.565	0.578
0.09	0.476	0.513	0.337	0.472	0.509	0.339	0.474	0.511	0.524
0.10	0.429	0.465	0.299	0.425	0.460	0.301	0.427	0.462	0.474
0.15	0.256	0.283	0.163	0.252	0.278	0.165	0.254	0.280	0.290
0.20	0.153	0.172	0.089	0.149	0.168	0.091	0.151	0.170	0.177
0.25	0.091	0.104	0.049	0.089	0.102	0.050	0.090	0.103	0.108
0.30	0.054	0.064	0.027	0.052	0.062	0.027	0.053	0.063	0.066
0.35	0.032	0.038	0.015	0.031	0.037	0.015	0.032	0.038	0.040
0.40	0.019	0.023	0.008	0.018	0.022	0.008	0.019	0.023	0.024
0.45	0.012	0.014	0.004	0.011	0.014	0.004	0.011	0.014	0.015
0.50	0.007	0.009	0.002	0.065	0.008	0.002	0.007	0.008	0.009
0.55	0.004	0.005	0.001	0.039	0.005	0.001	0.004	0.005	0.006
0.60	0.002	0.003	0.001	0.002	0.003	0.001	0.002	0.003	0.003

Having demonstrated the importance of selecting the proper time increment and mesh for Case (a) boundary conditions, we now present results for Cases (b) and (c). The analytical solution for Case (c) is

$$\theta(\xi,\tau) = 2\sum_{n=0}^{\infty} \frac{\lambda_n + H\sin\lambda_n}{\lambda_n^2 + H\sin^2\lambda_n}\sin(\lambda_n\xi)\,e^{-\lambda_n^2\tau}, \quad H = \frac{\beta L}{k}, \tag{5.5.13}$$

where λ_n are the roots of the transcendental equation

$$\lambda\cot\lambda + H = 0, \quad H = \frac{\beta L}{k}. \tag{5.5.14}$$

The first twelve roots of $\lambda\cos\lambda + H\sin\lambda = 0$ for $H = 2$ are

$$\begin{aligned} &\lambda_1 = 2.28893, \quad \lambda_2 = 5.08696, \quad \lambda_3 = 8.09614, \quad \lambda_4 = 11.1728 \\ &\lambda_5 = 14.2763, \quad \lambda_6 = 17.3933, \quad \lambda_7 = 20.5174, \quad \lambda_8 = 23.6463 \\ &\lambda_9 = 26.7779, \quad \lambda_{10} = 29.9119, \quad \lambda_{11} = 33.0741, \quad \lambda_{12} = 36.1834. \end{aligned} \tag{5.5.15}$$

The roots are very close (less than 0.15 percent error) to the value $\lambda_n = (2n-1)\pi/2$ for $n > 12$. The analytical solution for Case (b) boundary conditions is obtained from Eq. (5.5.13) by setting $H = 0$ and taking $\lambda_n = (2n-1)\pi/2$ (the roots of the equation $\lambda\cos\lambda = 0$).

Figures 5.5.4 and 5.5.5 show the solutions $\theta(1,\tau)$ for Cases (b) and (c), respectively, using four and eight linear (L) and quadratic (Q) element meshes. The numerical solutions are off initially (due to the conflict between the initial and boundary conditions), but get close to their respective analytical solutions with an increase in time. Of course, with mesh refinements (which require smaller time increments), finite element solutions converge to the analytical solutions.

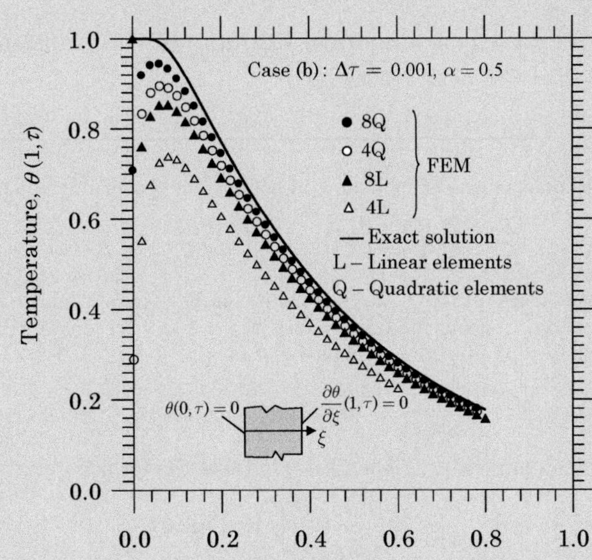

Fig. 5.5.4 Transient solutions in a plane wall for Case (b) as predicted by the Crank–Nicolson ($\alpha = 0.5$) scheme ($\Delta\tau = 0.001$).

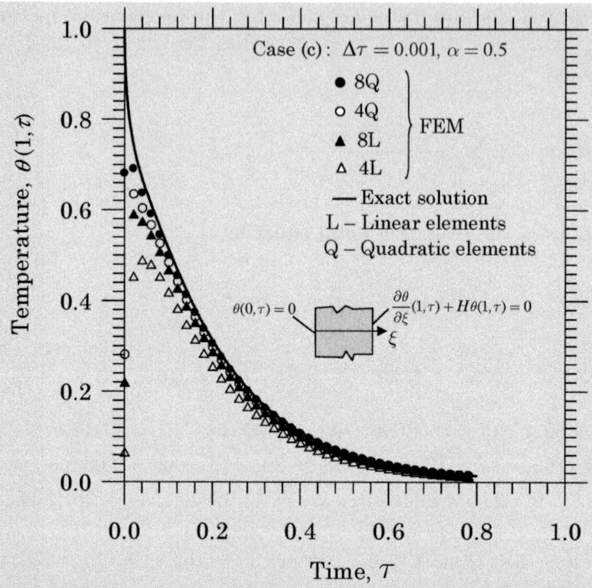

Fig. 5.5.5 Transient solutions in a plane wall for Case (c) as predicted by the Crank–Nicolson ($\alpha = 0.5$) scheme ($\Delta\tau = 0.001$).

5.5.2 Two-Dimensional Example

In this section, we present a numerical example of the transient analysis of a two-dimensional problem (see Reddy [1]).

Example 5.5.2

Consider two-dimensional transient heat flow in solids. The governing equation is given by

$$\rho c_p \frac{\partial T}{\partial t} - \frac{\partial}{\partial x}\left(k_{xx}\frac{\partial T}{\partial x}\right) - \frac{\partial}{\partial y}\left(k_{yy}\frac{\partial T}{\partial y}\right) = g(x, y), \tag{1}$$

where T is the temperature and (k_{xx}, k_{yy}) are the conductivities of an orthotropic medium. For the isotropic and homogeneous medium, we set $k_{xx} = k_{yy} = k$, constant. We can normalize the variables for a square plate of dimensions $a \times a$ as

$$\xi = \frac{x}{a} \quad \eta = \frac{y}{a} \quad \tau = \frac{k}{\rho c_p a^2} t, \quad \theta = T - T_0\,, \quad f(\xi, \eta) = \frac{a^2}{k} g(x(\xi), y(\eta)), \tag{2}$$

where T_0 is a reference temperature. Then Eq. (1) takes the form

$$\frac{\partial\theta}{\partial\tau} - \left(\frac{\partial^2\theta}{\partial\xi^2} + \frac{\partial^2\theta}{\partial\eta^2}\right) = f(\xi, \eta). \tag{3}$$

Determine the normalized temperature field $\theta(\xi, \eta, \tau)$ in the square plate when the plate boundaries are subjected to the following boundary conditions [see Fig. 5.5.6(a)]:

$$\left.\frac{\partial\theta}{\partial\xi}\right|_{(0,\eta,\tau)} = 0, \quad \left.\frac{\partial\theta}{\partial\eta}\right|_{(\xi,0,\eta)} = 0$$
$$\theta(1, \eta, \tau) = 0, \quad \theta(\xi, 1, \tau) = 0. \tag{4}$$

Assume that $f = 1$ and that the plate is initially at zero temperature:

$$\theta(\xi, \eta, 0) = 0 \quad \text{for all } (\xi, \eta) \text{ in } \Omega. \tag{5}$$

Figure 5.5.6(b) contains the finite element meshes to be used.

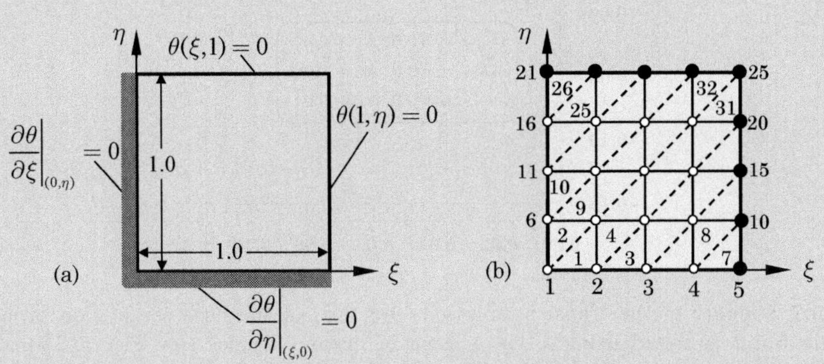

Fig. 5.5.6 (a) Domain and boundary and initial conditions, and (b) a typical 4×4 finite element mesh of linear triangles (with broken lines) or linear rectangles (without the broken lines) for the transient analysis.

Solution First we investigate the stability and accuracy of the forward difference scheme ($\alpha = 0.0$) and the accuracy of the Crank–Nicolson method ($\alpha = 0.5$). Since the Crank–Nicolson method is unconditionally stable, one can choose any value of $\Delta\tau$, but its accuracy will depend on the choice of $\Delta\tau$ (i.e., for large values of $\Delta\tau$ the solution may not be accurate). The forward difference scheme is conditionally stable [i.e., it is stable if $\Delta\tau < \Delta\tau_{cr}$, where $\Delta\tau_{cr}$ satisfies the condition in Eq. (5.2.23) for $\alpha = 0$]. When a 4×4 mesh of linear triangles is used, the maximum eigenvalue is $\lambda_{max} = 386.426$; hence, the critical time increment is

$$\Delta\tau_{cr} = \frac{2}{\lambda_{max}} = \frac{2}{386.426} = 0.00518, \tag{6}$$

whereas for 4×4 mesh of linear rectangles the critical time increment is $\Delta\tau_{cr} = 2/343.256 = 0.00583$.

The boundary conditions on the primary variable for the 4×4 of linear triangles or linear rectangles in the full domain are

$$U_5 = U_{10} = U_{15} = U_{20} = U_{21} = U_{22} = U_{23} = U_{24} = U_{25} = 0.0. \tag{7}$$

Beginning with the initial conditions $U_i^0 = 0$ ($i = 1, 2, \ldots, 25$), we solve the assembled set of equations associated with Eq. (5.3.7a). All numerical results presented herein are obtained with the 4×4 mesh of linear triangles shown in Fig. 5.5.6(b).

The forward difference scheme would be unstable, for example, for $\Delta\tau > 0.00518$ when 4×4 mesh of triangles is used. To illustrate this point, the equations are solved using $\alpha = 0$, $\Delta\tau = 0.01 > \Delta\tau_{cr} = 0.00518$ and $\alpha = 0.5$, $\Delta\tau = 0.01$. The Crank–Nicolson method gives a stable solution, while the forward difference scheme yields an unstable solution (i.e., the solution error grows unboundedly with time), as can be seen from Fig. 5.5.7. For $\Delta\tau = 0.005$ the forward difference scheme yields stable solution.

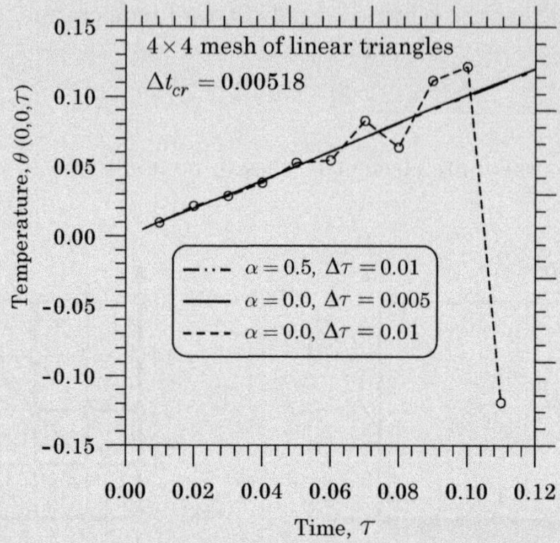

Fig. 5.5.7 Stability of the transient solutions $\theta(\xi, 0, \tau)$ of the heat conduction problem in **Example 5.5.2** analyzed using a 4×4 mesh of linear triangles (see Fig. 5.5.6) and the Crank–Nicolson ($\alpha = 0.5$) and forward difference ($\alpha = 0.0$) time integration schemes.

The Crank–Nicolson method gives a stable and accurate solution for even $\Delta\tau = 0.05$. The temperature $\theta(\xi, 0, \tau)$ versus ξ for various values of time are shown in Fig. 5.5.8(a). The steady state is reached at time $\tau = 1.0$ (a tolerance of $\epsilon = 10^{-3}$ is used to check the difference between the solutions of two consecutive times). The temperature $\theta(0, 0, \tau)$ versus time, predicted by the Crank–Nicolson method, is shown in Fig. 5.5.8(b). A comparison of the transient solution at $\tau = 1.0$ is presented in Table 5.5.2 with the steady state FEM, the FDM, and the analytical solutions. Table 5.5.3 contains FEM solutions for the temperature predicted by 4×4 meshes of triangles and rectangles and various values of $\Delta\tau$ and $\alpha = 0.5$.

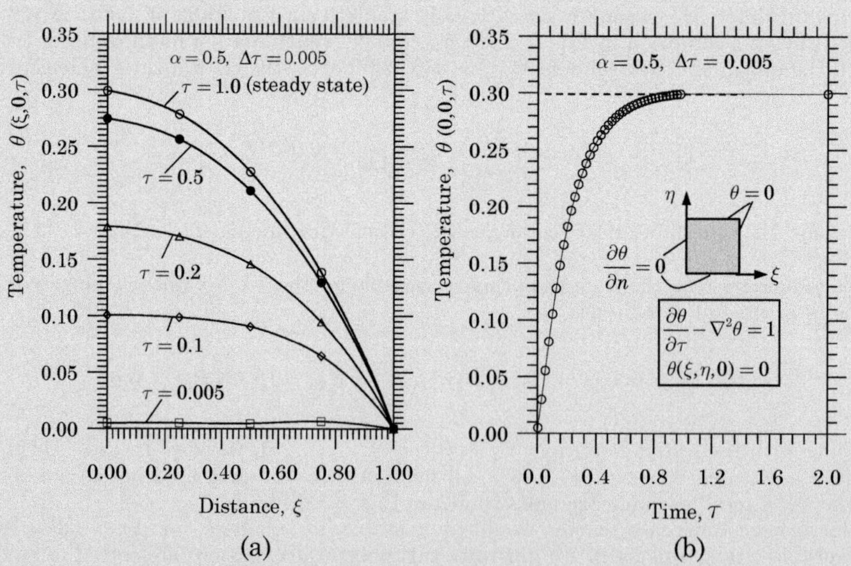

Fig. 5.5.8 (a) Variation of the temperature $\theta(\xi, 0, \tau)$ as a function of position ξ and time τ and (b) variation of the temperature $\theta(0, 0, \tau)$ as a function of time τ for the transient heat conduction problem of **Example 5.5.1** (4×4 mesh of linear triangles).

Table 5.5.2 Comparison of the solutions by the FDM and FEM with the exact solution of the heat conduction problem.

Node	Exact (steady)	FDM (steady)	Error	FEM (steady)	Error	FEM[a] at $\tau = 1.0$
1	0.2947	0.2911	0.0036	0.3013	−0.0066	0.2993
2	0.2789	0.2755	0.0034	0.2805	−0.0016	0.2786
3	0.2293	0.2266	0.0027	0.2292	0.0001	0.2278
4	0.1397	0.1381	0.0016	0.1392	0.0005	0.1385
5	0.0000	0.0000	0.0000	0.0000	0.0000	0.0000
7	0.2642	0.2609	0.0033	0.2645	−0.0003	0.2628
8	0.2178	0.2151	0.0027	0.2172	0.0006	0.2159
9	0.1333	0.1317	0.0016	0.1327	0.0006	0.1320
10	0.0000	0.0000	0.0000	0.0000	0.0000	0.0000
13	0.1811	0.1787	0.0024	0.1801	0.0010	0.1791
14	0.1127	0.1110	0.0017	0.1117	0.0010	0.1111
15	0.0000	0.0000	0.0000	0.0000	0.0000	0.0000
19	0.0728	0.0711	0.0017	0.0715	0.0013	0.0712
20	0.0000	0.0000	0.0000	0.0000	0.0000	0.0000
25	0.0000	0.0000	0.0000	0.0000	0.0000	0.0000

[a]Obtained with the Crank–Nicolson scheme with 4×4 mesh of linear triangles and $\Delta\tau = 0.005$.

Table 5.5.3 Comparison of transient solutions obtained using meshes of triangular and rectangular elements.

Time	Element[1]	Temperature along the line $\eta = 0$: $\theta(\xi, 0, \tau) \times 10$			
τ		$\xi = 0.0$	$\xi = 0.25$	$\xi = 0.5$	$\xi = 0.75$
0.1	T1	0.9758	0.9610	0.9063	0.7104
	R1	0.9684	0.9556	0.8956	0.6887
	T2	0.9928	0.9798	0.9168	0.6415
	R2	0.9841	0.9718	0.9020	0.6323
0.2	T1	1.8003	1.7238	1.4891	0.9321
	R1	1.7723	1.7216	1.4829	0.9367
	T2	1.7979	1.7060	1.4644	0.9462
	R2	1.7681	1.6990	1.4626	0.9469
0.3	T1	2.3130	2.1671	1.7961	1.1466
	R1	2.2747	2.1650	1.8084	1.1499
	T2	2.2829	2.1448	1.7943	1.1249
	R2	2.2479	2.1432	1.8018	1.1319
1.0	T1	2.9960	2.7871	2.2804	1.3843
	R1	2.9648	2.8053	2.3090	1.4059
	T2	2.9925	2.7862	2.2776	1.3849
	R2	2.9621	2.8037	2.3065	1.4053

[1]T1, triangular element mesh with $\Delta\tau = 0.1$; T2, triangular element mesh with $\Delta\tau = 0.05$; R1, rectangular element mesh with $\Delta\tau = 0.1$; R2, rectangular element mesh with $\Delta\tau = 0.05$. In all cases, $\alpha = 0.5$ and 4×4 mesh of linear triangles are used.

5.6 Summary

In this chapter, FEM (in space) and FDM (in time) approximations of the PDEs governing transient heat transfer have been discussed. The spatial approximation has been carried out, much the same way as for the steady state case, using the FEM. This step is known as the semidiscretization and it results in a set of ODEs among the nodal values in time. The time derivatives are then approximated using the alpha-family of approximation, which contains the backward, forward, and Crank–Nicolson schemes as special cases. The resulting equations are algebraic equations in terms of the nodal values at time t_{n+1} in terms of the known values at time t_n.

Numerical stability and accuracy of the approximation have been discussed and the stability criterion was established for the maximum time increment (above which the numerical scheme is unstable), called the critical time step (Δt_{cr}), in terms of the maximum eigenvalue (λ_{max}) of the discrete problem (i.e., for the chosen mesh). Since the eigenvalues of a discrete problem depend on the material properties as well as the mesh size, the stability criterion provides a restriction on the largest time step one can take for a given mesh size (measured in terms of the element length). All continuum problems (i.e., problems described by differential equations) have an infinite number of eigenvalues, whereas their discrete counterparts have a finite number of eigenvalues, the number being equal to the number of unspecified degrees of freedom of the discrete model. Thus, as the mesh is refined, additional eigenvalues from the infinite set of eigenvalues of the continuum problem are obtained and they are necessarily larger in magnitude than those already obtained with a coarser mesh. Thus, the value of λ_{max} increases with the mesh refinement and hence Δt_{cr} decreases. Therefore, mesh refinements do not automatically give stable solutions unless the associated time step is also reduced to that below Δt_{cr} for the given mesh. Although these comments are valid for conditionally stable schemes like the forward and central difference schemes, Δt_{cr} provides a estimate of the time step used even for stable schemes.

The size of the time increment can be determined by solving the corresponding (i.e., with the same but homogeneous boundary conditions and mesh) eigenvalue problem and using the formula $\Delta t_{cr} = 2/\lambda_{max}$, where λ_{max} is the largest eigenvalue. Since the λ_{max} increases with mesh refinements, Δt_{cr} decreases. With $\Delta t = (\Delta t_{cr})/10$ or smaller and refined meshes, one can get close to the actual solution. To determine whether the solution has converged, one must carry out a convergence study by using increasingly refined meshes. A conflict in the specification of the boundary and initial conditions can result in oscillatory behavior and larger error in the first few time steps; to reduce this error, one may use a very small time increment for the first few time steps. These ideas are illustrated with numerical examples of one- and two-dimensional problems.

Chapters 4 and 5 have provided a basic introduction to finite element analysis of ODEs and PDEs (in space and time) involving a single dependent unknown. The ideas and steps presented for the single variable problems can be extended to multivariable problems in multidimensions, as will be discussed in Chapter 6.

Problems

Additional examples and problems can be found in [1] and [6] to test and extend their understanding of the FEM as applied to linear heat transfer problems.

ONE-DIMENSIONAL PROBLEMS

5.1 Consider the PDE arising in connection with unsteady heat transfer in an insulated rod:

$$\frac{\partial T}{\partial t} - \frac{\partial}{\partial x}\left(kA\frac{\partial T}{\partial x}\right) = g \quad \text{for } 0 < x < L$$

$$T(0,t) = 0, \quad T(x,0) = T_0, \quad \left[kA\frac{\partial T}{\partial x} + \beta A(T - T_\infty) + \hat{Q}\right]\bigg|_{x=L} = 0.$$

Following the procedure outlined in Section 5.2, derive the weak form, the semidiscrete finite element model, and the fully discretized finite element equations for a typical element. Be sure to include the convection-type boundary condition in deriving the weak form.

5.2 Using a mesh of two linear elements and the semidiscrete finite element equations derived in **Problem 5.1**, determine the nodal temperatures as functions of time for the case in which $kA = 1$, $g = 0$, $T_0 = 1$, and $\hat{Q} = 0$.

5.3 Consider the normalized form of the equation governing heat transfer in a plane wall [see Eq. (5.5.4)]:

$$\frac{\partial \theta}{\partial \tau} - \frac{\partial^2 \theta}{\partial \xi^2} = 0, \quad 0 < \xi < 1, \quad \tau > 0 \tag{1}$$

subjected to the boundary (B.C.) and initial conditions (I.C.)

$$\text{B.C.: } \theta(0,\tau) = 1, \quad \theta(1,\tau) = 0; \quad \text{I.C.: } \theta(\xi,0) = 0. \tag{2}$$

Determine the finite element solution using a uniform mesh of 10 linear finite elements. Compare the results with the exact solution

$$\theta(\xi,\tau) = 1 - \xi - 2\sum_{n=1}^{\infty} \frac{\sin(\lambda_n \xi)}{\lambda_n} e^{-\lambda_n^2 \tau}, \quad \lambda_n = n\pi. \tag{3}$$

5.4 Consider the normalized form of the heat transfer equation in a plane wall [see Eq. (5.5.4) or Eq. (1) of **Problem 5.3**] with the following boundary and initial conditions:

$$\text{B.C.: } \theta(0,\tau) = 0, \quad \theta(1,\tau) = 0, \quad \theta(\xi,0) = \xi - 1. \tag{1}$$

Determine the finite element solution using two linear finite elements. Compare the results with the exact solution

$$\theta(\xi,\tau) = -2\sum_{n=1}^{\infty} \frac{\sin(\lambda_n \xi)}{\lambda_n} e^{-\lambda_n^2 \tau}, \quad \lambda_n = n\pi. \tag{2}$$

Determine the critical time increment and find the finite element solution for $\tau > \tau_0$, where τ_0 is the normalized time at which the normalized temperature reaches the steady state solution.

5.5 Consider the normalized form of the heat transfer equation in a plane wall [see Eq. (5.5.4) or Eq. (1) of **Problem 5.3**] with the following boundary and initial conditions:

$$\text{B.C.: } \frac{\partial \theta}{\partial \xi}\bigg|_{(0,\tau)} = -1, \quad \frac{\partial \theta}{\partial \xi}\bigg|_{(1,\tau)} = 0, \quad \theta(\xi,0) = 0. \tag{1}$$

Determine the finite element solution using a uniform mesh of 10 linear finite elements. Compare the results with the exact solution

$$\theta(\xi,\tau) = \tau + \frac{1}{3} - \xi + 0.5\xi^2 - 2\sum_{n=1}^{\infty} \frac{\cos(\lambda_n \xi)}{\lambda_n^2} e^{-\lambda_n^2 \tau}, \quad \lambda_n = n\pi. \tag{2}$$

5.6 Consider a thick-walled pipe whose inner radius is half its outer radius (unity). The pipe is initially at a uniform temperature (of unity). At time $t = 0$ the temperature of both inside and outside surfaces dropped to zero temperature. The normalized governing equation in the cylindrical coordinates is

$$\frac{\partial \theta}{\partial \tau} - \frac{1}{r}\frac{\partial}{\partial r}\left(r\frac{\partial \theta}{\partial r}\right) = 0, \quad 0.5 < r < 1, \quad \tau > 0, \tag{1}$$

subjected to the boundary and initial conditions

$$\text{B.C.: } \theta(0.5,\tau) = 0, \quad \theta(1,\tau) = 0, \quad \text{I.C.: } \theta(r,0) = 1. \tag{2}$$

Use (a) 10 linear finite elements and (b) 5 quadratic element to determine the finite element solution.

Two-Dimensional Problems

5.7 Determine the critical time increment for the transient analysis (with $\alpha \le \frac{1}{2}$) of the problem

$$\frac{\partial u}{\partial t} - \nabla^2 u = 1 \quad \text{in } \Omega; \quad u = 0 \quad \text{in } \Omega \text{ at } t = 0 \tag{1}$$

by determining the maximum eigenvalue of the problem

$$-\nabla^2 u = \lambda u \quad \text{in } \Omega; \quad u = 0 \quad \text{on } \Gamma. \tag{2}$$

The domain is a square of one unit. Use (a) one triangular element in the octant, (b) four linear triangular elements in the octant [see Fig. P5.7(b)], and (c) a 2×2 mesh of linear rectangular elements in a quadrant [see Fig. P5.7(c)]. Determine the critical time increment for the forward difference scheme. *Answer:* (a) $\lambda = 24$. (b) $\lambda_{max} = 305.549$.

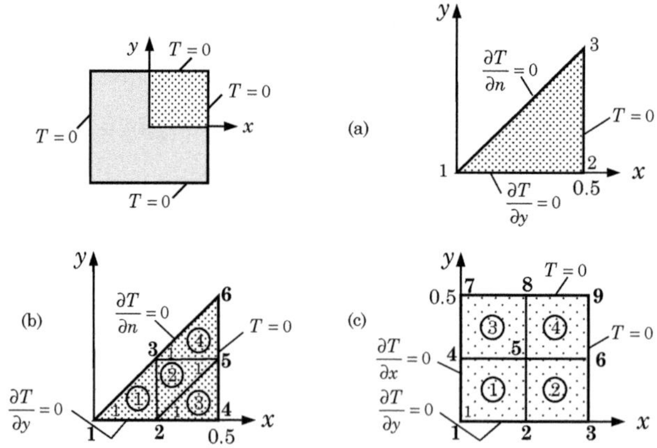

Fig. P5.7

6

Finite Element Analysis of Viscous Incompressible Flows

6.1 Governing Equations

The equations governing flows of Newtonian viscous incompressible fluids were reviewed in Chapter 2. The equations are revisited here, in the Cartesian component form, for the two-dimensional case (i.e., set $v_z = 0$ and the derivatives with respect to z to zero). The governing equations include the equations of motion and the conservation of mass equation:

$$\rho\left(\frac{\partial u}{\partial t} + u\frac{\partial u}{\partial x} + v\frac{\partial u}{\partial y}\right) + \frac{\partial p}{\partial x} - \mu\left[2\frac{\partial^2 u}{\partial x^2} + \frac{\partial}{\partial y}\left(\frac{\partial v}{\partial x} + \frac{\partial u}{\partial y}\right)\right] = \rho f_x \tag{6.1.1}$$

$$\rho\left(\frac{\partial v}{\partial t} + u\frac{\partial v}{\partial x} + v\frac{\partial v}{\partial y}\right) + \frac{\partial p}{\partial y} - \mu\left[\frac{\partial}{\partial x}\left(\frac{\partial v}{\partial x} + \frac{\partial u}{\partial y}\right) + 2\frac{\partial^2 v}{\partial y^2}\right] = \rho f_y \tag{6.1.2}$$

$$\frac{\partial u}{\partial x} + \frac{\partial v}{\partial y} = 0, \tag{6.1.3}$$

where (u, v) are the components of the velocity vector (i.e., $u = v_x = v_1$ and $v = v_y = v_2$), p is the hydrostatic pressure, (f_x, f_y) are the components of the body force vector, ρ is the density, and μ is the viscosity of the fluid.

The boundary conditions are of the form

$$u = \hat{u}, \quad v = \hat{v} \quad \text{on } \Gamma_v \tag{6.1.4a}$$

$$\left.\begin{aligned} t_x &\equiv \sigma_{xx} n_x + \sigma_{xy} n_y = \hat{t}_x \\ t_y &\equiv \sigma_{yx} n_x + \sigma_{yy} n_y = \hat{t}_y \end{aligned}\right\} \quad \text{on } \Gamma_\sigma, \tag{6.1.4b}$$

where $\hat{\mathbf{n}}$ is the unit normal to the boundary and Γ_v and Γ_σ are the nonoverlapping segments of the boundary such that their union is the total boundary Γ of the domain Ω.

In this study, we shall consider two different finite element models of Eqs. (6.1.1)–(6.1.3). The first one is a natural formulation in which the weak forms of Eqs. (6.1.1)–(6.1.3) are used to construct the finite element model. The resulting finite element model is termed the *velocity–pressure finite model* . The second model is based on the interpretation that the continuity equation (6.1.3) is an additional relation among the velocity components (u and v), and the constraint is included in the weak form in a least-squares sense, known as the *penalty method*, and the model is termed the *penalty-finite element model* (see [1, 3, 40–46]). In this case, the pressure p is eliminated from the primary computation and it is recovered through postcomputation. It is informative to note that the velocity–pressure formulation is that obtained by including the continuity condition in Eq. (6.1.3) as a constraint through the Lagrange multiplier method. The Lagrange multiplier turns out to be the negative of the pressure, p.

Sections 6.2 and 6.3 are concerned with mixed and penalty finite element models of the linearized Navier–Stokes equations; that is, the advective terms

$$v_x \frac{\partial v_x}{\partial x} + v_y \frac{\partial v_x}{\partial y}, \quad v_x \frac{\partial v_y}{\partial x} + v_y \frac{\partial v_y}{\partial y}$$

are assumed to be negligible. This amounts to modeling slow-moving laminar flows only. The nonlinear penalty finite element model of full Navier–Stokes equations is considered in Section 6.4.

Before leaving this introduction, we note that there are other types of finite element models of fluid flows that have been developed. Among them, finite element models based on the least-squares formulations of the governing equations have been advanced by Jiang [47], Bochev and Gunzburger [48], and Surana and Reddy and their colleagues [49–55]. This alternative approach is beyond the scope of this book and thus is not discussed here.

6.2 Velocity–Pressure Finite Element Model

6.2.1 Weak-Form Development

The starting point for the development of the finite element model is to construct the weak forms over a typical finite element Ω_e. Unlike in the previous chapters, we have three equations and three dependent unknowns (i.e., three degrees of freedom, u, v, and p). The three-step procedure to develop weak forms of the three equations is presented.

Next the weak forms of Eqs. (6.1.1) and (6.1.2), without the advective terms, are developed using the three-step procedure outlined in Section 4.3.3. We obtain

$$0 = \int_{\Omega^e} \left[\rho w_{1i} \frac{\partial u}{\partial t} + \frac{\partial w_{1i}}{\partial x} \left(2\mu \frac{\partial u}{\partial x} - p \right) + \mu \frac{\partial w_{1i}}{\partial y} \left(\frac{\partial u}{\partial y} + \frac{\partial v}{\partial x} \right) - \rho w_{1i} f_x \right] dxdy - \oint_{\Gamma^e} w_{1i} t_x \, ds \tag{6.2.1}$$

$$0 = \int_{\Omega^e} \left[\rho w_{2i} \frac{\partial v}{\partial t} + \frac{\partial w_{2i}}{\partial y} \left(2\mu \frac{\partial v}{\partial y} - p \right) + \mu \frac{\partial w_{2i}}{\partial x} \left(\frac{\partial u}{\partial y} + \frac{\partial v}{\partial x} \right) - \rho w_{2i} f_y \right] dxdy - \oint_{\Gamma^e} w_{2i} t_y \, ds, \tag{6.2.2}$$

where w_{1i} and w_{2i} are weight functions that have the meaning of velocity u and v, respectively (i.e., $w_{1i} \sim u$ and $w_{2i} \sim v$). This is dictated by the fact that $w_{1i}\, \rho f_x\, dxdy$ and $w_{2i}\, \rho f_y\, dxdy$ are expressions that constitute "power" (or rate of "work done") by the forces $\rho f_x\, dxdy$ and $\rho f_y\, dxdy$, respectively. Other combinations of products are not meaningful.

The weighted-integral statement of Eq. (6.1.3) over a typical element Ω^e is given by

$$0 = \int_{\Omega^e} Q_i \left(\frac{\partial u}{\partial x} + \frac{\partial v}{\partial y} \right) dxdy, \tag{6.2.3}$$

where Q_i are a set of weight functions. Since there is no advantage in transferring derivatives from v_x and v_y to the weight functions Q_i, the integral statement of the first-order equation is used to develop the finite element model. In fact, if one integrates the expression by parts, it will result in boundary expressions that will be in conflict with the classification of the variables as the *primary* and *secondary* variables. Since the coefficient of Q_i in the integrand of Eq. (6.2.3) represents the volume change, Q_i will have the meaning of the pressure variable (which acts towards the surface and causes volume change; i.e., $Q_i \sim p$).

6.2.2 Semidiscretized Finite Element Model

Since we are developing the weak-form Galerkin finite element model, the choice of the weight functions is restricted to the spaces of approximation functions used for the pressure and velocity fields. Suppose that the dependent variables (v_x, v_y, p) are approximated by expansions of the form

$$\begin{aligned} u(x,y,t) &= \sum_{j=1}^{m} \psi_j^{(1)}(x,y) u_j(t) \\ v(x,y,t) &= \sum_{j=1}^{m} \psi_j^{(1)}(x,y) v_j(t) \\ p(x,y,t) &= \sum_{j=1}^{n} \psi_j^{(2)}(x,y) p_j(t), \end{aligned} \tag{6.2.4}$$

where (u_j, v_j, p_j) are the nodal values of (u, v, p) and $\psi_j^{(1)}$ and $\psi_j^{(2)}$ are the two-dimensional interpolation functions discussed in Section 4.3.6. The reason for using the same interpolation functions $(\psi_j^{(1)})$ for both u and v is that they are components of the same vector.

Substitution of Eq. (6.2.4) for (u, v, p) and $w_{1i} = w_{2i} = \psi_i^{(1)}$ and $Q_i = \psi_i^{(2)}$ into Eqs. (6.2.1)–(6.2.3) yields the following semidiscrete finite element equations associated with momentum balance and mass conservation:

$$0 = \sum_{j=1}^{n} \int_{\Omega^e} \left[\rho\, \psi_i^{(1)} \psi_j^{(1)} \frac{du_j}{dt} + \mu \left(2 \frac{\partial \psi_i^{(1)}}{\partial x} \frac{\partial \psi_j^{(1)}}{\partial x} + \frac{\partial \psi_i^{(1)}}{\partial y} \frac{\partial \psi_j^{(1)}}{\partial y} \right) u_j \right] dxdy$$

$$+\sum_{j=1}^{n}\left[\int_{\Omega^e}\mu\frac{\partial\psi_i^{(1)}}{\partial y}\frac{\partial\psi_j^{(1)}}{\partial x}\,dxdy\right]v_j-\sum_{j=1}^{m}\left[\int_{\Omega^e}\frac{\partial\psi_i^{(1)}}{\partial x}\psi_j^{(2)}\,dxdy\right]p_j$$
$$-\int_{\Omega^e}\rho\psi_i^{(1)}f_x\,dxdy-\oint_{\Gamma^e}\psi_i^{(1)}t_x\,ds \tag{6.2.5a}$$

or

$$\mathbf{M}^{11}\dot{\mathbf{u}}+\mathbf{K}_\mu^{11}\mathbf{u}+\mathbf{K}_\mu^{12}\mathbf{v}-\mathbf{K}^{13}\mathbf{p}=\mathbf{F}^1, \tag{6.2.5b}$$

and

$$0=\sum_{j=1}^{n}\Bigg\{\left[\int_{\Omega^e}\rho\psi_i^{(1)}\psi_j^{(1)}\,dxdy\right]\frac{dv_j}{dt}$$
$$+\left[\int_{\Omega^e}\mu\left(2\frac{\partial\psi_i^{(1)}}{\partial y}\frac{\partial\psi_j^{(1)}}{\partial y}+\frac{\partial\psi_i^{(1)}}{\partial x}\frac{\partial\psi_j^{(1)}}{\partial x}\right)dxdy\right]v_j\Bigg\}$$
$$+\sum_{j=1}^{n}\left[\int_{\Omega^e}\mu\frac{\partial\psi_i^{(1)}}{\partial x}\frac{\partial\psi_j^{(1)}}{\partial y}\,dxdy\right]u_j-\sum_{j=1}^{m}\left[\int_{\Omega^e}\frac{\partial\psi_i^{(1)}}{\partial y}\psi_j^{(2)}\,dxdy\right]p_j$$
$$-\int_{\Omega^e}\rho\psi_i^{(1)}f_y\,dxdy-\oint_{\Gamma^e}\psi_i^{(1)}t_y\,ds. \tag{6.2.6a}$$

In matrix form, we have

$$\mathbf{M}^{22}\dot{\mathbf{v}}+\mathbf{K}_\mu^{21}\mathbf{u}+\mathbf{K}_\mu^{22}\mathbf{v}-\mathbf{K}^{23}\mathbf{p}=\mathbf{F}^2. \tag{6.2.6b}$$

The weighted-integral statement of the continuity equation is

$$0=-\sum_{j=1}^{n}\int_{\Omega^e}\psi_i^{(2)}\left(\frac{\partial\psi_j^{(1)}}{\partial x}u_j+\frac{\partial\psi_j^{(1)}}{\partial y}v_j\right)dxdy,\quad i=1,2,\ldots,m \tag{6.2.7a}$$

or

$$-\mathbf{K}^{31}\mathbf{u}-\mathbf{K}^{32}\mathbf{v}=\mathbf{0}. \tag{6.2.7b}$$

The coefficients of the matrices appearing in Eqs. (6.2.5b)–(6.2.7b) are defined as follows:

$$M_{ij}^{11}=M_{ij}^{22}=\int_{\Omega^e}\rho\psi_i^{(1)}\psi_j^{(1)}\,dxdy$$
$$(K_{ij}^{11})_\mu=\int_{\Omega^e}\mu\left(2\frac{\partial\psi_i^{(1)}}{\partial y}\frac{\partial\psi_j^{(1)}}{\partial y}+\frac{\partial\psi_i^{(1)}}{\partial x}\frac{\partial\psi_j^{(1)}}{\partial x}\right)dxdy$$
$$(K_{ij}^{12})_\mu=\int_{\Omega^e}2\mu\frac{\partial\psi_i^{(1)}}{\partial y}\frac{\partial\psi_j^{(1)}}{\partial x}\,dxdy,\quad K_{ij}^{13}=\int_{\Omega^e}\frac{\partial\psi_i^{(1)}}{\partial x}\psi_j^{(2)}\,dxdy$$
$$F_i^1=\int_{\Omega^e}\rho\psi_i^{(1)}f_x\,dxdy+\oint_{\Gamma^e}\psi_i^{(1)}t_x\,ds$$
$$(K_{ij}^{22})_\mu=\int_{\Omega^e}\mu\left(2\frac{\partial\psi_i^{(1)}}{\partial y}\frac{\partial\psi_j^{(1)}}{\partial y}+\frac{\partial\psi_i^{(1)}}{\partial x}\frac{\partial\psi_j^{(1)}}{\partial x}\right)dxdy$$

$$K_{ij}^{23} = \int_{\Omega^e} \frac{\partial \psi_i^{(1)}}{\partial y} \psi_j^{(2)}\, dxdy = K_{ji}^{32}$$

$$F_i^2 = \int_{\Omega^e} \rho \psi_i^{(1)} f_y \, dxdy + \oint_{\Gamma^e} \psi_i^{(1)} t_y \, ds. \tag{6.2.8}$$

We note that

$$\mathbf{K}_\mu^{21} = (\mathbf{K}_\mu^{12})^{\mathrm{T}}, \quad \mathbf{K}^{31} = (\mathbf{K}^{13})^{\mathrm{T}}, \quad \mathbf{K}^{32} = (\mathbf{K}^{23})^{\mathrm{T}}. \tag{6.2.9}$$

Equations (6.2.5b), (6.2.6b), and (6.2.7b) can be written symbolically in matrix form as

$$\begin{bmatrix} \mathbf{M}^{11} & \mathbf{0} & \mathbf{0} \\ \mathbf{0} & \mathbf{M}^{22} & \mathbf{0} \\ \mathbf{0} & \mathbf{0} & \mathbf{0} \end{bmatrix} \begin{Bmatrix} \dot{\mathbf{u}} \\ \dot{\mathbf{v}} \\ \dot{\mathbf{p}} \end{Bmatrix} + \begin{bmatrix} \mathbf{K}^{11} & \mathbf{K}^{12} & -\mathbf{K}^{13} \\ \mathbf{K}^{21} & \mathbf{K}^{22} & -\mathbf{K}^{23} \\ -\mathbf{K}^{31} & -\mathbf{K}^{32} & \mathbf{0} \end{bmatrix} \begin{Bmatrix} \mathbf{u} \\ \mathbf{v} \\ \mathbf{p} \end{Bmatrix} = \begin{Bmatrix} \mathbf{F}_1 \\ \mathbf{F}_2 \\ \mathbf{0} \end{Bmatrix}. \tag{6.2.10}$$

In single array form $\boldsymbol{\Delta} = \{\mathbf{u}\ \mathbf{v}\ \mathbf{p}\}^{\mathrm{T}}$, Eq. (6.2.9) can be expressed as

$$\mathbf{M}\dot{\boldsymbol{\Delta}} + \mathbf{K}\boldsymbol{\Delta} = \mathbf{F}. \tag{6.2.11}$$

The meaning of **M**, **K**, and **F** should be obvious from Eq. (6.2.10).

The two sets of interpolation functions used in Eq. (6.2.4) should be of the Lagrange type [i.e., Eq. (6.2.4) interpolate only the values of the functions, and not their derivatives]. There are two different finite elements associated with the two field variables (u, v, p), hence there are two different finite element meshes corresponding to the two variables over the same domain, Ω. If one of the meshes contains the other mesh as a subset, then we choose to display the first mesh and indicate the nodal degrees of freedom associated with the nodes of a typical element of the mesh.

As is clear from the weak forms, the finite element model requires only the C^0-continuous functions (i.e., only the function and not its derivatives are continuous across elements) to approximate velocities and pressure. Thus, any of the Lagrange and serendipity family of interpolation functions are admissible for the interpolation of the velocity and pressure fields.

The interpolation used for the pressure variable should be different from that used for the velocities, because the weak forms in Eqs. (6.2.1) and (6.2.2) contain the first derivatives of the velocity components (u, v) and no derivatives of the pressure p. In addition, the essential boundary conditions of the formulation do not include specification of the pressure; p enters the boundary conditions as a part of the natural boundary conditions. This implies that the pressure variable need not be carried as a variable that is continuous across interelement boundaries. These observations lead to the conclusion that the pressure variable should be interpolated with functions that are one order less than those used for the velocity field. Thus, (a) linear interpolation of the velocities and element-wise constant representation of the pressure and (b) quadratic interpolation of (u, v) and linear interpolation of p are admissible (see Fig. 6.2.1). Finite element models that use the same degree of interpolation of the velocities as well as pressure are known to give inaccurate results for the pressure (see Gresho and Sani [4] and references therein).

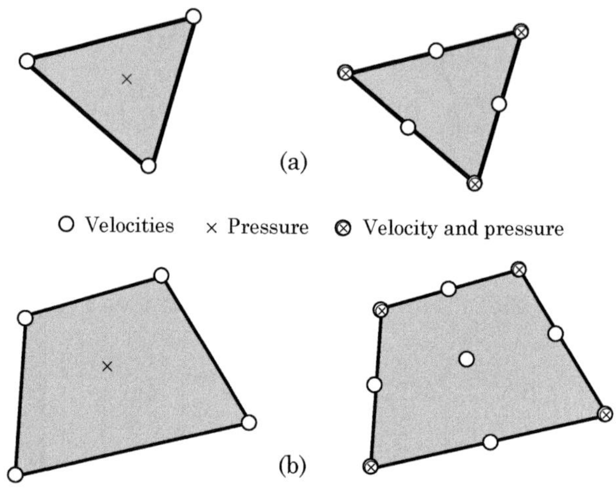

Fig. 6.2.1 Velocity–pressure finite element models with velocity and pressure degrees of freedom. Velocity and pressure are interpolated with different degrees of freedom.

6.2.3 Fully Discretized Equations

The general form of Eq. (6.2.10) is the same as Eq. (4.2.6), with **C** replaced by **M**. Therefore, the alpha-family of the time-approximation scheme discussed in Section 4.2.5 is readily applicable to the ODEs in Eq. (6.2.10). Therefore, we have

$$\hat{\mathbf{K}}\boldsymbol{\Delta}^{s+1} = \bar{\mathbf{K}}\boldsymbol{\Delta}^{s} + \bar{\mathbf{F}}^{s,s+1} \equiv \hat{\mathbf{F}}^{s,s+1}, \tag{6.2.12a}$$

where

$$\begin{aligned} \hat{\mathbf{K}} &= \mathbf{M} + a_1\mathbf{K}, \quad \bar{\mathbf{K}} = \mathbf{M} - a_2\mathbf{K} \\ \bar{\mathbf{F}}^{s,s+1} &= \left(a_1\mathbf{F}^{s+1} + a_2\mathbf{F}^{s}\right) \\ a_1 &= \alpha\,\Delta t, \quad a_2 = (1-\alpha)\Delta t, \end{aligned} \tag{6.2.12b}$$

and $0 \leq \alpha \leq 1$ defines various schemes:

$$\alpha = \begin{cases} 0, & \text{the forward difference scheme (conditionally stable),} \\ & \text{order of accuracy } = O(\Delta t); \\ \frac{1}{2}, & \text{the Crank–Nicolson scheme (stable),} \\ & \text{order of accuracy } = O(\Delta t)^2; \\ 1, & \text{the backward difference scheme (stable),} \\ & \text{order of accuracy } = O(\Delta t). \end{cases} \tag{6.2.13}$$

The conditionally stable schemes require use of the time step Δt that is less than Δt_{cr}:

$$\Delta t < \Delta t_{cr} \equiv \frac{2}{(1-2\alpha)\lambda_{max}}, \quad \alpha < \frac{1}{2}. \tag{6.2.14}$$

A few remarks concerning the pressure–velocity model, also known as the *mixed model*, are in order. An examination of the finite element model in Eq.

(6.2.10) shows that zeros appear on the matrix diagonals corresponding to the pressure variables. Direct equation solvers must use some type of pivoting strategy. The fact that the pressure does not appear explicitly in the continuity equation makes the system time-singular in the pressure and precludes the use of purely explicit time-integration methods.

The computer implementation of the mixed model is complicated by the fact that the element contains variable degrees of freedom at each node. For example, with the quadratic approximation of the velocity field and bilinear continuous approximation of the pressure, the element has three degrees of freedom (v_x, v_y, P) at the corner nodes and two degrees of freedom (v_x, v_y) at the midside and interior nodes. This complicates the calculation of element matrices as well as the assembly of element equations to form the global system of equations.

6.3 Penalty Finite Element Model

6.3.1 Weak Forms

The penalty function method is a method in which a variational problem with a constraint is reformulated as one without constraint by including the constraint into the variational statement. In the interest of simplicity, the theoretical details of the method are not outlined here. Interested readers may consult [1, 3] and [41–46]. The main feature of the penalty method is that a differential constraint is included in the weak form of the constrained problem in a weighted-integral sense, without bringing additional variables into the formulation.

Consider the weak forms, Eqs. (6.2.1) and (6.1.2), of the two momentum equations. Adding the two statements, we obtain

$$\begin{aligned}
0 &= \int_{\Omega^e} \left[\rho w_{1i}\frac{\partial v_x}{\partial t} + \frac{\partial w_{1i}}{\partial x}\left(2\mu\frac{\partial v_x}{\partial x} - p\right) + \mu\frac{\partial w_{1i}}{\partial y}\left(\frac{\partial u}{\partial y} + \frac{\partial v_y}{\partial x}\right) - \rho w_{1i} f_x\right] dxdy \\
&\quad - \oint_{\Gamma^e} w_{1i} t_x \, ds + \int_{\Omega^e} \left[\rho w_{2i}\frac{\partial v_y}{\partial t} + \frac{\partial w_{2i}}{\partial y}\left(2\mu\frac{\partial v_y}{\partial y} - p\right)\right. \\
&\quad \left. + \mu\frac{\partial w_{2i}}{\partial x}\left(\frac{\partial u}{\partial y} + \frac{\partial v_y}{\partial x}\right) - \rho w_{2i} f_y\right] dxdy - \oint_{\Gamma^e} w_{2i} t_y \, ds \\
&= \int_{\Omega^e} \left[\rho\left(w_{1i}\frac{\partial v_x}{\partial t} + w_{2i}\frac{\partial v_y}{\partial t}\right) + 2\mu\left(\frac{\partial w_{1i}}{\partial x}\frac{\partial v_x}{\partial x} + \frac{\partial w_{2i}}{\partial y}\frac{\partial v_y}{\partial y}\right)\right. \\
&\quad \left. + \mu\left(\frac{\partial w_{1i}}{\partial y} + \frac{\partial w_{2i}}{\partial x}\right)\left(\frac{\partial u}{\partial y} + \frac{\partial v_y}{\partial x}\right) - \left(\frac{\partial w_{1i}}{\partial x} + \frac{\partial w_{2i}}{\partial y}\right)p\right] dxdy \\
&\quad - \int_{\Omega^e} \rho\left(w_{1i} f_x + w_{2i} f_y\right) dxdy - \oint_{\Gamma^e} \left(w_{1i} t_x + w_{2i} t_y\right) ds. \qquad (6.3.1)
\end{aligned}$$

We note that the above expression is equal to the two statements in Eqs. (6.2.1) and (6.2.2), separated into two equations by terms containing w_{1i} and w_{2i}. Now we assume that the incompressibility condition in Eq. (6.1.3) is satisfied by the velocity components. Then the weight functions w_{1i} and w_{2i} must also satisfy the constraint condition

$$\frac{\partial w_{1i}}{\partial x} + \frac{\partial w_{2i}}{\partial y} = 0. \tag{6.3.2}$$

As a result, the expression containing P in Eq. (6.3.1) drops out. Now the (variational) problem with the constraint can be stated as: Find (v_x, v_y), which satisfy the constraint condition in Eq. (6.1.3), such that the following weak form holds:

$$\begin{aligned} 0 = &\int_{\Omega^e} \left[\rho \left(w_{1i} \frac{\partial v_x}{\partial t} + w_{2i} \frac{\partial v_y}{\partial t} \right) + 2\mu \left(\frac{\partial w_{1i}}{\partial x} \frac{\partial v_x}{\partial x} + \frac{\partial w_{2i}}{\partial y} \frac{\partial v_y}{\partial y} \right) \right. \\ &\left. + \mu \left(\frac{\partial w_{1i}}{\partial y} + \frac{\partial w_{2i}}{\partial x} \right) \left(\frac{\partial u}{\partial y} + \frac{\partial v_y}{\partial x} \right) \right] dxdy \\ &- \int_{\Omega^e} \rho \left(w_{1i} f_x + w_{2i} f_y \right) dxdy - \oint_{\Gamma^e} \left(w_{1i} t_x + w_{2i} t_y \right) ds. \end{aligned} \tag{6.3.3}$$

The penalty function method allows us to include the constraint, namely the incompressibility condition in Eq. (6.1.3), into the variational problem defined by Eq. (6.3.3) by adding the following *penalty term*:

$$\int_{\Omega^e} \gamma \left(\frac{\partial w_{1i}}{\partial x} + \frac{\partial w_{2i}}{\partial x} \right) \left(\frac{\partial v_x}{\partial x} + \frac{\partial v_y}{\partial y} \right) dxdy, \tag{6.3.4}$$

where γ, a positive parameter, is called the *penalty parameter*, whose value is preassigned. The larger the value of γ, the more accurately the constraint is satisfied. Obviously, it should not be too large to make other terms in the total expression be negligible. More on this is presented in the sequel. Thus, the modified weak form is given by

$$\begin{aligned} 0 = &\int_{\Omega^e} \left[\rho \left(w_{1i} \frac{\partial v_x}{\partial t} + w_{2i} \frac{\partial v_y}{\partial t} \right) + 2\mu \left(\frac{\partial w_{1i}}{\partial x} \frac{\partial v_x}{\partial x} + \frac{\partial w_{2i}}{\partial y} \frac{\partial v_y}{\partial y} \right) \right. \\ &\left. + \mu \left(\frac{\partial w_{1i}}{\partial y} + \frac{\partial w_{2i}}{\partial x} \right) \left(\frac{\partial u}{\partial y} + \frac{\partial v_y}{\partial x} \right) \right] dxdy \\ &+ \int_{\Omega^e} \gamma \left(\frac{\partial w_{1i}}{\partial x} + \frac{\partial w_{2i}}{\partial y} \right) \left(\frac{\partial v_x}{\partial x} + \frac{\partial v_y}{\partial y} \right) dxdy \\ &- \int_{\Omega^e} \rho \left(w_{1i} f_x + w_{2i} f_y \right) dxdy - \oint_{\Gamma^e} \left(w_{1i} t_x + w_{2i} t_y \right) ds \\ = &\int_{\Omega^e} \left\{ \rho \left(w_{1i} \frac{\partial v_x}{\partial t} + w_{2i} \frac{\partial v_y}{\partial t} \right) + \frac{\partial w_{1i}}{\partial x} \left[2\mu \frac{\partial v_x}{\partial x} + \gamma \left(\frac{\partial v_x}{\partial x} + \frac{\partial v_y}{\partial y} \right) \right] \right. \\ &+ \frac{\partial w_{2i}}{\partial y} \left[2\mu \frac{\partial v_y}{\partial y} + \gamma \left(\frac{\partial v_x}{\partial x} + \frac{\partial v_y}{\partial y} \right) \right] \\ &\left. + \mu \left(\frac{\partial w_{1i}}{\partial y} + \frac{\partial w_{2i}}{\partial x} \right) \left(\frac{\partial u}{\partial y} + \frac{\partial v_y}{\partial x} \right) \right\} dxdy \end{aligned}$$

$$- \int_{\Omega^e} \rho\, (w_{1i} f_x + w_{2i} f_y)\, dxdy - \oint_{\Gamma^e} (w_{1i} t_x + w_{2i} t_y)\, ds. \tag{6.3.5}$$

Equation (6.3.5) forms the basis of the penalty finite element model (i.e., it takes the place of the weak form). The statement in Eq. (6.3.5) is equivalent to the following two statements (i.e., separate into two expressions on the basis of terms containing w_{1i} and w_{2i}):

$$\begin{aligned} 0 = \int_{\Omega^e} \Big\{ & \rho w_{1i} \frac{\partial v_x}{\partial t} + \frac{\partial w_{1i}}{\partial x} \left[2\mu \frac{\partial v_x}{\partial x} + \gamma \left(\frac{\partial v_x}{\partial x} + \frac{\partial v_y}{\partial y} \right) \right] \\ & + \mu \frac{\partial w_{1i}}{\partial y} \left(\frac{\partial u}{\partial y} + \frac{\partial v_y}{\partial x} \right) \Big\} dxdy \\ & - \int_{\Omega^e} \rho w_{1i} f_x dxdy - \oint_{\Gamma^e} w_{1i} t_x \, ds, \end{aligned} \tag{6.3.6}$$

$$\begin{aligned} 0 = \int_{\Omega^e} \Big\{ & \rho w_{2i} \frac{\partial v_y}{\partial t} + \frac{\partial w_{2i}}{\partial y} \left[2\mu \frac{\partial v_y}{\partial y} + \gamma \left(\frac{\partial v_x}{\partial x} + \frac{\partial v_y}{\partial y} \right) \right] \\ & + \mu \frac{\partial w_{2i}}{\partial x} \left(\frac{\partial u}{\partial y} + \frac{\partial v_y}{\partial x} \right) \Big\} dxdy \\ & - \int_{\Omega^e} \rho w_{2i} f_y \, dxdy - \oint_{\Gamma^e} w_{2i} t_y \, ds. \end{aligned} \tag{6.3.7}$$

A close comparison of Eqs. (6.3.6) and (6.3.7) with Eqs. (6.2.1) and (6.2.2), shows that if we replace P in Eqs. (6.3.6) and (6.3.7) with

$$\begin{aligned} p_\gamma^e &= -\gamma_e \left(\frac{\partial v_x^e}{\partial x} + \frac{\partial v_y^e}{\partial y} \right) \\ &= -\gamma^e \sum_{j=1}^{m} \left(\frac{\partial \psi_j^{(1)}}{\partial x} v_x^j + \frac{\partial \psi_j^{(1)}}{\partial y} v_y^j \right), \end{aligned} \tag{6.3.8}$$

we arrive at Eqs. (6.3.6) and (6.3.7). Thus, in the penalty finite element model one can postcompute the pressure using Eq. (6.3.8) once the velocity field is known.

6.3.2 Finite Element Model

The penalty finite element model is obtained from Eqs. (6.3.6) and (6.3.7) by substituting finite element interpolation (6.2.4) for the velocity field, and setting $w_{1i} = w_{2i} = \psi_i^{(1)}$:

$$\begin{bmatrix} \mathbf{M}^{11} & \mathbf{0} \\ \mathbf{0} & \mathbf{M}^{22} \end{bmatrix} \begin{Bmatrix} \dot{\mathbf{v}}_x \\ \dot{v}_y \end{Bmatrix} + \begin{bmatrix} \mathbf{K}_\mu^{11} + \mathbf{K}_\gamma^{11} & \mathbf{K}_\mu^{12} + \mathbf{K}_\gamma^{12} \\ \mathbf{K}_\mu^{21} + \mathbf{K}_\gamma^{21} & \mathbf{K}_\mu^{22} + \mathbf{K}_\gamma^{22} \end{bmatrix} \begin{Bmatrix} \mathbf{v}_x \\ \mathbf{v}_y \end{Bmatrix} = \begin{Bmatrix} \mathbf{F}^1 \\ \mathbf{F}^2 \end{Bmatrix} \tag{6.3.9a}$$

or

$$\mathbf{M}\dot{\mathbf{\Delta}} + \mathbf{K}\mathbf{\Delta} = \mathbf{F}, \quad \mathbf{\Delta} = \{\mathbf{v}_x \; \mathbf{v}_y\}, \tag{6.3.9b}$$

where $M_{ij}^{11} = M_{ij}^{22}$, $(\mathbf{K}_{ij}^{\alpha\beta})_\mu$ for $\alpha, \beta = 1, 2$, and $\mathbf{F}^\alpha$ for $\alpha = 1, 2$ are the same as those defined in Eq. (6.2.8), and $(\mathbf{K}_{ij}^{\alpha\beta})_\gamma$ are defined by

$$
\begin{aligned}
(K_{ij}^{11})_\gamma &= \int_{\Omega^e} \gamma \frac{\partial \psi_i^{(1)}}{\partial x} \frac{\partial \psi_j^{(1)}}{\partial x} \, dx dy \\
(K_{ij}^{12})_\gamma &= \int_{\Omega^e} \gamma \frac{\partial \psi_i^{(1)}}{\partial x} \frac{\partial \psi_j^{(1)}}{\partial y} \, dx dy = (K_j^{21} i)_\gamma \\
(K_{ij}^{22})_\gamma &= \int_{\Omega^e} \gamma \frac{\partial \psi_i^{(1)}}{\partial y} \frac{\partial \psi_j^{(1)}}{\partial y} \, ds.
\end{aligned}
\tag{6.3.9c}
$$

The fully discretized finite element model of Eq. (6.3.9b) can be obtained as described in Section 6.2.3. As is clear from the weak forms, the penalty finite element model also requires only the C^0-continuous functions to approximate the velocities. The most commonly used penalty finite elements are shown in Fig. 6.3.1.

The numerical evaluation of the coefficient matrices appearing in Eq. (6.3.9b) requires special consideration. For the steady-state flows with constant material properties, Eq. (6.3.9b) is of the form

$$
(\mathbf{K}_v + \mathbf{K}_p)\, \boldsymbol{\Delta} = \mathbf{F}, \tag{6.3.10}
$$

where $\mathbf{K}_v$ is the contribution from the viscous terms and $\mathbf{K}_p$ is the contribution from the penalty terms (and depends on the penalty parameter γ) , which comes from the incompressibility constraint . In theory, as we increase the value of γ,

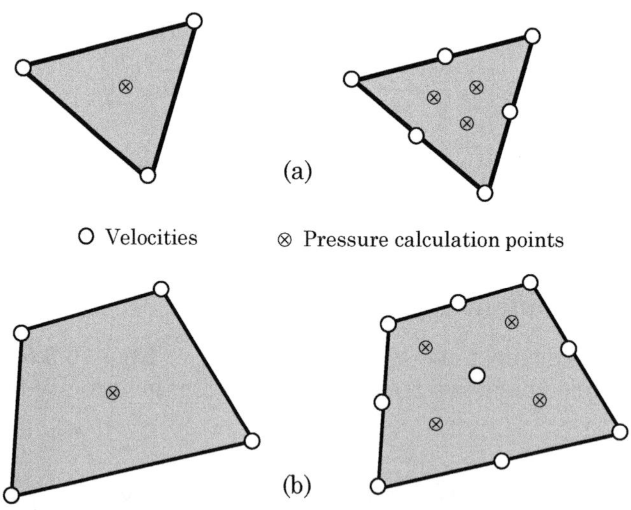

Fig. 6.3.1 Penalty finite elements with velocity degrees of freedom and Gauss points for the post-computation of the pressure.

the conservation of mass is satisfied more accurately. However, in practice, for some large value of γ, the contribution from the viscous terms would be negligibly small compared with the penalty terms. Thus, if $\mathbf{K}_p$ is a nonsingular (i.e., invertible) matrix, the solution of Eq. (6.3.9b) (after its assembly and imposition of boundary conditions) for a large value of γ is trivial, $\boldsymbol{\Delta} = \mathbf{0}$. While the solution satisfies the continuity equation, it does not satisfy the momentum equations. In this case the discrete problem (6.3.10) is said to be over-constrained or "locked." If $\mathbf{K}_p$ is singular, then the sum $(\mathbf{K}_v + \mathbf{K}_p)$ is nonsingular (because $\mathbf{K}_v$ is nonsingular), and a nontrivial solution to the problem is obtained.

The numerical problem described above is eliminated by proper evaluation of the integrals in $\mathbf{K}_v$ and $\mathbf{K}_p$. It is found that if the coefficients of $\mathbf{K}_p$ (i.e., penalty matrix coefficients) are evaluated using a numerical integration rule of an order less than that required to integrate them exactly, the finite element equations (6.3.10) give acceptable solutions for the velocity field. This technique of underintegrating the penalty terms is known in the literature as *reduced integration* and the finite element model is termed the *reduced integration penalty* (RIP) finite element model. For example, if a linear quadrilateral element is used to approximate the velocity field, the matrix coefficients $\mathbf{K}_v$ (as well as $\mathbf{M}$ for unsteady problems) are evaluated using the 2×2 Gauss quadrature, and $\mathbf{K}_p$ is evaluated using the one-point (1×1) Gauss quadrature. The one-point quadrature yields a singular $\mathbf{K}_p$. Therefore, Eq. (6.3.9b) can be solved because $(\mathbf{K}_v + \mathbf{K}_p)$ is nonsingular and can be inverted (after assembly and imposition of boundary conditions) to obtain a good finite element solution of the original problem. When a quadratic quadrilateral element is used, the 3×3 Gauss quadrature is used to evaluate $\mathbf{K}_v$, and $\mathbf{M}$, and the 2×2 Gauss quadrature is used to evaluate $\mathbf{K}_p$.

6.3.3 Postcomputation

For a plane two-dimensional flow, the stress components $(\sigma_{xx}, \sigma_{yy}, \sigma_{xy})$ can be computed using

$$
\begin{aligned}
\sigma_{xx}(x,y) &= 2\mu \sum_{j=1}^{n} \frac{\partial \psi_j^{(1)}}{\partial x} u_j - p_\gamma(x,y) \\
\sigma_{yy}(x,y) &= 2\mu \sum_{j=1}^{n} \frac{\partial \psi_j^{(1)}}{\partial y} v_j - p_\gamma(x,y) \\
\sigma_{xy}(x,y) &= \mu \sum_{j=1}^{n} \left(\frac{\partial \psi_j^{(1)}}{\partial y} u_j + \frac{\partial \psi_j^{(1)}}{\partial x} v_j \right),
\end{aligned}
\tag{6.3.11}
$$

where p_γ is calculated from Eq. (6.3.8). In the velocity–pressure model, p is calculated using Eq. (6.2.4). The derivatives of the interpolation functions in Eqs. (6.3.8) and (6.3.11) must be evaluated using the reduced integration. Thus, the stresses as well as the pressure are computed using the one-point Gauss rule for linear elements and with the 2×2 Gauss rule for the quadratic elements when rectangular elements are used. This is equivalent to using an interpolation for pressure that is one order less than the one used for the velocity field. The pressure computed using Eq. (6.3.8) at the reduced integration points is not always reliable and accurate. The pressures predicted using the linear elements are seldom acceptable. Quadratic elements are known to yield more reliable results. The choice of the penalty parameter is largely dictated by the ratio of the magnitude of the penalty terms to the viscous terms, the mesh, and the precision of the computer. Generally, a value of $\gamma = 10^4\mu$ to $\gamma = 10^{12}\mu$, where μ denotes the viscosity, gives good results.

6.3.4 Numerical Examples

In this section, we consider five numerical examples (see Reddy [1, 40] for additional information). Both velocity–pressure and penalty finite element models are evaluated in the context of slow flows (i.e., nonlinear effects are omitted). The first three examples deal with steady-state flows.

Example 6.3.1

Consider a steady flow between two parallel plates separated by a distance b (the plates are of length a and infinite into the plane of the paper), as shown in Fig. 6.3.2. Determine the velocity fields for the following two cases: (a) both walls are stationary and (b) one of the wall is stationary and the other wall is moving at a velocity of $u(x, b) = U_0$. Assume that the left side is subject to pressure P_L and the right side is subjected to a pressure of p_R ($p_L \neq p_R$).

Solution This is a parallel flow problem (in a parallel flow only one component of the velocity vector is nonzero, all fluid particles moving in one direction [35]), where only u is not zero and $\partial u/\partial x = 0$ by the continuity equation (6.2.3). Thus, $u = u(y)$. Then the momentum equations, (6.1.1) and (6.1.2), reduce to

$$\frac{dp}{dy} = 0, \quad -\mu\frac{d^2u}{dy^2} = -\frac{dp}{dx} = \frac{p_L - p_R}{a}, \tag{6.3.12}$$

where a is the horizontal distance between the sides. Therefore, this is a one-dimensional problem requiring only one element in the x-direction and several elements in the y-direction.

We discuss the details of the analysis for a 2×8 mesh of four-node linear (L4) elements for the penalty model, and 1×4 mesh of nine-node quadratic (Q9) elements for the velocity–pressure model. Figure 6.3.3 contains the meshes and the boundary conditions ($p_L = p = 1$, $p_R = 0$, $a = 1$, $b = 4$).

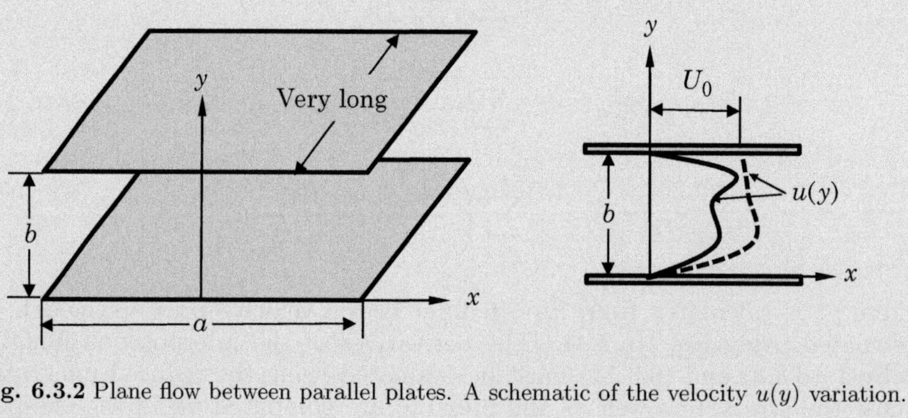

Fig. 6.3.2 Plane flow between parallel plates. A schematic of the velocity $u(y)$ variation.

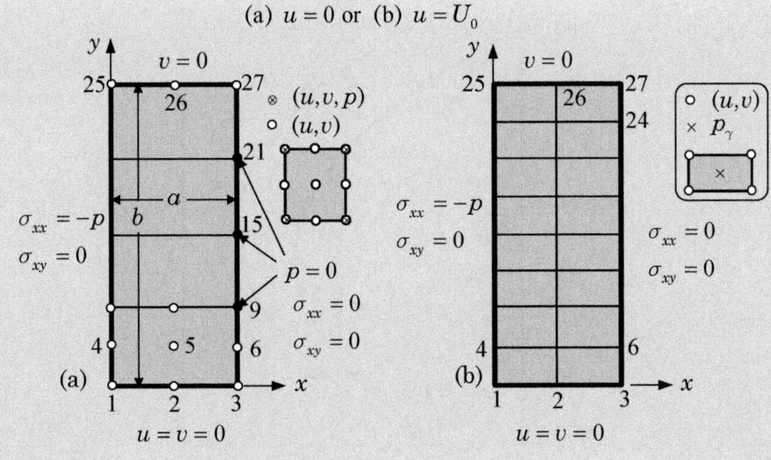

Fig. 6.3.3 Meshes used for the (a) velocity–pressure finite element model and (b) penalty finite element model of plane flow between parallel plates. A typical element of each model with nodal degrees of freedom is shown. In the case of the penalty finite element model, the pressure p_γ is post-computed using Eq. (6.3.8).

The specified boundary are:

$$\begin{aligned}
&\text{At } x=0: \;\; t_x = -\sigma_{xx} = -\Big(2\mu\frac{du}{dx} - p\Big) = p, \;\; t_y = -\sigma_{xy} = 0.\\
&\text{At } x=a: \;\; t_x = \sigma_{xx} = 0, \;\; t_y = \sigma_{xy} = 0.\\
&\text{At } y=0: \;\; t_x = -\sigma_{xy} = 0, \;\; t_y = -\sigma_{yy} = 0.\\
&\text{At } y=b: \;\; u = 0 \text{ or } U_0, \;\; t_y = \sigma_{yy} = 0.
\end{aligned} \tag{6.3.13}$$

One must calculate the forces corresponding to $p = 1$ at $x = 0$. For the linear element, we have $f_i = ph/2$, where $h = 4/8 = 0.5$ is the element height; and for the quadratic element they are $f_1 = ph/6$, $f_2 = 4ph/6$, and $f_3 = ph/6$, with $h = 4/4 = 1$. Thus, we have

$$F_4 = 0.666667, \;\; F_7 = 0.333333, \;\; F_{10} = 0.666667, \;\; F_{13} = 0.333333$$
$$F_{16} = 0.666667, \;\; F_{19} = 0.333333, \;\; F_{22} = 0.666667$$

for the velocity–pressure model, and

$$F_4 = 0.5, \;\; F_7 = 0.5, \;\; F_{10} = 0.5, \;\; F_{13} = 0.5, \;\; F_{16} = 0.5, \;\; F_{19} = 0.5, \;\; F_{22} = 0.5$$

for the penalty model. These values enter as boundary conditions on the secondary variables. The boundary conditions on the velocity field enter as boundary conditions on the primary variables.

Figure 6.3.4 contains the results of the analysis. The finite element solutions match with the analytical (or exact) solution ($\frac{dp}{dx} = -1$)

$$u(y) = \frac{y}{b}U_0 - \frac{b^2}{2\mu}\frac{y}{b}\Big(1 - \frac{y}{b}\Big)\frac{dp}{dx}. \tag{6.3.14}$$

The influence of the magnitude of the penalty parameter γ is also illustrated in the figure.

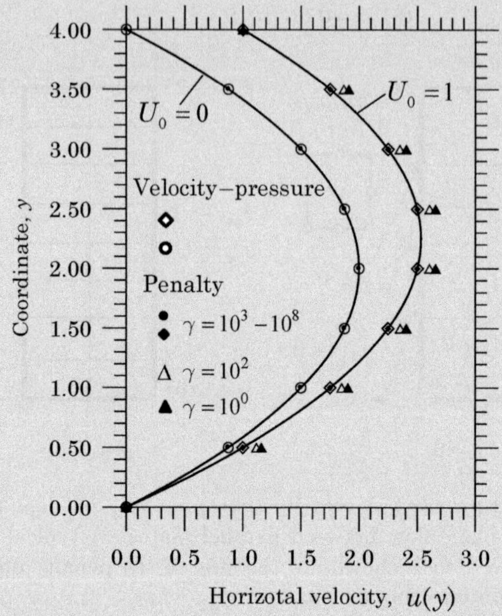

Fig. 6.3.4 Comparison of the finite element solutions with the exact solutions.

Example 6.3.2

Consider creeping flow induced by two parallel plates, with a viscous fluid between them, moving towards each other (see Fig. 6.3.5). When the length of the plates (into the plane of the paper) is very large compared with both the width and the distance between the plates, we have a case of plane flow. We wish to determine the velocity and pressure fields for a fixed distance between the plates, assuming that a state of plane flow exists. Exploit biaxial symmetry and use 12×8 mesh of nine-node quadratic elements to analyze the problem for velocity and pressure fields.

Solution Let V_0 be the velocity with which the two plates are moving toward each other, and let $2b$ and $2a$ denote, respectively, the distance between and the length of the plates at a particular instant of time. Because of the biaxial symmetry, it suffices to model only a quadrant of the domain.

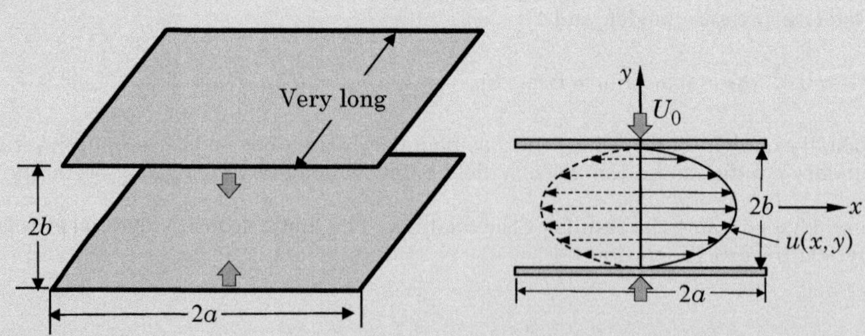

Fig. 6.3.5 Slow flow of viscous incompressible fluid between parallel plates.

The following nonuniform meshes (12×8 mesh of nine-node quadratic elements is shown in Fig. 6.3.6; the 24×16 mesh of linear elements is obtained by subdividing the 12×8 mesh of nine-node quadratic elements), with smaller elements near the free surface (i.e., at $x = a$), are used to approximate accurately the singularity in the shear stress at the point $(a, b) = (6, 2)$. The following subdivisions along the x and y coordinates are used:

$$\{DX\} = \{0.5\ 0.5\ 0.5\ 0.5\ 0.5\ 0.5\ 0.5\ 0.5\ 0.25\ 0.25\ 0.25\ 0.25\ 0.125\ 0.125\ 0.125\ 0.125\ 0.125\ 0.125\ 0.0625\ 0.0625\ 0.03125\ 0.03125\ 0.03125\ 0.03125\},$$

$$\{DY\} = \{0.0625\ 0.0625\ 0.0625\ 0.0625\ 0.25\ 0.25\ 0.25\ 0.25\ 0.125\ 0.125\ 0.125\ 0.125\ 0.0625\ 0.0625\ 0.0625\ 0.0625\}.$$

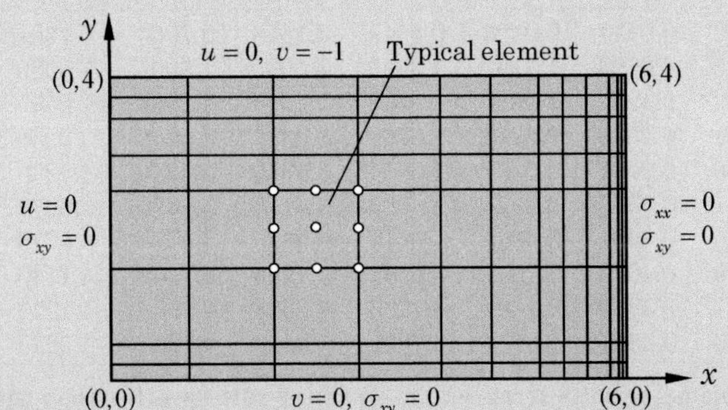

Fig. 6.3.6 Mesh and boundary conditions for the fluid squeezed between parallel plates.

The boundary conditions are also depicted in Fig. 6.3.6. In the velocity–pressure finite element model, it is necessary to specify the pressure at least at one node. In the present case, the node at $(x, y) = (a, 0)$ is specified to have zero pressure.

The problem at hand has no exact solution because of singularities at points $(x, y) = (a, b)$, $(x, y) = (a, 0)$, and $(x, y) = (0, b)$. An approximate analytical solution to this two-dimensional problem is provided by Nadai [56], and it is given by

$$u(x,y) = \frac{3U_0 x}{2b}\left(1 - \frac{y^2}{b^2}\right), \quad v(x,y) = -\frac{3U_0 y}{2b}\left(3 - \frac{y^2}{b^2}\right),$$
$$p(x,y) = \frac{3\mu U_0}{2b^3}(a^2 + y^2 - x^2). \tag{6.3.15}$$

The velocities $u(x, 0)$ obtained with the two finite element models compare well with the analytical solution, as shown in Table 6.3.1. The nine-node element gives very good results for both the penalty and mixed models. The influence of the penalty parameter on the accuracy of the solution is clear from the results. Whether the element is linear or quadratic, it is necessary to use a large value of the penalty parameter ($\gamma = 10^8$).

Figure 6.3.7 contains plots of the velocity $v_x(x, y)$ for $x = 2$, $x = 4$, and $x = 6$ obtained with velocity–pressure (mixed) and penalty finite element models, and Fig. 6.3.8 contains plots of pressure $P(x, y)$, for $y = y_0$, where y_0 is the y-coordinate of the Gauss point nearest to the centerline or top plate. The pressure is computed in the penalty model using Eq. (6.3.8) with the 2×2 Gauss rule for the quadratic rectangular element, whereas in the mixed model (as well as the analytical solution) it is computed at the nodes. If the pressure in the penalty model were computed using the full quadrature rule for rectangular elements, we would have obtained erroneous values. The oscillations in pressure computed nearest to the top plate are the result of the singularity in the boundary conditions at $(x, y) = (6, 2)$. Also, one should not think that the finite element solutions are in error when compared with Nadai's approximate solution; rather one may see that Nadai's solution [56] is not a bad approximation of the true solution.

Table 6.3.1 Comparison of finite element solutions $u(x,0)$ with the analytical solution for fluid squeezed between plates.

	penalty model						mixed model	Series solution
	$\gamma = 10^0$		$\gamma = 10^2$		$\gamma = 10^8$			
x	Four node	Nine[a] node	Four node	Nine node	Four node	Nine node		
1.00	0.0309	0.0311	0.6525	0.6514	0.7516	0.7500	0.7501	0.7500
2.00	0.0690	0.0694	1.3102	1.3080	1.5037	1.5005	1.5000	1.5000
3.00	0.1232	0.1237	1.9810	1.9776	2.2595	2.2549	2.2555	2.2500
4.00	0.2061	0.2068	2.6819	2.6769	3.0335	3.0268	3.0256	3.0000
4.50	0.2630	0.2637	3.0481	3.0417	3.4310	3.4228	3.4235	3.3750
5.00	0.3311	0.3317	3.2314	3.4007	3.8175	3.8076	3.8072	3.7500
5.25	0.3685	0.3691	3.4085	3.5619	4.9890	3.9778	3.9779	3.9375
5.50	0.4069	0.4073	3.7037	3.6941	4.1266	4.1146	4.1145	4.1250
5.75	0.4447	0.4450	3.7881	3.7783	4.2088	4.1966	4.1965	4.3125
6.00	0.4803	0.4807	3.8014	3.7927	4.2103	4.1995	4.2025	4.5000

[a] 2×2 Gauss rule for penalty terms and 3×3 Gauss rule for other than the penalty terms.

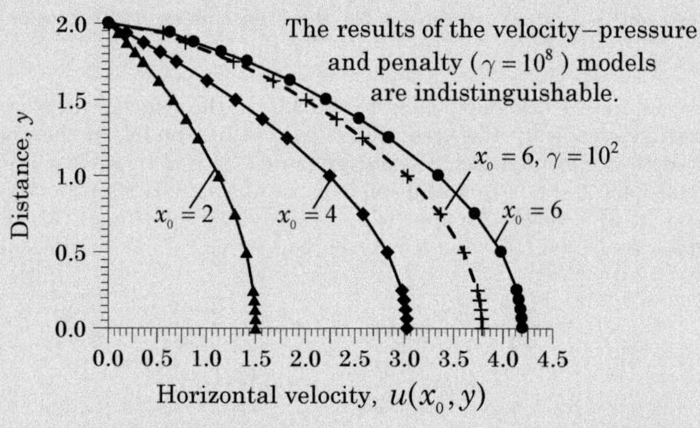

Fig. 6.3.7 Velocity fields for fluid squeezed between parallel plates.

Example 6.3.3

Consider the laminar flow of a viscous, incompressible fluid ($\mu = 1$ for computational purposes) in a unit square cavity bounded by three motionless walls (two side walls and the bottom) and the top (lid) moving at a constant velocity in its own plane, as depicted in Fig. 6.3.9. Singularities exist at each corner where the moving lid meets a fixed wall. Use the velocity–pressure and penalty finite element models to determine the velocity and pressure fields in the cavity and plot center-line velocities as a function of the distance for a 8×10 mesh of nine-node quadratic elements (and equivalent 16×20 mesh of bi-linear elements).

Solution At the singular points, namely at the top corners of the lid, we assume that $u(x, 1) = U_0 = 1.0$. The following subdivisions along the x and y coordinates are used for the 16×20 mesh of bi-linear elements (equivalent mesh of 8×10 meshes of nine-node quadratic elements are used in the velocity–pressure and penalty models):

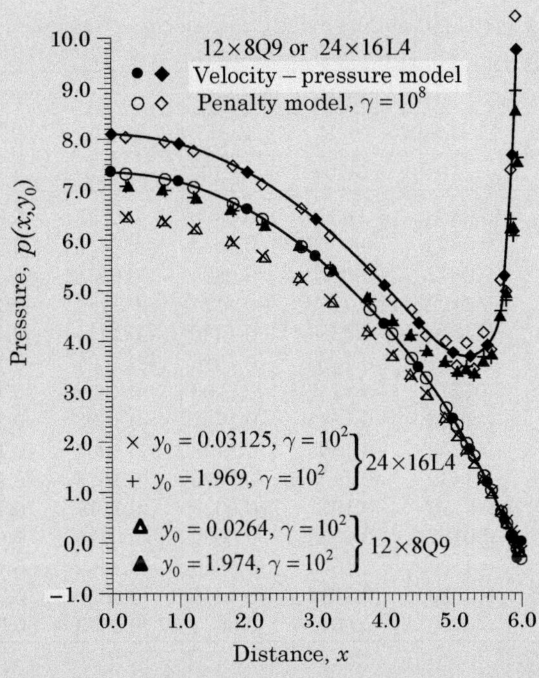

Fig. 6.3.8 Pressures for fluid squeezed between parallel plates.

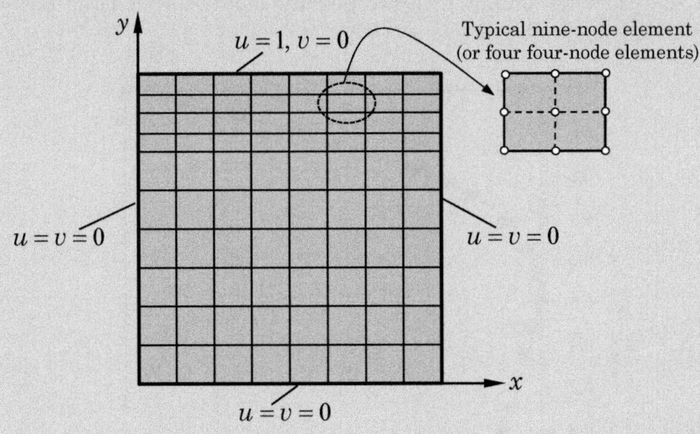

Fig. 6.3.9 Geometry, boundary conditions, and mesh for lid-driven cavity problem.

$$\{DX\} = \{0.0625\ 0.0625\ 0.0625\ 0.0625\ 0.0625\ 0.0625\ 0.0625\ 0.0625\ 0.0625\ 0.0625\ 0.0625\ 0.0625\ 0.0625\ 0.0625\ 0.0625\ 0.0625\},$$

$$\{DY\} = \{0.0625\ 0.0625\ 0.0625\ 0.0625\ 0.0625\ 0.0625\ 0.0625\ 0.0625\ 0.0625\ 0.0625\ 0.0625\ 0.0625\ 0.03125\ 0.03125\ 0.03125\ 0.03125\ 0.031250.031250.031250.03125\}.$$

The penalty parameter γ is taken to be $\gamma = 10^8$ to investigate its influence on the solution. The numerical values of the centerline velocity $u(0.5, y)$ are tabulated in Table 6.3.2. Low values of the penalty parameter adversely affect the accuracy of the solution but when $\gamma > 10^2$, the solution is relatively insensitive to the values of the penalty parameter, γ.

Table 6.3.2 Velocity $v_x(0.5, y)$ obtained with the velocity–pressure and penalty models.

	penalty model				mixed model
	Mesh: 16 × 10Q4		Mesh: 8 × 10Q9		
y	$\gamma = 10^2$	$\gamma = 10^8$	$\gamma = 10^2$	$\gamma = 10^8$	
0.0625	−0.0353	−0.0369	−0.0359	−0.0375	−0.0374
0.1250	−0.0633	−0.0663	−0.0644	−0.0675	−0.0675
0.1875	−0.0877	−0.0920	−0.0894	−0.0938	−0.0937
0.2500	−0.1102	−0.1159	−0.1121	−0.1181	−0.1180
0.3125	−0.1316	−0.1389	−0.1340	−0.1414	−0.1413
0.3750	−0.1511	−0.1603	−0.1536	−0.1631	−0.1629
0.4375	−01672	−0.1782	−0.1700	−0.1812	−0.1811
0.5000	−0.1756	−0.1889	−0.1785	−0.1923	−0.1921
0.5625	−0.1712	−0.1865	−0.1748	−0.1904	−0.1903
0.6250	−0.1449	−0.1625	−0.1492	−0.1673	−0.1672
0.6875	−0.0872	−0.1057	−0.0929	−0.1117	−0.1119
0.7500	0.0152	−0.0038	0.0082	−0.0115	−0.0118
0.8125	0.1730	0.1553	0.1646	0.1462	0.1458
0.8750	0.3926	0.3782	0.3841	0.3690	0.3694
0.9375	0.6725	0.6641	0.6662	0.6573	0.6594

The centerline velocities, $u(0.5, y)$ versus y and $v(x, 0.5)$ versus x, are shown in Figs. 6.3.10 and 6.3.11, respectively. The variation of pressure as a function of x is shown in Fig. 6.3.12. It is clear that the pressure computed in the penalty model through the post-computation (using the reduced integration) is not accurate.

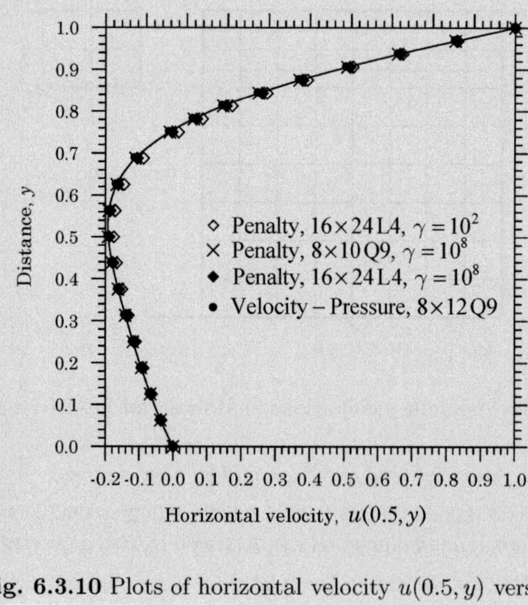

Fig. 6.3.10 Plots of horizontal velocity $u(0.5, y)$ versus y.

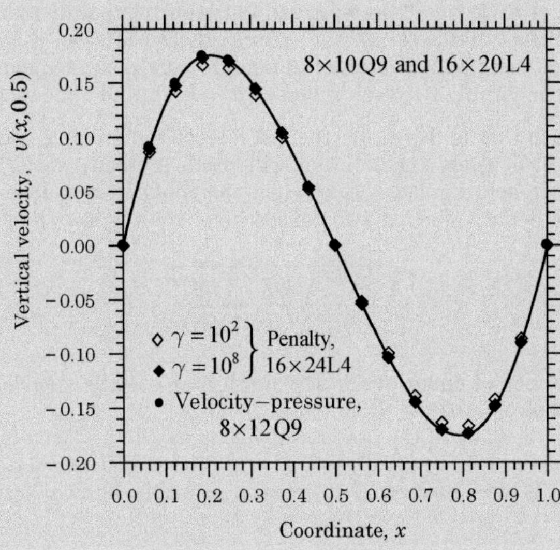

Fig. 6.3.11 Plots of vertical velocity $v(x, 0.5)$ versus x.

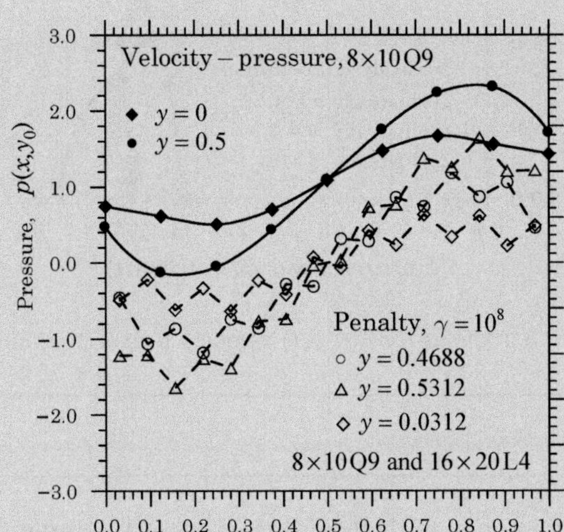

Fig. 6.3.12 Plots of pressure $p(x, y_0)$ versus x.

The next two examples deal with transient response of the problems in **Examples 6.3.2** and **6.3.3** using the penalty finite element model. The Crank–Nicolson scheme ($\alpha = 0.5$) is used to convert the ODEs to algebraic equations.

Example 6.3.4

Consider the problem of a viscous fluid squeezed between two parallel plates (see Fig. 6.3.5). Assuming that the initial conditions on the velocity field are zero, determine the transient solution of the problem using 12×8 mesh of nine-node quadratic elements (the same mesh as in **Example 6.3.2**), the Crank–Nicolson scheme ($\alpha = 0.5$), and time step $\Delta t = 0.1$.

Solution Since the solution is driven by the velocity of the moving plates, $v_y(b, x, t) = -1$, which is not a function of time, the solution will reach a steady state for some time. The solution is declared to reach a steady state when the root-mean-square (RMS) value of the difference between the nodal values at two consecutive times is less than a prescribed value:

$$\mathcal{E} = \sqrt{\frac{\sum_I^{\text{NEQ}} |U_I^{s+1} - U_I^s|^2}{\sum_I^{\text{NEQ}} |U_I^{s+1}|^2}} < \epsilon,$$

where NEQ is the number of equations in the mesh and ϵ is the specified tolerance. In the present example, we took $\epsilon = 10^{-3}$.

Figure 6.3.13 contains plots of the horizontal velocity $v_x(6, y, t)$ as a function of y for three different values of time: $t = 0.1$, $t = 0.2$, and $t = 2$. The transient solution reaches a steady state around $t = 2$ (for a difference of 10^{-3} between solutions at two consecutive time steps).

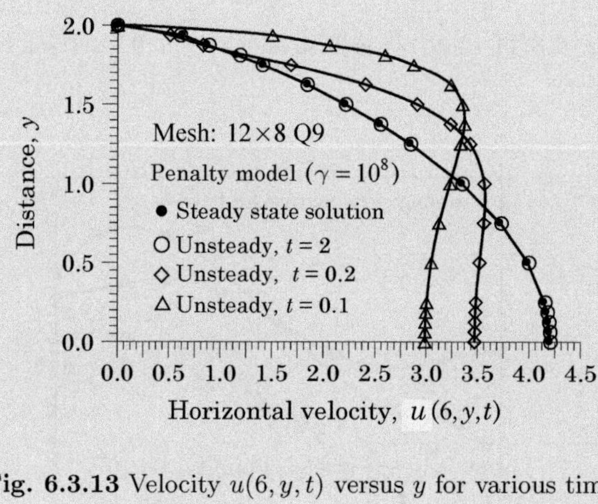

Fig. 6.3.13 Velocity $u(6, y, t)$ versus y for various times.

Example 6.3.5

Investigate the transient solution of the lid-driven cavity of **Example 6.3.3**. Use 8×10Q9 mesh of **Example 6.3.3**, the Crank–Nicolson scheme ($\alpha = 0.5$), and the penalty model with $\gamma = 10^8$, and obtain the transient solutions with two different time steps: $\Delta t = 0.01$ and $\Delta t = 0.001$.

Solution For the lid-driven problem, the specified lid velocity is independent of time. Therefore, the velocity field will reach a steady state at some point in time. The evolution of the horizontal velocity $u(0.5, y, t)$ versus y and the vertical velocity $v(x, 0.5, t)$ versus x with time t are shown in Figs. 6.3.14 and 6.3.15, respectively (using 8×10Q9 mesh and $\Delta t = 0.01$). The solution for velocity field reaches a steady state at $t = 0.1$ (the time at which the steady-state condition is declared with $\epsilon = 10^{-3}$). Graphically, the solution at $t = 0.1$ is already close to the steady-state solution (for the scale used in the figures).

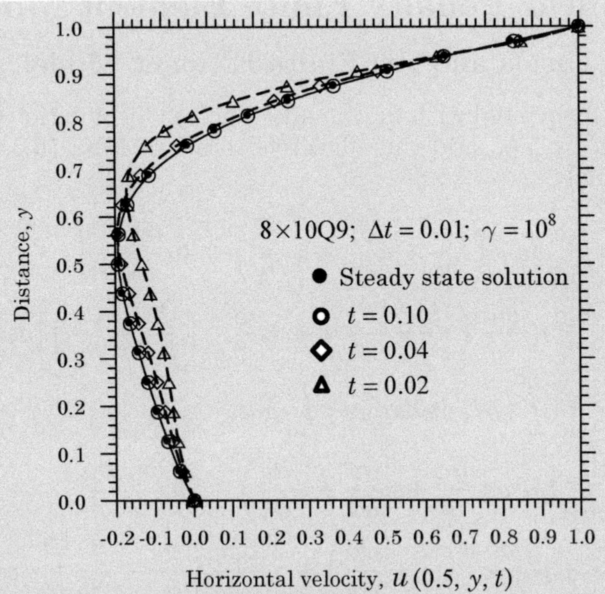

Fig. 6.3.14 The horizontal velocity field $u(0.5, y, t)$ at the center of the cavity as a function of y for various values of time t for the wall-driven cavity problem (8×10Q9 mesh). The flow field develops and reaches a steady state at $t = 0.1$.

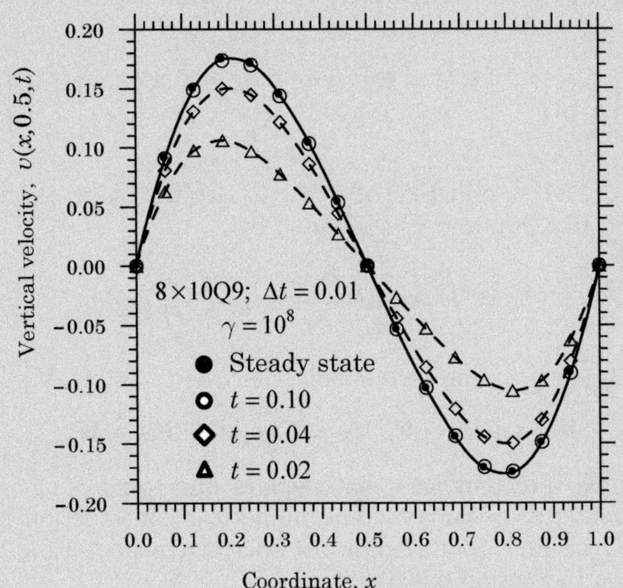

Fig. 6.3.15 The vertical velocity field $v(x, 0.5, t)$ at the center of the cavity as a function of x for various values of time t for the wall-driven cavity problem (8×10Q9 mesh). The flow field develops and reaches a steady state at $t = 0.1$.

6.4 Nonlinear Penalty Finite Element Model

6.4.1 Weak Forms and the Finite Element Model

The weak forms associated with the penalty function formulation of Eqs. (6.1.1)–(6.1.3) are given by [i.e., add the advective terms to Eqs. (6.3.6) and (6.3.7)] (see [1] for details)

$$
\begin{aligned}
0 = \int_{\Omega^e} \bigg[& \delta u\, \rho \left(\frac{\partial u}{\partial t} + u \frac{\partial u}{\partial x} + v \frac{\partial v}{\partial y} \right) + 2\mu \frac{\partial \delta u}{\partial x} \frac{\partial u}{\partial x} \\
& + \mu \frac{\partial \delta u}{\partial y} \left(\frac{\partial u}{\partial y} + \frac{\partial v}{\partial x} \right) + \gamma_e \frac{\partial \delta u}{\partial x} \left(\frac{\partial u}{\partial x} + \frac{\partial v}{\partial y} \right) \bigg] dx\, dy \\
& - \int_{\Omega^e} \rho f_x\, \delta u\, dx dy - \oint_{\Gamma^e} \delta u t_x\, ds
\end{aligned}
\tag{6.4.1a}
$$

$$
\begin{aligned}
0 = \int_{\Omega^e} \bigg[& \delta v\, \rho \left(\frac{\partial v}{\partial t} + u \frac{\partial v}{\partial x} + v \frac{\partial v}{\partial y} \right) + 2\mu \frac{\partial \delta v}{\partial y} \frac{\partial v}{\partial y} \\
& + \mu \frac{\partial \delta v}{\partial x} \left(\frac{\partial u}{\partial y} + \frac{\partial v}{\partial x} \right) + \gamma_e \frac{\partial \delta v}{\partial y} \left(\frac{\partial u}{\partial x} + \frac{\partial v}{\partial y} \right) \bigg] dx dy \\
& - \int_{\Omega^e} \rho f_y\, \delta v\; dx\, dy - \oint_{\Gamma^e} \delta u\, t_y\, ds,
\end{aligned}
\tag{6.4.1b}
$$

where $\delta u = w_{1i}$ and $\delta v = w_{2i}$ are the weight functions.

The penalty finite element model is obtained from Eqs. (6.4.1a) and (6.4.1b) by substituting the finite element interpolation from Eq. (6.2.4) for the velocity field, and $\delta u = \psi_i^e$ and $\delta v = \psi_i^e$:

$$
\begin{bmatrix} \mathbf{M}^{11} & \mathbf{0} \\ \mathbf{0} & \mathbf{M}^{11} \end{bmatrix} \begin{Bmatrix} \dot{\mathbf{u}} \\ \dot{\mathbf{v}} \end{Bmatrix} + \begin{bmatrix} \mathbf{K}_\mu^{11} + \mathbf{K}_\gamma^{11} + \mathbf{C} & \mathbf{K}_\mu^{12} + \mathbf{K}_\gamma^{12} \\ \mathbf{K}_\mu^{21} + \mathbf{K}_\gamma^{21} & \mathbf{K}_\mu^{22} + \mathbf{K}_\gamma^{22} + \mathbf{C} \end{bmatrix} \begin{Bmatrix} \mathbf{u} \\ \mathbf{v} \end{Bmatrix} = \begin{Bmatrix} \mathbf{F}^1 \\ \mathbf{F}^2 \end{Bmatrix},
\tag{6.4.2}
$$

where $\mathbf{K}_\mu^{\alpha\beta}$, $\mathbf{K}_\gamma^{\alpha\beta}$, $\mathbf{M}^{\alpha\beta}$, and $\mathbf{F}^\alpha$ $(\alpha, \beta = 1, 2)$ are the same as those defined in Eqs. (6.2.8) and (6.3.9c), and $\mathbf{C}$ is defined by

$$
C_{ij} = \int_{\Omega^e} \psi_i^e \left(U \frac{\partial \psi_j^e}{\partial x} + V \frac{\partial \psi_j^e}{\partial y} \right) dx dy.
\tag{6.4.3}
$$

Equation (6.4.2) can be expressed symbolically as ($\mathbf{v} = \{\mathbf{U}, \mathbf{V}\}^{\mathrm{T}}$)

$$
\mathbf{K}\mathbf{v} + \mathbf{M}\dot{\mathbf{v}} = \mathbf{F}, \quad \mathbf{K} = \mathbf{C}(\rho, \mathbf{v}) + \mathbf{K}_\mu + \mathbf{K}_\gamma,
\tag{6.4.4}
$$

where $\mathbf{K}_\mu$ and $\mathbf{K}_\gamma$ are matrices that depend only on μ and γ, respectively; whereas $\mathbf{C}$ depends on ρ and $\mathbf{v}$. The numerical construction of the penalty terms presented in Eq. (6.4.2) requires special consideration, the details of which were given in Section 6.3.2.

The fully discretized finite element model of Eq. (6.4.4) can be obtained as explained in Section 6.2.3. Application of the alpha-family of the time-approximation schemes discussed in Sections 5.2.5 and 6.2.3 to Eq. (6.4.4) yields

$$
\hat{\mathbf{K}}\boldsymbol{\Delta}^{s+1} = \bar{\mathbf{K}}\boldsymbol{\Delta}^{s} + \bar{\mathbf{F}}^{s,s+1} \equiv \hat{\mathbf{F}}^{s,s+1},
\tag{6.4.5a}
$$

where

$$
\begin{aligned}
\hat{\mathbf{K}} = \mathbf{M} + a_1\mathbf{K}, \quad \bar{\mathbf{K}} = \mathbf{M} - a_2\mathbf{K} \\
\bar{\mathbf{F}}^{s,s+1} = \left(a_1\mathbf{F}^{s+1} + a_2\mathbf{F}^s\right) \\
a_1 = \alpha\,\Delta t, \quad a_2 = (1-\alpha)\Delta t,
\end{aligned}
\tag{6.4.5b}
$$

where $0 \leq \alpha \leq 1$ defines various schemes:

$$
\alpha = \begin{cases}
0, & \text{the forward difference scheme, (conditionally stable),} \\
& \text{order of accuracy } = O(\Delta t); \\
\frac{1}{2}, & \text{the Crank–Nicolson scheme (stable),} \\
& \text{order of accuracy } = O(\Delta t)^2; \\
1, & \text{the backward difference scheme (stable),} \\
& \text{order of accuracy } = O(\Delta t).
\end{cases}
\tag{6.4.6}
$$

Recall that the conditionally stable schemes require use of the time step Δt that is less than Δt_{cr}:

$$
\Delta t < \Delta t_{cr} \equiv \frac{2}{(1-2\alpha)\lambda_{max}}, \quad \alpha < \frac{1}{2}. \tag{6.4.7}
$$

6.4.2 Tangent Matrix for the Penalty Finite Element Model

Numerical procedures used to solve nonlinear algebraic equations of the type in Eq. (6.4.5a) are iterative in nature, and they require *linearization* of the equations, as discussed in Section 3.2 (see Reddy [40]). The linearization amounts to evaluating the coefficient matrices containing nonlinear terms by using the solution from the immediate previous iteration. For the first iteration, one must begin with a presumed (or guess) solution vector. The solution obtained from the previous iteration is used to compute the coefficient matrix for the next iteration and the new solution is found. This iteration is continued until the normalized error between two consecutive iterations is less than a preassigned tolerance.

Two iterative procedures were outlined in Section 3.2 using finite element equations of the form

$$
\mathbf{K}^e(\boldsymbol{\Delta}^e)\boldsymbol{\Delta}^e = \mathbf{F}^e, \tag{6.4.8}
$$

where the coefficient matrix $\mathbf{K}^e$ is a function of the solution vector $\boldsymbol{\Delta}^e$ of unknown nodal velocities (u_i^e, v_i^e) of a typical element Ω^e. These include the Picard (or direct) iteration and Newton iteration methods. In the case of the Newton iteration scheme, we must compute the tangent coefficient matrix [see Eq. (3.2.23)] associated with the nonlinear equation in Eq. (6.4.8). Since element matrices are computed at the element level, we apply Newton's procedure at the element level, and then assemble these element equations to form the global equations.

The element equation for the Newton iteration procedure is

$$
\mathbf{T}^e(\boldsymbol{\Delta}^{(r-1)})\,\Delta\boldsymbol{\Delta}^e = -\mathbf{R}^e(\boldsymbol{\Delta}^{(r-1)}) \tag{6.4.9}
$$

and the tangent coefficient matrix at the element level is

$$T_{ij}^e \equiv \frac{\partial R_i^e}{\partial \Delta_j^e} = \frac{\partial}{\partial \Delta_j^e}\left(\sum_{m=1}^{n} K_{im}^e \Delta_m^e - F_i^e\right) = K_{ij}^e + \sum_{m=1}^{n} \frac{\partial K_{im}^e}{\partial \Delta_j^e}\Delta_m^e - \frac{\partial F_i^e}{\partial \Delta_j^e}. \tag{6.4.10}$$

Here $\mathbf{R}^e$ denotes the residual vector

$$\mathbf{R}^e = \mathbf{K}^e(\mathbf{\Delta}^e)\mathbf{\Delta}^e - \mathbf{F}^e. \tag{6.4.11}$$

When the vector $\mathbf{F}^e$ is independent of $\mathbf{\Delta}^e$, we have $\partial F_i^e/\partial \Delta_j^e = 0$.

To compute the tangent matrix coefficients for the penalty finite element model, we recall the reduced integration penalty finite element model of the steady two-dimensional flows of viscous incompressible fluids [see Eq. (6.4.2)] is

$$\begin{bmatrix} \mathbf{K}^{11} & \mathbf{K}^{12} \\ \mathbf{K}^{21} & \mathbf{K}^{22} \end{bmatrix} \begin{Bmatrix} \mathbf{u} \\ \mathbf{v} \end{Bmatrix} = \begin{Bmatrix} \mathbf{F}^1 \\ \mathbf{F}^2 \end{Bmatrix}, \tag{6.4.12}$$

where the element matrices $K_{ij}^{\alpha\beta}$ are defined by assuming that $\bar{u}$ and $\bar{v}$ in the advective terms

$$\rho\left(\bar{u}\frac{\partial u}{\partial x} + \bar{v}\frac{\partial u}{\partial y}\right), \quad \rho\left(\bar{u}\frac{\partial v}{\partial x} + \bar{v}\frac{\partial v}{\partial y}\right)$$

are known from the previous iteration:

$$\begin{aligned}
K_{ij}^{11} &= \int_{\Omega^e}\left[\mu\left(2\frac{\partial \psi_i}{\partial x}\frac{\partial \psi_j}{\partial x} + \frac{\partial \psi_i}{\partial y}\frac{\partial \psi_j}{\partial y}\right) + \gamma\frac{\partial \psi_i}{\partial x}\frac{\partial \psi_j}{\partial x}\right] dxdy \\
&\quad + \int_{\Omega^e} \rho\psi_i\left(\bar{u}\frac{\partial \psi_j}{\partial x} + \bar{v}\frac{\partial \psi_j}{\partial y}\right) dx\, dy \\
K_{ij}^{12} &= \int_{\Omega^e}\left(\mu\frac{\partial \psi_i}{\partial y}\frac{\partial \psi_j}{\partial x} + \gamma\frac{\partial \psi_i}{\partial x}\frac{\partial \psi_j}{\partial y}\right) dxdy, \quad K_{ij}^{21} = K_{ji}^{12} \\
K_{ij}^{22} &= \int_{\Omega^e}\left[\mu\left(\frac{\partial \psi_i}{\partial x}\frac{\partial \psi_j}{\partial x} + 2\frac{\partial \psi_i}{\partial y}\frac{\partial \psi_j}{\partial y}\right) + \gamma\frac{\partial \psi_i}{\partial y}\frac{\partial \psi_j}{\partial y}\right] dx\, dy \\
&\quad + \int_{\Omega^e} \rho\psi_i\left(\bar{u}\frac{\partial \psi_j}{\partial x} + \bar{v}\frac{\partial \psi_j}{\partial y}\right) dxdy \\
F_i^1 &= \int_{\Omega^e} \rho\,\psi_i f_x\, dx\, dy + \oint_{\Gamma^e} \psi_i t_x\, ds, \quad F_i^2 = \int_{\Omega^e} \rho\,\psi_i f_y\, dxdy + \oint_{\Gamma^e} \psi_i t_y\, ds.
\end{aligned} \tag{6.4.13}$$

The Newton iteration scheme associated with Eq. (6.4.12) can be cast in the same form as the original equation, Eq. (6.4.12) (the element label is omitted):

$$\begin{bmatrix} \mathbf{T}^{11} & \mathbf{T}^{12} \\ \mathbf{T}^{21} & \mathbf{T}^{22} \end{bmatrix} \begin{Bmatrix} \Delta\mathbf{u} \\ \Delta\mathbf{v} \end{Bmatrix} = -\begin{Bmatrix} \mathbf{R}^1 \\ \mathbf{R}^2 \end{Bmatrix}, \tag{6.4.14}$$

where $\mathbf{T}^{\alpha\beta}$ and $\mathbf{R}^{\alpha}$ are defined by

$$T_{ij}^{\alpha\beta} = \frac{\partial R^{\alpha}}{\partial \Delta_j^{\alpha}}, \quad R_i^{\alpha} = \sum_{\beta=1}^{2}\sum_{j=1}^{n} K_{ij}^{\alpha\beta}\Delta_j^{\beta}. \tag{6.4.15}$$

Here $\Delta_j^1 = u_j$ and $\Delta_j^2 = v_j$ $(j = 1, 2, \ldots, n)$. To compute the tangent matrix coefficients $T_{ij}^{\alpha\beta}$, we note that K_{ij}^{12} and K_{ij}^{21} are independent of the nodal values u_i and v_i. We have

$$
\begin{aligned}
T_{ij}^{11} &= K_{ij}^{11} + \sum_{k=1}^{n} \left[\frac{\partial}{\partial u_j} \int_{\Omega^e} \rho\psi_i \left(u\frac{\partial\psi_k}{\partial x} + v\frac{\partial\psi_k}{\partial y} \right) dx\, dy \right] v_{xk} \\
&= K_{ij}^{11} + \int_{\Omega^e} \rho\frac{\partial u}{\partial x}\, \psi_i\psi_j\, dxdy, \\
T_{ij}^{22} &= K_{ij}^{22} + \sum_{k=1}^{n} \left[\frac{\partial}{\partial v_j} \int_{\Omega^e} \rho\psi_i \left(u\frac{\partial\psi_k}{\partial x} + v\frac{\partial\psi_k}{\partial y} \right) dx\, dy \right] v_{yk} \\
&= K_{ij}^{22} + \int_{\Omega^e} \rho\frac{\partial v}{\partial y}\, \psi_i\psi_j\, dx\, dy, \\
T_{ij}^{12} &= K_{ij}^{12} + \int_{\Omega^e} \rho\frac{\partial u}{\partial y}\, \psi_i\psi_j\, dxdy, \\
T_{ij}^{21} &= K_{ij}^{21} + \int_{\Omega^e} \rho\frac{\partial v}{\partial x}\, \psi_i\psi_j\, dxdy.
\end{aligned} \tag{6.4.16}
$$

Some remarks on the Newton iteration scheme as applied to the solution of nonlinear equations $\mathbf{K}(\mathbf{U})\mathbf{U} = \mathbf{F}$ are in order. In the direct iteration method, the actual boundary conditions are applied to the assembled system during each iteration, prior to solving them. In the Newton iteration scheme, at the end of each iteration, we only compute the increment $\delta\boldsymbol{\Delta}$ to the known solution $\boldsymbol{\Delta}^{(r-1)}$. Therefore, if $\boldsymbol{\Delta}^{(r-1)}$ already satisfies the specified boundary conditions, then $\delta\boldsymbol{\Delta}$ must only satisfy the corresponding *homogeneous* boundary conditions so that $\boldsymbol{\Delta}^{(r)} = \boldsymbol{\Delta}^{(r-1)} + \delta\boldsymbol{\Delta}$ satisfies the specified boundary conditions. For example, if U_m is specified to be $\Delta_m = \alpha_m$, and the initial guess is such that $\Delta_m^{(0)} = \alpha_m$, then $\delta\Delta_m = 0$ so that $\Delta_m^{(1)} = \Delta_m^{(0)} + \delta\Delta_m = \alpha_m$. Alternatively, if the initial guess is such that $\Delta_m^{(0)} = 0$, then $\delta\Delta_m = \alpha_m$ for the first iteration so that $\Delta_m^{(1)} = \Delta_m^{(0)} + \delta\Delta_m = \alpha_m$, and the boundary condition on $\delta\Delta_m$ during the second iteration onwards should be $\delta\Delta_m = 0$ for all m.

6.4.3 Numerical Examples

In this section, a small number of flow problems will be solved using the finite element models developed herein are presented. Additional examples may be found in the book by Reddy and Gartling [3]. The examples presented herein were solved using the RIP and mixed finite element model. The objective of the first examples, which deal with linear solutions, is to evaluate the accuracy of the penalty and mixed finite element models by comparing with the available analytical or numerical results and to illustrate the effect of the penalty parameter on the accuracy of the solutions. The remaining examples are for Reynolds numbers ($Re = \rho v_0 a/\mu$) greater than unity (i.e. advective terms are included), and the results were obtained using the RIP finite element model.

For high Reynolds number flows, the total value of Re is divided into suitable increments, ΔRe, in much the same way the total load is divided in a structural problem. The converged nonlinear solution of the preceding Reynolds number is used as the initial guess in the first iteration of the next Reynolds number. In general, for very high Reynolds numbers under-relaxation may be necessary to achieve convergence. Recall that under-relaxation involves the use of a weighted average of velocities from the last two iterations

$$\bar{\mathbf{v}}^r = (1-\beta)\mathbf{v}^{(r)} + \beta\,\mathbf{v}^{(r-1)} \tag{6.4.17}$$

to compute the element matrix. Here β is the acceleration parameter, whose value is problem dependent. When $\beta = 0$ does not yield convergence, one may use $\beta = 0.25$ and numerically experiment to determine a value of β that gives convergent solution.

Here we consider a couple of examples involving the solution of the Navier–Stokes equations. First we revisit the lid-driven cavity flow.

Example 6.4.1

Lid-Driven Cavity Flow. Consider the lid-driven cavity problem of **Example 6.3.3** for non-linear analysis (i.e. solve Navier–Stokes equations for $Re > 0$). For the problem at hand, the Reynolds number ($Re = \rho v_0 a/\mu$) can be varied by changing the density ρ while keeping the viscosity μ and lid velocity v_0 fixed. Thus, take (in addition to the choice of the characteristic length $a = 1$) $\mu = 1$ and $v_0 = 1$ so that $Re = \rho$. Solve the problem for $Re = 250, 500$, and 750, using a uniform 8×8 mesh of linear elements as well as 4×4 mesh of nine-node quadratic elements, penalty parameter $\gamma = 10^8$, and convergence tolerance of $\epsilon = 10^{-3}$.

Also, present results for $Re = 5000$ and $Re = 10000$. using a nonuniform 18×20 mesh of linear quadrilateral elements. The following increments [$DX(I)$ and $DY(I)$] along x and y must be used:

$$\{DX\} = \{0.03125\ 0.03125\ 0.03125\ 0.03125\ 0.06250\ 0.06250\ 0.06250\ 0.06250\ 0.12500\ 0.12500\ 0.06250\ 0.06250\ 0.06250\ 0.06250\ 0.03125\ 0.03125\ 0.03125\ 0.03125\},$$

$$\{DY\} = \{0.03125\ 0.03125\ 0.03125\ 0.03125\ 0.06250\ 0.06250\ 0.06250\ 0.06250\ 0.06250\ 0.06250\ 0.06250\ 0.06250\ 0.06250\ 0.06250\ 0.06250\ 0.06250\ 0.03125\ 0.03125\ 0.03125\ 0.03125\}.$$

Plot (a) the horizontal velocity $v_x(0.5, y)$ as a function of y for the linear and nonlinear cases and (b) pressure $p(x, y)$ as a function of x at $y = 0.15625$ for the two different Reynolds numbers.

Solution In solving this problem, the mesh used should be such that the boundary layer thickness, which is of the order of $Re^{-\frac{1}{2}}$, near the walls is resolved. The numerical results obtained with the two meshes and an acceleration parameter $\beta = 0.25$ [see Eq. (6.4.12)], after some numerical tries, are presented in Table 6.4.1 for $Re = 250, 500$, and 750. The convergence may not be reached for $Re = 750$ if $\beta = 0$.

Figure 6.4.1 shows plots of the horizontal velocity along the cavity centerline obtained with 8×10 mesh of nine-node quadratic elements and 16×20 mesh of linear elements for $Re = 0$, 500, and 1000 ($\beta = 0.5$). The mesh grading used for the Q9 elements is as follows:

$$\{DX\} = \{0.125\ 0.125\ 0.125\ 0.125\ 0.125\ 0.125\ 0.125\ 0.125\},$$

$$\{DY\} = \{0.125\ 0.125\ 0.125\ 0.125\ 0.125\ 0.125\ 0.0625\ 0.0625\ 0.0625\ 0.0625\}.$$

Table 6.4.1 Velocity $v_x(0.5, y)$ obtained with linear and quadratic elements and for various values of the Reynolds number (values in parentheses are linear solutions).

y	Mesh: 8 × 8L4			Mesh: 4 × 4Q9		
Re →	250(9)[a]	500(9)	750(10)	250(10)	500(12)	750(20)
0.125	−0.0367 (−0.0579)	−0.0239	−0.0128	−0.0412 (−0.0615)	−0.0131	0.0146
0.250	−0.0688 (−0.0988)	−0.0502	−0.0320	−0.0851 (−0.1039)	−0.0520	0.0017
0.375	−0.0944 (−0.1317)	−0.0733	−0.0533	−0.1283 (−0.1393)	−0.1133	−0.0481
0.500	−0.0911 (−0.1471)	−0.0696	−0.0569	−0.1305 (−0.1563)	−0.1284	−0.1086
0.625	−0.0176 (−0.0950)	0.0043	0.0020	−0.0437 (−0.1118)	−0.0494	−0.0901
0.750	0.0469 (0.0805)	0.0414	0.0323	0.0753 (0.0481)	0.1042	0.0546
0.875	0.2616 (0.4500)	0.1712	0.1207	0.2833 (0.4186)	0.2137	0.1495

[a]The number in parentheses denotes the number of iterations taken for convergence.

For the linear elements the above lengths are subdivided. The results show excellent convergence with p refinement. The pressure and shear stress near to the moving lid obtained with the two meshes are shown in Figs. 6.4.2 and 6.4.3, respectively. Clearly, both pressure and shear stress exhibit oscillations.

The velocity $v_x(0.5, y)$ obtained with 18×20 mesh of linear quadrilateral elements, penalty parameter $\gamma = 10^{10}$, tolerance $\epsilon = 10^{-3}$ ($itmax = 30$), and acceleration parameter of $\beta = 0.25$ are presented in Table 6.4.2. Figures 6.4.4 and 6.4.5 contain plots of the velocity along the vertical centerline, $v_x(0.5, y)$, versus y for Reynolds numbers $Re = 0, 500$, and 1000 obtained with 8×10Q9 and 16×20L4 meshes.

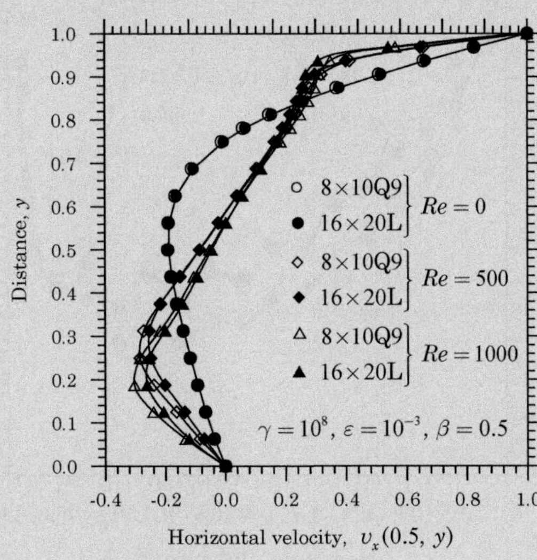

Fig. 6.4.1 Velocity $v_x(0.5, y)$ versus y for Reynolds numbers $Re = 0, 500$, and 1000 obtained with 8×10Q9 and 16×20L4 meshes (penalty parameter $\gamma = 10^{10}$, tolerance $\epsilon = 10^{-3}$, and the maximum number of iterations allowed, $itmax = 30$).

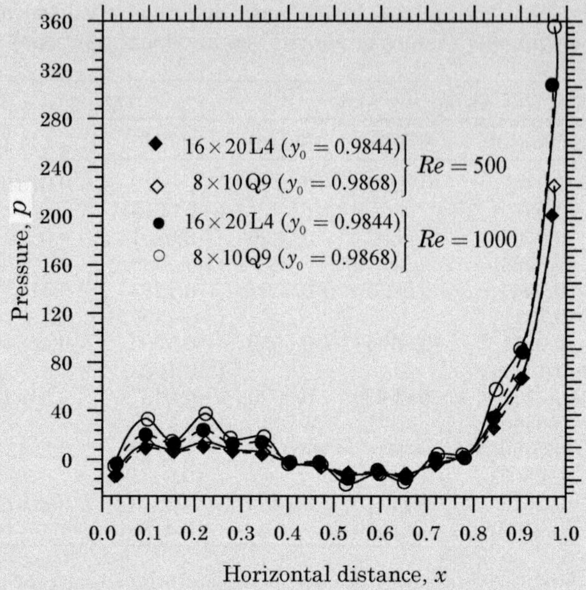

Fig. 6.4.2 Plots of pressure $p(x, y)$ along the top wall of the cavity using $8 \times 10Q9$ and $16 \times 20L4$ meshes (penalty parameter $\gamma = 10^{10}$, tolerance $\epsilon = 10^{-3}$, and the maximum number of iterations allowed, *itmax* = 30.

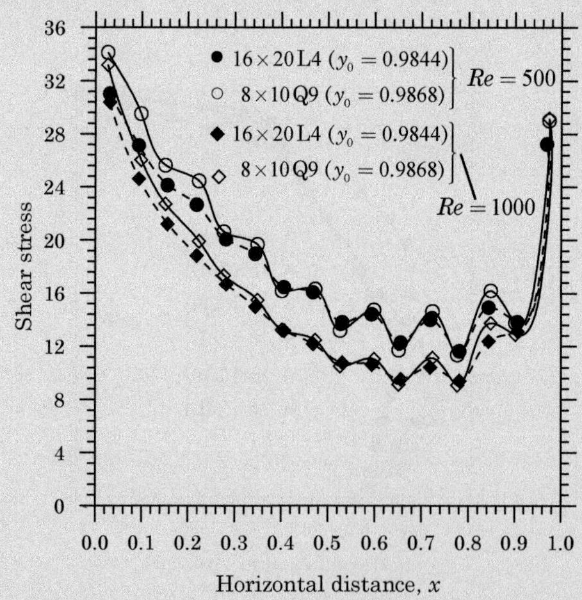

Fig. 6.4.3 Plots of shear stress $\sigma_{xy}(x, y_0)$ along the top wall of the cavity ($8 \times 10Q9$ and $16 \times 20L4$ meshes).

Table 6.4.2 Values of $v_x(0.5, y)$ as a function of y for $Re = 0$, 5×10^3, and $Re = 10^4$.

Node	y	$Re = 0$	$Re = 5 \times 10^3$	$Re = 10^4$
10	0.00000	0.00000	0.00000	0.00000
29	0.03125	−0.01961	−0.11364	−0.08990
48	0.06250	−0.03659	−0.16729	−0.12343
67	0.09375	−0.05176	−0.19152	−0.13295
86	0.12500	−0.06567	−0.19078	−0.12613
105	0.18750	−0.09120	−0.16006	−0.10247
124	0.25000	−0.11514	−0.11976	−0.08036
143	0.31250	−0.13828	−0.10475	−0.06799
162	0.37500	−0.16002	−0.06764	−0.04196
181	0.43750	−0.17854	−0.04226	−0.02620
200	0.50000	−0.18992	−0.01189	−0.00541
219	0.56250	−0.18839	0.01810	0.01477
238	0.62500	−0.16490	0.04819	0.03360
257	0.68750	−0.10850	0.08437	0.05980
276	0.75000	−0.00578	0.01148	0.07915
295	0.81250	0.15435	0.15809	0.10615
314	0.87500	0.37920	0.19178	0.13308
333	0.90625	0.51533	0.19835	0.12908
352	0.93750	0.66614	0.18587	0.12934
371	0.96875	0.82884	0.25227	0.13930
390	1.00000	1.00000	1.00000	0.10000

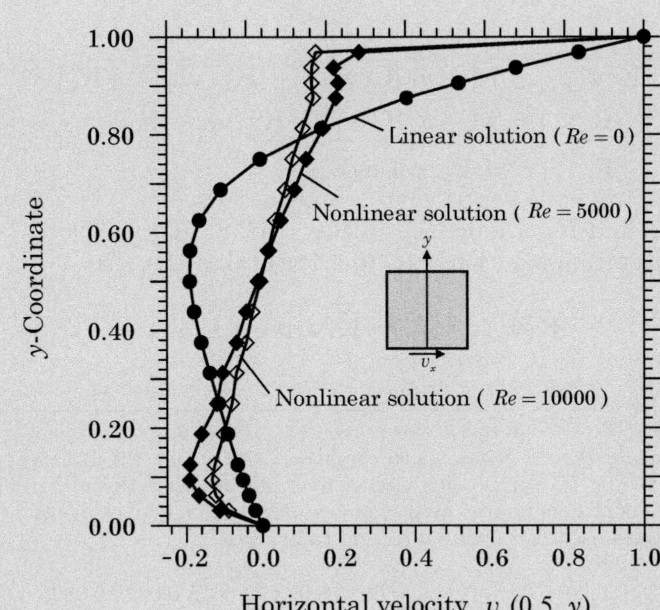

Fig. 6.4.4 Centerline horizontal velocity $u(0.5, y) = v_x(0.5, y)$ versus y for $Re = 5 \times 10^3$ and $Re = 10^4$.

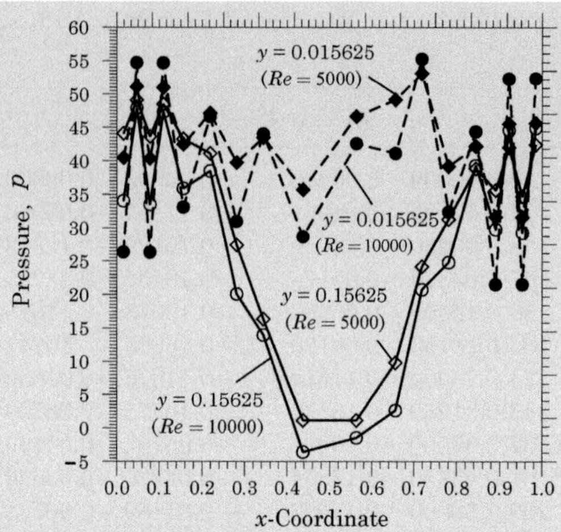

Fig. 6.4.5 Pressure $p(x, y)$ versus x for various values of y for $Re = 5 \times 10^3$ and $Re = 10^4$.

The weak-form penalty finite element models developed in Section 6.4 already include time-dependent terms [see Eqs. (6.4.2)–(6.4.4)]. The fully discretized equations in Eqs. (6.4.5a) and (6.4.5b) have the specific form

$$\hat{\mathbf{K}}(\mathbf{v}_{s+1})\mathbf{v}_{s+1} = \bar{\mathbf{K}}(\mathbf{v}_s)\mathbf{v}_s + \hat{\mathbf{F}}_{s,s+1}, \tag{6.4.18a}$$

where

$$\begin{aligned} \hat{\mathbf{K}}(\mathbf{v}_{s+1}) &= \mathbf{M} + a_1 \left[\mathbf{C}(\mathbf{v}_{s+1}) + \mathbf{K}_\mu(\mathbf{v}_{s+1}) + \mathbf{K}_\gamma\right] \\ \bar{\mathbf{K}}(\mathbf{v}_s) &= \mathbf{M} - a_2 \left[\mathbf{C}(\mathbf{v}_s) + \mathbf{K}_\mu(\mathbf{v}_s) + \mathbf{K}_\gamma\right] \\ \hat{\mathbf{F}}_{s,s+1} &= a_1 \mathbf{F}_{s+1} + a_2 \mathbf{F}_s, \end{aligned} \tag{6.4.18b}$$

and $\mathbf{M}$, $\mathbf{C}$, $\mathbf{K}_\mu$, and $\mathbf{K}_\gamma$ are defined in Eqs. (6.2.8) and (6.3.9c).

For direct iteration solution, Eq. (6.4.18a) takes the form

$$\hat{\mathbf{K}}(\mathbf{v}_{s+1}^r)\mathbf{v}_{s+1}^{r+1} = \bar{\mathbf{K}}(\mathbf{v}_s)\mathbf{v}_s + \hat{\mathbf{F}}_{s,s+1}. \tag{6.4.19}$$

Example 6.4.2

Transient analysis of the lid-driven cavity problem. Carry out the transient analysis, for $Re = 0$, $Re = 1000$, and $Re = 2500$, of a viscous fluid inside a wall-driven cavity (see Example 6.3.3). Use 16×20Q4 nonuniform mesh of four-node rectangular elements in the domain. Take the mesh size in each coordinate direction to be

$$\{DX\} = \{0.0625\ 0.0625\ 0.0625\ 0.0625\ 0.0625\ 0.0625\ 0.0625\ 0.0625\ 0.0625\ 0.0625\ 0.0625\ 0.0625\ 0.0625\ 0.0625\ 0.0625\ 0.0625\},$$

$$\{DY\} = \{0.0625\ 0.0625\ 0.0625\ 0.0625\ 0.0625\ 0.0625\ 0.0625\ 0.0625\ 0.0625\ 0.0625\ 0.0625\ 0.0625\ 0.03125\ 0.03125\ 0.03125\ 0.03125\ 0.03125\ 0.03125\ 0.03125\ 0.03125\}.$$

Use the Crank–Nicolson method ($\alpha = 0.5$) with two different time steps $\Delta t = 0.01$ and $\Delta t = 0.001$.

Solution Table 6.4.3 contains the velocity field $v_x(0.5, y, t) \times 10$ for times $t = 0.01$, 0.05, and 0.1 for $Re = 0$. The solution reaches the steady state ($\epsilon = 10^{-2}$) at time $t = 0.1$ when $\Delta t = 0.01$ is used. Figure 6.4.6 shows the evolution of $v_x(0.5, y, t)$.

Table 6.4.3 The horizontal velocity field $v_x(0.5, y, t) \times 10$ versus time t for the wall-driven cavity problem (16×20Q4 mesh).

y	$t = 0.01$ $\Delta t = 0.01$	$t = 0.01$ $\Delta t = 0.001$	$t = 0.05$ $\Delta t = 0.01$	$t = 0.05$ $\Delta t = 0.001$	$t = 0.10$ $\Delta t = 0.01$	Steady state
0.0625	−0.1342	−0.1953	−0.3103	−0.3247	−0.3655	−0.3688
0.1250	−0.1936	−0.3140	−0.5624	−0.5841	−0.6558	−0.6631
0.1875	−0.2314	−0.3940	−0.7888	−0.8163	−0.9108	−0.9198
0.2500	−0.2691	−0.4651	−1.0122	−1.0435	−1.1499	−1.1593
0.3125	−0.3157	−0.5475	−1.2346	−1.2746	−1.3802	−1.3886
0.3750	−0.3759	−0.6536	−1.4790	−1.5053	−1.5967	−1.6028
0.4375	−0.4516	−0.7902	−1.6964	−1.7151	−1.7793	−1.7820
0.5000	−0.5435	−0.9605	−1.8536	−1.8643	−1.8906	−1.8895
0.5625	−0.6465	−1.1577	−1.8846	−1.8878	−1.8700	−1.8652
0.6250	−0.7474	−1.3479	−1.7011	−1.6946	−1.6336	−1.6250
0.6875	−0.8097	−1.4428	−1.1889	−1.1653	−1.0700	−1.0572
0.7500	−0.7536	−1.1523	−0.2093	−0.1693	−0.0520	−0.0382
0.7813	−0.6325	−0.7744	0.5100	0.5471	0.6713	0.6820
0.8125	−0.4077	−0.1695	1.4014	1.4197	1.5467	1.5526
0.8438	−0.0054	0.7336	2.4885	2.4716	2.5918	2.5965
0.8750	0.6329	1.9318	3.7259	3.6716	3.7646	3.7824
0.9063	1.7000	3.5232	5.1185	5.0707	5.1198	5.1616
0.9375	3.3334	5.3837	6.5139	6.5756	6.6082	6.6410
0.9688	5.9470	7.5970	7.9975	8.2488	8.3805	8.2838

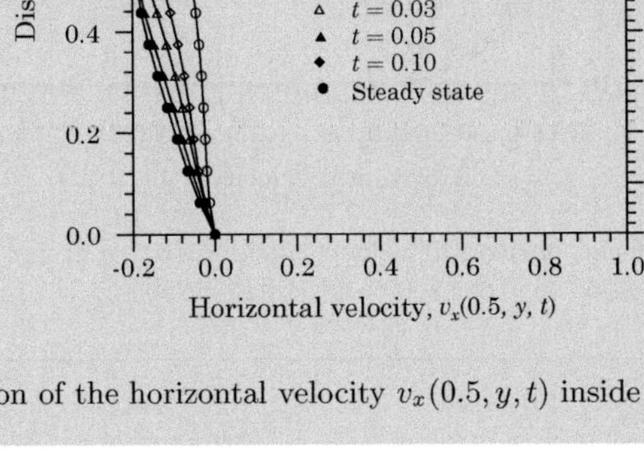

Fig. 6.4.6 Evolution of the horizontal velocity $v_x(0.5, y, t)$ inside a wall-driven cavity.

Next, we consider nonlinear transient analysis . The same mesh, namely, the mesh of 16×20 four-node quadrilateral elements, is used with $\Delta t = 0.001$. The transient response is calculated for Reynolds numbers $Re = 1,000$ and $Re = 2,500$ separately. Figures 6.4.7(a) and (b) show plots of the nonlinear steady-state and transient center horizontal velocity $v_x(0.5, y, t)$ versus y for various times.

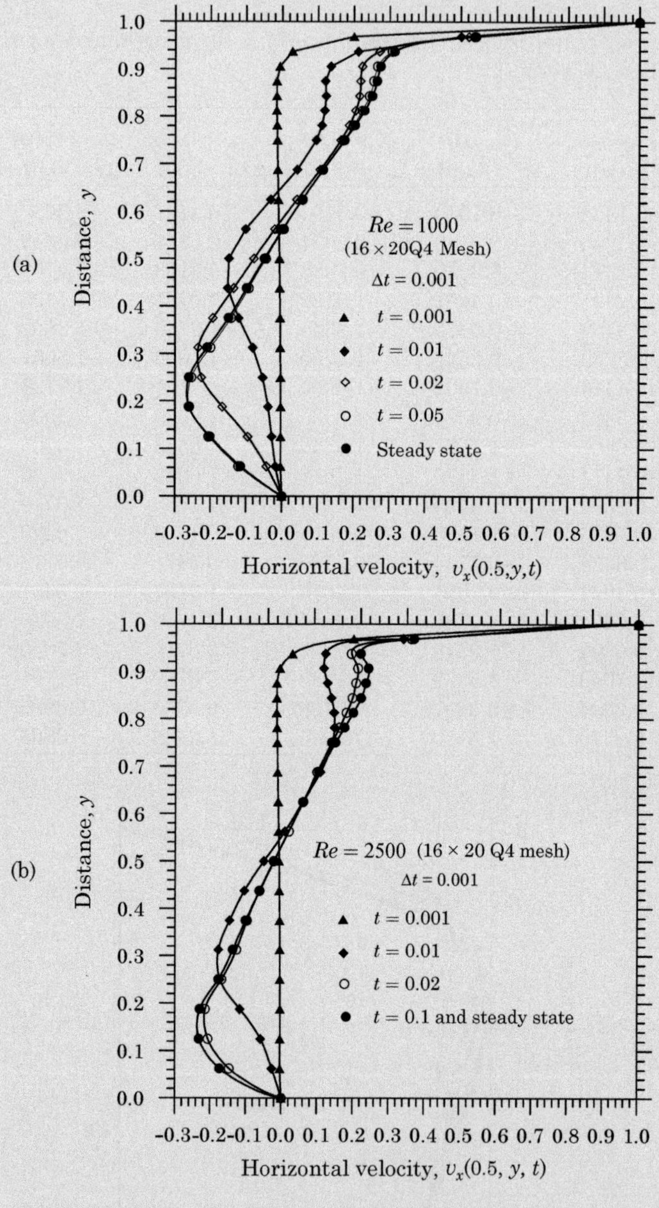

Fig. 6.4.7 Evolution of the horizontal velocity field $v_x(0.5, y, t)$ for the wall-driven cavity problem (nonlinear analysis). (a) Re=1000 and (b) Re=2500.

6.5 Summary

In this chapter two different finite element models of the equations governing two-dimensional flows of viscous, incompressible fluids have been presented. The first one is based on the velocity–pressure formulation and the second is based on the penalty function formulation. The velocity–pressure finite element model treats (u, v, p) as the primary nodal degrees of freedom and the penalty finite element model has (u, v) as the primary nodal degrees of freedom. In the penalty function method, the pressure is post-computed from the velocity field, and often the pressure is not accurate. The finite element adopted for the velocity–pressure finite element model is one in which the velocity components (u, v) are approximated using quadratic interpolation and the pressure p with bilinear interpolation (in a rectangular element). In the penalty finite element model, the pressure is eliminated through the use of the incompressibility condition (i.e., continuity equation, $\boldsymbol{\nabla} \cdot \mathbf{v} = 0$) as a constraint. The terms of the coefficient matrix that contain the penalty terms are evaluated using reduced integration. Both triangular and rectangular elements can be employed for the two models, but only the quadrilateral elements are used in the numerical computations presented herein.

Sections 6.2 and 6.3 were devoted to the finite element models of Stokes flows (i.e., flows in which the advective terms are assumed to be negligible). Several numerical examples of steady-state and transient solutions of Stokes flows of viscous incompressible fluids in two dimensions were presented with the help of FLOMXD2D and FLOPNL2D programs developed by the first author.

In Section 6.4, the nonlinear penalty finite element model of the Navier–Stokes equations (i.e., account for the advective terms) governing flows of viscous, incompressible, Newtonian fluids was presented. Solution of the nonlinear finite element equations using the Newton iteration as well as the Picard (or direct) iteration was discussed and the calculation of the coefficients of the tangent matrix was presented. It was noted that the element coefficient matrices of the two finite element models are unsymmetric due to the presence of the advective (nonlinear) terms. A number of examples of steady as well unsteady flow problems using the lid-driven cavity flow were presented.

Problems

6.1 For axisymmetric flows of viscous incompressible fluids (i.e., flow field is independent of the θ coordinate), we have

$$\rho\frac{\partial v_r}{\partial t} = \frac{1}{r}\frac{\partial}{\partial r}(r\sigma_{rr}) - \frac{\sigma_{\theta\theta}}{r} + \frac{\partial \sigma_{rz}}{\partial z} + f_r \tag{1}$$

$$\rho\frac{\partial v_z}{\partial t} = \frac{1}{r}\frac{\partial}{\partial r}(r\sigma_{rz}) + \frac{\partial \sigma_{zz}}{\partial z} + f_z \tag{2}$$

$$\frac{1}{r}\frac{\partial}{\partial r}(rv_r) + \frac{\partial v_z}{\partial z} = 0, \tag{3}$$

where v_r and v_z are the r- and z-components, respectively, of the velocity vector $\mathbf{v}$, and σ_{rr}, $\sigma_{\theta\theta}$, σ_{zz}, and σ_{rz} are the Cauchy stress components, which are related to the velocity gradients and pressure through the following equations:

$$\sigma_{rr} = -p + 2\mu\frac{\partial v_r}{\partial r}, \quad \sigma_{\theta\theta} = -p + 2\mu\frac{u}{r}$$
$$\sigma_{zz} = -p + 2\mu\frac{\partial v_z}{\partial z}, \quad \sigma_{rz} = \mu\left(\frac{\partial v_r}{\partial z} + \frac{\partial v_z}{\partial r}\right). \tag{4}$$

Develop the semidiscrete finite element model of the equations using the velocity–pressure (mixed) formulation.

6.2 Develop the semidiscrete finite element model of the equations in **Problem 6.1** using the penalty function formulation.

6.3 The equations governing unsteady slow flow of viscous, incompressible fluids in the (x, y)-plane can be expressed in terms of vorticity ζ and stream function ψ:

$$\rho\frac{\partial \omega_z}{\partial t} - \mu\nabla^2\omega_z = 0, \quad -2\omega_z - \nabla^2\psi = 0. \tag{1}$$

Develop the semidiscrete finite element model of the equations.

6.4 A slider bearing consists of a short sliding pad moving at a velocity $u = U_0$ relative to a stationary pad inclined at a small angle with respect to the stationary pad, and the small gap between the two pads is filled with a lubricant [see Fig. P6.4(a)]. If the upper pad is inclined to the base pad, a pressure distribution, in general, a function of x and y, is set up in the gap to support heavy loads normal to the base pad. Using the mesh of 16 × 8 mesh of 4-node quadrilateral elements shown in Fig. P6.4(b) and the penalty finite element model, determine the velocity and pressure distributions in the gap, and plot (a) horizontal velocity $u(x_0, y)$ as a function of y for $x_0 = 0, 0.18$ and $x_0 = 0.36$ and (b) pressure and shear stress as a function of x for $y = 0$ (see Reddy [1]).

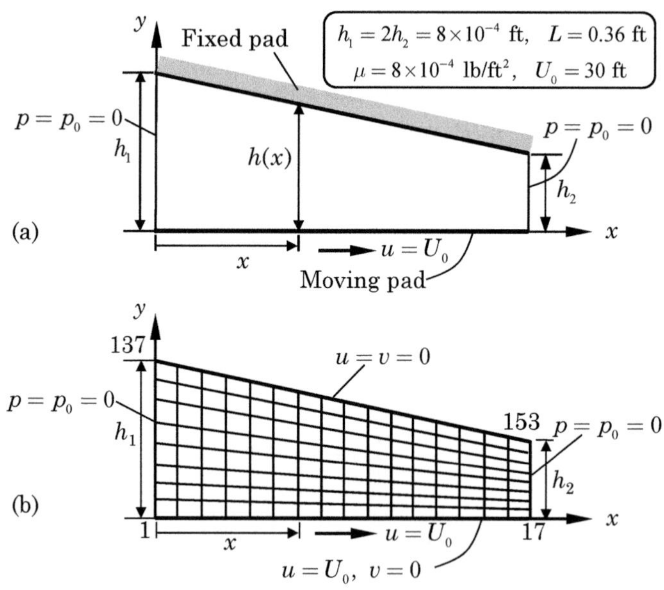

Fig. P6.4

6.5 Consider **Problem 6.1** with the advective (nonlinear) terms included. We have

$$\rho \frac{Dv_r}{Dt} = \frac{1}{r}\frac{\partial}{\partial r}(r\sigma_{rr}) - \frac{\sigma_{\theta\theta}}{r} + \frac{\partial \sigma_{rz}}{\partial z} + f_r \tag{1}$$

$$\rho \frac{Dv_z}{Dt} = \frac{1}{r}\frac{\partial}{\partial r}(r\sigma_{rz}) + \frac{\partial \sigma_{zz}}{\partial z} + f_z \tag{2}$$

$$0 = \frac{1}{r}\frac{\partial}{\partial r}(rv_r) + \frac{\partial v_z}{\partial z}, \tag{3}$$

where

$$\frac{Dv_r}{Dt} = \frac{\partial v_r}{\partial t} + v_r\frac{\partial v_r}{\partial r} + v_z\frac{\partial v_r}{\partial z}, \quad \frac{Dv_z}{Dt} = \frac{\partial v_z}{\partial t} + v_r\frac{\partial v_z}{\partial r} + v_z\frac{\partial v_z}{\partial z}. \tag{3}$$

Develop the semidiscrete mixed finite element model of the equations.

6.6 Develop the semidiscrete penalty finite element model of the equations in **Problem 6.5**.

6.7 Verify the tangent matrix coefficients in Eq. (6.4.16).

6.8 (a) Give the element coefficients $K_{ij}^{\alpha\beta}$ when $\partial v_x/\partial x$, $\partial v_x/\partial y$, $\partial v_y/\partial x$, and $\partial v_y/\partial y$ are treated as known from the previous iteration:

$$\rho\left(v_x\frac{\partial \bar{v}_x}{\partial x} + v_y\frac{\partial \bar{v}_x}{\partial y}\right), \quad \rho\left(v_x\frac{\partial \bar{v}_y}{\partial x} + v_y\frac{\partial \bar{v}_y}{\partial y}\right).$$

(b) Compute the tangent stiffness coefficients associated with the second (new) formulation [see Eq. (6.4.16) for the tangent stiffness coefficients of the first formulation].

Part III The Finite Volume Method

7

The Finite Volume Method: Diffusion Problems

7.1 Introduction

The present chapter deals with the FVM as applied to the diffusion equation in one and two dimensions. In the FVM, the domain Ω is divided into a collection of nonoverlapping subdomains, called *control volumes* and the collection is called a *mesh* or *grid*. A node is deployed at the geometric center of all internal control volumes. Nodes are the locations at which the dependent variables (unknowns) are calculated. The discrete equations in the FVM are derived by making the flux balance (in the integral sense) over each control volume in terms of the nodal values.

As in the FEM and other numerical methods [1], the FVM [7–12] has the following three basic steps.

(1) Grid (mesh) generation.
(2) Development of discretized equations of the governing equation for the entire grid, including the boundary conditions.
(3) Solution of the system of discretized equations.

These ideas will be illustrated first for the diffusion equation in this chapter, and then for other differential equations of heat transfer and fluid dynamics in the remainder of the book. We will go through the three steps in the context of the one- and two-dimensional problems of steady-state heat conduction (or diffusion) considered in Chapter 4.

7.2 One-Dimensional Problems

7.2.1 Governing Equations

In this section we will illustrate the development of the FVM using the one-dimensional heat flow equation, often called the *diffusion equation*. Consider a one-dimensional heat diffusion in thin rod of length L with a heat source S. The governing equation is [a special case of Eq. (4.2.1), with $a = k$, $b = c = 0$, and $g = S$]

$$-\frac{d}{dx}\left(k\frac{dT}{dx}\right) = S, \tag{7.2.1}$$

where T denotes temperature and k is the thermal conductivity. The associated boundary conditions are:

$$x = 0: \quad T = T_0; \qquad x = L: \quad T = T_L. \tag{7.2.2}$$

In this problem, the boundary condition specified at $x = 0$ and $x = L$ is of the *first kind* or *Dirichlet* type. Here we assume that the thermophysical properties (e.g., k) of the material are constant.

7.2.2 Grid Generation

This step involves dividing the domain into a set of subdomains, called control volumes (in the one-dimensional case they are just line segments), deploying nodes at the geometric center of each control volume, and systematically numbering the nodes. The control volumes, in general, need not be of the same size. As a matter of fact, control volumes are made smaller in regions of steep gradients (typically near boundaries or in the vicinity of discontinuities). Varying control volume sizes can be accomplished by varying the distance between the nodes. Grid or mesh generation is a subject in itself and is not the focus of this book; interested readers may consult Thompson et al. [57–59].

The domain (of length L) is divided into NCV, the number of control volumes, with nodes (open circles) positioned at the center of the control volumes (i.e., node-centered control volumes), as shown in Fig. 7.2.1, which corresponds to $NCV = 5$. The number of nodal points is equal to $NCV + 2$. Each control volume is numbered as CV-1, CV-2, and so on. The nodes are numbered sequentially from the left to the right. The solid circles denote the interfaces as well as boundary points while the open circles denote the nodes. The nodal values are denoted with capital letters. This method of grid generation in the context of FVM is termed "zero thickness control volume at the boundary." As stated in Chapter 1, the alternative approach is the "half control volume at the boundary," which is not discussed in this book; only the zero thickness control volume at the boundary is adopted.

The discretized (algebraic) equations for a typical internal control volume will be derived first. The discretized equations for the boundary control volumes are derived using a different treatment. For a uniform grid (i.e., all control volumes are of the same size), the length of each control volume is

$$\Delta x = \frac{L}{NCV}. \tag{7.2.3}$$

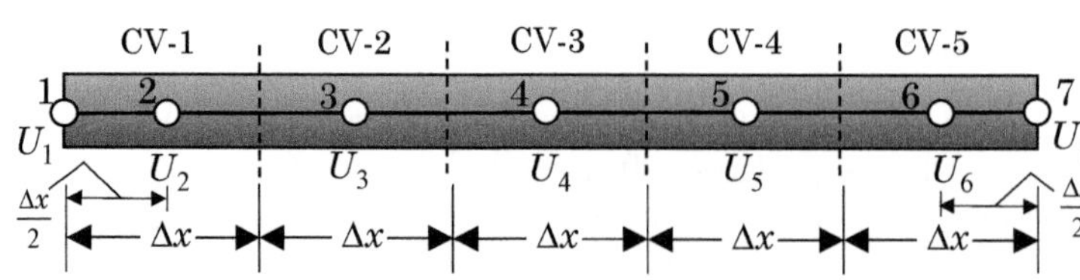

Fig. 7.2.1 Subdivision of the one-dimensional domain into a set of control volumes ($NCV = 5$). The open circles denote the nodes and vertical lines indicate interfaces between control volumes.

7.2.3 Development of Discretization Equations

7.2.3.1 Control Volume-Based Discretization

Consider a generic one-dimensional internal control volume, as shown in Fig. 7.2.2, where Δx represents length of the control volume and δx represents distance between nodes (called the "diffusion length") . The capital letters W, P, and E represent nodal points, with P referring to the center node of the control volume, W referring to the *W*est node, and E referring to the *E*ast node. The points w and e represent control volume interfaces on the left (*w*est) and right (*e*ast) of the control volume. For a uniform mesh, we have $\Delta x = \delta x_w = \delta x_e$, except for the first and last control volumes (see Patankar [7]).

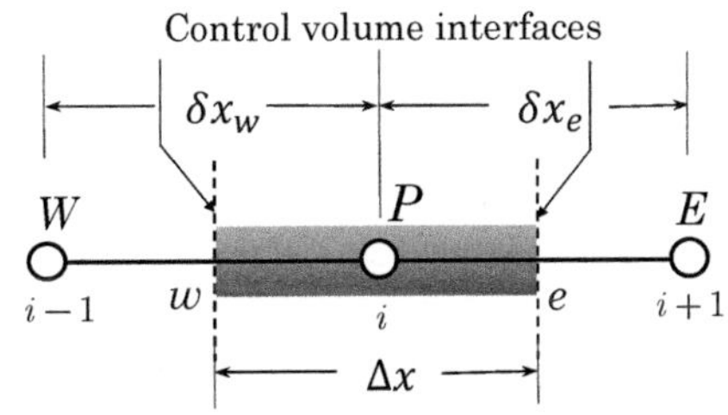

Fig. 7.2.2 A typical internal control volume for one-dimensional problems.

Integrating the governing differential equation, Eq. (7.2.1), over an internal control volume (occupying the domain between points w and e; see Fig. 7.2.2), we obtain

$$\int_{\mathrm{CV}} \left[\frac{d}{dx}\left(k\frac{dT}{dx}\right) + S \right] dx = \left[k\frac{dT}{dx} \right]_w^e + S_P \Delta x = 0, \tag{7.2.4}$$

where CV denotes the control volume; here we have approximated the integral of the source term S as (a feature that is common with the FVM):

$$\int_{\mathrm{CV}} S\, dx \approx S_P \Delta x, \tag{7.2.5}$$

where S_P is an average value. Other representations of the integrals (e.g., trapezoidal rule or Simpson's rule) may also be used. Carrying out the integration on the first term, we obtain the control-volume representation of the governing equation:

$$\left[\left(k\frac{dT}{dx}\right)_e - \left(k\frac{dT}{dx}\right)_w \right] + S_P \Delta x = J_w - J_e + S_P \Delta x = 0. \tag{7.2.6}$$

Here $J = -k\frac{dT}{dx}$ denotes the diffusive flux. Note that Eq. (7.2.6) is a flux balance equation, where the diffusive flux entering at the west face (J_w) and diffusive flux leaving from the east face (J_e) are balanced by the source term $S_P \Delta x$. In other words, the net flux is conserved *locally* in each control volume. Equation (7.2.6) forms the basis of obtaining the discretized (algebraic) equations for the model equation at hand.

7.2.3.2 Approximation of the First Derivatives across Control Volumes

The next step in deriving the discretized (algebraic) equations is to replace the derivatives appearing in Eq. (7.2.6) with suitable approximations. For simplicity as well as from practical point of view, we consider a uniform mesh of control volumes (i.e., $\delta x_e = \delta x_w = \Delta x$), as shown in Fig. 7.2.1. However, one should not use or represent δx and Δx interchangeably as δx represents diffusion length and Δx represents geometric length. To derive the approximation of the derivative at the eastern face (e), we expand T_P and T_E in the Taylor series about T_e (see Fig. 7.2.2):

$$T_P = T_e - \frac{\delta x}{2}\frac{dT}{dx}\bigg|_e + \frac{1}{2}\left(\frac{\delta x}{2}\right)^2 \frac{d^2T}{dx^2}\bigg|_e - \frac{1}{3!}\left(\frac{\delta x}{2}\right)^3 \frac{d^3T}{dx^3}\bigg|_e + \cdots \tag{7.2.7a}$$

$$T_E = T_e + \frac{\delta x}{2}\frac{dT}{dx}\bigg|_e + \frac{1}{2}\left(\frac{\delta x}{2}\right)^2 \frac{d^2T}{dx^2}\bigg|_e + \frac{1}{3!}\left(\frac{\delta x}{2}\right)^3 \frac{d^3T}{dx^3}\bigg|_e + \cdots . \tag{7.2.7b}$$

Subtracting Eq. (7.2.7a) from Eq. (7.2.7b) and solving for $(dT/dx)_e$, we obtain

$$\frac{dT}{dx}\bigg|_e = \frac{T_E - T_P}{\delta x} + \frac{(\delta x)^2}{24}\frac{d^3T}{dx^3}\bigg|_e + \cdots = \frac{T_E - T_P}{\delta x} + O(\delta x_e)^2. \tag{7.2.8}$$

Equation (7.2.8) is a central difference approximation of the first derivative dT/dx at the eastern face of the control volume. It is clear from Eq. (7.2.8) that the approximation is second-order accurate. Following the same procedure, we can derive the following expression for the western face (w) of the control volume:

$$\frac{dT}{dx}\bigg|_w \approx \frac{T_P - T_W}{\delta x_w}. \tag{7.2.9}$$

Note that in the approximate representations of the derivative (flux) we have made use of values of unknowns at two neighboring points. We could have used more than one unknown to represent the derivative of the variable at interfaces w and e, as discussed in the sequel. Since we use two nodes on either of the interfaces to represent the flux, it is referred to as *two-node formulation*.

7.2.3.3 Discretized Equation for an Interior Control Volume

We assume a piecewise-linear profile to represent the variation of temperature and estimate the derivative dT/dx at interfaces. One should note that flux representation on all control volume interfaces must remain the same to maintain the flux balance. For example, one should not represent flux at the east face (e) by a piecewise-linear profile and flux at west face (w) by a quadratic profile. Substituting the approximations in Eqs. (7.2.8) and (7.2.9) into Eq. (7.2.6), we obtain (see Fig. 7.2.2)

$$\left[\left(k_e \frac{T_E - T_P}{\delta x_e}\right) - \left(k_w \frac{T_P - T_W}{\delta x_w}\right)\right] + S_P \Delta x = 0 \tag{7.2.10a}$$

or

$$\left(\frac{k_w}{\delta x_w}\right)T_W - \left(\frac{k_w}{\delta x_w} + \frac{k_e}{\delta x_e}\right)T_P + \left(\frac{k_e}{\delta x_e}\right)T_E + S_P\,\Delta x = 0. \tag{7.2.10b}$$

The discretized equation in Eq. (7.2.10b) can be expressed as

$$-a_W T_W + a_P T_P - a_E T_E = b_P, \tag{7.2.11a}$$

where

$$a_W = \frac{k_w}{\delta x_w} = D_w, \quad a_E = \frac{k_e}{\delta x_e} = D_e, \quad a_P = a_W + a_E, \quad b_P = S_P\Delta x. \tag{7.2.11b}$$

Here, D_w and D_e are used to indicate diffusion strengths. Note that all thermophysical properties such as thermal conductivity, density, specific heat, viscosity, and the like are prescribed at nodal points, and not at the interface (w or e). For now we will assume that thermophysical properties remain constant, $k_e = k_w = k_P = k_W = k$. For variable thermophysical properties discussion will be deferred to a later section.

Equation (7.2.10a) can also be represented as,

$$a_P T_P = \sum_{nb} a_{nb} T_{nb} + b_P, \tag{7.2.12}$$

where nb stands for neighboring nodal points. We note that a_P includes contributions from the neighboring nodes through the coefficients a_W and a_E. We also note that the Scarborough criterion [36] is satisfied trivially for pure diffusion problems. Thus, the original differential equation is converted to a set of algebraic equations. Equations (7.2.11a) and (7.2.11b) are valid for nodes that are in control volumes CV-2 through CV-(N-1) (i.e., $P = 3, 4, \ldots, NCV$). The equations for the first (CV-1) and last (CV-N) control volumes must be derived using suitable approximations of $(dT/dx)_w$ in CV-1 and $(dT/dx)_e$ in CV-N, as discussed next.

7.2.3.4 Discretized Equations for Control Volumes CV-1 and CV-N

For CV-1, Eq. (7.2.6) takes the form ($P = 2$ and $E = 3$)

$$0 = -\left(k\frac{dT}{dx}\right)_w + k_e\frac{T_E - T_P}{\delta x_e} + S_P\Delta x, \tag{7.2.13}$$

where $(dT/dx)_e$ is replaced with the second-order accurate formula (7.2.8). On the left end of the domain either gradient of temperature $(dT/dx)_w$ is specified or temperature T_w is specified. In the latter case, $(dT/dx)_w$ is replaced with a suitable approximation.

For CV-N, Eq. (7.2.6) becomes ($P = NCV + 1$ and $W = NCV$)

$$0 = -k_w\frac{T_P - T_W}{\delta x_w} + \left(k\frac{dT}{dx}\right)_e + S_P\Delta x, \tag{7.2.14}$$

where $(dT/dx)_w$ is replaced with the second-order accurate formula (7.2.9). On the right end of the domain, either gradient of temperature $(dT/dx)_e$ is specified or temperature T_e is known. In the latter case, $(dT/dx)_e$ is suitably approximated.

When T is specified on the left end of the domain, we can use the first-order accurate representation of dT/dx on the west face, although it is inconsistent with the second-order accurate representation used on the east face (see **Problem 1.7** for a second-order accurate representation of the first derivative at a boundary). In this case, Eq. (7.2.10b) is valid with $T_W = T_0$, $\delta x_w = 0.5\Delta x$, and $\delta x_e = \Delta x$ [see Fig. 7.2.3(a)]. This approach is known as the "zero-thickness" control volume at the boundaries. Thus we have (for CV-1)

$$\left(\frac{k_w}{\delta x_w} + \frac{k_e}{\delta x_e}\right) T_P - \left(\frac{k_e}{\delta x_e}\right) T_E = S_P\,\Delta x + \left(\frac{k_w}{\delta x_w}\right) T_0 \tag{7.2.15}$$

or

$$a_P\,T_P - a_E\,T_E = b_P, \tag{7.2.16a}$$

where

$$a_W = \frac{k_w}{\delta x_w} = \frac{2k_w}{\Delta x},\quad a_E = \frac{k_e}{\delta x_e} = \frac{k_e}{\Delta x},\quad a_P = a_W + a_E,\quad b_P = S_P\,\Delta x + a_W T_0. \tag{7.2.16b}$$

Changing the subscripts into index notation (for $P = 2$, $W = 1$, and $E = 3$), we obtain

$$a_2\,T_2 - a_3\,T_3 = b_2, \tag{7.2.16c}$$

where $a_1 = a_W$, $a_2 = a_P$, $a_3 = a_E$, $T_1 = T_W = T_0$, $T_2 = T_P$, $T_3 = T_E$, and $b_P = b_2 = S_2\Delta x + a_1\,T_0$.

Next, we consider the last CV, namely, CV-N [see Fig. 7.2.3(b)] when T_E is specified. In this case, when using the first-order accurate approximation on the east face, Eq. (7.2.11a) is valid with $T_E = T_L$, $\delta x_e = 0.5\Delta x$ and $\delta x_w = \Delta x$. Thus we have (for CV-5)

$$\left(\frac{k_w}{\delta x_w} + \frac{k_e}{\delta x_e}\right) T_P - \left(\frac{k_w}{\delta x_w}\right) T_W = S_P\,\Delta x + \left(\frac{k_e}{\delta x_e}\right) T_L \tag{7.2.17}$$

or

$$a_P\,T_P - a_W\,T_W = b_P, \tag{7.2.18a}$$

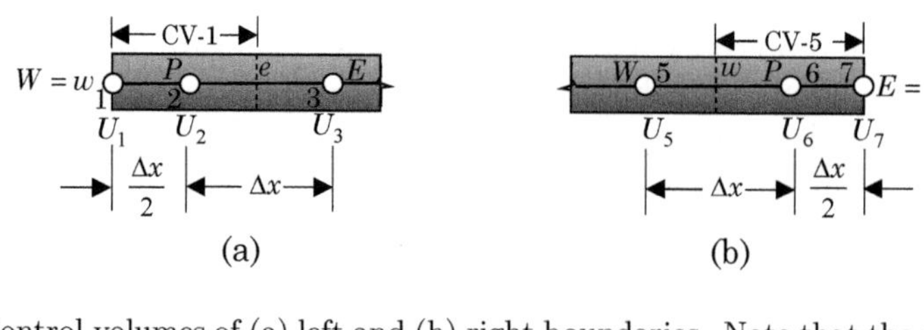

Fig. 7.2.3 Control volumes of (a) left and (b) right boundaries. Note that the control volumes of the boundary nodes have no dimension (hence, zero thickness).

where

$$a_W = \frac{k_w}{\delta x_w} = \frac{k_w}{\Delta x}, \quad a_E = \frac{k_e}{\delta x_e} = \frac{2k_e}{\Delta x}, \quad a_P = a_W + a_E, \quad b_P = S_P\,\Delta x + a_E T_L. \tag{7.2.18b}$$

Changing the subscripts into index notation ($P = 6$, $W = 5$, and $E = 7$), we obtain

$$a_6\,T_6 - a_5\,T_5 = b_6, \tag{7.2.18c}$$

where $a_5 = a_W$, $a_6 = a_P$, $a_7 = a_E$, $T_5 = T_W$, $T_6 = T_P$, $T_7 = T_E = T_L$, and $b_P = b_6 = S_6\Delta x + a_7\,T_L$.

7.2.3.5 Algebraic Equations for the Whole Mesh

For convenience in coding, the nodal values in Eqs. (7.2.11a), (7.2.16a), and (7.2.18a) can be expressed in terms of indices ($P = i$, $W = i-1$, and $E = i+1$) to generate the required system of linear algebraic equations (see Fig. 7.2.2):

$$a_P(i)T_i = a_W(i)T_{i-1} + a_E(i)T_{i+1} + b_P(i). \tag{7.2.19}$$

Equation (7.2.19) is valid for internal control volumes (node $i = 3$ to NCV). Therefore, for each node i the coefficients a_P, a_W, and a_E should be calculated and stored. Depending on the boundary conditions, the coefficients for boundary control volumes are modified. As an example, let $NCV = 5$ (i.e., there are seven nodes) and let the boundary conditions be of the Dirichlet type. The values of $a_W(i)$, $a_P(i)$, $a_E(i)$, and $b_P(i)$ [referring to Eq. (7.2.19)], for nodes $i = 2, 3, \ldots, 6$, are defined in Table 7.2.1, where the parameters $a_W(i)$, $a_E(i)$, and $a_P(i)$ appearing in the table are defined as follows:

$$a_W(i) = \frac{k_w}{\delta x_w}, \quad a_E(i) = \frac{k_e}{\delta x_e} \quad a_P(i) = a_W(i) + a_E(i), b_P(i) = S_i\Delta x + c, \tag{7.2.20}$$

where c is the additional term arising from the prescribed temperature boundary condition.

Table 7.2.1 Values of the coefficients $a_W(i)$, $a_P(i)$, $a_E(i)$, $b_P(i)$ for the five equations.

CV	i	T_2	T_3	T_4	T_5	T_6	b_P
1	2	$a_P(2)$	$-a_E(2)$	0	0	0	$S_2\,\Delta x$ $+a_W(2)T_0$
2	3	$-a_W(3)$	$a_P(3)$	$-a_E(3)$	0	0	$S_3\,\Delta x$
3	4	0	$-a_W(4)$	$a_P(4)$	$-a_E(4)$	0	$S_4\,\Delta x$
4	5	0	0	$-a_W(5)$	$a_P(5)$	$-a_E(5)$	$S_5\,\Delta x$
5	6	0	0	0	$-a_W(6)$	$a_P(6)$	$S_6\,\Delta x$ $+a_E(6)T_L$

Thus, we have the following five equations corresponding to nodes $i = 2, 3, \ldots, 6$.

$$\begin{aligned}
\text{For } i = 2: &\quad a_P(2)\,T_2 - a_E(2)\,T_3 = S_2\Delta x + a_W(2)\,T_0. \\
\text{For } i = 3: &\quad -a_W(3)\,T_2 + a_P(3)\,T_3 - a_E(3)\,T_4 = S_3\Delta x. \\
\text{For } i = 4: &\quad -a_W(4)\,T_3 + a_P(4)\,T_4 - a_E(4)\,T_5 = S_4\Delta x. \\
\text{For } i = 5: &\quad -a_W(5)\,T_4 + a_P(5)\,T_5 - a_E(5)\,T_6 = S_5\Delta x. \\
\text{For } i = 6: &\quad -a_W(6)\,T_5 + a_P(6)\,T_6 = S_6\Delta x + a_E(6)\,T_L.
\end{aligned} \tag{7.2.21}$$

The set of equations in Eq. (7.2.21) can be solved by any of the direct or iterative methods described in Chapter 3. It is evident that the above coefficient matrix is a tridiagonal matrix. Accordingly, we will use the TDMA to solve the set of equations. In any given row (equation) there are at most three nonzero coefficients corresponding to three unknowns and in the first and the last row (equation) there are only two nonzero coefficients corresponding to two unknowns. The TDMA or Thomas algorithm is the same as the Gauss elimination method specialized to a tridiagonal system of equations.

7.2.4 Neumann Boundary Condition: Prescribed Flux

In the illustrative problem we have used Dirichlet boundary conditions at both the left and right boundaries. Let us assume that the boundary condition at right boundary ($x = L$) is changed to a constant heat flux boundary. Now the problem statement becomes

$$\frac{d}{dx}\left(-k\frac{dT}{dx}\right) = S \tag{7.2.22a}$$

$$T(0) = T_0 \tag{7.2.22b}$$

$$-k\frac{dT}{dx}\bigg|_{x=L} = q_L, \tag{7.2.22c}$$

where q_L is a constant. Recall that for the last CV (CV-N), Eq. (7.2.6) becomes

$$0 = -k_w\frac{T_P - T_W}{\delta x_w} + \left(k\frac{dT}{dx}\right)_e + S_P\Delta x. \tag{7.2.23}$$

As we have mentioned earlier, the integral equation in the FVM is in flux balance form. As a result, any prescribed flux boundary condition can be directly applied into the semidiscretized form of the equation in FVM. In Eq. (7.2.23), the specified heat flux can be readily applied to the east face:

$$-k_w\frac{T_P - T_W}{\delta x_w} - q_L + S_P\Delta x = 0 \tag{7.2.24a}$$

$$\left(\frac{k_w}{\delta x_w}\right)T_P - \left(\frac{k_w}{\delta x_w}\right)T_W = S_P\,\Delta x - q_L \tag{7.2.24b}$$

$$-a_W T_W + a_P T_P - a_E T_E = b_P, \tag{7.2.24c}$$

where

$$a_W = \frac{k_w}{\delta x_w},\ a_E = 0,\ a_P = a_W + a_E,\ b_P = S_P \Delta x - q_L. \tag{7.2.24d}$$

In this way we obtain N equations for N number of control volumes. More specifically, for five control volumes we will have five equations involving five nodal temperatures (T_2 through T_6). Note that, when heat flux is specified, the temperature at the right boundary (T_7) does not enter into the system of equations. Once the equations for internal control volumes are solved, the temperature T_7 can be obtained from a suitable approximation of the flux term. Using a first-order approximation we get

$$-k_e \frac{T_7 - T_6}{\delta x_e} = q_L \ \rightarrow \therefore T_7 = T_6 - q_L \frac{\Delta x}{2k_e}. \tag{7.2.25a}$$

For zero Neumann (or zero heat flux) boundary condition $q_L = 0$, Eqs. (7.2.24c), and (7.2.24d) become,

$$\begin{aligned} &-a_W T_W + a_P T_P - a_E T_E = b_P \\ &a_W = \frac{k_w}{\delta x_w},\ a_E = 0,\ a_P = a_W + a_E,\ b_P = S_P \Delta x. \end{aligned} \tag{7.2.25b}$$

Also, the last term of Eq. (7.2.24d) vanishes for zero Neumann boundary condition and the temperature at the boundary becomes equal to the adjacent nodal temperature for the first order approximation of the first derivative term,

$$T_7 = T_6. \tag{7.2.26}$$

7.2.5 Mixed Boundary Condition: Convective Heat Flux

Let us assume that the heat flux at the right end of the rod varies as a function of the temperature difference between the right boundary and contacting medium. In this case, the boundary heat flux is not uniform and the problem statement becomes:

$$\frac{d}{dx}\left(-k\frac{dT}{dx}\right) = S \tag{7.2.27a}$$

$$T(0) = T_0 \tag{7.2.27b}$$

$$-k\left.\frac{dT}{dx}\right|_{x=L} = h_\infty (T_L - T_\infty), \tag{7.2.27c}$$

where T_∞ is the temperature of the contacting medium or ambient temperature and h_∞ is the convective heat transfer coefficient. Note that, heat flux depends on the temperature at the right boundary (T_L), which is not known. This is an example of mixed or Robin boundary condition. Here we will derive the discretized equation for the right-most control volume and show how the mixed boundary condition is treated in FVM. First, we apply the boundary condition

Eq. (7.2.27c) in the discretized equation for the last control volume (CV-N) in Eq. (7.2.6):

$$-k_w \frac{T_P - T_W}{\delta x_w} + \left(k \frac{dT}{dx}\right)_e + S_P \Delta x = 0 \tag{7.2.28a}$$

$$-k_w \frac{T_P - T_W}{\delta x_w} - h_\infty (T_L - T_\infty) + S_P \Delta x = 0. \tag{7.2.28b}$$

Since T_L is unknown, we need another equation to express T_L in terms of known quantities. We achieve that by discretizing the first derivative term in the boundary condition given in Eq. (7.2.27c):

$$-k \frac{dT}{dx}\bigg|_{x=L} = h_\infty (T_L - T_\infty) \tag{7.2.29a}$$

$$-k_e \frac{T_L - T_P}{\delta x_e} = h_\infty (T_L - T_\infty) \tag{7.2.29b}$$

$$T_L = \frac{h_\infty T_\infty + \frac{k_e}{\delta x_e} T_P}{h_\infty + \frac{k_e}{\delta x_e}} \tag{7.2.29c}$$

$$T_L - T_\infty = \frac{\frac{k_e}{\delta x_e}(T_P - T_\infty)}{h_\infty + \frac{k_e}{\delta x_e}}. \tag{7.2.29d}$$

Using Eq. (7.2.29d) in Eq. (7.2.28b), we get,

$$-k_w \frac{T_P - T_W}{\delta x_w} - \frac{h_\infty \frac{k_e}{\delta x_e}(T_P - T_\infty)}{h_\infty + \frac{k_e}{\delta x_e}} + S_P \Delta x = 0 \tag{7.2.30a}$$

$$\left(\frac{k_w}{\delta x_w} + \frac{h_\infty \frac{k_e}{\delta x_e}}{h_\infty + \frac{k_e}{\delta x_e}}\right) T_P - \left(\frac{k_w}{\delta x_w}\right) T_W = S_P \Delta x + \frac{h_\infty \frac{k_e}{\delta x_e} T_\infty}{h_\infty + \frac{k_e}{\delta x_e}}. \tag{7.2.30b}$$

Now we can express the above equation in our familiar form with a-coefficients:

$$-a_W T_W + a_P T_P - a_E T_E = b_P, \tag{7.2.31a}$$

where

$$a_W = \frac{k_w}{\delta x_w}, \; a_E = 0, \; a_P = a_W + a_E + R_{th}, \; b_P = S_P \Delta x + R_{th} T_\infty \tag{7.2.31b}$$

$$R_{th} = \frac{h_\infty \frac{k_e}{\delta x_e}}{h_\infty + \frac{k_e}{\delta x_e}}. \tag{7.2.31c}$$

Here, R_{th} acts as a thermal resistance for combined conduction and convection. This way we have a system of equations involving only the internal nodes or cell centers as unknowns. Once this system of equations is solved, we can compute the temperature at the right boundary using Eq. (7.2.29c):

$$T_7 = \frac{h_\infty T_\infty + \frac{k_e}{\delta x_e} T_6}{h_\infty + \frac{k_e}{\delta x_e}}. \tag{7.2.32}$$

Note that, for the zero heat flux boundary condition, $h_\infty = 0$, so R_{th} becomes zero; and Eq. (7.2.31b) transforms to Eq. (7.2.24c). Similarly, Eq. (7.2.32) becomes identical to Eq. (7.2.26):

$$T_7 = T_6. \tag{7.2.33}$$

7.2.6 Interface Properties

In the one-dimensional control volume shown in Fig. 7.2.2, the diffusion heat flux at the western interface is

$$q''_w = -k_w \left. \frac{dT}{dx} \right|_w \tag{7.2.34a}$$

$$q''_w = -k_w \frac{T_P - T_W}{\delta x_w}. \tag{7.2.34b}$$

As stated in an earlier discussion, all thermophysical properties are prescribed and stored at the nodes (W, P, and E). So thermal conductivity values are stored at nodes W, P, and E. But after the integration over a control volume, we need the value of thermal conductivity at the faces (i.e. w, e for one-dimensional problems). One may be tempted to use a simple arithmetic mean for a uniform grid when the interface w is midway between W and P, or a weighted arithmetic mean for a nonuniform grid

$$k_w = \frac{k_P + k_W}{2}. \tag{7.2.35}$$

This may work for a single material or fluid, but will produce inconsistent heat flux for a composite material where the thermophysical properties are discontinuous at the interface. For example, if we have a composite slab of one insulating material ($k_P \to 0$) and one highly conducting material ($k_W \neq 0$), the arithmetic mean formulation gives a nonzero heat flux at the interface ($q''_w \neq 0$), which is incorrect. The remedy is to use a harmonic mean instead of an arithmetic mean:

$$k_w = \frac{2}{\frac{1}{k_P} + \frac{1}{k_W}} = \frac{2k_P k_W}{k_P + k_W}. \tag{7.2.36}$$

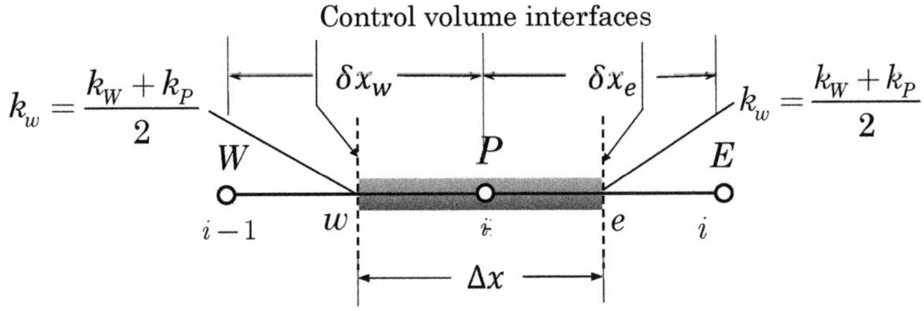

Fig. 7.2.4 A one-dimensional control volume for calculating interface properties k_w and k_e.

If $k_P \to 0$, then $k_w \to 0$ i.e., $q'' \to 0$, which is physically realistic. This discussion is based on the assumption that the interface w is midway between W and P. But the interface w need not be at the midpoint (e.g., for a nonuniform grid). If the distance between the west face w and node W is δx_i and the distance between the west face w and node P is δx_{i+1}, then $\delta x_w = \delta x_i + \delta x_{i+1}$. Since the heat flux on each side of interface w must be same, we get:

$$q''_w = \frac{T_W - T_w}{\delta x_i/k_W} = \frac{T_w - T_P}{\delta x_{i+1}/k_P} = \frac{T_W - T_P}{\delta x_i/k_W + \delta x_{i+1}/k_P} = \frac{T_W - T_P}{(\delta x_i + \delta x_{i+1})/k_w} \tag{7.2.37a}$$

$$\frac{\delta x_i + \delta x_{i+1}}{k_w} = \frac{\delta x_i}{k_W} + \frac{\delta x_{i+1}}{k_P} \tag{7.2.37b}$$

from which obtain

$$k_w = \frac{k_W k_P(\delta x_i + \delta x_{i+1})}{k_P \delta x_i + k_W \delta x_{i+1}}. \tag{7.2.37c}$$

For uniform grid we have $\delta x_i = \delta x_{i+1} = \delta x_w/2$, and k_w becomes,

$$k_w = \frac{2k_P k_W}{k_P + k_W}. \tag{7.2.38}$$

Similarly, we should use the harmonic mean method to estimate any other thermophysical properties arising out of a flux term, i.e., whenever the flux balance is considered at the interface with discontinuous properties, such as dynamic viscosity (μ), turbulent viscosity (μ_t), mass diffusivity (D), and the like. Since we are discussing a one-dimensional formulation, the weights in Eq. (7.2.21) are geometric lengths. For a two-dimensional formulation the weights will be areas, and for three-dimensional formulation the weights will be volumes.

7.2.7 Numerical Examples

Example 7.2.1

Consider the linear differential equation

$$-\frac{d^2u}{dx^2} = g_0 \cos x, \quad 0 < x < 1 \tag{1}$$

with the following two cases of boundary conditions (**Case 1** is of BC1 type; and **Case 2** is of BC2 type with $b_L = 0$ and $q_L = 0$):

$$\textbf{Case 1:}\quad u(0) = u(1) = 0; \quad \textbf{Case 2:}\quad u(0) = \left.\frac{du}{dx}\right|_{x=1} = 0. \tag{2}$$

Determine the finite volume numerical solution of the problem for $g_0 = 10$ using uniform meshes and various formulations discussed in this section and compare the results with the exact solutions:

$$\textbf{Case 1:}\quad u(x) = g_0\left[-(1-\cos x) + x(1-\cos 1)\right];$$

$$q(x) \equiv -\frac{du}{dx} = g_0\left[\sin x - (1-\cos 1)\right]. \tag{3}$$

$$\textbf{Case 2:}\quad u(x) = g_0\left[-(1-\cos x) + x\sin 1\right];$$

$$q(x) \equiv -\frac{du}{dx} = g_0\left(\sin x - \sin 1)\right). \tag{4}$$

Solution Equation (1) is a special case of the model equation in Eq. (7.2.1), with $k = 1$ and $S(x) = g_0 \cos x$. We begin with a mesh of four control volumes (i.e., $NCV = 4$ and $\Delta x = 0.25$) and six nodes ($NCV + 2$) in the zero-thickness formulation, as shown in Fig. 7.2.5. The source term $g(x) = g_0 \cos x$ was evaluated using the trapezoidal rule within each control volume (i.e., $F_P = 0.5\Delta x[g(x_w) + g(x_e)]$).

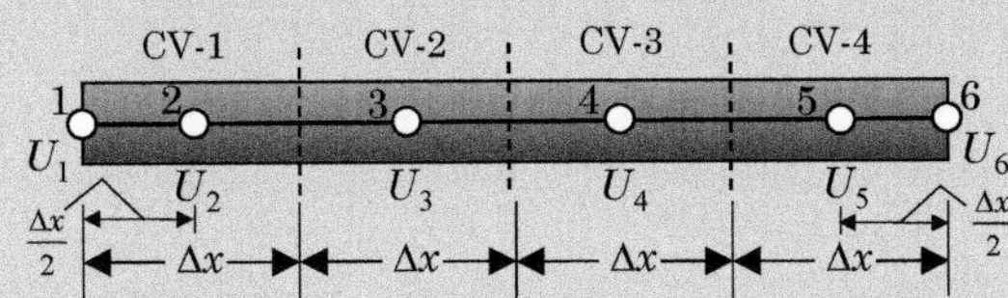

Fig. 7.2.5 A uniform mesh of four ($N = 4$) subdivisions used in the zero-thickness control volume formulation. There are four control volumes, $NCV = N$ and 6 nodes.

Case 1 [$u(0) = u(1) = 0$]

The four algebraic equations for the first-order accurate zero-thickness formulation are [the first equation is obtained using Eqs. (7.2.16a) and (7.2.16b) with $\delta x_w = \Delta x/2$ and $\delta x_e = \Delta x$; the second and third equations from Eqs. (7.2.11a) and (7.2.11b); and the fourth equation from Eqs. (7.2.18a) and (7.2.18b)]:

$$\begin{aligned} -8U_1 + 12U_2 - 4U_3 &= 2.4611 \\ -4U_2 + 8U_3 - 4U_4 &= 2.3081 \\ -4U_3 + 8U_4 - 4U_5 &= 2.0116 \\ -4U_4 + 12U_5 - 8U_6 &= 1.5900, \end{aligned} \tag{5}$$

whose solution is ($U_1 = U_6 = 0$): $U_2 = 0.5686, U_3 = 1.0906, U_4 = 1.0356$, and $U_5 = 0.4777$.

The four algebraic equations for the second-order accurate zero-thickness formulation are [the first equation is obtained using Eq. (7.2.13) and the result from **Problem 1.7**; the second and third equations are obtained from Eqs. (7.2.11a) and (7.2.11b); and the fourth equation is obtained from Eq. (7.2.14) and **Problem 1.7**]:

$$\begin{aligned} -\tfrac{32}{3}U_1 + 16U_2 - \tfrac{16}{3}U_3 &= 2.4611 \\ -4U_2 + 8U_3 - 4U_4 &= 2.3081 \\ -4U_3 + 8U_4 - 4U_5 &= 2.0116 \\ -\tfrac{16}{3}U_4 + 16U_5 - \tfrac{32}{3}U_6 &= 1.5900 \end{aligned} \tag{6}$$

whose solution is ($U_1 = U_6 = 0$): $U_2 = 0.4951, U_3 = 1.0240, U_4 = 0.9757$, and $U_5 = 0.4246$.

The numerical values obtained with various number of subdivisions are compared in Table 7.2.2, where the underlined numbers are the nodal values and others are *linearly* interpolated values (we note that the use of linear interpolation for the second-order accurate scheme is not correct). The table also contains gradient of the solution at the left end ($x = 0$) and the right end ($x = 1$) of the domain (the last two rows). The FEM solutions are obtained using (nonuniform) meshes of linear elements that correspond to the FVM meshes. The FVM solutions are exact at the nodes for this problem [i.e., when a is a constant, $c = 0$, and g is any function; see Eq. (7.2.3)], independent of the number of elements.

Table 7.2.2 Comparison of the finite volume solutions with the exact solution (u) of: $-\frac{d^2u}{dx^2} = 10\cos x;\quad 0 < x < 1;\quad u(0) = u(1) = 0.$

		Finite volume method				Finite element method	
x	Exact	FVM1(4)	FVM2(4)	FVM1(8)	FVM2(8)	FEM(5)	FEM(9)
0.0625	0.2678	0.2843	0.2476	0.2866	0.2676	0.2483	0.2678
0.1250	0.4966	0.5686	0.4951	0.4953	0.4768	0.4966	0.4772
0.1875	0.6867	0.6991	0.6273	0.7040	0.6861	0.6297	0.6867
0.2500	0.8384	0.8296	0.7595	0.8362	0.8187	0.7628	0.8194
0.3125	0.9522	0.9601	0.8917	0.9683	0.9513	0.8959	0.9522
0.3750	1.0289	1.0906	1.0239	1.0263	1.0098	1.0289	1.0108
0.4375	1.0693	1.0769	1.0119	1.0842	1.0682	1.0174	1.0693
0.5000	1.0743	1.0631	0.9998	1.0715	1.0560	1.0058	1.0572
0.5625	1.0450	1.0494	0.9878	1.0588	1.0438	0.9943	1.0450
0.6250	0.9827	1.0356	0.9757	0.9802	0.9657	0.9827	0.9669
0.6875	0.8888	0.8961	0.8380	0.9015	0.8875	0.8451	0.8888
0.7500	0.7646	0.7567	0.7002	0.7626	0.7491	0.7075	0.7503
0.8125	0.6119	0.6172	0.5624	0.6237	0.6107	0.5699	0.6119
0.8750	0.4323	0.4777	0.4246	0.431	0.4187	0.4323	0.4198
0.9375	0.2277	0.2388	0.2123	0.2387	0.2266	0.2162	0.2277
$-q_w$	4.5970	5.3696	4.5764	5.0000	4.5929	4.5970	4.5970
q_e	3.8177	4.3517	3.7944	4.0648	3.8108	3.8177	3.8177

Figures 7.2.6 and 7.2.7 contains a comparison of the numerical solutions obtained from the zero-thickness formulation with the exact solutions $u(x)$ and $q(x)$. Meshes of four and eight control volume solutions are presented with the first-order and second-order accurate schemes. As expected, the second-order accurate scheme performs better than the first-order accurate scheme, although with mesh refinements they all converge to the exact solutions. The first-order accurate solution for coarse meshes has more error in u as well as q (especially at the boundary).

Case 2 $\left[u(0) = \frac{du}{dx}\Big|_{x=1} = 0\right]$

For this case the five algebraic equations for the first-order accurate zero-thickness formulation are [the first equation is obtained using Eqs. (7.2.16a) and (7.2.16b); the second and third equations are obtained from Eqs. (7.2.11a) and (7.2.11b); the fourth equation is determined from Eqs. (7.2.18a) and (7.2.18b); and the fifth equation is obtained from Eq. (7.2.33) with $h_\infty = 0$ and $q_L = 0$]:

$$\begin{aligned}
-8U_1 + 12U_2 - 4U_3 &= 2.4611\\
-4U_2 + 8U_3 - 4U_4 &= 2.3081\\
-4U_3 + 8U_4 - 4U_5 &= 2.0116\\
-4U_4 + 4U_5 &= 1.5900\\
-U_5 + U_6 &= 0.0,
\end{aligned} \tag{7}$$

whose solution is ($U_1 = 0$):

$$U_2 = 1.0463,\quad U_3 = 2.5238,\quad U_4 = 3.4242,\quad U_5 = U_6 = 3.8217. \tag{8}$$

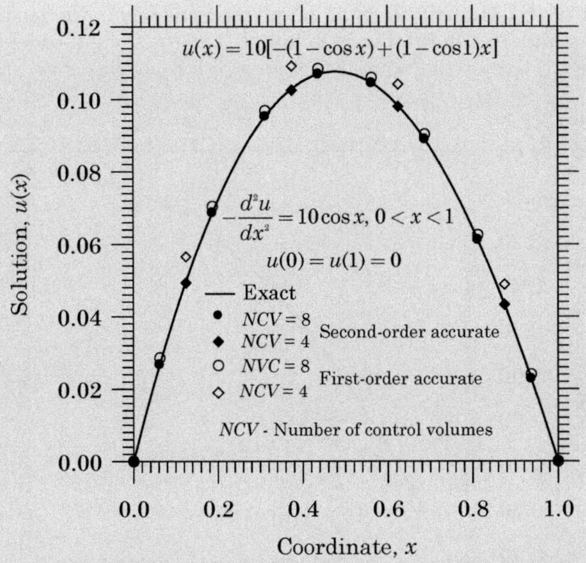

Fig. 7.2.6 Comparison of FVM solutions for $u(x)$ with the exact solution of Case 1.

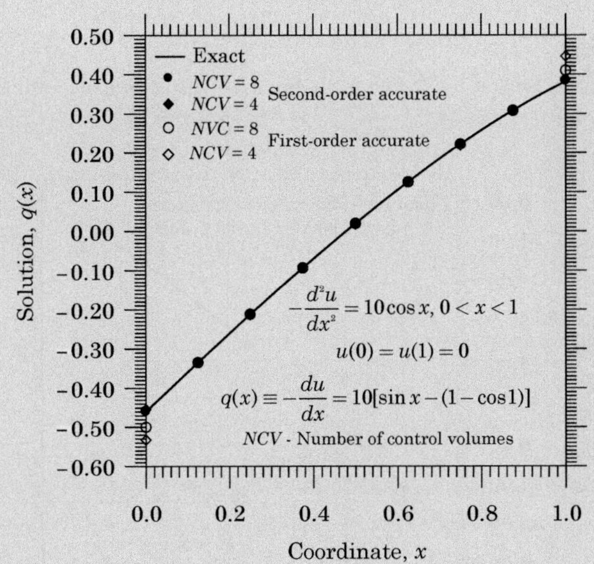

Fig. 7.2.7 Comparison of FVM solutions for $q(x)$ with the exact solution of Case 1.

The finite element solution at the same points is $U_2 = 0.9738$, $U_3 = 2.4606$, $U_4 = 3.3688$, $U_5 = 3.7728$, and $u_6 = 3.8177$. which coincides with the exact solution. Once again, recall that the finite element solution is exact at the nodes for any number of elements.

Figures 7.2.8 and 7.2.9 contains a comparison of the numerical solutions obtained from the zero-thickness formulation with the exact solutions $u(x)$ and $q(x)$ for meshes of four and eight control volumes and the first-order and second-order accurate schemes. Both first-order and second-order accurate FVM schemes yield numerical solutions that converge to the exact solution with mesh refinements.

This completes the FVM analysis of an inhomogeneous second-order differential equation with two different boundary conditions. It is found that the second-order accurate representation of the first derivative at the boundary points results in better accuracy.

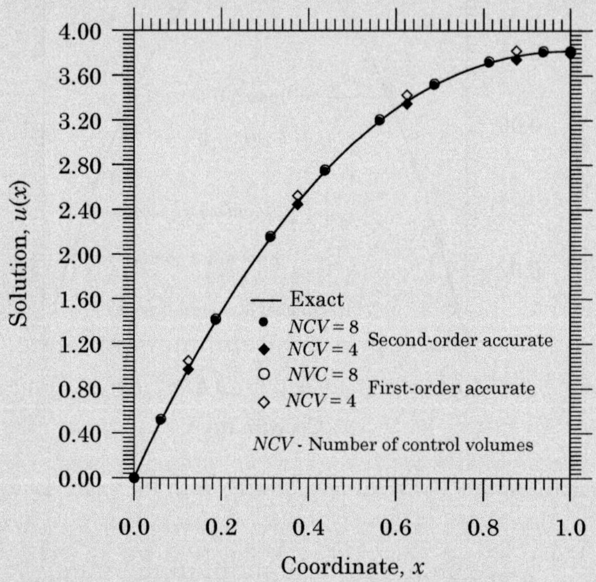

Fig. 7.2.8 Comparison of FVM solutions for $u(x)$ with the exact solution of Case 2.

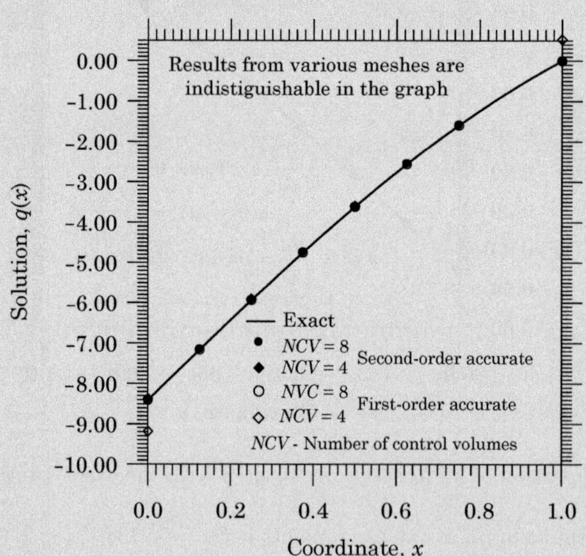

Fig. 7.2.9 Comparison of FVM solutions for $q(x)$ with the exact solution of Case 2.

Example 7.2.2

Compute the finite volume solution of one-dimensional heat conduction in a rod given in **Example 4.2.1**. Determine the temperature distribution, $u = T - T_\infty$, using uniform meshes of 5, 10, and 20 control volumes. Compute the maximum error in the nodal values for each case.

Solution The governing equation of the problem is the same as that in Eq. (7.2.3). For the sake of simplicity, we divide throughout Eq. (7.2.3) by $a = kA$ so that $a = 1$, $c = P\beta/kA = 400$, and $\beta/k = 2$.

Here we consider, as in **Example 4.2.1**, the following two cases of boundary conditions:

$$\textbf{Case 1:}\ x = 0: \quad u = u_0; \quad x = L: \quad \frac{du}{dx} = 0. \tag{1}$$

$$\textbf{Case 2:}\ x = 0: \quad u = u_0; \quad x = L: \quad \frac{du}{dx} + (\beta/k)u = 0, \tag{2}$$

where $L = 0.05$m is the length of the rod. We begin with a uniform mesh of five control volumes (see Fig. 7.2.10) to show the details of the actual computation. The control volume length is: $\Delta x = L/5 = 0.05/5 = 0.01$ m. For internal control volumes, the diffusion length is $\delta x_e = \delta x_w = \Delta x = 0.01$ m. The diffusion length for the left boundary is $\delta x_w = \Delta x/2 = 0.005$ m, and for the right boundary it is $\delta x_e = \Delta x/2 = 0.005$ m. Therefore, the diffusion coefficients for internal control volumes are:

$$a_W = \frac{1.0}{\delta x_w} = \frac{1.0}{0.01} = 100.0, \quad a_E = \frac{1.0}{\delta x_e} = \frac{1.0}{0.01} = 100.0. \tag{3}$$

At boundaries, the diffusion coefficients are [a_E(left) = 100.0 and a_W(right) = 100.0]

$$a_W(\text{left}) = \frac{1.0}{\delta x_w} = \frac{1.0}{0.005} = 200.0, \quad a_E(\text{right}) = \frac{1.0}{\delta x_e} = \frac{1.0}{0.005} = 200.0. \tag{4}$$

Since the source term is zero, we have $F_P = 0$. We can now calculate the a-coefficients and the discretized equations for each case of boundary conditions [noting that $a_W(i) = 100$, $a_E(i) = 200 + \frac{\beta}{k}$, $a_P(i) = 304$, $u_0 = 300$, and $\beta/k = 2$; see Eq. (7.2.20)].

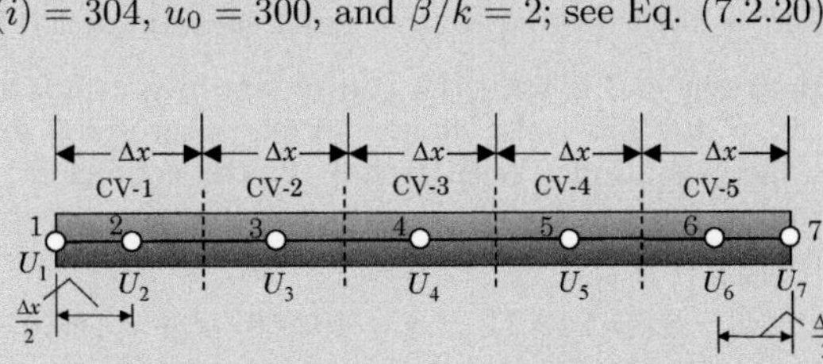

Fig. 7.2.10 A uniform mesh of five ($N = 5$) subdivisions used in the zero-thickness control volume formulation. There are five control volumes, $NCV = N$ and 7 nodes.

Case 1 For Case 1, the right boundary is that of zero heat flow , $\frac{du}{dx} = 0$ (i.e., $\beta = 0$, insulated). The coefficients a_W, a_E, and a_P for the six equations based on the *first-order accurate approximation*, are:

$$\{a_W\} = [200\ 100\ 100\ 100\ 100\ 200] \tag{5}$$

$$\{a_E\} = [100\ 100\ 100\ 100\ 200\ 0] \tag{6}$$

$$\{a_P\} = [304\ 204\ 204\ 204\ 304\ 200]\,. \tag{7}$$

Note that a_P for the last equation contains β/k (which is zero for Case 1) in place of $c\Delta x$. Hence, the discretized equations for the mesh of five control volumes are:

$$
\begin{aligned}
304\,U_2 - 100\,U_3 &= 200\,u_0 \\
-100\,U_2 + 204\,U_3 - 100\,U_4 &= 0 \\
-100\,U_3 + 204\,U_4 - 100\,U_5 &= 0 \\
-100\,U_4 + 204\,U_5 - 100\,U_6 &= 0 \\
-100\,U_5 + 304\,U_6 - 200\,U_7 &= 0 \\
-200\,U_6 + 200\,U_7 &= 0.
\end{aligned} \tag{8}
$$

These linear equations can be solved using the Gauss elimination or the Gauss–Seidel iteration. The solution of these equations using the Gauss elimination is obtained as (the solution is obtained at $x = 0.005, 0.015, 0.025, 0.035, 0.045$, and 0.005)

$$U_2 = 277.29,\quad U_3 = 242.95,\quad U_4 = 218.33,\quad U_5 = 202.45,\quad U_6 = 194.66,\quad U_7 = 194.66. \tag{9}$$

The six equations based on the *second-order accurate approximation* are:

$$
\begin{aligned}
404\,U_2 - 133.333\,U_3 &= 266.667\,u_0 \\
-100\,U_2 + 204\,U_3 - 100\,U_4 &= 0 \\
-100\,U_3 + 204\,U_4 - 100\,U_5 &= 0 \\
-100\,U_4 + 204\,U_5 - 100\,U_6 &= 0 \\
-133.333\,U_5 + 404\,U_6 - 266.667\,U_7 &= 0 \\
-33.333\,U_5 - 300U_6 + 266.667\,U_7 &= 0,
\end{aligned} \tag{10}
$$

whose solution is

$$U_2 = 278.57,\quad U_3 = 244.08,\quad U_4 = 219.35,\quad U_5 = 203.39,\quad U_6 = 195.67,\quad U_7 = 194.59. \tag{11}$$

The FEM solution at the same points is

$$U_2^F = 278.57,\quad U_3^F = 243.90,\quad U_4^F = 219.05,\quad U_5^F = 203.03,\quad U_6^F = 195.18,\quad U_7^F = 194.21, \tag{12}$$

and the exact solution at the same points is

$$u_2^e = 278.62,\quad u_3^e = 244.03,\quad u_4^e = 219.23,\quad u_5^e = 203.23,\quad u_6^e = 195.39.\quad u_7^e = 194.42. \tag{13}$$

As noted before, the FEM is not expected to give exact solution at the nodes when $c \neq 0$.

Case 2 In this case, the right end is with the convection heat transfer boundary condition ($\beta/k = 2$), $(du/dx) + 2u = 0$. The only change to the discretized set of equations is the coefficient of U_7 in the last equation. With $\beta/k = 2$, the coefficient becomes 202 for the first-order accurate formulation and 268.667 for the second-order accurate formulation. The solution of the resulting equations for the first-order accurate formulation is

$$U_2 = 276.11,\quad U_3 = 239.38,\quad U_4 = 212.23,\quad U_5 = 193.56,\quad U_6 = 182.64,\quad U_7 = 180.84, \tag{14}$$

whereas the second-order accurate formulation gives

$$U_2 = 277.35,\quad U_3 = 240.51,\quad U_4 = 213.24,\quad U_5 = 194.50,\quad U_6 = 183.54,\quad U_7 = 180.82. \tag{15}$$

The finite element solution at the same points is

$$U_2^F = 277.40,\quad U_3^F = 240.34,\quad U_4^F = 212.96,\quad U_5^F = 194.16,\quad U_6^F = 183.18,\quad U_7^F = 180.47 \tag{16}$$

and exact solution at the same points is

$$u_2^e = 277.44,\quad u_3^e = 240.46,\quad u_4^e = 213.13,\quad u_5^e = 194.35,\quad u_6^e = 183.37,\quad u_7^e = 180.66. \tag{17}$$

Figure 7.2.11 shows a comparison of numerical solutions obtained with various FVM formulations and the exact solution. In all cases, the FVM2 and FEM solutions (shown with the symbol +) are the closest to the exact solution.

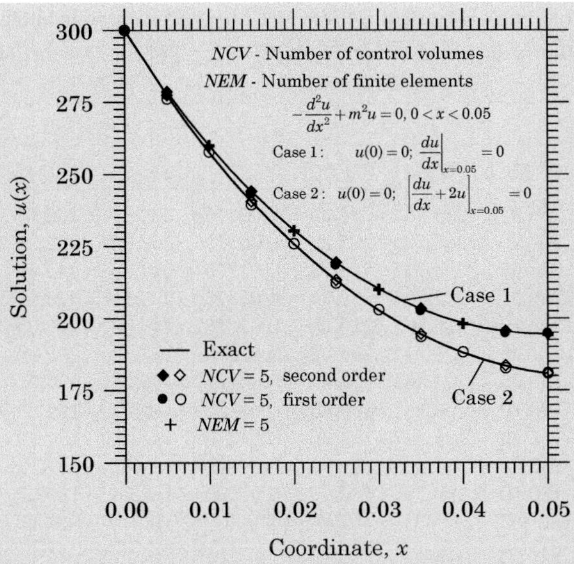

Fig. 7.2.11 Comparison of various FVM solutions for $u(x)$ with the exact solution of the problem in **Example 7.2.2**.

The finite volume solutions obtained with various numbers of control volumes (NCV) are compared with the solutions obtained with the *same nodal locations* using linear finite elements and the exact solution for Case 1 in Table 7.2.3 and for Case 2 in Table 7.2.2. In trying to match the nodal locations of the FVM, one is required to make the FEM mesh to be nonuniform at the boundaries. Consequently, the equivalent finite element meshes will have one extra element. The tables also contain the boundary flux at the left end (where the temperature is specified). Clearly, both the FVM and FEM solutions converge to the exact solution with mesh refinements.

Table 7.2.3 Comparison of the FVM, FEM, and exact solutions for Case 1:
$-\frac{d^2u}{dx^2}+400u=0;\quad 0<x<L=0.05;\quad u(0)=0;\quad \left.\frac{du}{dx}\right|_{x=L}=0.$

	$NCV=10$					$NCV=20$			
x^a	Exact	FVM1	FVM2	FEM	x^a	Exact	FVM1	FVM2	FEM
0.0025	288.95	288.59	288.94	288.94	0.00375	283.69	283.61	283.69	283.69
0.0075	268.98	268.67	268.99	268.96	0.00875	264.42	264.34	264.42	264.41
0.0125	251.71	251.42	251.73	251.68	0.01375	247.79	247.72	247.80	247.78
0.0175	236.95	236.70	236.98	236.92	0.01875	233.64	233.58	233.65	233.63
0.0225	224.57	224.34	224.61	224.53	0.02375	221.83	221.77	221.84	221.82
0.0275	214.44	214.22	214.48	214.39	0.02875	212.24	212.19	212.25	212.23
0.0325	206.45	206.25	206.49	206.39	0.03375	204.77	204.73	204.79	204.76
0.0375	200.52	200.33	200.57	200.47	0.03875	199.36	199.31	199.37	199.34
0.0425	196.61	196.42	196.66	196.55	0.04375	195.94	195.89	195.95	195.92
0.0475	194.66	194.48	194.71	194.60	0.04875	194.48	194.43	194.49	194.46
0.0500	194.42	194.48	194.47	194.36	0.00500	194.42	194.43	194.43	194.40
$-q_x(0)$	4570	4755	4568	4424	$-q_x(0)$	4570	4666	4569	4496

[a]Results for only alternate values of x are listed in the table.

Table 7.2.4 Comparison of the FVM, FEM, and exact solutions for Case 2: $-\frac{d^2u}{dx^2} + 400u = 0, \quad 0 < x < 0.05; \; u(0) = 0; \; \frac{du}{dx}\Big|_{x=0.05} + 2u(0.05) = 0.$

	NCV = 10					NCV = 20			
x^a	Exact	FVM1	FVM2	FEM	x^a	Exact	FVM1	FVM2	FEM
0.0025	288.36	288.01	288.35	288.35	0.00375	282.81	282.73	282.81	282.81
0.0075	267.22	266.90	267.22	267.20	0.00875	262.36	262.28	262.36	262.35
0.0125	248.75	248.47	248.77	248.72	0.01375	244.53	244.46	244.53	244.52
0.0175	232.77	232.51	232.80	232.74	0.01875	229.15	229.08	229.16	229.14
0.0225	219.12	218.89	219.16	219.08	0.02375	216.06	216.00	216.07	216.05
0.0275	207.67	207.45	207.71	207.62	0.02875	205.13	205.08	205.14	205.12
0.0325	198.29	198.09	198.33	198.24	0.03375	196.26	196.21	196.27	196.25
0.0375	190.90	190.70	190.94	190.85	0.03875	189.35	189.30	189.36	189.34
0.0425	185.41	185.23	185.46	185.36	0.04375	184.33	184.29	184.35	184.32
0.0475	181.79	181.61	181.84	181.74	0.04875	181.17	181.12	181.18	181.15
0.0500	180.66	180.70	180.71	180.61	0.05000	180.66	180.67	180.67	180.64
$-q_x(0)$	4804	4989	4803	4658	$-q_x(0)$	4804	4900	4803	4730

[a]Results for only alternate values of x are listed in the table.

7.2.8 Axisymmetric Problems

7.2.8.1 Governing Equation

Here we consider steady heat transfer in axisymmetric problems; we note that the geometry, material properties, heat source, and boundary conditions should exhibit the same axisymmetry. In particular, here we consider a long circular cylinder of radius R_0 and made of an isotropic material, with heat source and boundary conditions independent of the radial coordinate (θ). In such cases, the problem becomes one dimensional with temperature $T = T(r)$ varying only with r, requiring us to consider a circular disc of unit thickness. The governing differential equation is given by [see Eq. (4.2.31) and Fig. 4.2.4]

$$-\frac{1}{r}\frac{d}{dr}\left(kr\frac{dT}{dr}\right) = g(r), \quad 0 < r < R_0, \tag{7.2.39}$$

where g is internal heat generation per unit length. We note that the heat flux in the axisymmetric problems is related to the temperature gradient by (Fourier's law)

$$q(r) = -k\frac{dT}{dr}. \tag{7.2.40}$$

In addition, the heat flux is zero [more precisely, $rq(r) = 0$] at $r = 0$ for a solid cylinder. In the following we assume that k and g are constants.

7.2.8.2 Discretization over an Internal Control Volume

Next, the integral statement of Eq. (7.2.39) over a typical internal control volume occupying the west r_w and east r_e interfaces (see Fig. 7.2.12) is developed.

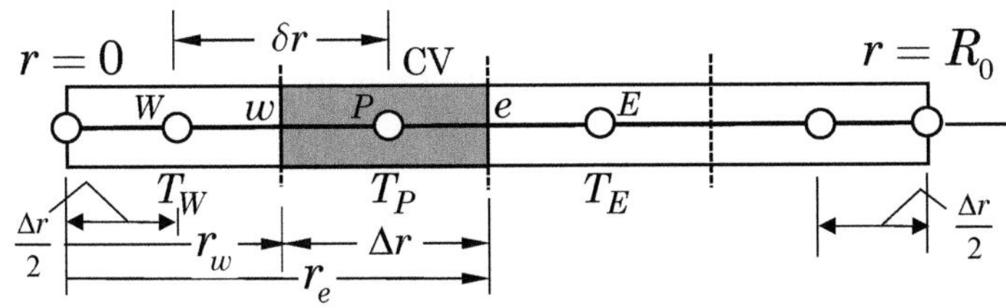

Fig. 7.2.12 Subdivision of the one-dimensional axisymmetric domain into a set of control volumes (similar to Fig. 7.2.1, except that x is replaced with r).

The area element of the axisymmetric geometry is $r\,dr\,d\theta$, but integration with respect to θ gives 2π as a multiplier to the whole equation, and hence we drop it.

Integrating the governing differential equation, Eq. (7.2.39) (assuming that g is a constant g_0) over an internal control volume (occupying the domain between points w and e; see Fig. 7.2.12), we obtain

$$\begin{aligned} 0 &= \int_{r_w}^{r_e} \left[-\frac{1}{r}\frac{d}{dr}\left(kr\frac{dT}{dr}\right) - g(r) \right] r dr = -\left[kr\frac{dT}{dr}\right]_w^e - F_P\delta r \\ &= -kr\frac{dT}{dr}\bigg|_e + kr\frac{dT}{dr}\bigg|_w - F_P\delta r, \end{aligned} \tag{7.2.41a}$$

where $F_P = g_0(r_w + r_e)/2$:

$$\int_{r_w}^{r_e} rg_0\,dr = \frac{g_0}{2}\left(r_e^2 - r_w^2\right) = \frac{g_0}{2}(r_e + r_w)(r_e - r_w) \equiv F_P\delta r. \tag{7.2.41b}$$

Using the usual approximations [see Eqs. (7.2.8) and (7.2.9)] to replace the derivative dT/dr at the west and east faces, we obtain

$$0 = -kr_e\frac{T_E - T_P}{\delta r_e} + kr_w\frac{T_P - T_W}{\delta r_w} - F_P\delta r,$$

or

$$a_P T_P = a_W T_W + a_E T_E + b_P, \tag{7.2.42a}$$

with

$$a_W = \frac{kr_w}{\delta r_w}, \quad a_E = \frac{kr_e}{\delta r_e}, a_P = a_W + a_E, \quad b_P = 0.5g_0(r_w + r_e)\delta r. \tag{7.2.42b}$$

7.2.8.3 Discretization of the Boundary Conditions

As before, we consider the zero-thickness approximations for the first and last control volumes. For the axisymmetric problems of the solid cylinder we consider here, the point $r = 0$ has specified flux. Hence, Eq. (7.2.22a) for the first control volume becomes

$$\begin{aligned} 0 &= -kr\frac{dT}{dr}\bigg|_e - \int_0^{r_e} rg_0\,dr \\ &= -kr_e\frac{T_E - T_P}{\delta r_e} - \frac{g_0 r_e^2}{2} \end{aligned} \tag{7.2.43}$$

or

$$a_P T_P = a_W T_W + a_E T_E + b_P \tag{7.2.44a}$$

with

$$a_W = 0, \quad a_E = \frac{k r_e}{\delta r_e}, a_P = a_W + a_E, \quad b_P = 0.5 g_0 r_e^2. \tag{7.2.44b}$$

The zero flux boundary condition at $r = 0$ gives

$$0 = q(0) = -k \frac{T_2 - T_1}{0.5\delta r} \quad \rightarrow \quad T_1 = T_2. \tag{7.2.45}$$

The right end of the domain can have specified temperature, specified flux, or specified convection boundary condition of the type

$$k \frac{dT}{dr} + \beta(T - T_\infty) = \hat{q}, \tag{7.2.46}$$

where β is the heat transfer coefficient and T_∞ is the ambient temperature. We consider this convection boundary condition, as it includes the specified flux as a special case (when $\beta = 0$). In this case, Eq. (7.2.37a) for the last control volume takes the form

$$\begin{aligned} 0 &= kr \frac{dT}{dr}\bigg|_w + \beta(T_P - T_\infty) - \hat{q} - \int_{r_w}^{R_0} r g_0 \, dr \\ &= k r_w \frac{T_P - T_w}{\delta r_w} + \beta(T_P - T_\infty) - \hat{q} - \frac{g_0}{2}\left(R_0^2 - r_w^2\right) \end{aligned} \tag{7.2.47}$$

or

$$a_P T_P = a_W T_W + a_E T_E + b_P \tag{7.2.48a}$$

with

$$a_W = \frac{k r_w}{\delta r_w}, \quad a_E = 0, a_P = a_W + a_E + \beta, \quad b_P = \frac{g_0}{2}\left(R_0^2 - r_w^2\right) + \beta T_\infty + \hat{q}. \tag{7.2.48b}$$

Equation (7.2.46) applied to the boundary point $r = R$ yields

$$k \frac{T_N - T_{N-1}}{0.5\Delta r} + \beta(T_N - T_\infty) = \hat{q}, \tag{7.2.49}$$

where N is the last node number.

Example 7.2.3

Consider a *long*, homogeneous, isotropic solid cylinder of outside radius $R_0 = 0.01$ m, thermal conductivity $k = 20$ W/(m·°C), a constant rate of internal heat generation $g_0 = 2 \times 10^8$ W/m^3. Consider the case where the boundary surface at $r = R_0$ is maintained at a constant temperature $T_0 = 100\,°$C. Use the FVM to calculate the temperature distribution $T(r)$ and heat flux $q(r) = -k dT/dr$ in the cylinder using various uniform meshes and compare the results with the FEM results from **Example 4.2.2**.

Solution We begin with a uniform mesh of five control volumes (i.e., $\Delta r = \delta r = \delta r_e = \delta r_w = 0.002$). The discretized equations for the mesh of five control volumes are (the condition of zero flux at $r = 0$ yields $T_1 = T_2$):

$$\begin{aligned} 20\,T_2 - 20\,T_3 &= 400 \\ -20\,T_2 + 60\,T_3 - 40\,T_4 &= 1200 \\ -40\,T_3 + 100\,T_4 - 60\,T_5 &= 2000 \\ -60\,T_4 + 140\,T_5 - 80\,T_6 &= 2800 \\ -80\,T_5 + 280\,T_6 - 200\,T_7 &= 23600, \end{aligned} \tag{1}$$

where $T_7 = 100$. The solution of these equations is

$$T_1 = 350, \quad T_2 = 350, \quad T_3 = 330, \quad T_4 = 290, \quad T_5 = 230, \quad T_6 = 150. \tag{2}$$

The FEM solution with six linear elements (with element lengths of 0.001, 0.002, 0.002, 0.002, 0.002, and 0.001 to match the nodal locations of the FVM) is (in °C)

$$T_1 = 354.35, \quad T_2 = 351.02, \quad T_3 = 329.35, \quad T_4 = 288.52, \quad T_5 = 227.96, \quad T_6 = 147.54. \tag{3}$$

The exact solution at the same locations is

$$T_1 = 350.00, \quad T_2 = 347.50, \quad T_3 = 327.50, \quad T_4 = 287.50, \quad T_5 = 227.50, \quad T_6 = 147.50. \tag{4}$$

The FVM solutions obtained using 5, 10, and 20 control volumes and corresponding (i.e., matching nodal locations) meshes of 6, 11, and 21 linear finite elements are compared with the exact solution in Table 7.2.5.

Table 7.2.5 Comparison of the finite volume, finite element, and exact solutions for temperature T in a radially symmetric circular disc. The underlined numbers are nodal values and others are interpolated values.

	Six nodes		Twelve nodes		Twenty two nodes		
$\frac{r}{R_0}$	FVM	FEM	FVM	FEM	FVM	FEM	Exact
0.000	<u>0.2500</u>	<u>0.2544</u>	<u>0.2500</u>	<u>0.2514</u>	<u>0.2500</u>	<u>0.2504</u>	0.2500
0.050	0.2500	0.2527	<u>0.2500</u>	<u>0.2505</u>	0.2494	0.2495	0.2494
0.100	<u>0.2500</u>	<u>0.2510</u>	0.2475	0.2479	0.2475	0.2476	0.2475
0.150	0.2450	0.2456	<u>0.2450</u>	<u>0.2451</u>	0.2444	0.2444	0.2444
0.200	0.2400	0.2402	0.2400	0.2400	0.2400	0.2400	0.2400
0.250	0.2350	0.2348	<u>0.2350</u>	<u>0.2349</u>	0.2344	0.2344	0.2344
0.275	0.2325	0.2321	0.2312	0.2312	<u>0.2312</u>	<u>0.2312</u>	0.2311
0.350	0.2200	0.2191	<u>0.2200</u>	<u>0.2198</u>	0.2194	0.2193	0.2194
0.425	0.2050	0.2038	0.2050	0.2047	<u>0.2050</u>	<u>0.2049</u>	0.2048
0.550	0.1750	0.1734	<u>0.1750</u>	<u>0.1746</u>	0.1744	0.1743	0.1744
0.625	0.1525	0.1507	0.1525	0.1521	<u>0.1525</u>	<u>0.1524</u>	0.1523
0.750	0.1100	0.1079	<u>0.1100</u>	<u>0.1095</u>	0.1094	0.1093	0.1094
0.825	0.0800	0.0777	0.0800	0.0794	<u>0.0800</u>	<u>0.0799</u>	0.0798
0.950	0.0250	0.0238	<u>0.0250</u>	<u>0.0244</u>	0.0244	0.0242	0.0244
0.975	0.0125	0.0119	0.0125	0.0122	<u>0.0125</u>	<u>0.0123</u>	0.0123

Convergence of the numerical solutions, $\bar{T} = (T - T_0)k/g_0R_0^2$, to the exact solution with an increasing number of elements is clear from Fig. 7.2.13, where the finite element solutions are not shown because they would have coincided with the corresponding finite volume solutions. Figure 7.2.14 shows plots of $\bar{Q}(r) = -krdT/dr/(R_0g_0)$ versus $\bar{r} = r/R_0$. Interestingly, the FVM solutions for all meshes produce the exact solution, which is not the case with the FEM.

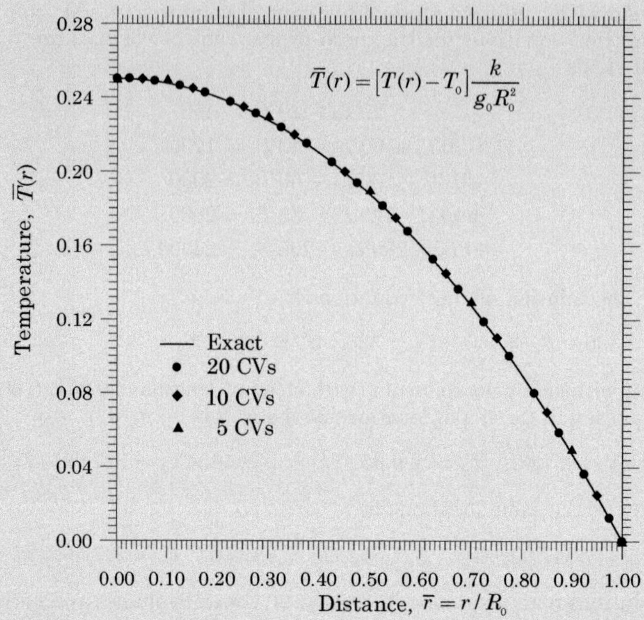

Fig. 7.2.13 Comparison of the finite volume and finite element solutions with the exact solution, $\bar{T}$, for heat transfer in a radially symmetric problem; CV stands for control volume.

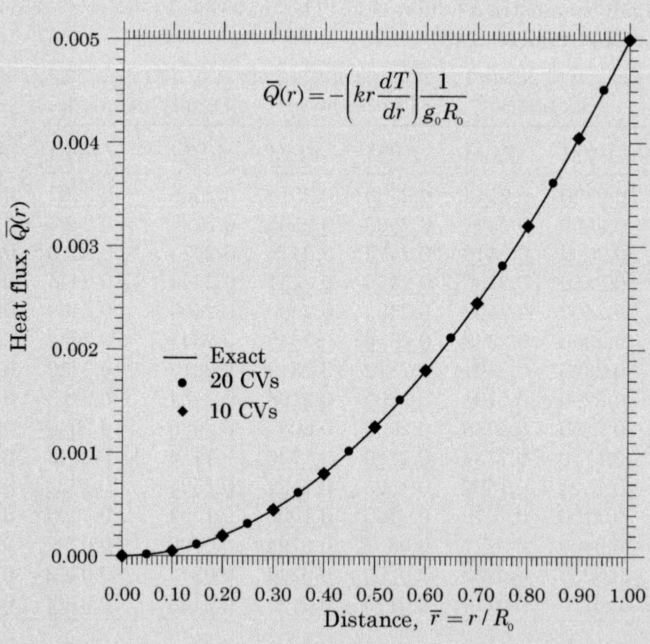

Fig. 7.2.14 Comparison of the finite volume and finite element solutions with the exact solution for the temperature gradient $\bar{Q}(r)$ in a radially symmetric problem; CV stands for control volume.

7.3 Two-Dimensional Diffusion

7.3.1 Model Equation

In this section we focus on the FVM applied to a model PDE in two dimensions, which governs steady heat transfer in a solid medium. The thickness of the solid medium is assumed to be small so that we can neglect the flow of heat in the thickness direction. For simplicity, we will consider the two-dimensional diffusion equation in rectangular Cartesian coordinates.

The steady-state temperature distribution in an orthotropic solid medium with internal heat source S (W/m^2) is governed by the following PDE [see Eq. (4.3.1)]:

$$-\left[\frac{\partial}{\partial x}\left(k\frac{\partial T}{\partial x}\right)+\frac{\partial}{\partial y}\left(k\frac{\partial T}{\partial y}\right)\right]=S, \tag{7.3.1}$$

where T is the temperature and k is the thermal conductivity of the solid medium. For the sake of simplicity of introducing the FVM, we shall consider a rectangular domain of dimensions $a \times b$ (see Fig. 7.3.1). The boundary conditions associated with Eq. (7.3.1) are:

$$\begin{gathered} T(a,y)=T_0, \quad T(x,b)=T_1, \\ -k\frac{\partial T}{\partial x}\bigg|_{y=0}=q_s, \quad -k\frac{\partial T}{\partial y}\bigg|_{x=0}=\beta\left(T-T_\infty\right), \end{gathered} \tag{7.3.2}$$

where T_0, and T_1 are the specified temperatures, q_s is specified heat flux normal to the boundary, β is the heat transfer coefficient, and T_∞ is the ambient temperature.

Prior to proceeding with the solution process the following observations are appropriate:

(1) The given equation is a second-order linear PDE assuming that 'k' is constant.

(2) The given PDE is nonhomogeneous.

(3) The given PDE is elliptic ($A = 0$, $B = C = 1$, $B^2 - 4AC = -4 < 0$).

(4) The PDE exhibits an elliptic or two-way street behavior in both x- and y-directions.

(5) Boundary conditions along $x = a$, and $y = b$ are of Dirichlet type or boundary conditions of the first kind. These boundary conditions are nonhomogeneous as T_0 and T_1 are not equal to zero.

(6) The boundary condition along $y = 0$ is a Neumann type or second kind.

(7) The boundary condition along $x = 0$ is a Robin or mixed boundary condition.

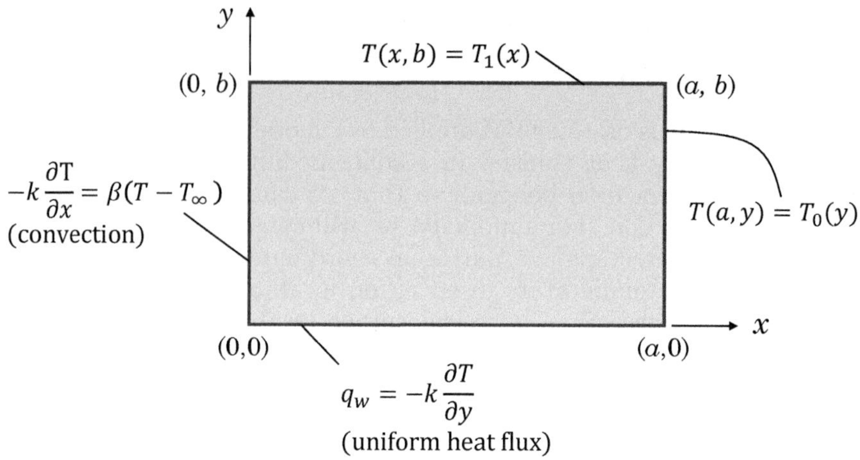

Fig. 7.3.1 Two-dimensional heat transfer in a rectangular domain with prescribed temperatures, heat flux, or convection on the boundary.

7.3.2 Grid Generation

As in the case of one-dimensional problems, we begin with the domain discretization. A mesh of rectangular control volumes is used to cover the domain. Then a node is placed at the geometric center of each control volume. In general, the control volume sizes or the distance between nodal points need not be uniform throughout the domain. However, for illustrative purposes we consider a uniform mesh of $NCV_x \times NCV_y = 6 \times 4$ (i.e., six control volumes in the x-direction and four control volumes in the y-direction) control volumes, as shown in Fig. 7.3.2. In addition to the nodes at the center of each control volume, nodes are placed on the boundary of the domain in such a way that they are aligned with the rows and columns of the internal nodes (as in the one-dimensional case). The nodes in the mesh are labelled in an orderly fashion, as shown in Fig. 7.3.2.

In theory, the node labeling can be quite arbitrary, but it only makes it difficult for the reader to follow the discussion. In addition, more importantly, if the grid points were not numbered in order, the resulting coefficient matrix associated with the discretized equations would not be in a desired pattern, for example, the penta-diagonal matrix or tridiagonal matrix, making the solution process time- consuming. The corner nodes $[(1,1),(8,1),(1,6),(8,6)]$ are multi-valued points, but will not enter into any calculations of the solution procedure. Once the converged solution is reached, one may represent the corner point value as the arithmetic mean of its neighbors, for example, $T_{1,1} = 0.5(T_{1,2} + T_{2,1})$. This is called "numerical decoration," and has no impact on the solution inside the domain whatsoever.

The geometric lengths (Δx and Δy) are to be defined independently of the diffusion lengths (δx and δy). Even though for internal control volumes the two lengths may be equal, the diffusion lengths for control volumes along the boundaries will be different than those for the internal control volumes. So, $\Delta x = a/NCV_x$ and $\Delta y = b/NCV_y$.

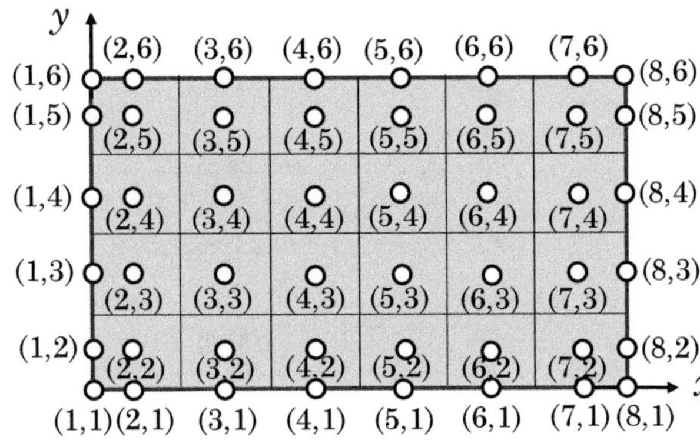

Fig. 7.3.2 A typical grid with numbered grid points. The node number of a mesh point located at (i, j)th location can be calculated using the formula $n = i + (j - 1) * (NCV_x + 2)$, where NCV_x is the number of control volumes along the x-axis (e.g., $NCV_x = 6$).

7.3.3 Discretization of the Model Equation

The next step is to discretize the differential equation by requiring that it is satisfied in an integral sense over a typical control volume (see Fig. 7.3.3). To keep the discussion simple, we will assume that the thermophysical properties (e.g., density, thermal conductivities, and film conductance) of the material remain constant. In addition, the source term S is also assumed to be a constant for now. Thus, Eq. (7.3.1) is a second-order linear PDE with constant coefficients.

Consider the integral of Eq. (7.3.1) over a control volume (see Fig. 7.3.3)

$$\int_{CV} \left[\frac{\partial}{\partial x}\left(k\frac{\partial T}{\partial x}\right) + \frac{\partial}{\partial y}\left(k\frac{\partial T}{\partial y}\right) + S \right] dxdy = 0. \tag{7.3.3}$$

Since

$$\int_{CV} (\cdot)\, dxdy = \int_0^{\Delta x}\int_0^{\Delta y} (\cdot)\, dxdy, \tag{7.3.4}$$

we obtain

$$\int_0^{\Delta y} \left[k\frac{\partial T}{\partial x}\bigg|_e - k\frac{\partial T}{\partial x}\bigg|_w \right] dy + \int_0^{\Delta x} \left[k\frac{\partial T}{\partial y}\bigg|_n - k\frac{\partial T}{\partial y}\bigg|_s \right] dx + \int_0^{\Delta x}\int_0^{\Delta y} S\, dxdy = 0, \tag{7.3.5}$$

where the subscripts e, w, s, and n refer to east, west, south, and north interfaces (i.e., boundaries), respectively, of the typical control volume, as shown in Fig. 7.3.3.

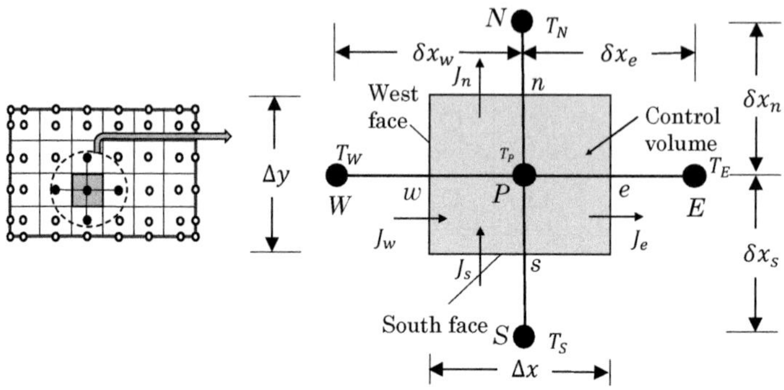

Fig. 7.3.3 A two-dimensional control volume with diffusion and control volume lengths.

Next, consider the following components of the heat flux vector (we already made the assumption that ρ, c, k, and S are constants):

$$J_x = -k\frac{\partial T}{\partial x}, \qquad J_y = -k\frac{\partial T}{\partial y}. \tag{7.3.6}$$

In the FVM, one assumes that J_x is independent of y on the east and west faces and J_y is independent of x on the south and north faces (an approximation). In other words, the diffusive fluxes are assumed to be locally one-dimensional. Under these assumptions, Eq. (7.3.5) takes the form

$$\left[k\frac{\partial T}{\partial x}\bigg|_e - k\frac{\partial T}{\partial x}\bigg|_w\right]\Delta y + \left[k\frac{\partial T}{\partial y}\bigg|_n - k\frac{\partial T}{\partial y}\bigg|_s\right]\Delta x + S\Delta x\,\Delta y = 0. \tag{7.3.7}$$

Then we use a piecewise-linear approximation to represent the fluxes at the interface [see Eqs. (7.2.8) and (7.2.9)] and replace the derivatives to obtain

$$k_e\frac{(T_E - T_P)}{\delta x_e}\Delta y - k_w\frac{(T_P - T_W)}{\delta x_w}\Delta y + k_n\frac{(T_N - T_P)}{\delta y_n}\Delta x$$
$$-k_s\frac{(T_P - T_S)}{\delta y_s}\Delta x + S\Delta x\,\Delta y = 0 \tag{7.3.8a}$$
$$D_e(T_E - T_P) - D_w(T_P - T_W) + D_n(T_N - T_P)$$
$$-D_s(T_P - T_S) + S\Delta x\,\Delta y = 0, \tag{7.3.8b}$$

where diffusion strength terms D are

$$D_e = \frac{k_e\Delta y}{\delta x_e}, \quad D_w = \frac{k_w\Delta y}{\delta x_w}, \quad D_n = \frac{k_n\Delta x}{\delta y_n}, \quad D_s = \frac{k_s\Delta x}{\delta y_s}. \tag{7.3.9}$$

Equations (7.3.8a) and (7.3.8b) can be expressed in the alternative form as

$$a_P T_P = a_W T_W + a_E T_E + a_S T_S + a_N T_N + b_P^T, \tag{7.3.10a}$$

where

$$a_W = D_w, \quad a_E = D_e, \quad a_S = D_s, \quad a_N = D_n$$
$$a_P = a_W + a_E + a_S + a_N, \quad b_P^T = S\,\Delta x\Delta y. \tag{7.3.10b}$$

Therefore, we have

$$T_P = \frac{1}{a_P}\left[a_W T_W + a_E T_E + a_S T_S + a_N T_N + b_P^T\right]. \tag{7.3.11}$$

7.3.4 Discrete Equations for Control Volumes and Nodes on the Boundary

Equation (7.3.11) is valid for any internal node. The formula is to be modified, as we did in the one-dimensional case, for nodes and control volumes near the boundary. In the two-dimensional problems, we only use the first-order accurate formulation for the nodes next to the boundary. In addition, boundary conditions are implemented at the boundary nodes following the procedures used in the case of one-dimensional problems (by equating the nodal values to the specified temperature or imposing the zero-thickness control volume concept for gradient boundary conditions). For illustrative purposes, we consider different cases shown in Fig. 7.3.2 to write the discretized equations, when the temperature at the boundary node is specified and *not* specified.

Note that both eastern (a, y) and northern (x, b) boundaries are subjected to Dirichlet boundary conditions. Therefore, the development of the discretization equations for control volumes along these boundaries will be similar. Here we will illustrate the development of discretization equation for control volumes along the eastern boundary. For the eastern boundary we consider a representative control volume around the node (7,4) as shown in Fig. 7.3.2(a). Integrating the given PDE over this control volume and rearranging the terms we obtain

$$a_P T_P = a_W T_W + a_E T_E + a_S T_S + a_N T_N + b_P, \tag{7.3.12a}$$

where

$$a_W = \frac{k_w}{\delta x_w}\Delta y,\quad a_E = 2\frac{k_e}{\delta x_e}\Delta y,\quad a_S = \frac{k_s}{\delta y_s}\Delta x,\quad a_N = \frac{k_n}{\delta y_n}\Delta x,$$
$$a_P = a_W + a_E + a_S + a_N,\quad b_P^T = S_T\,\Delta x \Delta y. \tag{7.3.12b}$$

Referring to the index notation:

$$a_P(7,4)T_{7,4} = a_W(7,4)T_{6,4} + a_E(7,4)T_{8,4} + a_S(7,4)T_{7,3} + a_N(7,4)T_{7,5} + b_{7,4}^T. \tag{7.3.12c}$$

Note that in Eq. (7.3.12c) $T_{8,4}$ is known due to the given boundary condition.

Next, we develop the discretization equations for control volumes along the southern boundary $(y = 0)$, which is subjected to uniform heat flux, that is, the Neumann boundary condition. To begin with, we will develop the discretization equation for control volume (6,2) along the southern boundary. As described in Section 7.3.3, we integrate the governing equation around the control volume and obtain the flux balance form. Then we replace the flux at the south face of the control volume with the given heat flux, $-k\frac{\partial T}{\partial y}|_s = q_s$. We discretize the fluxes at other faces by using a two-node formulation, that is, assuming a piecewise-linear profile for the temperature:

$$\int_{CV}\left[\frac{\partial}{\partial x}\left(k\frac{\partial T}{\partial x}\right)+\frac{\partial}{\partial y}\left(k\frac{\partial T}{\partial y}\right)\right]dV+\int_{CV}S_T dV=0 \qquad (7.3.13a)$$

$$k\frac{\partial T}{\partial x}\bigg|_w^e \Delta y+k\frac{\partial T}{\partial y}\bigg|_s^n \Delta x+S_T\Delta x\Delta y=0 \qquad (7.3.13b)$$

$$\left[k\frac{\partial T}{\partial x}\bigg|_e - k\frac{\partial T}{\partial x}\bigg|_w\right]\Delta y+\left[k\frac{\partial T}{\partial y}\bigg|_n - k\frac{\partial T}{\partial y}\bigg|_s\right]\Delta x+S_T\Delta x\Delta y=0 \qquad (7.3.13c)$$

$$k_e\left(\frac{T_{7,2}-T_{6,2}}{\delta x_e}\right)\Delta y-k_w\left(\frac{T_{6,2}-T_{5,2}}{\delta x_w}\right)\Delta y$$
$$+k_n\left(\frac{T_{6,3}-T_{6,2}}{\delta y_n}\right)\Delta x+q_s\Delta x+S_T\Delta x\Delta y=0 \qquad (7.3.13d)$$

$$T_{6,2}\left(\frac{k_w\Delta y}{\delta x_w}+\frac{k_e\Delta y}{\delta x_e}+\frac{k_n\Delta x}{\delta y_n}\right)=q_s\Delta x+\frac{k_w\Delta y}{\delta x_w}T_{5,2}+\frac{k_e\Delta y}{\delta x_e}T_{7,2}$$
$$+\frac{k_n\Delta x}{\delta y_n}T_{6,3}+S_T\Delta x\Delta y \qquad (7.3.13e)$$

$$a_P^T(6,2)T_{6,2}=a_W^T(6,2)T_{5,2}+a_E^T(6,2)T_{7,2}+a_N^T(6,2)T_{6,3}+b_P^T(6,2), \qquad (7.3.13f)$$

where

$$a_W^T(6,2)=\frac{k_w\Delta y}{\delta x_w},\quad a_E^T(6,2)=\frac{k_e\Delta y}{\delta x_e},\quad a_S^T(6,2)=0$$
$$a_N^T(6,2)=\frac{k_n\Delta x}{\delta y_n},\quad b_P^T(6,2)=q_s\Delta x+S_T\Delta x\Delta y \qquad (7.3.13g)$$
$$a_P^T(6,2)=a_W^T(6,2)+a_E^T(6,2)+a_S^T(6,2)+a_N^T(6,2).$$

Equation (7.3.13f) is incorporated into the system of equations and can be solved using a Gauss–Seidel method, a line-by-line method, or the strongly implicit procedure (section 3.3.3). Once $T_{6,2}$ is obtained then $T_{6,1}$ is calculated outside of the matrix inversion process:

$$-k_s\left(\frac{T_{6,2}-T_{6,1}}{\delta y/2}\right)=q_s,\quad T_{6,2}-T_{6,1}=\frac{-q_s}{2k_s}\delta y,\quad T_{6,1}=T_{6,2}+\frac{q_s}{2k_s}\delta y. \qquad (7.3.14)$$

Note that, if the southern boundary is insulated, that is, subjected to the zero heat flux boundary condition, then $q_s=0$, and the second equation of Eq. (7.3.14) gives

$$T_{6,1}=T_{6,2}. \qquad (7.3.15)$$

Next, we develop the discretization equation for control volume (2,4) at the western boundary ($x=0$), which is subjected to the convection boundary condition. For this we again first integrate the governing equation around the control volume and apply the convective heat flux boundary condition for the west face, that is, $-k\frac{\partial T}{\partial x}|_w = \beta(T_{1,4} - T_\infty)$. However, unlike the given heat flux boundary condition, the right-hand side of this equation is unknown, since $T_{1,4}$ is not given. So we discretize the flux at the boundary using a first-order approximation and express the temperature at the boundary node in terms of temperature at the inner node, and the ambient temperature, which is given:

$$-k\frac{\partial T}{\partial x}\bigg|_w = \beta(T_{1,4} - T_\infty) \tag{7.3.16a}$$

$$-k_w\frac{T_{1,4} - T_{2,4}}{\delta x/2} = \beta(T_{1,4} - T_\infty) \tag{7.3.16b}$$

$$T_{1,4}\left(\beta + \frac{2k_w}{\delta x}\right) = \frac{2k_w}{\delta x}T_{2,4} + \beta T_\infty \tag{7.3.16c}$$

$$T_{1,4} = \frac{\frac{2k_w}{\delta x}T_{2,4} + \beta T_\infty}{\beta + \frac{2k_w}{\delta x}} \tag{7.3.16d}$$

$$\text{or} \quad T_{1,4} - T_\infty = \frac{\frac{2k_w}{\delta x}(T_{2,4} - T_\infty)}{\beta + \frac{2k_w}{\delta x}}. \tag{7.3.16e}$$

We then replace the convective flux term in the discretized equation with Eq. (7.3.16e). Integrating the given PDE over a control volume with node 2,4:

$$\int_{CV}\left[\frac{\partial}{\partial x}\left(k\frac{\partial T}{\partial x}\right) + \frac{\partial}{\partial y}\left(k\frac{\partial T}{\partial y}\right)\right]dV + \int_{CV} S_T dV = 0 \tag{7.3.17a}$$

$$k\frac{\partial T}{\partial x}\bigg|_w^e \Delta y + k\frac{\partial T}{\partial y}\bigg|_s^n \Delta x + S_T\Delta x\Delta y = 0 \tag{7.3.17b}$$

$$\left[k\frac{\partial T}{\partial x}\bigg|_e - k\frac{\partial T}{\partial x}\bigg|_w\right]\Delta y + \left[k\frac{\partial T}{\partial y}\bigg|_n - k\frac{\partial T}{\partial y}\bigg|_s\right]\Delta x + S_T\Delta x\Delta y = 0 \tag{7.3.17c}$$

$$k_e\left(\frac{T_{3,4} - T_{2,4}}{\delta x_e}\right)\Delta y + \beta(T_{1,4} - T_\infty)\Delta y + k_n\left(\frac{T_{2,5} - T_{2,4}}{\delta y_n}\right)\Delta x$$
$$-k_s\left(\frac{T_{2,4} - T_{2,3}}{\delta y_s}\right)\Delta x + S_T\Delta x\Delta y = 0 \tag{7.3.17d}$$

$$k_e\left(\frac{T_{3,4} - T_{2,4}}{\delta x_e}\right)\Delta y + \frac{\beta\Delta y\frac{2k_w}{\delta x}}{\beta + \frac{2k_w}{\delta x}}(T_{2,4} - T_\infty) + k_n\left(\frac{T_{2,5} - T_{2,4}}{\delta y_n}\right)\Delta x$$
$$-k_s\left(\frac{T_{2,4} - T_{2,3}}{\delta y_s}\right)\Delta x + S_T\Delta x\Delta y = 0 \tag{7.3.17e}$$

$$T_{2,4}\left(\frac{k_e\Delta y}{\delta x_e}+\frac{k_s\Delta x}{\delta y_s}+\frac{k_n\Delta x}{\delta y_n}-\frac{\beta\Delta y\frac{2k_w}{\delta x}}{\beta+\frac{2k_w}{\delta x}}\right)$$
$$=\frac{k_e\Delta y}{\delta x_e}T_{3,4}+\frac{k_s\Delta x}{\delta y_s}T_{2,3}+\frac{k_n\Delta x}{\delta y_n}T_{2,5}$$
$$-\frac{\beta\Delta y\frac{2k_w}{\delta x}}{\beta+\frac{2k_w}{\delta x}}T_\infty+S_T\Delta x\Delta y \tag{7.3.17f}$$

$$a_P^T(2,4)T_{2,4}=a_E^T(3,4)T_{2,4}+a_S^T(2,4)T_{2,3}+a_N^T(2,4)T_{2,5}+b_P^T(2,4), \tag{7.3.17g}$$

where

$$a_W^T(2,4)=0,\quad a_E^T(2,4)=\frac{k_e\Delta y}{\delta x_e},\quad a_S^T(2,4)=\frac{k_s\Delta x}{\delta y_s}$$
$$a_N^T(2,4)=\frac{k_n\Delta x}{\delta y_n},\quad b_P^T(2,4)=-\frac{\beta\Delta y\frac{2k_w}{\delta x}}{\beta+\frac{2k_w}{\delta x}}T_\infty+S_T\Delta x\Delta y \tag{7.3.18}$$
$$a_P^T(2,4)=a_W^T(2,4)+a_E^T(2,4)+a_S^T(2,4)+a_N^T(2,4)-\frac{\beta\Delta y\frac{2k_w}{\delta x}}{\beta+\frac{2k_w}{\delta x}}.$$

Once we solve for the internal node temperature $T_{2,4}$ we can calculate the boundary node temperature $T_{1,4}$ by using Eq. (7.3.16d). Note that, for the zero flux condition ($\beta=0$) on the west face, the above condition reduces to

$$T_{1,4}=T_{2,4}. \tag{7.3.19}$$

Similar equations can be written for other boundary nodes and control volumes on the boundary. One can write a computer program to implement the ideas developed here and solve steady two-dimensional heat transfer problems using the FVM. Here we revisit the problem of **Example 4.3.2**.

Example 7.3.1

Consider steady-state heat conduction in an isotropic rectangular solid (see **Example 4.3.2** and Fig. 7.3.4). The dimensions of the solid region are $3a\times 2a$. Use the FVM to determine the temperature distribution using (a) 3×2 control volumes, (b) 6×4 control volumes, and (c) 12×8 control volumes.

Solution The steady-state heat conduction equation with no source or sink term is given by

$$\frac{\partial}{\partial x}\left(k\frac{\partial T}{\partial x}\right)+\frac{\partial}{\partial y}\left(k\frac{\partial T}{\partial y}\right)=0. \tag{1}$$

The finite volume discretized equation for a interior nodes (has the following form:

$$a_PT_P=a_WT_W+a_ET_E+a_ST_S+a_NT_N+b_P \tag{2}$$
$$a_W=D_w,\quad a_E=D_e,\quad a_S=D_s,\quad a_N=D_n,\quad a_P=a_W+a_E+a_S+a_N,\quad b_P=0 \tag{3}$$

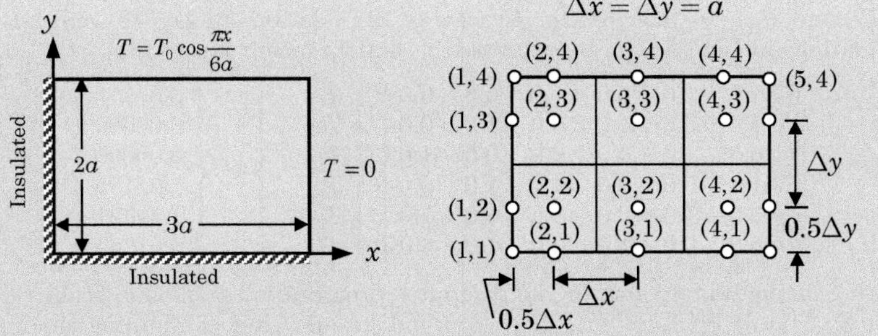

Fig. 7.3.4 Domain and 3 × 2 uniform mesh of control volumes and associated nodes.

and D_e, D_w, D_s, and D_n are the diffusion coefficients, which are calculated for a given mesh, as described next.

First, we use a mesh of 3 × 2 control volumes. The number of nodes in the x- and y-directions are: $itmax = 5$ and $jtmax = 4$ (see Fig. 7.3.4). Then $\Delta x = 3a/3 = a$ and $\Delta y = 2a/2 = a$. For the internal control volumes the diffusion lengths are:

$$\delta x_w = \delta x_e = \Delta x = a, \quad \delta y_s = \delta y_n = \Delta y = a. \tag{4}$$

For the boundary control volumes the diffusion lengths are:

$$\delta x_w(\text{left}) = \delta x_e(\text{right}) = \Delta x/2 = 0.5a, \quad \delta y_s(\text{bottom}) = \delta y_n(\text{top}) = \Delta y/2 = 0.5a \tag{5}$$

Thus, the diffusion coefficients for the internal control volumes are:

$$\begin{aligned} D_w &= \frac{k\Delta y}{\delta x_w} = \frac{k \times a}{a} = k, \quad D_e = \frac{k\Delta y}{\delta x_e} = \frac{k \times a}{a} = k \\ D_s &= \frac{k\Delta x}{\delta y_s} = \frac{k \times a}{a} = k, \quad D_n = \frac{k\Delta x}{\delta y_n} = \frac{k \times a}{a} = k. \end{aligned} \tag{6}$$

Similarly, the diffusion coefficients for the boundary control volumes are:

$$\begin{aligned} D_w(\text{left}) &= \frac{k\Delta y}{\delta x_w} = \frac{k \times a}{0.5a} = 2k, \quad D_e(\text{right}) = \frac{k\Delta y}{\delta x_e} = \frac{k \times a}{0.5a} = 2k \\ D_s(\text{bottom}) &= \frac{k\Delta x}{\delta y_s} = \frac{k \times a}{0.5a} = 2k, \quad D_n(\text{top}) = \frac{k\Delta x}{\delta y_n} = \frac{k \times a}{0.5a} = 2k. \end{aligned} \tag{7}$$

For the problem at hand, the zero flux conditions, $\partial T/\partial x = 0$ at $x = 0$ and $\partial T/\partial y = 0$ at $y = 0$, give, for the 3 × 2 mesh shown in Fig. 7.3.4, the following relations [see Eqs. (7.3.13a)–(7.3.16b)]:

$$T_{(1,1)} = T_{(2,1)} = T_{(1,2)} = T_{(2,2)}, \quad T_{(3,1)} = T_{(3,2)}, \quad T_{(4,1)} = T_{(4,2)}, \quad T_{(1,3)} = T_{(2,3)}. \tag{8}$$

Thus, the only unknowns are $T_{(2,2)}$, $T_{(3,2)}$, $T_{(4,2)}$, $T_{(2,3)}$, $T_{(3,3)}$, and $T_{(4,3)}$. The six equations needed to solve for the six unknowns are provided by equations of the form in Eq. (7.3.10a). We have (noting that $\Delta x = \Delta y = a$)

$$\begin{aligned} 6kT_{(2,2)} &= k\left(2T_{(1,2)} + T_{(3,2)}\right) + k\left(2T_{(2,1)} + T_{(2,3)}\right) + ga^2 \\ 5kT_{(3,2)} &= k\left(T_{(2,2)} + T_{(4,2)}\right) + k\left(2T_{(3,1)} + T_{(3,3)}\right) + ga^2 \\ 6kT_{(4,2)} &= k\left(T_{(3,2)} + 2T_{(5,2)}\right) + k\left(2T_{(4,1)} + T_{(4,3)}\right) + ga^2 \\ 6kT_{(2,3)} &= k\left(2T_{(1,3)} + T_{(3,3)}\right) + k\left(2T_{(2,4)} + T_{(2,2)}\right) + ga^2 \\ 5kT_{(3,3)} &= k\left(T_{(2,3)} + T_{(4,3)}\right) + k\left(2T_{(3,4)} + T_{(3,2)}\right) + ga^2 \\ 6kT_{(4,3)} &= k\left(T_{(3,3)} + 2T_{(5,3)}\right) + k\left(2T_{(4,4)} + T_{(4,2)}\right) + ga^2, \end{aligned} \tag{9}$$

where $T_{(5,1)} = T_{(5,2)} = T_{(5,3)} = T_{(5,4)} = 0$, $T_{(1,4)} = T_0$, $T_{(2,4)} = 0.965926T_0$, $T_{(3,4)} = 0.707107T_0$, and $T_{(4,4)} = 0.258819T_0$. In view of the relations in Eq. (8) and known values, the six equations in Eq. (9) can be expressed in matrix form (noting that $g = 0$) as

$$\begin{bmatrix} 2.0 & -1.0 & 0.0 & -1.0 & 0.0 & 0.0 \\ -1.0 & 3.0 & -1.0 & 0.0 & -1.0 & 0.0 \\ 0.0 & -1.0 & 4.0 & 0.0 & 0.0 & -1.0 \\ -1.0 & 0.0 & 0.0 & 4.0 & -1.0 & 0.0 \\ 0.0 & -1.0 & 0.0 & -1.0 & 5.0 & -1.0 \\ 0.0 & 0.0 & -1.0 & 0.0 & -1.0 & 6.0 \end{bmatrix} \begin{Bmatrix} T_{(2,2)} \\ T_{(3,2)} \\ T_{(4,2)} \\ T_{(2,3)} \\ T_{(3,3)} \\ T_{(4,3)} \end{Bmatrix} = \begin{Bmatrix} 0.0 \\ 0.00000 \\ 0.00000 \\ 1.93185 \\ 1.41421 \\ 0.51764 \end{Bmatrix}. \tag{10}$$

The ordering of the nodal value in the computer program is as follows: $(1,1) \to 1; (2,1) \to 2; (3,1) \to 3; (4,1) \to 4; (5,1) \to 5; (1,2) \to 6$ and so on. After eliminating the known nodal values, we have six unknowns (T_6, T_7, T_8, T_{12}, T_{13}, T_{14}), where $(2,2) \to 6; (3,2) \to 7; (4,2) \to 8; (2,3) \to 12; (3,3) \to 13; (4,3) \to 14$. Solution of Eq. (10) gives

$$\begin{aligned} &T_{(2,2)} = 0.6145T_0, \quad T_{(3,2)} = 0.4499T_0, \quad T_{(4,2)} = 0.1647T_0, \\ &T_{(2,3)} = 0.7792T_0, \quad T_{(3,3)} = 0.5704T_0, \quad T_{(4,3)} = 0.2088T_0, \end{aligned} \tag{11}$$

whereas the exact solution at the same locations is

$$\begin{aligned} &T(0.5a, 0.5a) = 0.6244T_0, \quad T(1.5a, 0.5a) = 0.4571T_0, \quad T(2.5a, 0.5a) = 0.1673T_0, \\ &T(0.5a, 1.5a) = 0.7995T_0, \quad T(1.5a, 1.5a) = 0.5853T_0, \quad T(2.5a, 1.5a) = 0.2142T_0. \end{aligned} \tag{12}$$

The FVM results obtained with three different meshes, 3×2, 6×4, and 12×8, are compared in Table 7.3.1 with the results obtained using corresponding finite element meshes

Table 7.3.1 Comparison of temperatures $T(x, 0)/T_0$ at nodes, obtained using various number of control volume and finite element meshes with the exact solution.

	3×2		6×4		12×8		
x	FVM	FEM	FVM	FEM	FVM	FEM	Exact
0.000	0.6145	0.6128	0.6224	0.6219	0.6243	0.6242	0.6249
0.125	−−	−−	−−	−−	0.6243	0.6215	0.6235
0.250	−−	−−	0.6224	0.6113	−−	0.6188	0.6195
0.375	−−	−−	−−	−−	0.6136	0.6108	0.6129
0.500	0.6145	0.5718	−−	0.6007	−−	0.6029	0.6036
0.625	−−	−−	−−	−−	0.5924	0.5900	0.5917
0.750	−−	−−	0.5800	0.5697	−−	0.5766	0.5773
0.875	−−	−−	−−	−−	0.5611	0.5586	0.5604
1.000	−−	0.5307	−−	0.5386	−−	0.5405	0.5412
1.125	−−	−−	−−	−−	0.5202	0.5179	0.5196
1.250	−−	−−	0.4981	0.4892	−−	0.4952	0.4958
1.375	−−	−−	−−	−−	0.4704	0.4683	0.4698
1.500	0.4499	0.4186	−−	0.4398	−−	0.4413	0.4419
1.625	−−	−−	−−	−−	0.4125	0.4106	0.4120
1.750	−−	−−	0.3822	0.3754	−−	0.3800	0.3804
1.875	−−	−−	−−	−−	0.3476	0.3460	0.3472
2.000	−−	0.3064	−−	0.3110	−−	0.3121	0.3124
2.125	−−	−−	−−	−−	0.2767	0.2755	0.2764
2.250	−−	−−	0.2402	0.2360	−−	0.2388	0.2391
2.375	−−	−−	−−	−−	0.2011	0.2002	0.2009
2.500	0.1647	0.1532	−−	0.1610	−−	0.1615	0.1617
2.625	−−	−−	−−	−−	0.1220	0.1215	0.1219
2.750	−−	−−	0.0819	0.0805	−−	0.0815	0.0816
2.875	−−	−−	−−	−−	0.0409	0.0473	0.0409

of linear rectangular elements (see Table 3.3.1) and the exact values. The underlined numbers denote linearly interpolated values between the nodal values of the finite element meshes used. The results are independent of the values of k and a for this problem (because the internal heat generation and specified boundary fluxes are zero). It is not possible to compare the nodal values of the temperature obtained from different FVM meshes because as the mesh is refined, the nodal locations change. It is possible to obtain the finite element solution at the mesh points of the finite volume mesh but one is required to use a nonuniform mesh of finite elements (e.g., a 4×3 nonuniform mesh of linear finite elements with $DX(I) = a/2, a, a, a/2$ and $DY(I) = a/2, a, a/2$ will have the same nodal locations as in the 3×2 uniform mesh of control volumes).

Figure 7.3.5 shows the numerical convergence of the finite volume and finite element solutions to the exact solution as meshes are refined.

The post-computation of heat fluxes on the boundaries $x = 3a$ and $y = 2a$ can be carried out using the finite difference approximation as in one-dimensional problems. For example, the flux normal to the boundary at $(x, y) = (1.5a, y = 2a)$ is $(\Delta y = a)$

$$q_y(1.5a, 2a) = -k\frac{\partial T}{\partial y}\bigg|_{(x=1.5a,y=2a)} = -k\left(\frac{T_{(3,4)} - T_{(3,3)}}{0.5\Delta y}\right). \tag{13}$$

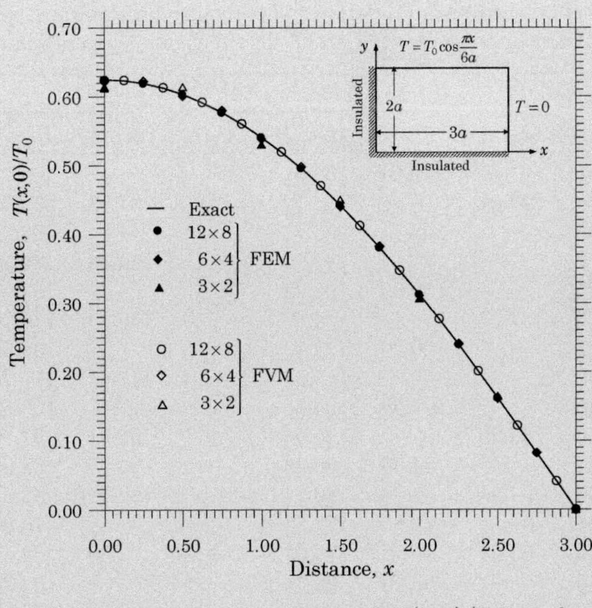

Fig. 7.3.5 Plots of numerical and analytical solutions $T(x, 0)/T_0$ as a function of x for $y = 0$.

Example 7.3.2

Consider a rectangular cooling fin with its base, $x = 0$, maintained at 300 °C, and the remaining surfaces exposed to a specified ambient temperature (see **Example 4.3.3** and Fig. 7.3.6). Use the FVM to determine the temperature distribution using the data and discretizations used in **Example 4.3.3** and compare the results.

Solution For the 4×1 mesh in the symmetric (about $y = 1$ cm) half shown in Fig. 7.3.6, we can use equations of the type in Eq. (7.3.13a) at nodes on the $y = 0.01$ m boundary, Eq. (7.3.14) at nodes 2, 3, 4, 5, 6, and 12, and Eqs. (7.3.16a) and (7.3.16b) for control

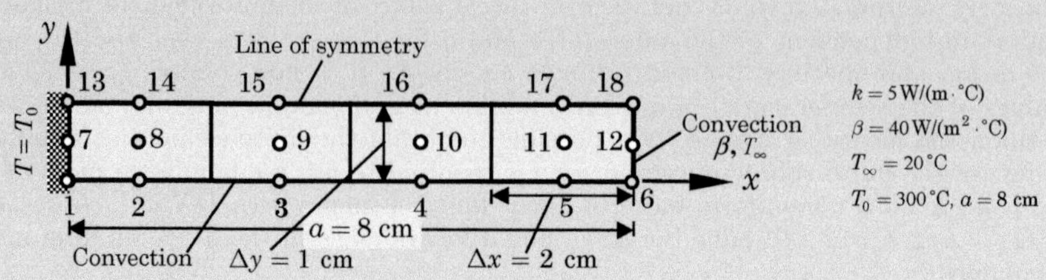

Fig. 7.3.6 Domain and 4×1 uniform mesh of control volumes and associated nodes.

volumes with center nodes 8, 9, 10, and 11 (twice). Table 7.3.2 contains numerical values of the temperature $T(x, 0.01)$ obtained with various meshes of the FVM and FEM. The underlined values denote the linearly interpolated values between the nodal values. Figure 7.3.7 contains plots of the temperature along the horizontal center line of the fin. Clearly, the FVM solution converges to the FEM solution as the mesh is refined. The inaccuracy of the FVM solution is attributed to the gross approximation of the convection boundary condition.

Table 7.3.2 Nodal temperatures $T(x, 0.01)$ of the cooling fin obtained using various finite volume and finite element meshes (see Fig. 7.3.6 for the geometry and boundary conditions).

	FVM			FEM		
x	4×1	8×1	16×2	4×1	8×1	16×2
0.005	—	261.14	265.57	—	—	266.71
0.010	224.21	—	234.12	—	237.13	235.75
0.015	—	202.71	206.71	—	—	208.31
0.020	—	—	182.90	185.04	184.79	184.52
0.025	—	158.90	162.28	—	—	163.93
0.030	137.97	—	144.48	—	146.24	146.16
0.035	—	126.20	129.13	—	—	130.86
0.040	—	—	115.95	116.20	117.64	117.73
0.045	—	101.99	104.66	—	—	106.52
0.050	89.45	—	95.06	—	96.81	97.00
0.055	—	84.35	86.93	—	—	88.98
0.060	—	—	80.14	81.05	82.09	82.31
0.065	—	71.85	74.54	—	—	76.87
0.070	63.23	—	70.03	—	72.30	72.54
0.075	—	63.50	66.56	—	—	69.23
0.080	58.57	61.07	63.80	63.96	66.66	66.89

Example 7.3.3

Consider an example of the manufactured solution method in Section 1.5.2. The heat diffusion equation in a two-dimensional rectangular domain (see Fig. 7.3.8) in the absence of source/sink term is

$$\frac{\partial^2 T}{\partial x^2} + \frac{\partial^2 T}{\partial y^2} = 0. \tag{7.3.20}$$

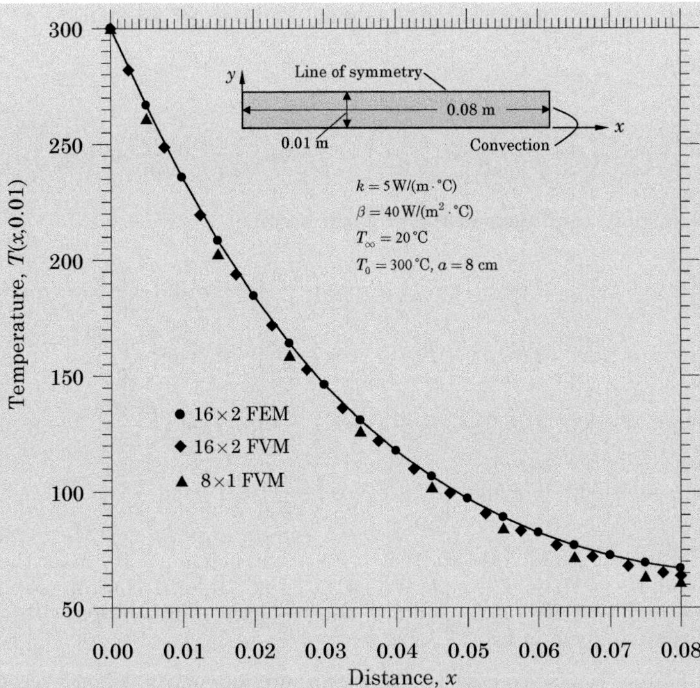

Fig. 7.3.7 Plots of numerical solutions $T(x, 0.01)$ as a function of x for $y = 0.01$ m.

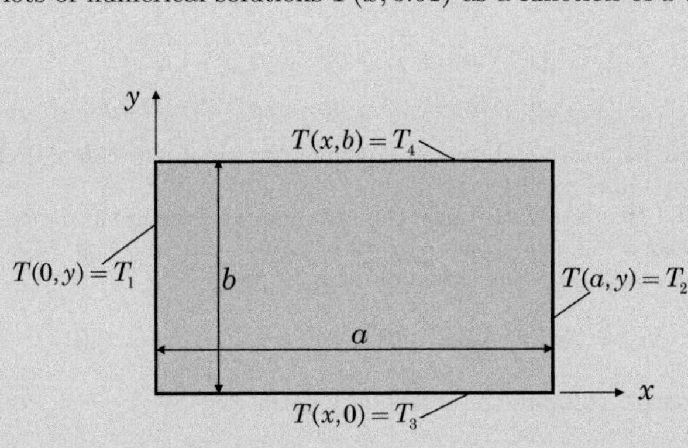

Fig. 7.3.8 Heat diffusion in a thin rectangular plate with specified temperature on all sides (i.e., Dirichlet boundary conditions).

The prescribed boundary conditions are:

$$T(0,y) = T_1, \quad T(a,y) = T_2, \quad T(x,0) = T_3, \quad T(x,b) = T_4. \tag{7.3.21}$$

The proposed manufactured solution must be continuous in the domain and must survive the highest order differentiation in the given differential equation. So we propose a manufactured solution of the form

$$T_m(x,y) = T_1 \cos\left(\frac{\pi x}{2a}\right) + T_2 \sin\left(\frac{\pi x}{2a}\right) + T_3 \cos\left(\frac{\pi y}{2b}\right) + T_4 \sin\left(\frac{\pi y}{2b}\right). \tag{7.3.22}$$

By substituting the above solution into the original PDE we get an MDE:

$$\frac{\partial^2 T_m}{\partial x^2} + \frac{\partial^2 T_m}{\partial y^2} = S_T \tag{7.3.23a}$$

$$S_T = -\frac{\pi^2}{4}\left[\frac{T_1}{a^2}\cos\left(\frac{\pi x}{2a}\right) + \frac{T_2}{a^2}\sin\left(\frac{\pi x}{2a}\right) + \frac{T_3}{b^2}\cos\left(\frac{\pi y}{2b}\right) + \frac{T_4}{b^2}\sin\left(\frac{\pi y}{2b}\right).\right] \tag{7.3.23b}$$

The modified boundary conditions of the problem become:

$$\text{On } x = 0 \text{ side:} \quad T(0, y) = T_1 + T_3\cos\left(\frac{\pi y}{2b}\right) + T_4\sin\left(\frac{\pi y}{2b}\right). \tag{7.3.24a}$$

$$\text{On } x = a \text{ side:} \quad T(a, y) = T_2 + T_3\cos\left(\frac{\pi y}{2b}\right) + T_4\sin\left(\frac{\pi y}{2b}\right). \tag{7.3.24b}$$

$$\text{On } y = 0 \text{ side:} \quad T(x, 0) = T_3 + T_1\cos\left(\frac{\pi x}{2a}\right) + T_2\sin\left(\frac{\pi x}{2a}\right). \tag{7.3.24c}$$

$$\text{On } y = b \text{ side:} \quad T(x, b) = T_4 + T_1\cos\left(\frac{\pi x}{2a}\right) + T_2\sin\left(\frac{\pi x}{2a}\right). \tag{7.3.24d}$$

Solve the MDE using the FVM. Take width $a = 0.1$ m and height $b = 0.1$ m for the rectangular domain. Assume $T_1 = T_2 = T_3 = T_4 = 100\,^\circ$C. Use 10×10 control volumes, 20×20 control volumes, and 40×40 control volumes and compare your solution with the proposed manufactured solution given in Eq. (7.3.22).

Solution The finite volume discretized equation for interior control volumes has the following form:

$$a_P T_P = a_W T_W + a_E T_E + a_S T_S + a_N T_N + b_P, \tag{2}$$

where

$$\begin{aligned} a_W &= D_w, \quad a_E = D_e, \quad a_S = D_s, \quad a_N = D_n, \\ a_P &= a_W + a_E + a_S + a_N, \quad b_P = S_T \Delta x \Delta y, \end{aligned} \tag{3}$$

and D_e, D_w, D_s, and D_n are the diffusion coefficients, which are calculated in a similar way as described in the previous example problems.

For a mesh of 10×10 control volumes, the number of nodes in the x- and y-directions are: $itmax = 12$ and $jtmax = 12$. Then $\Delta x = a/10 = 0.01$ m and $\Delta y = b/10 = 0.1/10 = 0.01$ m. For the internal control volumes the diffusion lengths are:

$$\delta x_w = \delta x_e = \Delta x = 0.01, \quad \delta y_s = \delta y_n = \Delta y = 0.01. \tag{4}$$

For the boundary control volumes the diffusion lengths are

$$\delta x_w(\text{left}) = \delta x_e(\text{right}) = \Delta x/2 = 0.005, \quad \delta y_s(\text{bottom}) = \delta y_n(\text{top}) = \Delta y/2 = 0.005. \tag{5}$$

Thus the diffusion coefficients for the internal control volumes are:

$$\begin{aligned} D_w &= \frac{\Delta y}{\delta x_w} = \frac{0.01}{0.01} = 1, \quad D_e = \frac{\Delta y}{\delta x_e} = \frac{0.01}{0.01} = 1 \\ D_s &= \frac{\Delta x}{\delta y_s} = \frac{0.01}{0.01} = 1, \quad D_n = \frac{\Delta x}{\delta y_n} = \frac{0.01}{0.01} = 1. \end{aligned} \tag{6}$$

Similarly, the diffusion coefficients for the boundary control volumes are:

$$\begin{aligned} D_w(\text{left}) &= \frac{\Delta y}{\delta x_w} = \frac{0.01}{0.005} = 2, \quad D_e(\text{right}) = \frac{\Delta y}{\delta x_e} = \frac{0.01}{0.005} = 2, \\ D_s(\text{bottom}) &= \frac{\Delta x}{\delta y_s} = \frac{0.01}{0.005} = 2, \quad D_n(\text{top}) = \frac{\Delta x}{\delta y_n} = \frac{0.01}{0.005} = 2. \end{aligned} \tag{7}$$

The modified boundary conditions for the problem are all Dirichlet boundary conditions. However, these boundary conditions vary as a function of x and/or y, as shown in Eqs. (7.3.24a)-(7.3.24d). Once we incorporate the a-coefficients and appropriate boundary conditions, we obtain a system of equations (10 in this case), which can be solved by the Gauss–Seidel scheme or line-by-line TDMA method. Figure 7.3.9 shows the centerline temperature distributions ($x = a/2$) calculated using the FVM with uniform meshes 10×10, 20×20, and 40×40 control volumes and the manufactured (analytical) solution. The FVM solutions for all three grid sizes agrees very well with the analytical solution, and the maximum difference is less than one percent thus *verifying* the FVM solution.

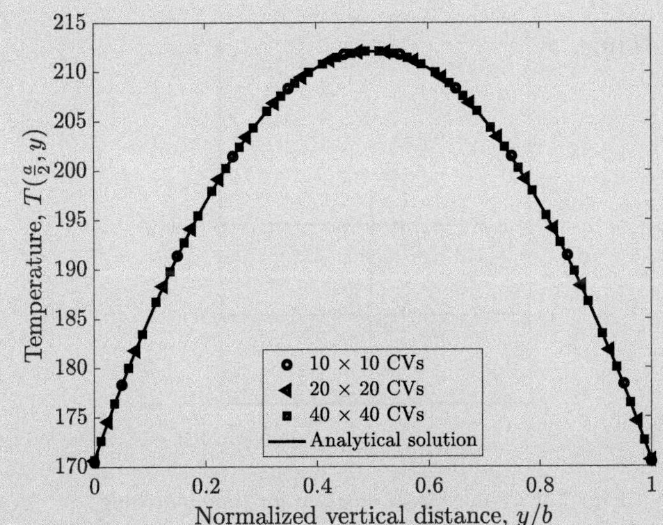

Fig. 7.3.9 Comparison of finite volume solution with the (analytical) solution. The temperature profiles along vertical distance are taken at centerline ($x = a/2$).

7.4 Unsteady Problems

7.4.1 One-Dimensional Problems

7.4.1.1 Model Equation

Consider the following equation governing one-dimensional unsteady heat transfer in an isotropic medium:

$$\rho c \frac{\partial T}{\partial t} = \frac{\partial}{\partial x}\left(k \frac{\partial T}{\partial x}\right) + S, \quad 0 < x < L, \quad t > 0, \tag{7.4.1}$$

where k denotes the thermal conductivity, ρ is the mass density, c is the specific heat, and S is the internal heat generation per unit length, all of which can be, in general, known functions of position x and time t. However, for simplicity of presentation, we assume that all geometric as well as material properties are independent of t and x. The given PDE can also be reconciled using the energy balance over a control volume.

Observations The given PDE (7.4.1) is parabolic since $B^2 - 4AC = 0$. The problem exhibits one-way-street or parabolic behavior in time t and two-way-street or elliptic behavior in spatial coordinate x. While one is able to march in time at each time step, one must solve a set of simultaneous equations to satisfy the boundary conditions. One could also look at this problem as exhibiting initial value behavior in time t and exhibiting boundary value behavior in x. Hence, this problem is an initial value-boundary value problem. The conceptual diagram (see Fig. 7.4.1) shows the way information propagates in the domain.

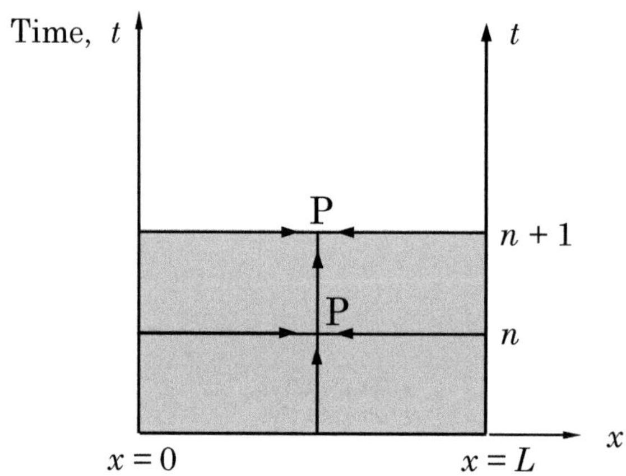

Fig. 7.4.1 Conceptual diagram for time marching

As shown in Fig. 7.4.1, the solution process marches in time (t) while satisfying boundary conditions at each time step. The temperature at point P at the $(n+1)$th time step is influenced by temperature at the nth time step or at earlier time steps depending on the formulation while it receives information from both boundaries $x = 0$ and $x = L$.

As in the case of steady-state heat transfer of Section 7.2, the domain $0 < x < L$ is discretized into a set of control volumes using the zero-thickness formulation. The PDE is satisfied in an integral sense over a typical one-dimensional control volume (see Fig. 7.2.2). A detailed description of this figure is not given here, as details have already been presented in Section 7.2.

7.4.1.2 Explicit Formulation

Recall from Section 4.4 that the explicit formulation is one in which all the nonzero elements are on the diagonal of the coefficient matrix and there no inversion of the discretized equations is required. Integrating the given PDE over the control volume and in the time interval t and $t + \Delta t$:

$$\int_t^{t+\Delta t} \int_{CV} \rho c \frac{\partial T}{\partial t}\, dx\, dt = \int_t^{t+\Delta t} \int_{CV} \left[\frac{\partial}{\partial x} \left(k \frac{\partial T}{\partial x} \right) \right] dx\, dt + \int_t^{t+\Delta t} \int_{CV} S\, dx\, dt \tag{7.4.2}$$

or

$$\rho c\left(T_P^{n+1} - T_P^n\right)\Delta x = \int_t^{t+\Delta t}\left[k\frac{\partial T}{\partial x}\bigg|_e - k\frac{\partial T}{\partial x}\bigg|_w + S_P\Delta x\right]dt. \tag{7.4.3}$$

Now the right-hand side term can be evaluated at the nth time step, or the $(n+1)$th time step, or with a combination of both. When the right-hand side is evaluated at the nth time step, the formulation is *explicit*:

$$\rho c\left(T_P^{n+1} - T_P^n\right)\Delta x = \left[k\frac{\partial T}{\partial x}\bigg|_e - k\frac{\partial T}{\partial x}\bigg|_w\right]^n \Delta t + S_P\Delta x\,\Delta t \tag{7.4.4}$$

$$\left(\frac{T_P^{n+1} - T_P^n}{\Delta t}\right)\rho c\,\Delta x = \left[\frac{k_e}{\delta x_e}(T_E - T_P) - \frac{k_w}{\delta x_w}(T_P - T_W)\right]^n + S_P\Delta x \tag{7.4.5}$$

$$\left(\frac{T_P^{n+1} - T_P^n}{\Delta t}\right)\rho c\Delta x = \left[D_e(T_E - T_P) - D_w(T_P - T_W)\right]^n + S_P\Delta x. \tag{7.4.6}$$

The diffusion strength D is defined as $k/\delta x$. Rearranging Eq. (7.4.6) we obtain

$$a_P T_P^{n+1} = a_W T_W^{(n)} + a_E T_E^n + (a_P^0 - a_E - a_W)T_P^n + b, \tag{7.4.7a}$$

where

$$a_E = D_e = \frac{k_e}{\delta x_e}, \quad a_W = D_w = \frac{k_w}{\delta x_w} \tag{7.4.7b}$$

$$b = S_P\,\Delta x, \quad a_P^0 = \frac{\rho c_v \Delta x}{\Delta t}, \quad a_P = a_P^0. \tag{7.4.7c}$$

For the Gauss–Seidel iterative method of solution, the only unknown in Eq. (7.4.7a) is T_P^{n+1}, that is, the solution at the previous time step and the solution at other points for the current step are known, the equation can be put in a nested do-loop and go through a point-by-point solution to calculate T_P^{n+1}.

Starting with initial condition $T(x_i, 0) = T_i^0$, the algorithm in Eq. (7.4.7a) is used to march in time until the specified time limit or steady state is reached. As discussed in Section 4.4, explicit schemes are conditionally stable, requiring satisfaction of a stability criterion, which relates the time step (Δt) to the grid size (Δx). As long as Δt meets the stability criterion in Eq. (4.2.23) (for $\alpha = 0$), the solution will be numerically stable; however, for accurate solution, Δt should be considerably smaller that the critical time step Δt_{cr}.

7.4.1.3 Implicit Formulation

In the implicit formulation, the right side of Eq. (7.4.3) is evaluated at $t + \Delta t = t^{n+1}$. Then we have

$$\rho c\left(T_P^{n+1} - T_P^n\right)\Delta x = \left[k\frac{\partial T}{\partial x}\bigg|_e - k\frac{\partial T}{\partial x}\bigg|_w\right]^{n+1} \Delta t + S_P\Delta x\,\Delta t \tag{7.4.8}$$

$$\left(\frac{T_P^{n+1} - T_P^n}{\Delta t}\right)\rho c\,\Delta x = \left[\frac{k_e}{\delta x_e}(T_E - T_P) - \frac{k_w}{\delta x_w}(T_P - T_W)\right]^{n+1} + S_P\Delta x \tag{7.4.9}$$

$$\left(\frac{T_P^{n+1} - T_P^n}{\Delta t}\right)\rho c\Delta x = \Big[D_e(T_E - T_P) - D_w(T_P - T_W)\Big]^{n+1} + S_P\Delta x. \quad (7.4.10)$$

Equation (7.4.10) can be rearranged to read:

$$a_P T_P^{n+1} = a_W T_W^{n+1} + a_E T_E^{n+1} + a_P^0 T_P^n + b, \quad (7.4.11a)$$

where

$$a_E = D_e = \frac{k_e}{\delta x_e} \quad (7.4.11b)$$

$$a_W = D_w = \frac{k_w}{\delta x_w} \quad (7.4.11c)$$

$$b = S_P\,\Delta x \quad (7.4.11d)$$

$$a_P^0 = \frac{\rho c_v \Delta x}{\Delta t} \quad (7.4.11e)$$

$$a_P = a_W + a_E + a_P^0. \quad (7.4.11f)$$

The resulting equation is of the form $AX = C$, where A is the coefficient matrix, X is the vector containing the unknown variables (nodal temperatures in the present case), and C is the vector of known terms. We will later introduce a procedure called the *line-by-line method* to solve the set of algebraic equations. Unlike explicit formulation (where no inversion of the coefficient matrix A is involved, because it is diagonal), fully implicit formulation is numerically stable independent of the size of the time step Δt. However, greater accuracy of the solution does require (as illustrated in Chapter 5) smaller time steps as well as refined meshes.

7.4.2 Two-Dimensional Problems

7.4.2.1 Model Equation

In this section we focus on the finite volume method applied to a model partial differential equation in two dimensions, which governs unsteady heat diffusion. For the purpose of illustration heat diffusion (conduction) in a solid is considered. The thickness of the solid medium is assumed to be small so that we can neglect any flow of heat in the thickness direction. For simplicity, we consider an isotropic medium with a rectangular geometry.

The unsteady temperature distribution in an isotropic solid with internal heat source S (W/m^2) is governed by the following partial differential equation [see Eq. (1) of **Example 5.5.2** with $k_{xx} = k_{yy} = k$]:

$$\rho c\frac{\partial T}{\partial t} - \left[\frac{\partial}{\partial x}\left(k\frac{\partial T}{\partial x}\right) + \frac{\partial}{\partial y}\left(k\frac{\partial T}{\partial y}\right)\right] = S, \quad (7.4.12)$$

where T is the temperature and k is the thermal conductivity of the solid medium. Much of the remaining development follows the formulation presented in the previous section for the unsteady one-dimensional case, while keeping in mind that the spatial domain is two-dimensional.

The next step is to discretize the differential equation by requiring that it is satisfied in an integral sense over a typical control volume. Equation (7.4.12) displays parabolic behavior in time (t) and elliptic behavior in space (x and y). To keep the discussion simple we will assume that the thermophysical properties (e.g., density, thermal conductivity, film conductance, and specific heat) of the material remain constant. In addition, the source term S is also assumed to be a constant for now. Thus Eq. (7.4.12) is a second order, linear, partial differential equation with constant coefficients.

As before, we discretize the first derivative of the field variable T with respect to time t using the forward difference approximation:

$$\dot{T}_i(\tau) \equiv \left[\frac{\partial T_i}{\partial t}\right]_{t=\tau} \approx \frac{T_i^{n+1} - T_i^n}{\Delta t}, \tag{7.4.13}$$

where τ is equal to either t^n or t^{n+1}. As discussed earlier, if the left side (time derivative of T_i) is evaluated at the $t = t^{n+1}$ then it is the backward difference scheme and if it is evaluated at $t = t^n$ it is the forward difference scheme; if the left side is equal to the average of $\dot{T}_i^n$ and $\dot{T}_i^{n+1}$, then it is known as the Crank–Nicolson scheme.

7.4.2.2 Explicit Formulation

Consider the integral of Eq. (7.4.12) over a control volume (see Fig. 7.3.4)

$$0 = \int_{CV} \rho c \frac{\partial T}{\partial t}\, dx dy - \int_{CV} \left[\frac{\partial}{\partial x}\left(k\frac{\partial T}{\partial x}\right) + \frac{\partial}{\partial y}\left(k\frac{\partial T}{\partial y}\right) + S\right] dx dy \tag{7.4.14}$$

or

$$\int_{CV} \rho c \frac{\partial T}{\partial t}\, dx dy - \int_0^{\Delta y} \left[k\frac{\partial T}{\partial x}\bigg|_e - k\frac{\partial T}{\partial x}\bigg|_w\right] dy - \int_0^{\Delta x} \left[k\frac{\partial T}{\partial y}\bigg|_n - k\frac{\partial T}{\partial y}\bigg|_s\right] dx$$
$$- \int_0^{\Delta x}\int_0^{\Delta y} S\, dx dy = 0. \tag{7.4.15}$$

Next, we assume (in addition to the assumptions made concerning the spatial derivatives of T) that $\dot{T}$ is independent of x and y. Then Eq. (7.4.15) takes the form

$$0 = \rho c \frac{\partial T}{\partial t} \Delta x \Delta y - \left[k\frac{\partial T}{\partial x}\bigg|_e - k\frac{\partial T}{\partial x}\bigg|_w\right] \Delta y - \left[k\frac{\partial T}{\partial y}\bigg|_n - k\frac{\partial T}{\partial y}\bigg|_s\right] \Delta x$$
$$- S\Delta x\, \Delta y. \tag{7.4.16}$$

Next we invoke the forward difference approximation of the time derivative (i.e., evaluate $\dot{T}$ at $\tau = t^n$) and obtain

$$\rho c \left(\frac{T_P^{n+1} - T_P^n}{\Delta t}\right) \Delta x\, \Delta y = \Big[D_e(T_E - T_P) - D_w(T_P - T_W) + D_n(T_N - T_P)$$
$$- D_s(T_P - T_S) + S_P \Delta x\, \Delta y\Big]^n. \tag{7.4.17}$$

Rearranging Eq. (7.4.17) we obtain

$$a_P T_P^{n+1} = a_W T_W^n + a_E T_E^n + a_S T_S^n + a_N T_N^n + T_P^n[a_P^0 - (a_W + a_E + a_S + a_N)] + b, \tag{7.4.18a}$$

where

$$a_E = D_e = \frac{k_e \Delta y}{\delta x_e} \tag{7.4.18b}$$

$$a_W = D_w = \frac{k_w \Delta y}{\delta x_w} \tag{7.4.18c}$$

$$a_N = D_n = \frac{k_n \Delta x}{\delta y_n} \tag{7.4.18d}$$

$$a_S = D_s = \frac{k_s \Delta x}{\delta y_s} \tag{7.4.18e}$$

$$a_P^0 = \frac{\rho c \Delta x \Delta y}{\Delta t} \tag{7.4.18f}$$

$$b = S_P \Delta x\, \Delta y \tag{7.4.18g}$$

$$a_P = a_P^0. \tag{7.4.18h}$$

Again, for the Gauss–Seidal iteration method of solving the linear equations, the only unknown in Eq. (7.4.18a) is T_P^{n+1}, the equation can be put in a nested do-loop and go through a point-by-point solution to calculate T_P^{n+1}.

We now shift to the index notation by making the following correspondence:

$$P \to it, jt, \quad W \to it-1, j, \quad E \to it+1, j, \quad S \to it, jt-1, \quad N \to it, jt+1. \tag{7.4.19}$$

The subscript t refers to temperature. The loops are as follows: loop over $jt = 2, jtmax - 1$ and loop over $it = 2, itmax - 1$. Then Eq. (7.4.18a) takes the following form suitable for computation:

$$T_{it,jt}^{n+1} = \frac{1}{A_{it,jt}} \Big[T_{it,jt}^n [A_{it,jt} - (D_w + D_e + D_s + D_n)] + S\Delta x \Delta y + D_w T_{it-1,jt}^n + D_e T_{it+1,jt}^{(n)} + D_s T_{it,jt-1}^n + D_n T_{it,jt+1}^n \Big]. \tag{7.4.20}$$

Starting with initial condition $T(x_i, y_j, 0) = T_{it,jt}^0$, the algorithm in Eq. (7.4.20) is used to march in time until the specified time limit or steady state is reached. As for a given point (it, jt), the points to the left and below are known at the $(n+1)$th time step, but we have chosen not to use it. One option is to use the information at the $(n+1)$th time step for grid points to the left and below the grid point (it, jt). This technique is similar to the Gauss–Seidel iteration method of solving linear simultaneous equations. Conceptually, the time-marching scheme is shown in Fig. 7.4.2 for a 5×5 uniform mesh in a square domain.

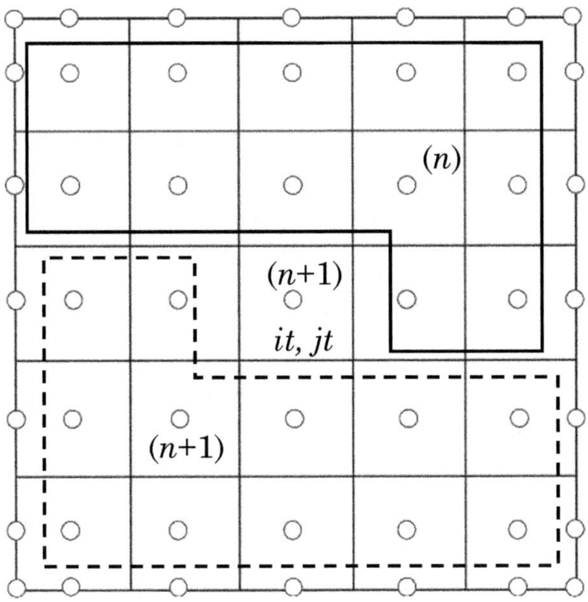

Fig. 7.4.2 Time marching for explicit method.

7.4.2.3 Implicit Formulation

In the implicit formulation, the time derivative in Eq. (7.4.14) is evaluated at $t + \Delta t = t^{n+1}$ (i.e., the backward difference scheme is used). Then we have

$$\rho c\left(\frac{T_P^{n+1} - T_P^n}{\Delta t}\right)\Delta x\,\Delta y = \Big[D_e(T_E - T_P) - D_w(T_P - T_W) + D_n(T_N - T_P) - D_s(T_P - T_S) + S_P\Delta x\,\Delta y\Big]^{n+1}. \quad (7.4.21)$$

The equation can be rearranged to read as:

$$a_P T_P^{n+1} = a_W T_W^{n+1} + a_E T_E^{n+1} + a_S T_S^{n+1} + a_N T_N^{n+1} + a_P^0 T_P^n + b_P, \quad (7.4.22a)$$

where

$$a_W = D_w, \quad a_E = D_e, \quad a_S = D_s, \quad a_N = D_n$$
$$a_P = a_W + a_E + a_S + a_N + a_P^0, \quad b_P = S_P\Delta x\Delta y. \quad (7.4.22b)$$

The above set of linear algebraic equations at each time step can be solved by the *line-by-line method* (discussed in Chapter 3) and marching in time is continued until a preset (i.e., final) time is reached or steady state is attained.

Next, we consider two numerical examples involving one- and two-dimensional time-dependent problems of transient heat transfer (see **Examples 5.5.1** and **5.5.2**).

7.4.3 Numerical Examples

Example 7.4.1

Suppose a plane wall, initially at temperature T_0, is suddenly temperature on the left side is changed to temperature T_∞ (see Fig. 7.4.3). Determine the time history of the temperature field in the wall when the right side is: (a) subjected to temperature T_∞ or (b) insulated (i.e., solve the problem for two different sets of boundary conditions). Use explicit formulation with five control volumes and plot the dimensionless temperature $\theta(\xi_0, \tau)$ as a function of the dimensionless time τ, where $\xi_0 = 0.5$ for the first set of boundary conditions and $\xi_0 = 1$ for the second set of boundary conditions. The time step for a stable and accurate solution can be taken as $\Delta\tau = 0.001$.

Solution This example is the same as that considered in **Example 5.5.1**. As discussed there, we introduce the following normalization of the temperature T, time t, and coordinate x [see Eq. (5.5.3)]:

$$\xi = \frac{x}{L}, \quad \tau = \frac{k}{\rho c_p L^2} t, \quad \theta = \frac{T - T_\infty}{T_0 - T_\infty} \tag{1}$$

so that the governing equation and the boundary and initial conditions become [see Eqs. (5.5.4) and (5.5.5)]:

$$\frac{\partial \theta}{\partial \tau} - \frac{\partial^2 \theta}{\partial \xi^2} = 0, \quad 0 < \xi < 1, \quad \tau > 0 \tag{2}$$

and

$$\text{(a)} \;\; \theta(\xi, 0) = 0, \;\; \theta(\xi, 1) = 0; \quad \text{(b)} \;\; \theta(\xi, 0) = 0, \;\; \left[\frac{\partial \theta}{\partial \xi}\right]_{\xi=1} = 0. \tag{3}$$

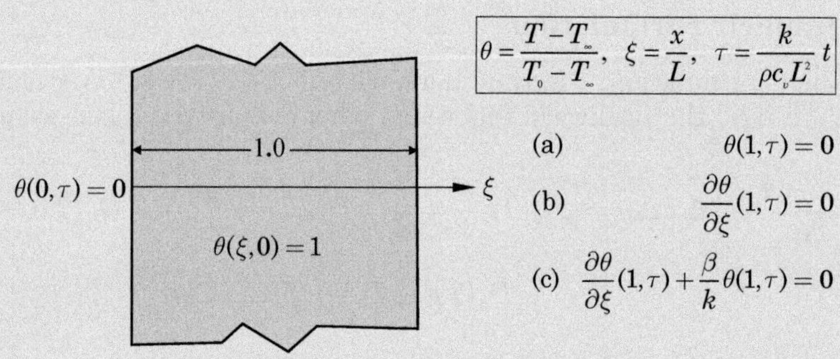

Fig. 7.4.3 A plane wall initially at temperature $\theta(\xi, 0) = 1$ is subjected to two different sets of boundary conditions.

For the , explicit formulation we use Eqs. (7.4.7a)–(7.4.7c). For Case (a) of the boundary conditions, the initial conditions for the seven nodal values should be

$$\theta_1 = 0.0, \quad \theta_2 = \theta_3 = \theta_4 = \theta_5 = \theta_6 = 1.0, \quad \theta_7 = 0.0. \tag{4}$$

For Case (b) of boundary conditions, the initial conditions are

$$\theta_1 = 0.0, \quad \theta_2 = \theta_3 = \theta_4 = \theta_5 = \theta_6 = \theta_7 = 1.0. \tag{5}$$

Owing to the explicit nature of Eq. (7.4.7a), one can solve for each nodal value θ_i ($i = 2, 3, 4, 5, 6$) for every time step, marching in time. The explicit scheme has a stability requirement that places a restriction on the size of the time step. For example, $\Delta t = 0.02$

results in an oscillatory (and unstable) response, whereas $\Delta t = 0.01$ gives a stable solution for the uniform mesh of five control volumes. Figure 7.4.4 shows the transient response for Case (a), where plots of the dimensionless temperature $\theta(0.5, \tau)$ versus the normalized time τ are shown. Both FVM and FEM give very good results for $\Delta t = 0.001$. Similar results are shown in Fig. 7.4.5 for Case (b) boundary conditions.

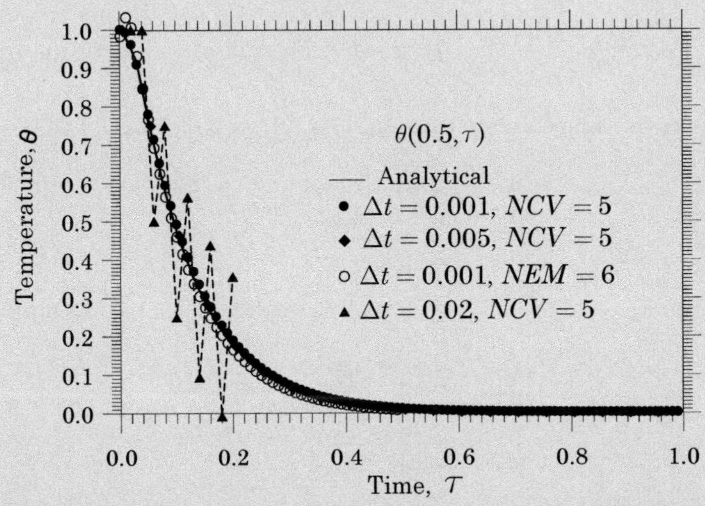

Fig. 7.4.4 Plots of normalized temperature $\theta(0.5, \tau)$ versus normalized time τ for plane wall initially at temperature $\theta(\xi, 0) = 1$ and subjected to Case (a) boundary conditions.

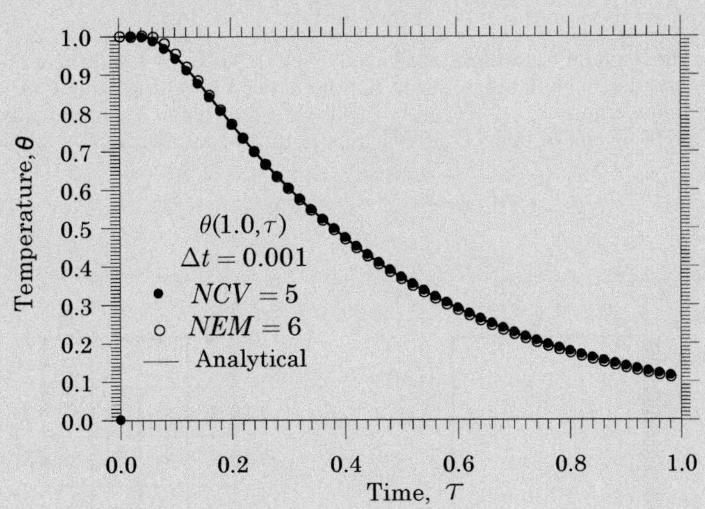

Fig. 7.4.5 Plots of normalized temperature $\theta(1.0, \tau)$ versus normalized time τ for a plane wall initially at temperature $\theta(\xi, 0) = 1$ and subjected to Case (b) boundary conditions.

Example 7.4.2

Consider the two-dimensional transient heat flow problem of **Example 5.5.2**. The governing equation is given by Eq. (7.4.12). As in **Example 5.5.2**, we can normalize the variables for a square plate of dimension $a \times a$ as

$$\xi = \frac{x}{a}, \quad \eta = \frac{y}{a}, \quad \tau = \frac{k}{\rho c_p a^2} t, \quad \theta = T - T_0, \quad f(\xi, \eta) = \frac{a^2}{k} g(x(\xi), y(\eta)), \tag{1}$$

where T_0 is a reference temperature. Then Eq. (7.4.12) takes the form

$$\frac{\partial \theta}{\partial \tau} - \left(\frac{\partial^2 \theta}{\partial \xi^2} + \frac{\partial^2 \theta}{\partial \eta^2} \right) = f(\xi, \eta). \tag{2}$$

Determine the normalized temperature field $\theta(\xi, \eta, \tau)$ in the square plate when the plate boundaries are subjected to the following boundary conditions [see Fig. 7.4.6(a)]:

$$\left.\frac{\partial \theta}{\partial \xi}\right|_{(0,\eta,\tau)} = 0, \quad \left.\frac{\partial \theta}{\partial \eta}\right|_{(\xi,0,\eta)} = 0$$
$$\theta(1, \eta, \tau) = 0, \qquad \theta(\xi, 1, \tau) = 0. \tag{3}$$

Assume that $f = 1$ and that the plate is initially at zero temperature:

$$\theta(\xi, \eta, 0) = 0 \quad \text{for all } (\xi, \eta) \text{ in } \Omega. \tag{4}$$

To begin with, use the 4×4 mesh shown in Fig. 7.4.6(b) with $\Delta\tau = 10^{-3}$ and then obtain solutions for refined meshes.

Solution For the 4×4 uniform mesh, we have $NCV_x = NCV_y = 4$ and there are 36 nodes in the mesh. We use the single label notation for the nodes by writing a node in location (i, j) to be the nth node, where $n = i + (j - 1) * (NCV_x + 2)$. Thus, a node located in position (6, 2), for example, becomes node 12. With this notation, we identify the known boundary values to be

$$\theta_6 = \theta_{12} = \theta_{18} = \theta_{24} = \theta_{30} = \theta_{31} = \cdots = \theta_{36} = 0. \tag{5}$$

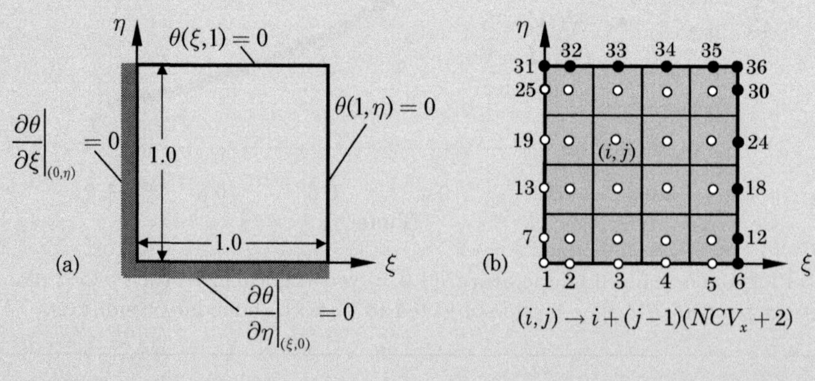

Fig. 7.4.6 (a) Domain, and boundary and initial conditions. (b) A 4×4 uniform mesh.

Thus there are 36 − 11 = 25 nodal unknowns. Using the explicit formulation,

$$\{\theta\}^{n+1} = \{\theta\}^n + \frac{\Delta\tau}{\Delta\xi\,\Delta\eta}\Big(\{F\} - [K]\{\theta\}^n\Big), \tag{6}$$

where $[K]$ is the coefficient matrix and $\{F\}$ is the source vector due to f (both are the same as those in the steady-state case), we march in time with $\{\theta\}^0 = \{0\}$. The solution obtained with the 4×4 mesh is found to be in significant error when compared with the FEM solutions obtained in **Example 5.5.2** with the 4×4 mesh and Crank–Nicolson (CN) or the forward difference (FD) scheme. With a mesh of 8×8 and $\Delta\tau = 10^{-3}$, the solution improves. However, the solution obtained with a 16×16 mesh with $\Delta\tau = 10^{-3}$ is found to be unstable. When $\Delta\tau = 10^{-4}$ is used, the 16×16 mesh gives nearly as accurate value of temperature as the FEM. These results are presented in Fig. 7.4.7. The inaccuracy of the FVM solutions may have to do with the assumption of $\dot{u}$ being constant in the control volume.

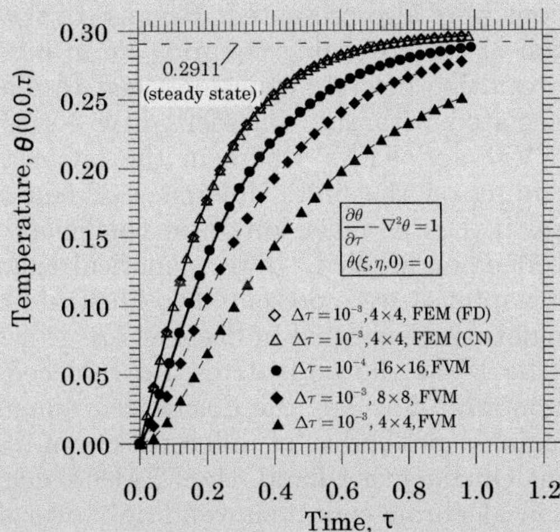

Fig. 7.4.7 Finite element and FVM solutions of the heat conduction problem in **Example 7.4.2** analyzed using various meshes and time increments.

7.5 Summary

This chapter was dedicated to the introduction of the FVM applied to differential equations governing one- and two-dimensional diffusion problems involving a single unknown. The FVM is introduced in this chapter for the one- and two-dimensional steady-state heat transfer.

The FVM is based on the idea of dividing the domain of the problem into a finite set of nonoverlapping subdomains, called *control volumes*. At the geometric center of each control volume a node (or grid point) is deployed, where the dependent variables are calculated. In addition, nodes at the boundary points are identified. Thus, in a one-dimensional problem using zero-thickness formulation, for NCV number of control volumes, there will be a total of $NCV + 2$ nodes, including two nodes associated with the boundaries. Then the spatial integral of the given differential equation over each control volume is set to zero (i.e., satisfy the global form of the energy balance equation). After carrying out the allowed spatial integration(s), the spatial derivatives are replaced by difference quotients derived from Taylor series approximations. The resulting algebraic equations involve the values of the dependent variable at the nodes of the control volume as well as those of the neighboring control volumes. Additional algebraic equations are derived for the boundary nodes using the boundary conditions.

At the end of the discretization, one will have exactly the same number of algebraic equations as there are unknown nodal values. Typically, resulting algebraic equations are in a *tridiagonal* form. A number of numerical examples, using the same problems as those considered in Chapter 4 by the FEM, were solved using the FVM and the numerical solutions obtained with the two methods were compared with the analytical solutions.

The finite volume formulations of initial value-boundary value problems in one and two dimensions were also presented. Because of the assumptions made during the satisfaction of the governing equation in an integral sense over the control volume, the resulting discrete equations lead to diagonal mass matrix (i.e., the matrix associated with the time-derivative term). This is a major difference between FVM and FEM, where in the latter the mass matrix is not diagonal. Depending on the finite difference scheme used, the resulting equations can be explicit (i.e., no inversion of the coefficient matrix to solve the final equations is needed) or implicit. Two numerical examples, the same as those considered in Chapter 4, were presented to illustrate the use of the finite volume explicit formulations developed in Section 7.4.

In summary, in the FVM the derivatives are replaced with *desired* (i.e., depends on the person who develops the discretized equations) Taylor series approximations, bringing a possibility of different sets of discretized equations for the same problem. On the other hand, the FEM uses element-wise approximations that result in algebraic equations valid only over an element. Consequently, one must "assemble" (i.e., put together) the algebraic equations from the various elements to form the global system of equations. This feature of the FEM can be seen both as an advantage and as a disadvantage. The advantage is largely computational in that the element calculations can be parallelized (and they are unique). The disadvantage is that the element structure creates derivative discontinuity across the element interfaces, and flux like variables (which involve the computation of the derivatives of the primary variable) are nonunique at the element interfaces, where only functions are made continuous (i.e., C^0 continuous approximations are used) across the element interfaces. In contrast, in the FVM the fluxes are computed at the interfaces of the control volumes, and therefore they are unique and continuous. Recently, a numerical method, termed *the dual mesh control domain method* (DMCDM), that brings several desirable features of both the FVM and FEM has been proposed by the first author (see Reddy et al. [60–62]). In the DMCDM, the domain is represented with a primal mesh of finite elements (defining the approximations used for the dependent variables) and a dual mesh that is superimposed on the primal mesh such that the nodes of the primal mesh are at the center of the dual mesh of control domains. Then the governing equation is satisfied in an integral sense over the control domain, as in the FVM, resulting directly in a set of global equations in terms of the nodal values of the primary variables.

Problems

One-Dimensional Problems

7.1 Show that the four algebraic relations associated with the second-order accurate finite volume formulation (FVM2) of the problem in **Example 7.2.1**, **Case 1**, for the mesh shown in Fig. 7.2.5(a) are:

$$\begin{aligned}-10.6667\,U_1 + 16\,U_2 - 5.3333\,U_3 &= 0.2413\\ -4\,U_2 + 8\,U_3 - 4\,U_4 &= 0.2308\\ -4\,U_3 + 8\,U_4 - 4\,U_5 &= 0.2012\\ -5.3333\,U_4 + 16\,U_5 - 10.6667\,U_6 &= 0.1689.\end{aligned}$$

7.2 Show that the three algebraic relations associated with the half control volume formulation (FVMH) of the problem in **Example 7.2.1** for the mesh shown in Fig. 7.2.5(b) are

$$\begin{aligned}-4\,U_1 + 8\,U_2 - 4\,U_3 &= 2.4034\\ -4\,U_2 + 8\,U_3 - 4\,U_4 &= 2.1768\\ -4\,U_3 + 8\,U_4 - 4\,U_5 &= 1.8150.\end{aligned}$$

7.3 Consider the following pair of first-order equations:

$$-\frac{dv}{dx} + cu - f = 0, \quad -\frac{du}{dx} + \frac{v}{a} = 0, \quad a \neq 0, \tag{1}$$

where u and v are the dependent unknowns. Develop the finite volume model of the pair of equations keeping both u and v as unknowns.

7.4 Solve **Problem 4.6** using the FVM.

7.5 Solve **Problem 4.7** using the FVM.

7.6 Solve **Problem 4.8** using the FVM.

Two-Dimensional Problems

7.7 Consider the PDE governing heat transfer in an axisymmetric geometry

$$-\frac{1}{r}\frac{\partial}{\partial r}\left(rk_{rr}\frac{\partial T}{\partial r}\right) - \frac{\partial}{\partial z}\left(k_{zz}\frac{\partial T}{\partial z}\right) = f(r,z), \tag{1}$$

where $(k_{rr},\, k_{zz})$ and f are the conductivities and internal heat generation per unit volume, respectively. Develop the finite volume model of Eq. (1).

7.8 Solve **Problem 4.12** using the FVM.

7.9 Solve **Problem 4.18** using the FVM.

8

The Finite Volume Method: Advection–Diffusion Problems

8.1 Introduction

General equations describing transport of momenta and energy by advection–diffusion was given in Chapter 2 (see, also, **Example 4.2.3**) and will not be repeated here. It is important to note that the entire finite volume formulation is based on local one-dimensional representation in each coordinate direction. The flux crossing a control volume surface is represented using a one-dimensional formulation. So the flux in the x-direction is completely represented in terms of x.

The steady-state advection–diffusion equation in one dimension with no source (or sink) term can be expressed in general form as

$$\rho u \frac{d\phi}{dx} - \frac{d}{dx}\left(\Gamma \frac{d\phi}{dx}\right) = 0, \tag{8.1.1}$$

where ρ is the density of the fluid, u is the velocity in the flow direction, and $\Gamma(= k/c_p)$ is the diffusion coefficient. The variable ϕ is transported by advection–diffusion and the one-dimensional advection–diffusion flux is given by

$$J = \rho u \phi - \Gamma \frac{d\phi}{dx}. \tag{8.1.2}$$

We can express Eqs. (8.1.1)–(8.1.2) in dimensionless form by assuming that Γ and ρu are nonzero constants. Let us introduce the following dimensionless quantities

$$x^* = \frac{x}{\delta x}, \quad P = \frac{\rho\, u\, \delta x}{\Gamma}, \tag{8.1.3}$$

where δx is the grid characteristic length and P is known as the cell *Péclet number*[1] and x^* is the dimensionless distance. Then Eq. (8.1.1) becomes

$$P \frac{d\phi}{dx^*} - \frac{d^2\phi}{dx^{*2}} = \frac{dJ^*}{dx^*} = 0, \tag{8.1.4}$$

where the dimensionless advection–diffusion flux is

$$J^* = P\,\phi - \frac{d\phi}{dx^*}. \tag{8.1.5}$$

[1]The reader is asked to be aware of the fact that P used for denoting a generic point (only appears in figures showing control volumes) is different from P used here for Péclet number; we could have adopted Pe, as we did in Chapter 4, for Péclet number but it would become very cumbersome to add subscripts to Pe for east face (e), west face (w), and so on.

8.2 Discretization of the Advection–Diffusion Flux

8.2.1 General Discussion

As in Chapter 7, we divide the one-dimensional domain into a set of nonoverlapping control volumes, each centered around a node, as shown in Fig. 8.2.1. Consider two adjacent nodes i and $i+1$ on either side of an interface as shown in Fig. 8.2.2. In the present two-node formulation (i.e., one node on either side of an interface) the general variable ϕ at the interface is represented as a weighted average of ϕ at the adjacent nodes, namely, ϕ_i and ϕ_{i+1}, and the derivative of $d\phi/dx^*$ at the interface is replaced in terms of the difference $\phi_{i+1} - \phi_i$. Thus, we have

$$\phi = \alpha\phi_i + (1-\alpha)\phi_{i+1}, \quad \frac{d\phi}{dx^*} = \beta(\phi_{i+1} - \phi_i), \tag{8.2.1}$$

where α and β are known parameters, which may be taken as functions of the cell Peclet number ($P = \rho U \delta x/\Gamma$). Then J^* becomes

$$\begin{aligned} J^* &= P[\alpha\phi_i + (1-\alpha)\phi_{i+1}] - \beta(\phi_{i+1} - \phi_i) \\ &= (\alpha P + \beta)\phi_i - (\alpha P + \beta - P)\phi_{i+1} \end{aligned} \tag{8.2.2}$$

or

$$J^* = B\phi_i - A\phi_{i+1}, \quad B(P) = \alpha P + \beta, \quad A(P) = \alpha P + \beta - P. \tag{8.2.3}$$

Symbols B and A are chosen to indicate "behind" and "ahead" of the interface, respectively (see Patankar [7]). From Eq. (8.2.3) it is evident that

$$B(P) = A(P) + P. \tag{8.2.4}$$

If the flow direction is reversed, we have

$$B \to A, \quad A \to B$$

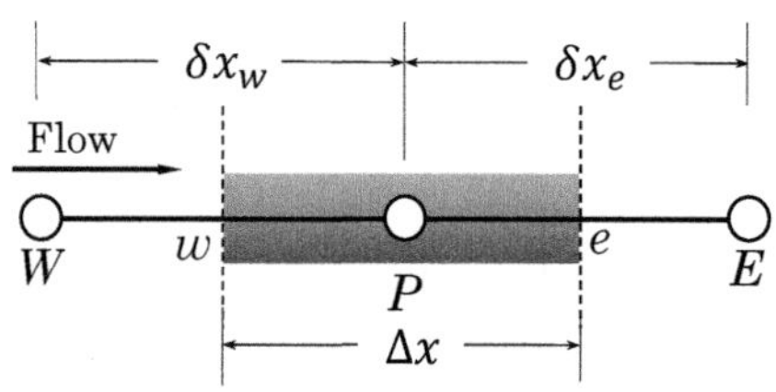

Fig. 8.2.1 One-dimensional internal control volume.

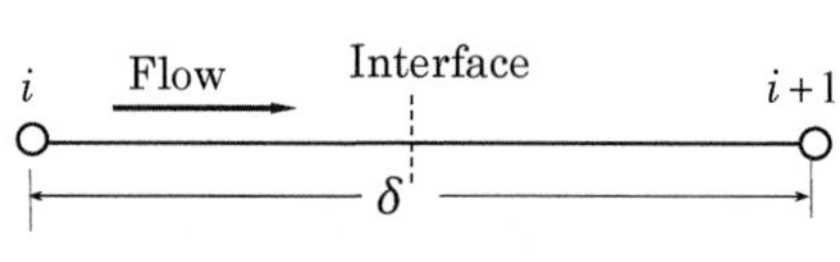

Fig. 8.2.2 Nodes i and $i+1$.

and

$$B(-P) = A(P), \quad A(-P) = B(P). \tag{8.2.5}$$

Equation (8.2.3) is regarded as a general two-node formulation for a nondimensional flux J^* across an interface.

Next, consider the expression

$$A(P) = A(|P|) + [\![-P, 0]\!], \tag{8.2.6}$$

where the operator $[\![\cdot,\cdot]\!]$ denotes the greater of the two quantities. For $P > 0$, Eq. (8.2.6) reduces to

$$A(P) = A(|P|). \tag{8.2.7}$$

For $P < 0$, let $P = -P^*$, where $P^* > 0$. Then

$$A(-P^*) = A(|-P^*|) + [\![P^*, 0]\!], \quad B(P^*) = A(P^*) + P^*, \tag{8.2.8}$$

which is true from Eq. (8.2.4). Thus, Eq. (8.2.3) is verified to be true for both positive and negative values of P. Similarly, for both positive and negative values of P, we have

$$B(P) = A(|P|) + [\![P, 0]\!]. \tag{8.2.9}$$

8.2.2 A General Two-Node Formulation

We can express the one-dimensional incompressible flow in terms of the advection–diffusion flux J as (where $F = \rho u$ is assumed to be a constant)

$$\frac{dJ}{dx} = 0. \tag{8.2.10}$$

The FVM formulation of Eq. (8.2.10) is presented next.

Integrating Eq. (8.2.10) over the control volume shown in Fig. 8.2.1, we obtain

$$J_e - J_w = 0, \tag{8.2.11}$$

where e and w denote the east and west interface locations respectively. Noting that $J_e = (\Gamma_e/\delta x_e)J_e^*$ and $J_w = (\Gamma_w/\delta x_w)J_w^*$ and using Eq. (8.2.3) [i.e., $J_e^* = B_e\phi_P - A_e\phi_E$ and $J_w^* = B_w\phi_W - A_w\phi_P$] in Eq. (8.2.11), we obtain

$$\frac{\Gamma_w}{\delta x_w}B_w\phi_W - \frac{\Gamma_w}{\delta x_w}A_w\phi_P = \frac{\Gamma_e}{\delta x_e}B_e\phi_P - \frac{\Gamma_e}{\delta x_e}A_e\phi_E. \tag{8.2.12}$$

Now, let $\Gamma/\delta x = D$. Here D signifies the *diffusion strength*. Then Eq. (8.2.12) can be written as

$$D_wB_w\phi_W - D_wA_w\phi_P = D_eB_e\phi_P - D_eA_e\phi_E \tag{8.2.13}$$

or

$$a_P\phi_P = a_W\phi_W + a_E\phi_E, \tag{8.2.14a}$$

where $B_w = B(P_w)$, $A_w = A(P_w)$, and so on, and

$$a_W = D_wB_w, \quad a_E = D_eA_e, \quad a_P = D_wA_w + D_eB_e. \tag{8.2.14b}$$

Making use of the relation from Eqs. (8.2.6) and (8.2.9) in Eq. (8.2.14a), we obtain

$$\begin{aligned} a_W &= D_w\Big[A(|P_w|) + [\![P_w, 0]\!]\Big] \\ a_E &= D_e\Big[A(|P_e|) + [\![-P_e, 0]\!]\Big] \\ a_P &= D_w\Big[A(|P_w|) + [\![-P_w, 0]\!]\Big] + D_e\Big[A(|P_e|) + [\![P_e, 0]\!]\Big]. \end{aligned} \tag{8.2.15}$$

Defining the *strength of advection (or flow strength)*, F as

$$F = PD = \frac{\rho u \delta x}{\Gamma}\frac{\Gamma}{\delta x} = \rho u, \tag{8.2.16}$$

Eq. (8.2.15) can be written as

$$\begin{aligned} a_W &= D_w A(|P_w|) + [\![F_w, 0]\!] \\ a_E &= D_e A(|P_e|) + [\![-F_e, 0]\!] \\ a_P &= D_w A(|P_w|) + D_e A(|P_e|) + [\![-F_w, 0]\!] + [\![F_e, 0]\!]. \end{aligned} \tag{8.2.17}$$

Combining the three relations in Eq. (8.2.17), one obtains

$$a_P = a_W + a_E + (F_e - F_w).$$

Since $F_e = F_w$ because of continuity, we have

$$a_P = a_W + a_E, \tag{8.2.18a}$$

where

$$\begin{aligned} a_W &= D_w A(|P_w|) + [\![F_w, 0]\!] \\ a_E &= D_e A(|P_e|) + [\![-F_e, 0]\!] \\ a_P &= a_W + a_E + (F_e - F_w) = a_W + a_E. \end{aligned} \tag{8.2.18b}$$

The function $A(|P|)$ depends on the specific formulation (e.g., central difference scheme or upwind scheme). The next step is to identify $A(|P|)$ for each specific formulation.

8.2.3 Central Difference Approximation

Here we develop the central difference approximation of the steady one-dimensional advection–diffusion equation for a variable ϕ, without a source or a sink term ($S = 0$),

$$\frac{d}{dx}(\rho u \phi) = \frac{d}{dx}\left(\Gamma \frac{d\phi}{dx}\right). \tag{8.2.19}$$

Integrating Eq. (8.2.19) over a control volume (see Fig. 8.2.1), we obtain

$$(\rho u \phi)_e - (\rho u \phi)_w = \left(\Gamma \frac{d\phi}{dx}\right)_e - \left(\Gamma \frac{d\phi}{dx}\right)_w. \tag{8.2.20}$$

The values of ϕ at the interface are replaced by the average ϕ values at the adjacent nodes on either side of the interface [i.e., $\phi_e = (\phi_P + \phi_E)/2$] and the derivative is replaced using a piecewise-linear profile [i.e., $(d\phi/dx)_e = (\phi_E - \phi_P)/\delta x_e$]. Thus, we have

$$F_e \frac{(\phi_P + \phi_E)}{2} - F_w \frac{(\phi_W + \phi_P)}{2} = \frac{\Gamma_e}{\delta x_e}(\phi_E - \phi_P) - \frac{\Gamma_w}{\delta x_w}(\phi_P - \phi_W). \quad (8.2.21)$$

Replacing $\Gamma/\delta x$ with D and rearranging the above equation, we arrive at

$$\phi_P \left(\frac{F_e}{2} + D_e - \frac{F_w}{2} + D_w \right) = \phi_W \left(\frac{F_w}{2} + D_w \right) + \phi_E \left(-\frac{F_e}{2} + D_e \right). \quad (8.2.22)$$

Thus, the flux variable ϕ at node P (i.e., ϕ_P) can is related to the adjacent nodes W and E by the following equation:

$$a_P \phi_P = a_W \phi_W + a_E \phi_E. \quad (8.2.23a)$$

Setting the source term S_P to zero,

$$\begin{aligned} a_W &= D_w + \frac{F_w}{2} \\ a_E &= D_e - \frac{F_e}{2} \\ a_P &= D_w - \frac{F_w}{2} + D_e + \frac{F_e}{2}. \end{aligned} \quad (8.2.23b)$$

Substituting for D_w and D_e from the first two relations in Eq. (8.2.23b) into the expression for a_P in Eq. (8.2.23b), we arrive at

$$a_P = a_W + a_E + F_e - F_w = a_W + a_E \quad (8.2.24)$$

because, by continuity, we have $F_e - F_w = 0$.

Comparing the a_E and a_W from Eq. (8.2.23b) with those in Eq. (8.2.18b) ($F = DP$), for the central difference scheme we have (here P is the Péclet number)

$$A(|P|) = 1 - 0.5|P|. \quad (8.2.25)$$

In the central difference scheme, depending on the values of F and D, there is a possibility that the coefficients will be negative, leading to physically unrealistic solutions. Also, it might violate the Scarborough criterion [36]. For example, if $F_e > 2D_e$ ($P_e > 2$, where P_e is the Péclet number at the east face) then a_E will be negative. When diffusion is zero, $a_P = 0$, which prevents one from using a point-by-point solution technique (see Patankar [7] and Minkowycz [37]).

8.2.4 Upwind Scheme

The motivation for the upwind scheme is that for $P_e > 2$ (P_e denotes the Péclet number at the east face) the upstream node influence on the node under consideration is much more than the downstream node. In the central difference scheme the effect of the downstream variable is considered; this could lead to physically unrealistic results. One could reduce the grid size such that $Pe < 2$ to make the central difference scheme valid, but indiscriminate refinement of the grid size will lead to excessive round-off error.

In the upwind scheme the flux variable ϕ at the interface is set equal to the value of ϕ at the immediate upstream node; that is, if $F_e > 0$, $\phi_e = \phi_P$, and if $F_e < 0$, $\phi_e = \phi_E$. This conditional statement can be written as

$$F_e\phi_e = \phi_P \llbracket F_e, 0 \rrbracket - \phi_E \llbracket -F_e, 0 \rrbracket. \tag{8.2.26}$$

Similarly,

$$F_w\phi_w = \phi_W \llbracket F_w, 0 \rrbracket - \phi_P \llbracket -F_w, 0 \rrbracket. \tag{8.2.27}$$

The derivative is replaced by using a piecewise-linear profile as in the case of the central difference scheme. Equation (8.2.20) can be expressed in terms of $F = \rho u$ [also see Eq. (8.2.21)]:

$$F_e\phi_e - F_w\phi_w = D_e(\phi_E - \phi_P) - D_w(\phi_P - \phi_W). \tag{8.2.28}$$

Using Eqs. (8.2.26) and (8.2.27) in Eq. (8.2.28), we obtain

$$\begin{aligned}\phi_P \llbracket F_e, 0 \rrbracket - \phi_E \llbracket -F_e, 0 \rrbracket + \phi_P \llbracket -F_w, 0 \rrbracket - \phi_W \llbracket F_w, 0 \rrbracket& \\ = D_e(\phi_E - \phi_P) - D_w(\phi_P - \phi_W).&\end{aligned} \tag{8.2.29}$$

Rearranging, we obtain

$$\begin{aligned}\phi_P \left(\llbracket F_e, 0 \rrbracket + \llbracket -F_w, 0 \rrbracket + D_e + D_w\right) = \phi_W \left(D_w + \llbracket F_w, 0 \rrbracket\right)& \\ + \phi_E \left(D_e + \llbracket -F_e, 0 \rrbracket\right)&\end{aligned} \tag{8.2.30}$$

or

$$a_P\phi_P = a_W\phi_W + a_E\phi_E, \tag{8.2.31a}$$

where

$$\begin{aligned}a_W &= D_w + \llbracket F_w, 0 \rrbracket \\ a_E &= D_e + \llbracket -F_e, 0 \rrbracket \\ a_P &= D_w + De + \llbracket F_e, 0 \rrbracket + \llbracket -F_w, 0 \rrbracket.\end{aligned} \tag{8.2.31b}$$

From Eq. (8.2.31b) we can write

$$a_P = a_W + a_E + (F_e - F_w) = a_W + a_E. \tag{8.2.32}$$

Comparing the expressions for a_W and a_E in Eq. (8.2.31b) with those in Eq. (8.2.18b) we find that $A(|P|) = 1$ for the upwind scheme.

8.2.5 Exponential Scheme

The exponential scheme is developed using the exact solution to the one-dimensional advection–diffusion equation over the domain, $0 < x < L$ [i.e., Eq. (8.2.19)] subjected to the following boundary conditions:

$$\phi(0) = \phi_0; \quad \phi(L) = \phi_L. \tag{8.2.33}$$

The exact solution is of the form

$$\phi(x) = C_1 + C_2 e^{Px/L},$$

where P is the Péclet number and C_1 and C_2 are constants of integration that are determined using the boundary conditions in Eq. (8.2.33). We obtain

$$\phi(x) = \frac{e^P \phi_0 - \phi_L + e^{Px/L}(\phi_L - \phi_0)}{e^P - 1}. \tag{8.2.34}$$

After rearranging (subtract ϕ_0 from both sides), we can write the solution $\phi(x)$ as

$$\frac{\phi(x) - \phi_0}{\phi_L - \phi_0} = \frac{e^{Px/L} - 1}{e^P - 1}. \tag{8.2.35}$$

It can easily be verified that $\phi(x)$ in Eq. (8.2.35) satisfies the boundary conditions $\phi(0) = \phi_0$ and $\phi(L) = \phi_L$. When $P = 0$ (i.e., $\rho u \to 0$), advection is absent and Eq. (8.2.19) becomes a pure diffusion equation. Letting $P \to 0$ in Eq. (8.2.35) and using L'Hôpital's rule, we obtain $\phi(x)$ as a linear function:

$$\begin{aligned}\frac{\phi(x) - \phi_0}{\phi_L - \phi_0} &= \lim_{P \to 0} \frac{e^{Px/L} - 1}{e^P - 1} \\ &= \frac{x}{L}.\end{aligned} \tag{8.2.36}$$

On the other hand, if $P \to \infty$ (i.e., $\Gamma \to 0$), diffusion is absent and Eq. (8.2.19) becomes a pure advection equation. It can be shown (again using L'Hôpital's rule) that $\phi(x) \to \phi_L$ as $P \to -\infty$ and $\phi \to \phi_0$ as $P \to \infty$. When $|P|$ is large, $d\phi/dx$ is zero at $x = L/2$. However, the upwind scheme always calculates the diffusion term from a linear profile $\phi(x) \sim x$ and thus over-estimates diffusion at large values of $|P|$ (see Patankar [7]). Hence, a discretization scheme based on the exact solution would not have these pitfalls. A discretization scheme based on the exact solution is called the *exponential scheme*.

Integrating the one-dimensional steady advection–diffusion equation over the control volume, we obtain [see Eq. (8.2.20)]

$$F_e \phi_e - F_w \phi_w = \left(\Gamma \frac{d\phi}{dx}\right)_e - \left(\Gamma \frac{d\phi}{dx}\right)_w. \tag{8.2.37}$$

Making use of the exact solution in Eq. (8.2.35), but applied to the domains between $W - P$ and $P - E$ (see Fig. 8.2.1) with end values (ϕ_W, ϕ_P) and

(ϕ_P, ϕ_E), respectively, we can write

$$\phi_e = \phi_P + (\phi_E - \phi_P)\frac{e^{P_e x_e/\delta x_e} - 1}{e^{P_e} - 1} \tag{8.2.38a}$$

$$\phi_w = \phi_W + (\phi_P - \phi_W)\frac{e^{P_w x_w/\delta x_w} - 1}{e^{P_w} - 1} \tag{8.2.38b}$$

$$\left.\frac{d\phi}{dx}\right|_e = \frac{P_e}{\delta x_e}\frac{e^{P_e x_e/\delta x_e}(\phi_E - \phi_P)}{e^{P_e} - 1} \tag{8.2.38c}$$

$$\left.\frac{d\phi}{dx}\right|_w = \frac{P_w}{\delta x_w}\frac{e^{P_w x_w/\delta x_w}(\phi_P - \phi_W)}{e^{P_w} - 1}. \tag{8.2.38d}$$

Substituting Eqs. (8.2.38a)–(8.2.38d) into Eq. (8.2.37) and simplifying,

$$\phi_P\left(F_e + \frac{F_e}{e^{P_e} - 1} + \frac{F_w}{e^{P_w} - 1}\right) = \phi_W\left(F_w + \frac{F_w}{e^{P_w} - 1}\right) + \phi_E\left(\frac{F_e}{e^{P_e} - 1}\right). \tag{8.2.39}$$

Therefore, we can write the above equation in the following familiar form:

$$a_P\phi_P = a_W\phi_W + a_E\phi_E, \tag{8.2.40a}$$

where

$$\begin{aligned} a_W &= F_w + \frac{F_w}{e^{P_w} - 1} \\ a_E &= \frac{F_e}{e^{P_e} - 1} \\ a_P &= F_e + \frac{F_e}{e^{P_e} - 1} + \frac{F_w}{e^{P_w} - 1}. \end{aligned} \tag{8.2.40b}$$

From Eq. (8.2.40b) it is clear that

$$a_P = a_W + a_E + (F_e - F_w) = a_W + a_E. \tag{8.2.41}$$

The next step is to identify $A(|P|)$ for the exponential scheme. Consider the first equation in Eq. (8.2.40b):

$$\begin{aligned} a_W &= F_w + \frac{F_w}{e^{P_w} - 1} \\ &= \frac{F_w}{e^{P_w} - 1} + [\![F_w, 0]\!] - [\![-F_w, 0]\!] \\ &= D_w\left(\frac{P_w}{e^{P_w} - 1} - [\![-P_w, 0]\!]\right) + [\![F_w, 0]\!]. \end{aligned} \tag{8.2.42}$$

The function $A(|P|)$ for the exponential scheme will be determined by comparing Eq. (8.2.42) with a_W in Eq. (8.2.18b). To start with

$$A(P_w) = \frac{P_w}{e^{P_w} - 1} - [\![-P_w, 0]\!]. \tag{8.2.43}$$

We have the following relations:

$$\text{For } P_w > 0: \quad A(P_w) = \frac{P_w}{e^{P_w} - 1} \tag{8.2.44a}$$

$$\text{For } P_w > 0: \quad A(|P_w|) = \frac{|P_w|}{e^{|P_w|} - 1} \tag{8.2.44b}$$

$$\text{For } P_w < 0: \quad A(P_w) = \frac{P_w}{e^{P_w} - 1} + P_w = \frac{P_w e^{P_w}}{e^{P_w} - 1} \tag{8.2.44c}$$

$$\text{For } P_w < 0: \quad A(|P_w|) = \frac{-|P_w| e^{-|P_w|}}{e^{-|P_w|} - 1} = \frac{|P_w|}{e^{|P_w|} - 1}. \tag{8.2.44d}$$

Thus, it is verified for both positive and negative values of P_w that we have

$$A(|P|) = \frac{|P|}{e^{|P|} - 1}. \tag{8.2.45}$$

8.2.6 Hybrid Scheme

The hybrid scheme was originally developed by Spalding [63]. This method can be best understood by examining a_E/D_e for the exponential scheme. Consider a plot of $A(|P|) = a_E/D_e$ as a function of $|P|$ as shown in Fig. 8.2.3 (see Patankar [7]). One can make the following observations: $a_E/D_e \to 0$ as $P_e \to \infty$, $a_E/D_e \to -P_e$ as $P_e \to -\infty$, and $a_E/D_e = 1 - P_e/2$ is tangent to the exponential scheme at $P_e = 0$. As $P_e \to \infty$, $a_E/D_e \to 0$ implies that at large values of the Péclet number Pe the downstream neighbor has no influence on the upstream neighbor, which is consistent with the physics. From these observations one can replace the exponential scheme with three different straight lines, that is,

$$\begin{aligned} P_e < -2: &\quad \frac{a_E}{D_e} = -P_e \\ -2 \le P_e \le 2: &\quad \frac{a_E}{D_e} = 1 - \frac{P_e}{2} \\ P_e > 2: &\quad \frac{a_E}{D_e} = 0. \end{aligned} \tag{8.2.46}$$

The above three straight lines can be combined into an operator form as

$$a_E = D_e \left[\!\left[-P_e, 1 - \frac{P_e}{2}, 0 \right]\!\right]. \tag{8.2.47}$$

or

$$a_E = \left[\!\left[-F_e, D_e - \frac{F_e}{2}, 0 \right]\!\right]. \tag{8.2.48}$$

In the range $-2 \le Pe \le 2$ the hybrid scheme reduces to the central difference scheme.

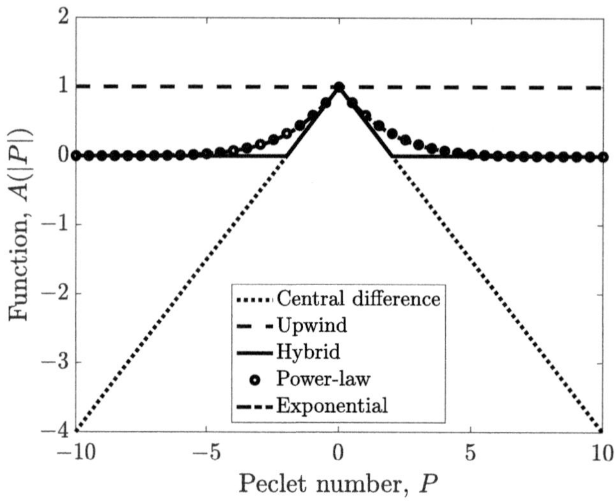

Fig. 8.2.3 Variation of function $A(|P|)$ with Peclet number, Pe.

The next step is to compare the expression in Eq. (8.2.48) with a_E for a general formulation to identify $A(|P|)$ for the hybrid scheme. In order to do that, add and subtract $[\![-F_e, 0]\!]$ from Eq. (8.2.48) and obtain

$$\begin{aligned} a_E &= D_e[\![-P_e, (1-\frac{P_e}{2}), 0]\!] - [\![-F_e, 0]\!] + [\![-F_e, 0]\!] \\ &= D_e\{[\![-P_e, (1-\frac{P_e}{2}), 0]\!] - [\![-P_e, 0]\!]\} + [\![-F_e, 0]\!]. \end{aligned} \tag{8.2.49}$$

Comparing Eq. (8.2.49) with a_E in Eq. (8.2.18b) we get

$$A(P_e) = [\![-P_e, (1-\frac{P_e}{2}), 0]\!] - [\![-P_e, 0]\!]. \tag{8.2.50}$$

Therefore, we have

$$A(P_e) = 1 - \frac{P_e}{2}, \qquad \text{for } 0 \leq |P_e| \leq 2, \tag{8.2.51a}$$

$$= 0, \qquad \text{for } |P_e| > 2. \tag{8.2.51b}$$

Thus, we have

$$A(|P_e|) = [\![0, 1 - 0.5|P_e|]\!] \tag{8.2.52}$$

for the hybrid scheme.

8.2.7 Power–Law Scheme

It can be seen from Fig. 8.2.3 that the maximum difference between the hybrid and exponential scheme is at $P_e = \pm 2$. Switching to the upwind scheme for $P_e > 2$ is premature. A better approximation to the exponential scheme was proposed by Patankar [7]. In the power–law scheme the exponential scheme is approximated by the following curves,

$$P_e < -10: \quad a_E/D_e = -P_e \tag{8.2.53a}$$

$$-10 \le P_e \le 0: \quad a_E/D_e = (1+0.1P_e)^5 - P_e \tag{8.2.53b}$$

$$0 \le P_e \le 10: \quad a_E/D_e = (1-0.1P_e)^5 \tag{8.2.53c}$$

$$P_e > 10: \quad a_E/D_e = 0. \tag{8.2.53d}$$

Combining the above expressions in the operator form one can write

$$a_E = D_e\left[\!\left[0, \left(1 - 0.1\frac{|F_e|}{D_e}\right)^5\right]\!\right] + \left[\!\left[-F_e, 0\right]\!\right]. \tag{8.2.54}$$

Comparing Eq. (8.2.54) with the a_E for a generalized formulation one gets, for the power–law, the following

$$A(|P|) = \left[\!\left[0, \left(1 - 0.1|P|\right)^5\right]\!\right]. \tag{8.2.55}$$

The power–law is the preferred scheme as it provides an economic alternative to the exponential scheme. Furthermore, the exponential scheme has the same limitation of formulation as other schemes as being locally one-dimensional. The functions for various two-node formulations are presented in Table 8.2.1. In solving two- and three-dimensional flow and heat transfer problems one will use the power–law scheme several times. A two-dimensional laminar flow and heat transfer problem with Dirichlet boundary conditions and a mesh of 5×5 control volumes will require the calculation of 100 a-coefficients for U, 100 a-coefficients for V, and 100 a-coefficients for temperature. Therefore, for this case, the power law will be used 300 times per iteration. So, for a mesh of 100×100 control volumes, the number will be $100 * 100 * 4 * 3 = 120\,000$ per iteration. Thus, to optimize CPU resources, the readers are encouraged to multiply P five times as opposed to using the exponential function.

Table 8.2.1 $A(|P|)$ Function for two-node formulations

Scheme	$A(\|P\|)$ function
Central difference	$1 - 0.5\|P\|$
Upwind	1
Hybrid	$[\![0, 1 - 0.5\|P\|]\!]$
Exponential	$\frac{\|P\|}{e^{\|P\|}-1}$
Power–law	$[\![0, (1 - 0.1\|P\|)^5]\!]$

8.2.8 A Three-Node Formulation: QUICK Scheme

The Quadratic Upstream Interpolation for Convective Kinetics scheme, called the QUICK scheme, was developed by Leonard [64]. The motivation for using the QUICK scheme is its superior accuracy compared with the upwind scheme. In this scheme the interface values are represented by using a quadratic interpolation function. The quadratic function connects three adjacent nodes. Two of them are upstream of the interface and one is downstream. Figure 8.2.4 shows a one-dimensional control volume wherein three adjacent nodes (two upstream and one downstream) are available to represent ϕ values at the interfaces w and e.

Consider a quadratic function

$$\phi(x) = c_1 + c_2 x + C_3 x^2. \tag{8.2.56}$$

If the distance between two adjacent nodes is δx and the origin is placed at node W then

$$\begin{aligned} x &= 0, & \phi &= \phi_W \\ x &= \delta X, & \phi &= \phi_P \\ x &= -\delta x, & \phi &= \phi_{WW}. \end{aligned} \tag{8.2.57}$$

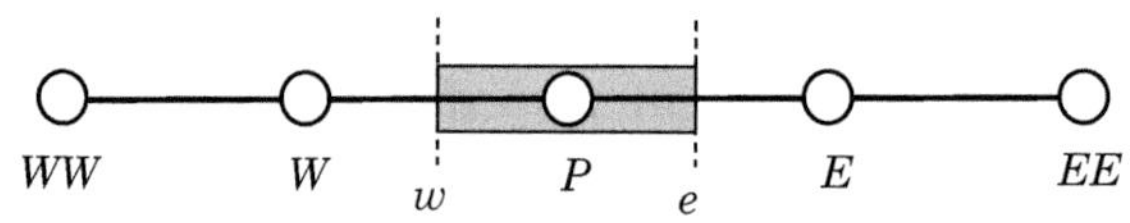

Fig. 8.2.4 One-dimensional control volume for a three-node formulation.

Using conditions from Eq. (8.2.57) in Eq. (8.2.56), one can solve for c_1, c_2, and c_3 and obtain

$$\phi(x) = \phi_W + \frac{(\phi_P - \phi_{WW})}{2\delta x}x + \frac{(\phi_P + \phi_{WW} - 2\phi_W)}{2\delta x^2}x^2. \tag{8.2.58}$$

The flux ϕ_W at the western face of the control volume, when $U_w > 0$ can be obtained by substituting $x = \delta x/2$,

$$\phi_w = \tfrac{3}{8}\phi_P + \tfrac{6}{8}\phi_W - \tfrac{1}{8}\phi_{WW}. \tag{8.2.59}$$

Similarly, when $U_e > 0$,

$$\phi_e = \tfrac{3}{8}\phi_E + \tfrac{6}{8}\phi_P - \tfrac{1}{8}\phi_W. \tag{8.2.60}$$

Integrating one-dimensional steady state advection–diffusion equation over the control volume results in

$$F_e\phi_e - F_w\phi_w = \left(k\frac{d\phi}{dx}\right)_e - \left(k\frac{d\phi}{dx}\right)_w. \tag{8.2.61}$$

The diffusion (derivative) terms in Eq. (8.2.61) at w and e will be obtained by differentiating the quadratic expression for ϕ [see Eq. (8.2.56)], and substituting the appropriate values of x at w and e:

$$\left(k\frac{d\phi}{dx}\right)_w = D_w(\phi_P - \phi_W). \tag{8.2.62a}$$

$$\left(k\frac{d\phi}{dx}\right)_e = D_e(\phi_E - \phi_P). \tag{8.2.62b}$$

where $D_w = k_w/\delta x_w$ and $D_e = k_e/\delta x_e$. Substituting for flux (ϕ) and derivative values at the interface into Eq. (8.2.61).

$$F_e\left(\tfrac{3}{8}\phi_E + \tfrac{6}{8}\phi_P - \tfrac{1}{8}\phi_W\right) - F_w\left(\tfrac{3}{8}\phi_P + \tfrac{6}{8}\phi_W - \tfrac{1}{8}\phi_{WW}\right) = D_e(\phi_E - \phi_P) - D_w(\phi_P - \phi_W). \quad (8.2.63)$$

Rearranging Eq. (8.2.63),

$$a_P\phi_P = a_W\,\phi_W + a_E\,\phi_E + a_{WW}\,\phi_{WW}, \quad (8.2.64a)$$

where

$$a_W = D_w + \tfrac{6}{8}F_w + \tfrac{1}{8}F_e \quad (8.2.64b)$$

$$a_E = D_e - \tfrac{3}{8}F_e \quad (8.2.64c)$$

$$a_{WW} = -\tfrac{1}{8}F_w \quad (8.2.64d)$$

$$a_P = a_W + a_E + a_{WW} + (F_e - F_w). \quad (8.2.64e)$$

When $F_w < 0$ and $F_e < 0$, the equation for ϕ_P becomes

$$a_P\phi_P = a_W\,\phi_W + a_E\,\phi_E + a_{EE}\,\phi_{EE}, \quad (8.2.65a)$$

where

$$a_W = D_w + \tfrac{3}{8}F_w \quad (8.2.65b)$$

$$a_E = D_e - \tfrac{6}{8}F_e - \tfrac{1}{8}F_w \quad (8.2.65c)$$

$$a_{EE} = \tfrac{1}{8}F_e \quad (8.2.65d)$$

$$a_P = a_W + a_E + a_{EE} + (F_e - F_w). \quad (8.2.65e)$$

For both negative and positive values of F_w and F_e, Eqs. (8.2.64a) and (8.2.65a) can be combined as

$$a_P\phi_P = a_W\,\phi_W + a_E\,\phi_E + a_{WW}\,\phi_{WW} + a_{EE}\,\phi_{EE} + (F_e - F_w), \quad (8.2.66a)$$

where

$$a_W = D_w + [\![F_w, 0]\!]\left(\tfrac{6}{8} + \tfrac{1}{8}\frac{F_e}{F_w}\right) + [\![-F_w, 0]\!]\left(-\tfrac{3}{8}\right) \quad (8.2.66b)$$

$$a_{WW} = [\![F_w, 0]\!]\left(-\tfrac{1}{8}\right) \quad (8.2.66c)$$

$$a_E = D_e + [\![-F_e, 0]\!]\left(\tfrac{6}{8} + \tfrac{1}{8}\frac{F_w}{F_e}\right) + [\![F_e, 0]\!]\left(-\tfrac{3}{8}\right) \quad (8.2.66d)$$

$$a_{EE} = [\![-F_e, 0]\!]\left(-\tfrac{1}{8}\right). \quad (8.2.66e)$$

As stated earlier, the QUICK scheme has superior accuracy compared with the upwind scheme. In terms of the Taylor series the truncation error is third order, which is the main advantage. The disadvantage is that the immediate coefficients a_W and a_E are not guaranteed to be positive. The far coefficients a_{WW} and a_{EE} are negative and this may lead to unstable solutions for certain values of the cell Peclet number, P. Further details can be found elsewhere (see, e.g., Leonard [64] and Versteeg and Malalasekera [8]).

8.2.9 A Numerical Example

Here we consider the same advection–diffusion problem as that considered in **Example 4.2.3**.

Example 8.2.1

Solve the normalized advection–diffusion equation

$$P\frac{d\phi}{dx} - \frac{d^2\phi}{dx^2} = 0, \quad 0 < x < 1 \tag{1}$$

with the following boundary conditions:

$$\phi(0) = 1, \quad \phi(1) = 0. \tag{2}$$

Use a uniform mesh of five control volumes. Compare the numerical solutions obtained with various schemes with the exact solution for Péclet number of $P = 30$. The exact solution of the problem is given by [set $L = 1$, $\phi_0 = 1$, and $\phi_L = 0$ in Eq. (8.2.35)]

$$\phi(x) = 1 - \frac{e^{Px} - 1}{e^P - 1} = \frac{1 - e^{P(x-1)}}{1 - e^{-P}}. \tag{3}$$

Solution The mesh of five control volumes is shown in Fig. 8.2.5. The control volume size is $\delta_e = \delta_w = \Delta x = 0.2$. The parameters D_e and D_w are (with $F = P_e$ and $k = 1$),

$$D_e = \frac{k_e}{\delta x_e} = \frac{1}{\Delta x} = 0.5, \quad D_w = D_e. \tag{4}$$

For the central difference scheme, for example, the parameters a_E, a_W, and a_P become

$$a_E = D_e - 0.5P, \quad a_W = D_w + 0.5P, \quad a_P = D_e + D_w. \tag{5}$$

The a-coefficients for control volumes 1 and 5 are determined as follows:

$$\int_w^e \left[P\frac{dT}{dx} - \frac{d}{dx}\left(\frac{dT}{dx}\right)\right] dx = PT_e - PT_w - \left.\frac{dT}{dx}\right|_e + \left.\frac{dT}{dx}\right|_w$$
$$= P\frac{T_3 + T_2}{2} - PT_1 - \frac{T_3 - T_2}{\Delta x} + \frac{T_2 - T_1}{0.5\Delta x} \tag{6}$$

$$\int_w^e \left[P\frac{dT}{dx} - \frac{d}{dx}\left(\frac{dT}{dx}\right)\right] dx = -P\frac{T_6 + T_5}{2} + PT_7 - \frac{T_7 - T_6}{0.5\Delta x} + \frac{T_6 - T_5}{\Delta x}. \tag{7}$$

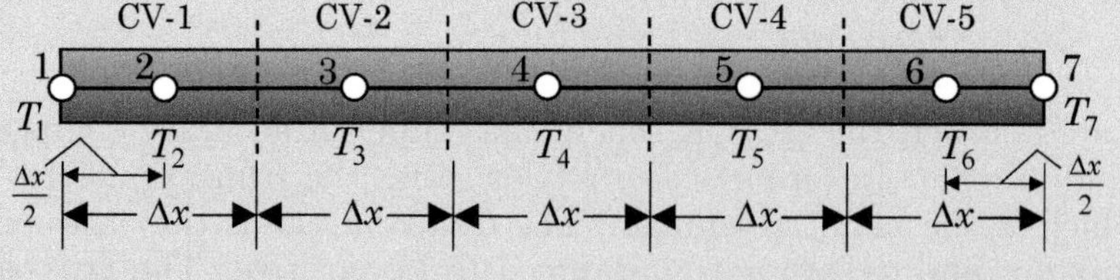

Fig. 8.2.5 Subdivision of the one-dimensional domain into a set of five control volumes (i.e., $NCV = 5$).

Table 8.2.2 contains numerical solutions obtained for the Péclet number of $P = 30$ with various schemes. The power–law scheme seems be the best when compared with the analytical solution. Table 8.2.3 contains a comparison between the upwind and central difference (CD) schemes for $P = 2$. Note that the finite element solution (using six linear finite elements to match the nodal locations of the FVM) gives exactly the same results as the central difference approximation.

Table 8.2.2 Comparison between different schemes ($P = 3$ and $NCV = 5$).

x	Exact	Upwind	CD	power–law	Hybrid	FEM
0.0	1.0000	1.0000	1.0000	1.0000	1.0000	1.0000
0.1	1.0000	0.9999	1.0909	1.0000	1.0000	1.0150
0.3	1.0000	0.9993	0.7273	1.0000	1.0000	0.9774
0.5	1.0000	0.9949	1.4545	1.0000	1.0000	1.0526
0.7	0.9999	0.9643	0.0000	0.9999	1.0000	0.9023
0.9	0.9502	0.7500	2.9091	0.9469	0.7500	1.2030
1.0	0.0000	0.0000	0.0000	0.0000	0.0000	0.0000
$e_{max}(\%)$	——	20.02	195.89	0.3264	20.02	26.604

Table 8.2.3 Comparison between exact solution, solutions of the upwind and central difference schemes, and the finite element solution for Péclet number $P = 2$.

x	Exact	Upwind	CD	FEM
0.0	1.0000	1.0000	1.0000	1.0000
0.1	0.9653	0.9559	0.9697	0.9661
0.3	0.8713	0.8499	0.8787	0.8730
0.5	0.7310	0.7017	0.7422	0.7333
0.7	0.5218	0.4941	0.5374	0.5238
0.9	0.2096	0.2034	0.2303	0.2095
1.0	0.0000	0.0000	0.0000	0.0000
$e_{max}(\%)$	——	2.9388	2.0691	0.3116

Figures 8.2.6 and 8.2.7 show the plots of the temperature versus x as predicted by various schemes and the FEM for $P = 30$. The power–law scheme performs the best, followed by the hybrid scheme. Figure 8.2.8 shows the plots of the temperature versus x as predicted by the central difference scheme, the upwind scheme, and the FEM for $P = 2$. For this case, the central difference scheme performs better than the upwind scheme.

8.3 Numerical Diffusion

In this section the concept of artificial viscosity or numerical diffusion will be introduced through an example. However, for detailed analysis readers are advised to refer to [65].

The effect of viscosity is to smooth the flow. Likewise, the numerical viscosity smooths the numerical solution. The source of numerical viscosity is the truncation error in the discretization equation. Eventually, we solve a set of discretization equations which implicitly contain truncation errors. These truncation error terms behave like additional viscosity terms.

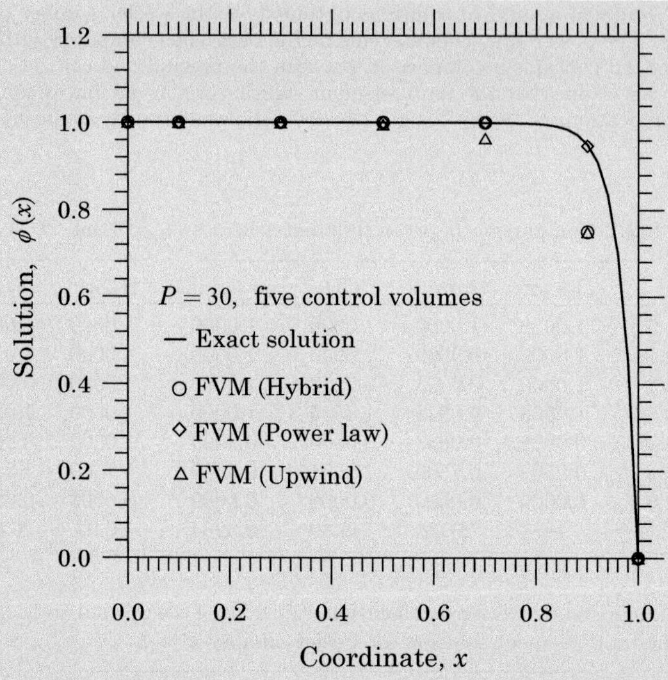

Fig. 8.2.6 Numerical solutions of the one-dimensional advection–diffusion equation obtained with the hybrid, power–law, and upwind schemes ($P = 30$).

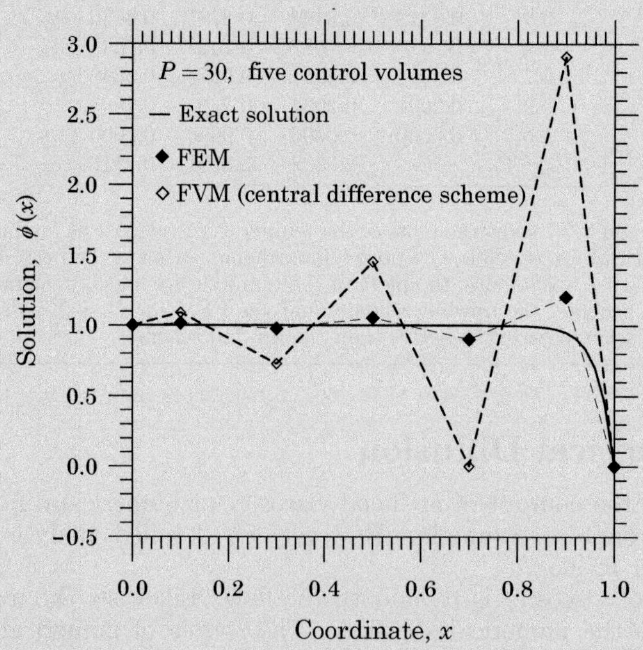

Fig. 8.2.7 Numerical solution of the one-dimensional advection–diffusion equation obtained with the central difference formula and the FEM ($P = 30$).

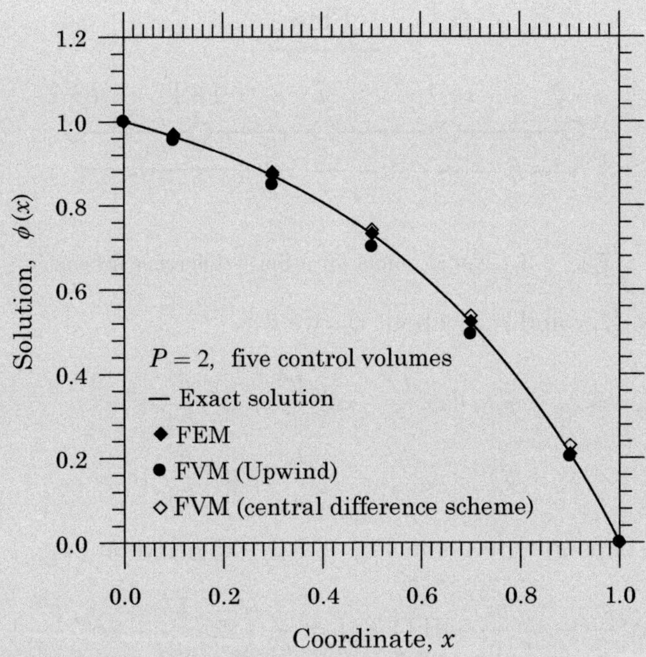

Fig. 8.2.8 Numerical solutions of the one-dimensional advection–diffusion equation with the upwind and central difference schemes and the FEM ($P = 2$).

The example we consider here is a one-dimensional advection–diffusion problem:

$$P\frac{d\phi}{dx} = \frac{d^2\phi}{dx^2} \tag{8.3.1a}$$

$$x = 0, \ \phi = 0 \tag{8.3.1b}$$

$$x = 1, \ \phi = 1, \tag{8.3.1c}$$

where, ϕ is a general nondimensional flux variable transported by advection–diffusion, P is the Péclet number, and x is nondimensional distance.

In the context of the two-node formulation we will represent the advection term (first derivative) by the *upwind difference* approximation and the diffusion term (second derivative) by a second-order central difference scheme. Consider the grid shown in Fig. 8.3.1 for the finite difference approximation and assume that $P > 0$ (flow is from left to right):

$$P\frac{\phi_i - \phi_{i-1}}{h} = \frac{\phi_{i+1} + \phi_{i-1} - 2\phi_i}{h^2}. \tag{8.3.2}$$

The next step is to use the Taylor series expansion of ϕ_{i+1} and ϕ_{i-1} about ϕ_i to recover the original differential equation and to identify the truncation error.

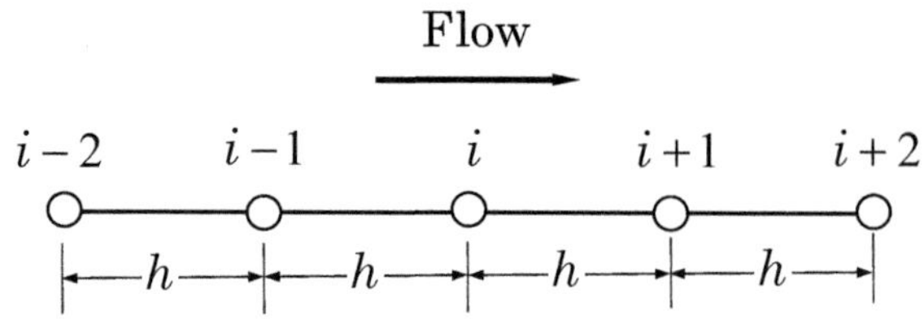

Fig. 8.3.1 Nodal points for a finite difference scheme.

Expanding ϕ_{i+1} and ϕ_{i-1} about ϕ_i, we get

$$\phi_{i+1} = \phi_i + \phi_i' h + \phi_i'' \frac{h^2}{2} + \phi_i''' \frac{h^3}{6} + \phi_i'''' \frac{h^4}{24} + \cdots \tag{8.3.3a}$$

$$\phi_{i-1} = \phi_i - \phi_i' h + \phi_i'' \frac{h^2}{2} - \phi_i''' \frac{h^3}{6} + \phi_i'''' \frac{h^4}{24} + \cdots . \tag{8.3.3b}$$

Replacing the expanded terms in Eq. (8.3.2), we arrive at

$$\frac{P}{h}\left(\phi_i' h - \phi_i'' \frac{h^2}{2} + \phi_i''' \frac{h^3}{6} + \text{H.O.T.}\right) = \frac{1}{h^2}\left(2\phi_i'' \frac{h^2}{2} + 2\phi_i'''' \frac{h^4}{24} + \text{H.O.T.}\right). \tag{8.3.3c}$$

Therefore, we obtain

$$P\frac{d\phi}{dx} = \frac{d^2\phi}{dx^2} + \frac{h}{2}\frac{d^2\phi}{dx^2} + \frac{h^2}{12}\frac{d^4\phi}{dx^4} + \text{H.O.T.} \tag{8.3.3d}$$

$$P\frac{d\phi}{dx} = \left(1 + \frac{h}{2}\right)\frac{d^2\phi}{dx^2} + O(h^2). \tag{8.3.3e}$$

If the truncation term were to be zero then we would have recovered the original ODE, implying that solving the finite difference approximation (discretization equation) is the same as the original ODE. But the finite difference approximation has introduced an additional diffusion-like term $[(h/2)(d^2\phi/dx^2)]$. This added diffusion term introduces additional viscosity, which artificially smooths or diffuses the numerical solution. Thus one is unable to capture sharp gradients in the solution process, for example capturing the shock front. There are several strategies reported in the literature to mitigate the effects of artificial diffusion. One obvious strategy is to add a term in the original ODE to counter the artificial diffusion term that arises due to numerical approximation. Furthermore, based on this example, one can always minimize the effect of artificial diffusion by keeping the value of the step size (h) as small as possible. Of course, one should be mindful of the round-off errors. As shown in the work of Rhie and Chow [66] (see, also, [9]), when body-fitted coordinates are used additional terms come into the finite difference or finite volume approximations. To mitigate this effect, effort should be made to align one of the grid lines along the main flow directions. Numerical schemes that are highly diffusive cannot be used for direct numerical simulation (DNS), where accurate solution of the Navier–Stokes equations are required. For this reason, higher-order finite difference schemes and spectral methods are preferred for DNS.

From the FVM perspective, a short discussion is in order. Recall the a_W coefficient for the central difference scheme,

$$a_W = D_w A(|P_w|) + [\![F_w, 0]\!], \quad A(|P_w|) = 1 - 0.5|P_w|. \tag{8.3.4}$$

For $F_w > 0$ and $P_w > 0$, we have

$$a_W = D_w \left(1 - \frac{P_w}{2}\right) + [\![F_w, 0]\!]. \tag{8.3.5}$$

For the upwind scheme $A(|P_w|) = 1$. So, the a_W-coefficient becomes,

$$a_W = D_w + [\![F_w, 0]\!] \tag{8.3.6a}$$

$$= D_w - \frac{F_w}{2} + \frac{F_w}{2} + [\![F_w, 0]\!] \tag{8.3.6b}$$

$$= D_w \left(1 - \frac{P_w}{2}\right) + \frac{F_w}{2} + [\![F_w, 0]\!]. \tag{8.3.6c}$$

Comparing Eqs. (8.3.5) and (8.3.6c) we can see that the diffusion term in the a_W-coefficient of the upwind scheme is augmented by $F_w/2$. In other words, the upwind scheme diffuses more than the central difference approximation. This is an issue that one needs to pay attention to when we have flows with large values of the Péclet number.

8.4 Steady Two-Dimensional Problems

Formulation of the two-dimensional advection–diffusion equation using the FVM is illustrated in this section by considering the two-dimensional laminar flow of a viscous fluid in a rectangular cavity of dimensions $a \times b$ (see Fig. 8.4.1).

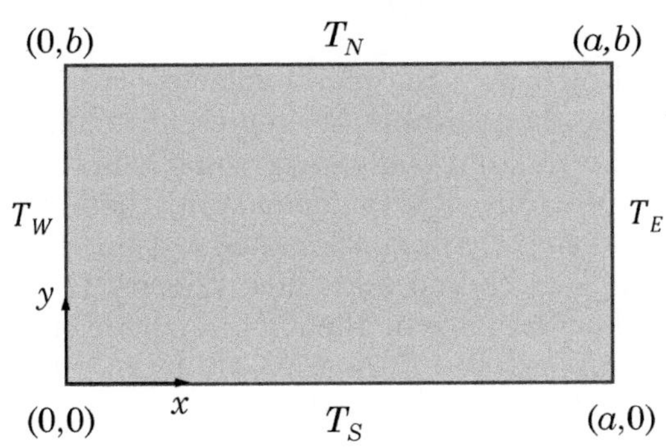

Fig. 8.4.1 A two-dimensional rectangular domain with specified temperatures on its boundary.

In the problem considered here, a viscous fluid is trapped in the rectangular cavity. The north, east, south, and west sides of the cavity are maintained at uniform temperatures T_N, T_E, T_S, and T_W respectively. Here, velocities are assumed to be known *a priori*, either through computation or through experimental measurements. So in this example we are interested only in a scalar variable (ϕ) transported by advection and diffusion.

The flux balance over a two-dimensional control volume (see Fig. 8.4.2) is given by ($T = \phi$)

$$\frac{\partial J_x}{\partial x} + \frac{\partial J_y}{\partial y} = S, \tag{8.4.1}$$

where

$$J_x = \rho u \phi - \Gamma \frac{\partial \phi}{\partial x}, \quad J_y = \rho v \phi - \Gamma \frac{\partial \phi}{\partial y}. \tag{8.4.2}$$

Here u and v are the x- and y-components of the velocity vector, k denotes the conductivity divided by c_p, the specific heat at constant pressure, and S is the source divided by c_p.

Integrating the flux balance equation, Eq. (8.4.1), over the control volume:

$$\int\int_{CV} \left(\frac{\partial J_x}{\partial x} + \frac{\partial J_y}{\partial y} - S \right) dxdy = 0. \tag{8.4.3}$$

As discussed earlier, a local one-dimensional formulation is used to represent the flux at each interface. Using a two-node formulation and integrating the above differential equation, we get

$$a_P \phi_P = a_W \phi_W + a_E \phi_E + a_S \phi_S + a_N \phi_N + S_P, \tag{8.4.4a}$$

where

$$a_W = D_w A(|P_w|) + [\![F_w, 0]\!] \tag{8.4.4b}$$
$$a_E = D_e A(|P_e|) + [\![-F_e, 0]\!] \tag{8.4.4c}$$
$$a_P = D_w A(|P_w|) + D_e A(|P_e|) + [\![-F_w, 0]\!] + [\![F_e, 0]\!] \tag{8.4.4d}$$

and S_P is the integral of S over the rectangular control volume. In the problem considered here, the flux variable (ϕ) represents temperature, $\phi = T$, and $\Gamma = k/c$, where k is the thermal conductivity and c is the specific heat of the fluid. Note that the function $A(|P|)$ depends on the expression one chooses to represent solution to the one-dimensional advection–diffusion equation. So, once the function $A(|P|)$ is chosen and since velocity (u and v) and pressure (p) fields are assumed to be known, then the a-coefficients can be calculated and that leaves the temperature (T) as the only unknown. Therefore, the discretization equation for temperature field becomes

$$a_P T_P = a_W T_W + a_E T_E + a_S T_S + a_N T_N + S_P. \tag{8.4.5}$$

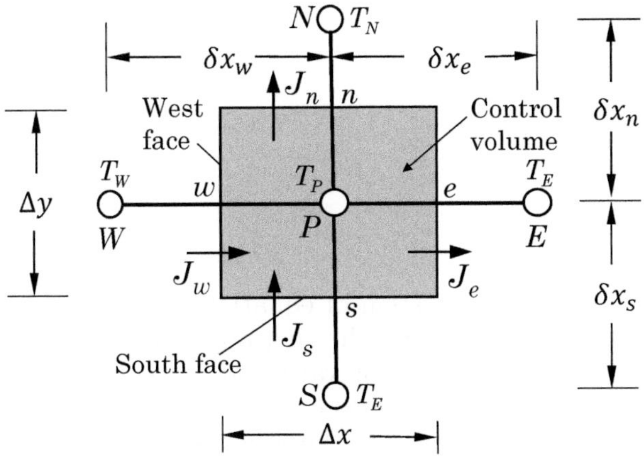

Fig. 8.4.2 A two-dimensional control volume with diffusion and control volume lengths Δx and Δy and interface fluxes (J_s, J_n, J_w, and J_e) on all four sides of the control volume.

For illustration purposes we consider a uniform mesh of 5×5 control volumes, as shown in Fig. 8.4.3. Numbering the mesh points as indicated in Fig. 8.4.3, we have $itmax = 7$ and $jtmax = 7$. Since values of the dependent variable (T) are specified along all four boundaries, there are $(itmax - 2) \times (jtmax - 2)$ unknowns. If the boundary conditions are specified as a function of derivatives and/or value of the unknowns, then the number of unknowns will increase. In this case, the equations for the unknowns can be written as follows.

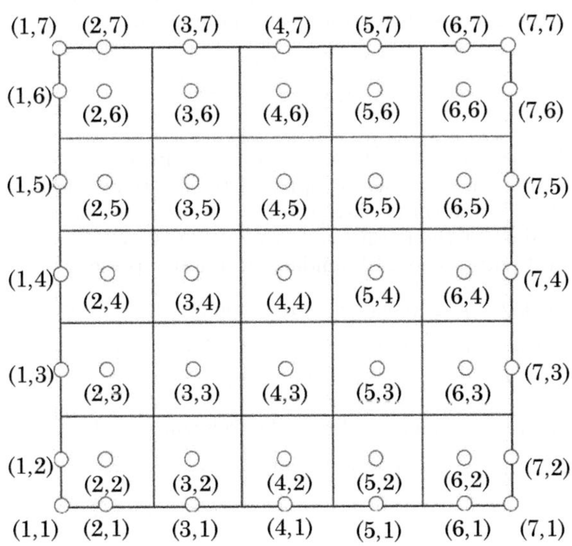

Fig. 8.4.3 A uniform mesh of 5 × 5 control volumes and numbering of the mesh points.

for loop $j = 2, jtmax - 1$
for loop $i = 2, itmax - 1$

$$\begin{aligned} a_P\, T(i,j) = a_W\, T(i-1,j) + a_E\, T(i+1,j) + a_S\, T(i,j-1) \\ + a_N\, T(i,j+1) + S_P \end{aligned}$$

end for loop i
end for loop j.

The above penta-diagonal system of equations has to be solved. If velocity and pressure fields are specified and the thermophysical properties remain constant, then the a-coefficients will not vary and are stationary. In such cases, the penta-diagonal system of equations can be solved by a direct method, such as the Gauss-elimination method. On the other hand, if one has to account for the nonlinearity through a-coefficients, then iterative procedures such as the *line-by-line method* or *strongly implicit procedure* are used.

Example 8.4.1

Solve the normalized convection–diffusion equation

$$P\left(\frac{\partial \phi}{\partial x} + \frac{\partial \phi}{\partial y}\right) - \left(\frac{\partial^2 \phi}{\partial x^2} + \frac{\partial^2 \phi}{\partial y^2}\right) = 0, \quad \text{in } 0 < x < 1,\ 0 < y < 1 \tag{1}$$

with the following boundary conditions:

$$\phi(x,0) = \frac{1 - e^{(x-1)P}}{1 - e^{-P}}, \quad \phi(0,y) = \frac{1 - e^{(y-1)P}}{1 - e^{-P}}, \quad \phi(x,1) = 0, \quad \phi(1,y) = 0. \tag{2}$$

Use 5×5 uniform mesh of control volumes. Compare the numerical solutions obtained with various schemes with the exact solution for Péclet number of $P = 10$. The exact solution of the problem is given by

$$\phi(x,y) = \frac{1 - e^{(x-1)P}}{1 - e^{-P}} \frac{1 - e^{(y-1)P}}{1 - e^{-P}}. \tag{3}$$

Solution The problem is one of Dirichlet boundary conditions (i.e., the dependent variable ϕ is specified on all boundary segments, as opposed to specifying the flux). Here we consider 5×5, 10×10, and 20×20 uniform meshes to solve the problem. In each case, we must input the known boundary values at the mesh points (nodes). For example, in the case of the 5×5 mesh, there are 49 total nodes, and 24 boundary nodes are specified as follows:

$$\begin{aligned} &T_1 = 1.0,\ T_2 = 0.99992,\ T_3 = 0.99913,\ T_4 = 0.99331,\ T_5 = 0.95026,\ T_6 = 0.63215, \\ &T_7 = 0.0,\ T_8 = 0.99992,\ T_{15} = 0.99913,\ T_{22} = 0.99331,\ T_{29} = 0.95026,\ T_{36} = 0.63215, \\ &T_{14} = T_{21} = T_{28} = T_{35} = T_{36} = T_{37} = \cdots = T_{49} = 0. \end{aligned} \tag{4}$$

Using the central difference scheme, we obtain the following results for $P = 10$:

$$\begin{aligned} &T_{16} = 0.9995,\ T_{17} = 1.0019,\ T_{18} = 0.9946,\ T_{19} = 1.0435,\ T_{20} = 0.6730, \\ &T_{23} = 0.9957,\ T_{24} = 0.9946,\ T_{25} = .9709,\ T_{26} = 1.0422,\ T_{27} = 0.6861. \end{aligned} \tag{5}$$

The exact solution is

$$\begin{aligned} &T(0.1,0.3) = 0.9991,\ T(0.3,0.3) = 0.9983,\ T(0.5,0.3) = 0.9925,\ T(0.7,0.3) = 0.9494 \\ &T(0.9,0.3) = 0.6316,\ T(0.1,0.5) = 0.9932,\ T(0.3,0.5) = 0.9925,\ T(0.5,0.5) = 0.9867 \\ &T(0.7,0.5) = 0.9439,\ T(0.9,0.5) = 0.6279. \end{aligned} \tag{6}$$

Clearly, the central difference scheme has oscillations. Figure 8.4.4 contains results obtained using 10×10 and 20×20 meshes. The central difference scheme exhibits significant numerical convection. Tables 8.4.1–8.4.3 contain numerical results obtained with 5×5 mesh using central difference, upwind, and power–law schemes, respectively.

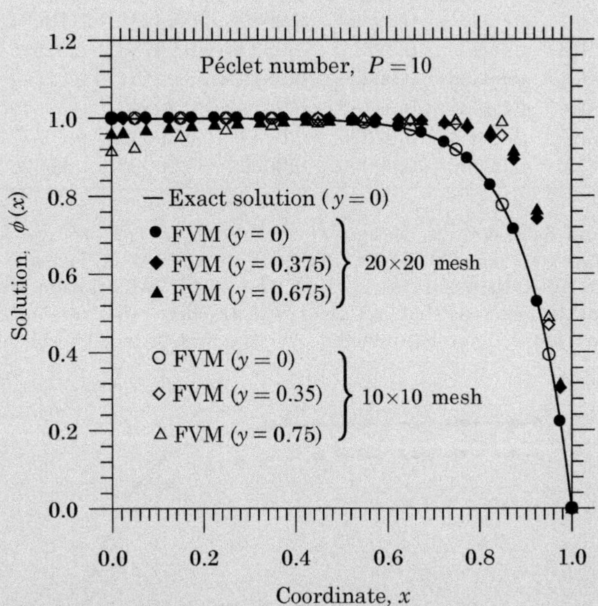

Fig. 8.4.4 Numerical results obtained with the central difference scheme for the advection–diffusion problem described by Eqs. (1) and (2).

Table 8.4.1 Solution obtained using the central difference scheme ($P = 10$).

x/y	0.0	0.1	0.3	0.5	0.7	0.9	1.0
0.0	1.00000	0.99992	0.99913	0.99331	0.95026	0.63215	0.00000
0.1	0.99992	0.99992	0.99945	0.99576	0.96846	0.63889	0.00000
0.3	0.99913	0.99940	0.99942	0.99759	0.98303	0.64877	0.00000
0.5	0.99331	0.99534	0.99738	0.99749	0.99026	0.65561	0.00000
0.7	0.95026	0.96528	0.98133	0.98941	0.98983	0.65818	0.00000
0.9	0.63215	0.63702	0.64734	0.65470	0.65781	0.43867	0.00000
1.0	0.00000	0.00000	0.00000	0.00000	0.00000	0.00000	0.00000

Table 8.4.2 Solution obtained using the upwind scheme ($P = 10$)

x/y	0.0	0.1	0.3	0.5	0.7	0.9	1.0
0.0	1.00000	0.99992	0.99913	0.99331	0.95026	0.63215	0.00000
0.1	0.99992	0.99914	0.99600	0.98193	0.91140	0.58045	0.00000
0.3	0.99913	0.99600	0.98818	0.96469	0.87538	0.54168	0.00000
0.5	0.99331	0.98193	0.96469	0.93180	0.83313	0.50761	0.00000
0.7	0.95026	0.91140	0.87538	0.83313	0.73587	0.44410	0.00000
0.9	0.63215	0.58045	0.54168	0.50761	0.44410	0.26646	0.00000
1.0	0.00000	0.00000	0.00000	0.00000	0.00000	0.00000	0.00000

Table 8.4.3 Solution obtained using the power–law scheme ($P = 10$).

x/y	0.0	0.1	0.3	0.5	0.7	0.9	1.0
0.0	1.00000	0.99992	0.99913	0.99331	0.95026	0.63215	0.00000
0.1	0.99992	0.99984	0.99903	0.99306	0.94931	0.63084	0.00000
0.3	0.99913	0.99903	0.99817	0.99206	0.94783	0.62936	0.00000
0.5	0.99331	0.99306	0.99206	0.98583	0.94154	0.62492	0.00000
0.7	0.95026	0.94931	0.94783	0.94154	0.89896	0.59649	0.00000
0.9	0.63215	0.63084	0.62936	0.62492	0.59649	0.39572	0.00000
1.0	0.00000	0.00000	0.00000	0.00000	0.00000	0.00000	0.00000

Figures 8.4.5 and 8.4.6 contain plots of the solution $\phi(x, y_0)$ versus x for different values of y_0 (i.e., $y_0 = 0.35$ and $y_0 = 0.75$) as predicted by the power–law and upwind schemes, respectively, with a 10×10 mesh. It is clear that the FVM solution with the power–law scheme is very accurate, whereas that obtained with the first order upwind scheme is in error. Therefore, the power law scheme is preferred over the first-order upwind scheme.

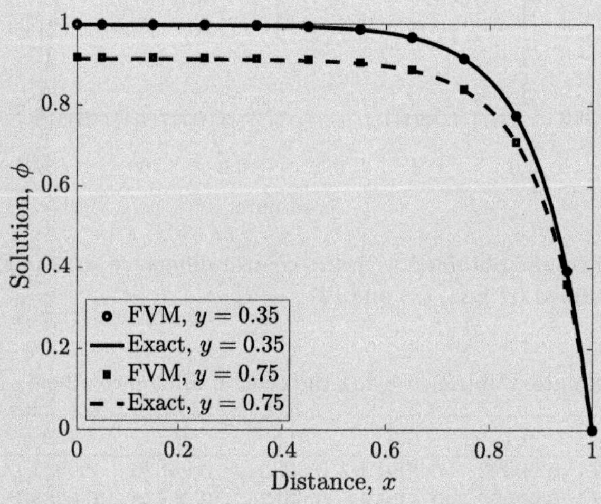

Fig. 8.4.5 Numerical results obtained for $\phi(x)$ as a function of the coordinate x with the power–law scheme for the convection–diffusion problem described by Eq. (1) of **Example 8.4.1** The FVM solution with the 10×10 mesh is compared with the exact solution.

8.5 Summary

In this chapter, the finite volume formulations of steady-state, one-dimensional and two-dimensional advection–diffusion problems were presented. The model equations used are those from diffusion heat transfer in the presence of flow. The flow variables like velocities are assumed to be known and remain constant along with the thermophysical properties. Various schemes such the central difference, upwind, hybrid, and power–law schemes have been discussed in the context of solving advection–diffusion problems.

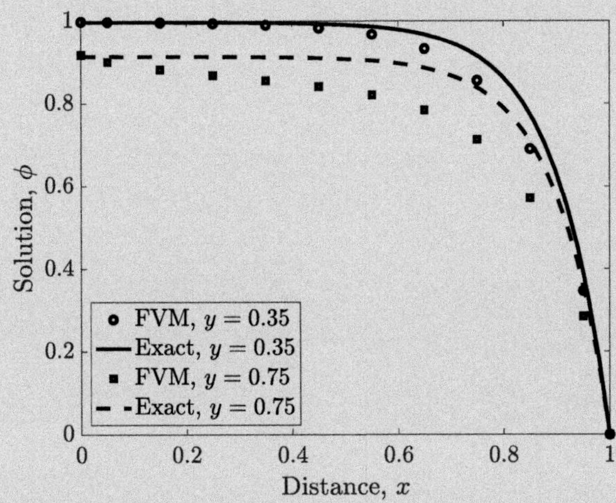

Fig. 8.4.6 Numerical results obtained for $\phi(x)$ as a function of the coordinate x with the upwind scheme for the advection–diffusion problem described by Eq. (1) of **Example 8.4.1**. The FVM solution with the 10×10 mesh is compared with the exact solution.

9

Finite Volume Methods for Viscous Incompressible Flows

9.1 Governing Equations

In this chapter, we will focus on solving the PDEs governing laminar flows of viscous incompressible fluids using the FVM (this chapter is a counterpart of Chapter 6 on FEM, where velocity–pressure and penalty finite element models of two dimensional flows of viscous incompressible fluids were presented) . These equations are expressed in terms of the primitive variable, namely, the velocity field and the pressure. To begin with, we will consider isothermal flows (flows without the presence of the temperature effect), and demonstrate the use of the FVM for two-dimensional laminar flows of viscous incompressible fluids. Then cases of non-isothermal flows with both forced convection and natural convection will be considered in the sequel.

The equations governing laminar, non-isothermal, three-dimensional flows of viscous incompressible fluids are provided by the physical principles, namely, conservation of mass, balance of linear and angular momenta, and balance of energy (see Section 2.4 for a summary of the equations). The governing equations can be expressed in terms of the velocity field ($\mathbf{v}$) and pressure (p) as:

$$\text{continuity:} \qquad \boldsymbol{\nabla}\cdot\mathbf{v}=0 \tag{9.1.1}$$

$$\text{momentum:} \quad \underbrace{\rho\frac{\partial \mathbf{v}}{\partial t}}_{\text{Unsteady term}} + \underbrace{\rho\mathbf{v}\cdot(\boldsymbol{\nabla}\mathbf{v})}_{\text{Advection}} = \underbrace{-\boldsymbol{\nabla}p}_{\text{Pressure drop}} + \underbrace{\boldsymbol{\nabla}\cdot(\mu\boldsymbol{\nabla}\mathbf{v})}_{\text{Viscous diffusion}} + \underbrace{S_v}_{\text{Source term}} \tag{9.1.2}$$

$$\text{energy:} \quad \underbrace{\rho c\frac{\partial T}{\partial t}}_{\text{Unsteady term}} + \underbrace{\mathbf{v}.(\boldsymbol{\nabla}T)}_{\text{Advection}} = \underbrace{\boldsymbol{\nabla}\cdot(-k\boldsymbol{\nabla}T)}_{\text{Diffusion}} + \underbrace{S_T}_{\text{Source term}} \; . \tag{9.1.3}$$

Assuming that thermophysical properties (i.e., ρ, c, and μ) of the fluid remain constant, the vector forms of governing equations are simplified to describe two-dimensional laminar steady flows. The velocity components are denoted here as $v_1 = u$ and $v_2 = v$. Then Eqs. (9.1.1)–(9.1.3) become (after rearranging terms and making use of the of the incompressibility condition $\nabla \cdot \mathbf{v} = 0$):

$$\text{continuity:} \qquad \frac{\partial u}{\partial x} + \frac{\partial v}{\partial y} = 0 \tag{9.1.4}$$

$$x\text{-momentum:} \qquad \rho \left(u\frac{\partial u}{\partial x} + v\frac{\partial u}{\partial y} \right) = -\frac{\partial p}{\partial x} + \mu \left(\frac{\partial^2 u}{\partial x^2} + \frac{\partial^2 u}{\partial y^2} \right) + S_u \tag{9.1.5}$$

$$y\text{-momentum:} \qquad \rho \left(u\frac{\partial v}{\partial x} + v\frac{\partial v}{\partial y} \right) = -\frac{\partial p}{\partial y} + \mu \left(\frac{\partial^2 v}{\partial x^2} + \frac{\partial^2 v}{\partial y^2} \right) + S_v \tag{9.1.6}$$

$$\text{energy:} \qquad \rho c_P \left(u\frac{\partial T}{\partial x} + v\frac{\partial T}{\partial y} \right) = \frac{\partial}{\partial x}\left(-k\frac{\partial T}{\partial x} \right) + \frac{\partial}{\partial y}\left(-k\frac{\partial T}{\partial y} \right) + S_T. \tag{9.1.7}$$

The boundary conditions for two-dimensional flows consist of specifying the velocity components or associated tractions and temperature or heat flux normal to the boundary:

$$\text{specify: } \mathbf{v} \text{ or } \mathbf{t} \equiv \hat{\mathbf{n}} \cdot \boldsymbol{\sigma}; \quad \text{specify: } T \text{ or } q_n \equiv \hat{\mathbf{n}} \cdot \mathbf{q}, \tag{9.1.8}$$

where $\hat{\mathbf{n}}$ is the unit vector normal to the boundary, $\boldsymbol{\sigma}$ is the Cauchy stress tensor, $\mathbf{t}$ is the stress vector, and $\mathbf{q}$ is the heat flux vector (with q_n being the flux normal to the boundary). In this chapter we will primarily focus on the solution strategies of the momentum equations as the energy equation can be solved using the methods described in Chapter 8.

Some observations on the governing equations are in order. In these equations, (x, y) are the independent coordinates and $(u,\ v,\ p)$ are the dependent variables (unknowns). Thus, there are three coupled differential equations in three unknowns. Mathematically, the continuity equation is hyperbolic and the two momentum equations are elliptic, which will have certain implications on the propagation of information. We note that the continuity equation does not contain the pressure variable, p, while it appears in the momentum equations as a gradient. Clearly, the boundary conditions do not involve specifying p by itself; it is a part of the tractions (t_x, t_y). Thus, for a given flow system, both velocity and the pressure gradient cannot be independently specified. Also, one can determine the pressure only within an arbitrary constant. In other words, for a given flow system, the momentum flux is transported by advection–diffusion phenomena, while the pressure is a wave phenomenon. To determine the absolute pressure, one has to specify the pressure at some point of the domain with respect to which other pressure values are measured.

The momentum equations are *owned* by the velocities (u, v); one may think of p as the owner of the continuity equation but it does not appear in the continuity equation. Since pressure manifests as a relative variable it may be convenient to eliminate pressure and solve for the velocity field first, as was done in Chapter 6 using the penalty function method. As an alternative to the penalty function method to eliminate the pressure, one can use the vorticity–stream function approach, which is discussed in Section 9.6.

9.2 The Velocity–Pressure Formulation

9.2.1 Introduction

Here we present the finite volume formulation to solve the continuity and momentum equations in two dimensions, Eqs. (9.1.4)–(9.1.6). As discussed in the previous chapters, the first step is to discretize the domain, which involves dividing the domain into a set of nonoverlapping, node-centered, control volumes, and systematically numbering them. The approximation of the first derivatives such as $\partial p/\partial x$, $\partial p/\partial y$, $\partial u/\partial x$, $\partial u/\partial y$, and so on, pose some unique challenges, as discussed by Patankar [7].

A wavy pressure field is made to look like a uniform pressure field and continuity is satisfied between alternate nodes both of which are unacceptable. Therefore, a staggered-grid approach is adopted, wherein velocity and pressure fields are stored at staggered locations as shown in Fig. 9.2.1.

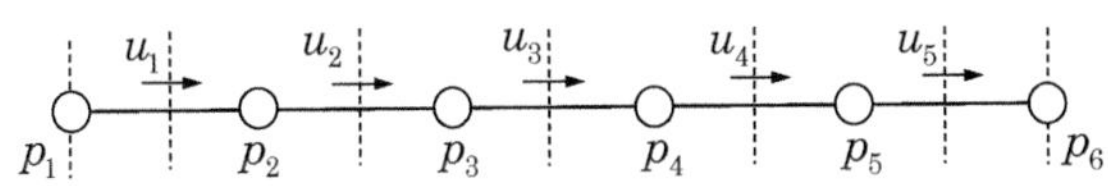

Fig. 9.2.1 Pressure and velocity in staggered positions for a one-dimensional flow field.

Here we acknowledge that the staggered-grid approach is not the only way to overcome challenges associated with representing first derivatives in numerical form. There are other strategies such as *momentum interpolation* or collocated-grid approach [66]. We will discuss both the staggered-grid approach and the collocated-grid approach in this chapter. In the staggered-grid approach, velocity and pressure variables are calculated at staggered locations as shown in the 5×5 control volume system in Fig. 9.2.2. Here we have used the notation

$$ipcv = jpcv = 5, \ ipmax = ipcv + 2, \ jpmax = jpcv + 2,$$

$$iumax = ipcv + 1, \ jumax = jpcv + 2,$$

$$ivmax = ipcv + 2, \ jvmax = jpcv + 1.$$

Assuming that all control volumes are of uniform size:

$$\Delta x = L/ipcv, \ \Delta y = L/jpcv.$$

Note that the diffusion lengths (δx and δy) have to be calculated as the distance between two adjacent nodes. The diffusion lengths (δx) for the transport of the x-momentum flux is the distance between two adjacent u-velocity nodes. Similarly, diffusion lengths (δy) for the transport of the y-momentum flux is the distance between two adjacent v-velocity nodes.

In Fig. 9.2.2, the locations of unknowns u, v, and p in the staggered-grid framework are identified. In the three figures shown in Fig. 9.2.3, locations where pressure, u-velocity, and v-velocity are calculated are specified. These three figures show p-, u-, and v- control volumes, respectively.

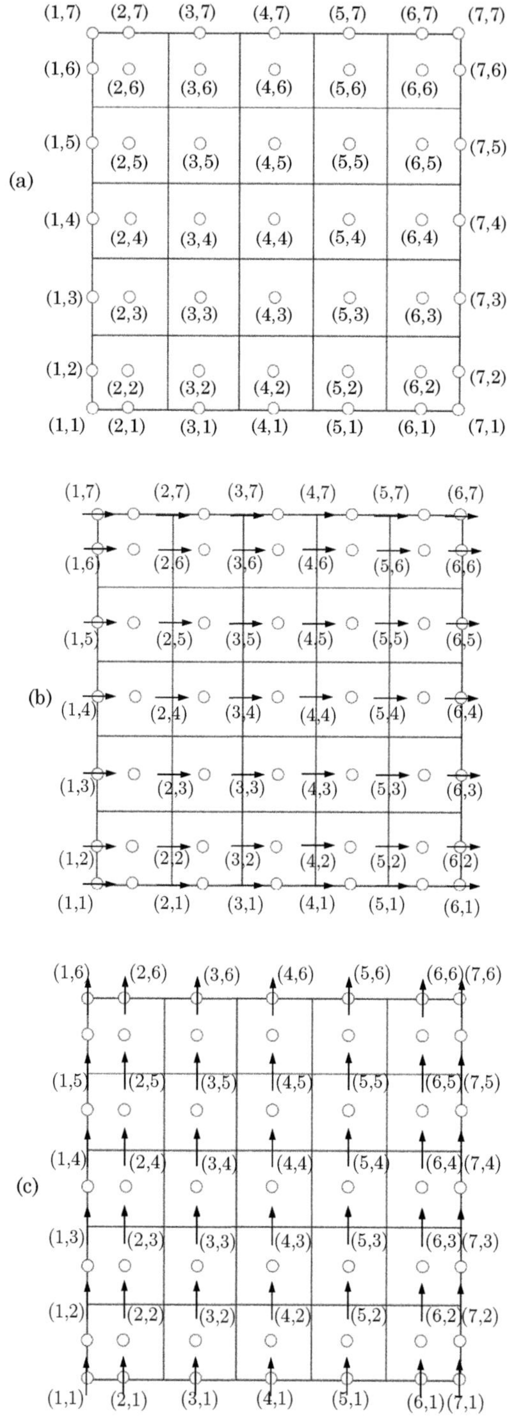

Fig. 9.2.2 (a) Pressure, (b) u-velocity and (c) v-velocity nodal positions in a 5×5 control volume system for a staggered-grid approach.

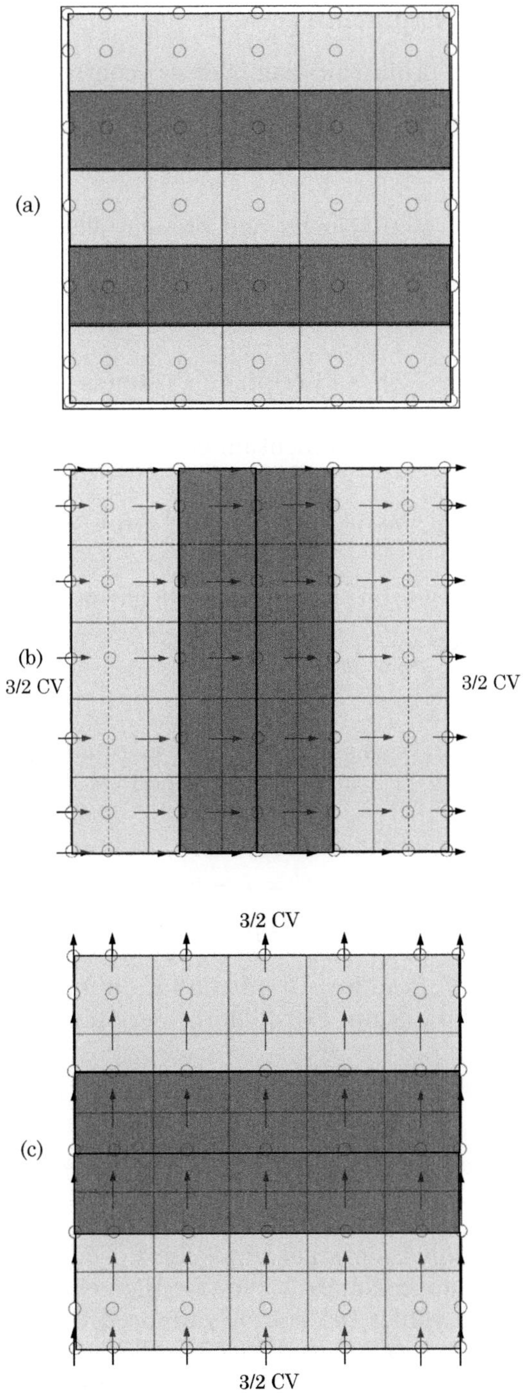

Fig. 9.2.3 (a) Pressure, (b) u-velocity and (c) v-velocity control volumes (denoted CV) in a 5×5 control volume system for a staggered-grid approach; CV stands for control volume.

9.2.2 Discretized Equations

Integrating the x-momentum equation over a u-control volume we obtain:

$$a^u_{iu,ju} u_{iu,ju} = \sum_{nb} a^u_{nb} u_{nb} + A_u \left(p_{iu,ju} - p_{iu+1,ju}\right) + b^u_{iu,ju}. \tag{9.2.1}$$

We will start with a guessed velocity and pressure fields u^*, v^*, p^*:

$$a^u_{iu,ju} u^*_{iu,ju} = \sum_{nb} a^u_{nb} u^*_{nb} + A_u \left(p^*_{iu,ju} - p^*_{iu+1,ju}\right) + b^u_{iu,ju}. \tag{9.2.2}$$

In Eqs. (9.2.1) and (9.2.2), subscript "nb" stands for neighborhood points (W, E, S, N) and $A_u = \Delta y \times 1$.

Similarly, by integrating y-momentum equation over a v-control volume we get

$$a^v_{iv,jv} v_{iv,jv} = \sum_{nb} a^v_{nb} v_{nb} + A_v \left(p_{iv,jv} - p_{iv,jv+1}\right) + b^v_{iv,jv}. \tag{9.2.3}$$

So the discretized equation for approximate or guessed values of u^*, v^*, p^* is:

$$a^v_{iv,jv} v^*_{iv,jv} = \sum_{nb} a^v_{nb} v^*_{nb} + A_v \left(p^*_{iv,jv} - p^*_{iv,jv+1}\right) + b^v_{iv,jv}. \tag{9.2.4}$$

In Eqs. (9.2.4) and (9.2.3), $A_v = \Delta x \times 1$. If u, v, and p are actual solutions, then the corrections to u^*, v^*, and p^* are defined as:

$$u' = u - u^* \tag{9.2.5a}$$

$$v' = v - v^* \tag{9.2.5b}$$

$$p' = p - p^*. \tag{9.2.5c}$$

Subtracting Eq. (9.2.2) from Eq. (9.2.1) and Eq. (9.2.4) from Eq. (9.2.3) and invoking the above relationship, Eq. (9.2.5), we obtain

$$a^u_{iu,ju} u'_{iu,ju} = \sum_{nb} a^u_{nb} u'_{nb} + A_u \left(p'_{iu,ju} - p'_{iu+1,ju}\right) \tag{9.2.6}$$

$$a^v_{iv,jv} v'_{iv,jv} = \sum_{nb} a^v_{nb} v'_{nb} + A_v \left(p'_{iv,jv} - p'_{iv,jv+1}\right). \tag{9.2.7}$$

First, we need to solve a system of algebraic equations, Eqs. (9.2.2) and (9.2.4), to predict the approximate values for the velocities u and v. Then, to correct these predicted values (u^* and v^*), we need to solve another system of algebraic equations, Eqs. (9.2.6) and (9.2.7). Clearly, this predictor–corrector method is not computationally efficient since at each iteration we are solving four different systems of equations of the type $Ax = b$. To reduce the computational effort, the velocity corrections can be made local as opposed to making a global correction. In other words, we can assume that the velocity corrections depend only on the pressure difference in the control volume but do not depend on the neighboring velocities. This method is famously called the *Semi-Implicit*

Pressure-Linked Equation or SIMPLE algorithm [67]. Here, we drop the first terms ($\sum_{nb} a^u_{nb} u'_{nb}$ and $\sum_{nb} a^v_{nb} v'_{nb}$) in the right-hand side of the Eqs. (9.2.6) and (9.2.7). Since the velocity correction is not global the formulation is no longer fully implicit, but semi-implicit, hence the term "semi-implicit" in the title of the SIMPLE algorithm. Once converged, $u' = v' = 0$, to obtain the final solution, it does not matter if the formulations for u' and v' are not implicit and accurate. However, the rate of convergence for the predictor–corrector method can be slower. But the overall computation time may still be smaller for the semi-implicit procedure compared with solving two extra set of implicit equations. The expressions for u' and v' can therefore be written as:

$$u'_{iu,ju} = \frac{A_u}{a^u_{iu,ju}} \left(p'_{iu,ju} - p'_{iu+1,ju}\right) \tag{9.2.8}$$

$$v'_{iv,jv} = \frac{A_v}{a^V_{iv,jv}} \left(p'_{iv,jv} - p'_{iv,jv+1}\right). \tag{9.2.9}$$

Let $d^u_{iu,ju} = \frac{A_u}{a^u_{iu,ju}}$ and $d^v_{iv,jv} = \frac{A_v}{a^v_{iv,jv}}$. Therefore we get:

$$u'_{iu,ju} = d^u_{iu,ju} \left(p'_{iu,ju} - p'_{iu+1,ju}\right) \tag{9.2.10}$$

$$v'_{iv,jv} = d^v_{iv,jv} \left(p'_{iv,jv} - p'_{iv,jv+1}\right). \tag{9.2.11}$$

In the above equations, the right-hand side consists of pressure-correction variables which are still unknown. Now the next step is to develop an equation for the pressure-correction (p'). Integrating the continuity equation for incompressible flow over a p-control volume we get,

$$[(\rho u)_e - (\rho u)_w]\Delta y + [(\rho v)_n - (\rho v)_s]\Delta x = 0. \tag{9.2.12}$$

Substituting for $u = u^* + u'$ and $v = v^* + v'$ and incorporating Eqs. (9.2.10) and (9.2.11), we obtain the following algebraic form of the pressure-correction equation:

$$\begin{aligned} a^{p'}_{ip,jp} p'_{ip,jp} &= a^{p'}_{ip-1,jp} p'_{ip-1,jp} + a^{p'}_{ip+1,jp} p'_{ip+1,jp} \\ &\quad + a^{p'}_{ip,jp-1} p'_{ip,jp-1} + a^{p'}_{ip,jp+1} p'_{ip,jp+1} + b^{p'}_{ip,jp}, \end{aligned} \tag{9.2.13}$$

where

$$a^{p'}_{ip-1,jp} = \rho_w d^u_{ip,jp} \Delta y$$

$$a^{p'}_{ip+1,jp} = \rho_e d^u_{ip+1,jp} \Delta y$$

$$a^{p'}_{ip,jp-1} = \rho_s d^v_{ip,jp} \Delta x$$

$$a^{p'}_{ip,jp+1} = \rho_n d^v_{ip,jp+1} \Delta x$$

$$b^{p'}_{ip,jp} = [(\rho u)_e - (\rho u)_w]\Delta y + [(\rho v)_n - (\rho v)_s]\Delta x.$$

The pressure-correction in the SIMPLE algorithm is global and the source term b^p for the pressure-correction equation will be zero at convergence due to the conservation of mass flux.

This brings us to the next question, that is, the boundary conditions for the pressure-correction variable (p'). There can be two different scenarios and we are going to discuss each in the following paragraphs:

Case 1 When pressure is specified at the boundary. When pressure (p) is specified at the boundary, there is no need to correct the pressure. So we can simply set $p' = 0$ at the boundary.

Case 2 When velocity is specified at the boundary. When velocities (u, v) are specified at the boundary, the velocity correction terms are zero: $u' = v' = 0$. Then, from Eqs. (9.2.10) and (9.2.11) we obtain:

$$p'_{iu,ju} = p'_{iu+1,ju} \tag{9.2.14}$$

$$p'_{iv,jv} = p'_{iv,jv+1}. \tag{9.2.15}$$

This relationship also means that at the boundary $\frac{\partial p'}{dx} = 0$ and $\frac{\partial p'}{dy} = 0$. This can be achieved by setting the linking a-coefficients in the pressure-correction equation to zero.

Momentum equations (9.2.2) and (9.2.4) and the pressure-correction equation (9.2.13) can be written in a general form as:

$$a_P\phi_P = a_W\phi_W + a_E\phi_E + a_S\phi_S + a_N\phi_N + b_\phi. \tag{9.2.16}$$

Here subscripts are mapped to corresponding u, v, or p variable as shown in the following table.

Node	u-index	v-index	p-index
P	iu, ju	iv, jv	ip, jp
W	$iu-1, ju$	$iv-1, jv$	$ip-1, jp$
E	$iu+1, ju$	$iv+1, jv$	$ip+1, jp$
S	$iu, ju-1$	$iv, jv-1$	$ip, jp-1$
N	$iu, ju+1$	$iv, jv+1$	$ip, jp+1$

In general terms, to solve a two-dimensional incompressible flow problem using the SIMPLE algorithm, one has to invert a penta-diagonal matrix for u, v, and p' at each iteration. As discussed in Chapters 3 and 7, the solution for these three equations at each iteration can be obtained by using a *line-by-line method* or a *strongly implicit procedure*, which is a combination of the TDMA and the Gauss–Seidel scheme. To avoid numerical instability an under-relaxation factor ω is introduced. When implicit in the x-direction, the formulation leads to the following.

Sweep from bottom to top and top to bottom:

$$-a_W\phi_W^{k+1} + \frac{a_P}{\omega}\phi_P^{k+1} - a_E\phi_E^{k+1} = a_S\phi_S^{k+1} + a_N\phi_N^k + b_P^\phi + a_P\phi_P^k\left(\frac{1}{\omega} - 1\right), \tag{9.2.17}$$

where variables with superscript k represent values from the previous iteration (known value).

Sweep from left to right and right to left:

$$-a_S\phi_S^{k+1} + \frac{a_P}{\omega}\phi_P^{k+1} - a_N\phi_N^{k+1} = a_W\phi_W^{k+1} + a_E\phi_E^k + b_P^\phi + a_P\phi_P^k\left(\frac{1}{\omega} - 1\right). \tag{9.2.18}$$

It is recommended that in each iteration lines are swept at least once in all four directions (bottom to top, top to bottom, left to right, and right to left). For classroom exercises, one sweep per iteration in all four directions would suffice. For exploratory research problems, multiple sweeps of iterations may be necessary.

The relaxation factors for u and v (ω_u and ω_v) are usually around 0.5. For periodic flow problems, smaller relaxation factor values are preferred (e.g., $\omega_u = \omega_v = 0.2$). While sweeping, the p' equation is not under-relaxed; however, only a fraction ω_p of p' is added to the estimated value of p^*. That is

$$p = p^* + \omega_p p'. \tag{9.2.19}$$

Typically $\omega_p = 0.8$ will work for classroom exercise problems.

Sequence of the SIMPLE Algorithm

(1) Guess pressure field ($p^*_{ip,jp}$).

(2) Using the guessed pressure field, solve for u^* and v^* [see Eqs. (9.2.2) and (9.2.4)].

(3) Calculate $d^u_{iu,ju}$ and $d^v_{iv,jv}$.

(4) Solve the pressure-correction (p') equation (9.2.13).

(5) Calculate velocity corrections $u'_{iu,ju}$ and $v'_{iv,jv}$ using Eqs. (9.2.10) and (9.2.11).

(6) Correct velocity and pressure fields:

$$p_{ip,jp} = p^*_{ip,jp} + \omega_p p'_{ip,jp} \tag{9.2.20}$$
$$u_{iu,ju} = u^*_{iu,ju} + u'_{iu,ju} \tag{9.2.21}$$
$$v_{ip,jp} = v^*_{iv,jv} + v'_{iv,jv}. \tag{9.2.22}$$

(7) Solve for any other scalar quantities, T, K, ϵ, C, and the like.

(8) Check for convergence.

Programming Call Sequence

```
call geometry
call property
call initial_conditions
```

```
call boundary_conditions
call reset_pressure
call ucoefficient
call usolve
call vcoefficient
call vsolve
call pccoefficient
call pcsolve
call ucorrection
call vcorrection
call pcorrection
call covergence
call print_results
```

9.2.2.1 Algorithms of the SIMPLE Family

There are several algorithms that are closely related to the SIMPLE algorithm, such as the SIMPLEC and SIMPLER algorithms. We will discuss only the SIMPLEC algorithm here.

SIMPLEC Algorithm SIMPLEC stands for SIMPLE Consistent and was developed by Van Doormal and Raithby [68]. This method is a modification of the SIMPLE algorithm. The terms $\sum_{nb} a_{nb}u'_{nb}$ and $\sum_{nb} a_{nb}v'_{nb}$ in Eqs. (9.2.6) and (9.2.7) were dropped in the SIMPLE algorithm. This approximation is inconsistent as all of these terms are of similar magnitude. These terms were dropped to obtain equations for velocity corrections strictly in terms of pressure-corrections. If we had not dropped $\sum_{nb} a_{nb}u'_{nb}$ and $\sum_{nb} a_{nb}v'_{nb}$, then the pressure-correction equation for each control volume would contain eight 'u' neighbors, eight 'v' neighbors, and four p' neighbors for two-dimensional problems, rendering the equation unmanageable. Also, this amounts to solving the exact momentum equations. By dropping $\sum_{nb} a_{nb}u'_{nb}$ and $\sum_{nb} a_{nb}v'_{nb}$, the pressure-correction equation is simplified since the p' equation for each control volume contains only four of its neighbors for two-dimensional problems. Also, note that the p' equation need not be accurate since it is only a vehicle to get a converged solution and p' will be zero, once the converged velocity field is obtained. The success of SIMPLE algorithm depends on how close the under-relaxation factor (ω_p) is to the optimum value $(p = p^* + \omega_p p')$.

The SIMPLEC algorithm removes the need for under-relaxing the pressure field. The solution cost in using SIMPLEC is equivalent to or better than that obtained by the optimal value of α_p in the SIMPLE algorithm (Van Doormal and Raithby [68]). The motivation for the SIMPLEC algorithm is to introduce a consistent approximation.

Consider a relaxed form of the x-momentum equation:

$$\frac{a^u_{iu,ju} u_{iu,ju}}{\omega_u} = \sum_{nb} a^u_{nb} u_{nb} + A_u(p_{iu,ju} - p_{iu+1,ju}) + b^u_{iu,ju} + a^u_{iu,ju} u^0_{iu,ju}\left(\frac{1}{\omega_u} - 1\right) \tag{9.2.23a}$$

$$\frac{a^v_{iv,jv}v_{iv,jv}}{\omega_v} = \sum_{nb} a^v_{nb}v_{nb} + A_v(p_{iv,jv} - p_{iv,jv+1}) + b^v_{iv,jv} + a^v_{iv,jv}v^0_{iv,jv}\left(\frac{1}{\omega_v} - 1\right). \tag{9.2.23b}$$

For an intermediate solution (u^*, v^*, p^*) these equations become:

$$\frac{a^u_{iu,ju}u^*_{iu,ju}}{\omega_u} = \sum_{nb} a^u_{nb}u^*_{nb} + A_u(p^*_{iu,ju} - p^*_{iu+1,ju}) + b^u_{iu,ju} + a^u_{iu,ju}u^0_{iu,ju}\left(\frac{1}{\omega_u} - 1\right) \tag{9.2.24a}$$

$$\frac{a^v_{iv,jv}v^*_{iv,jv}}{\omega_v} = \sum_{nb} a^v_{nb}v^*_{nb} + A_v(p^*_{iv,jv} - p^*_{iv,jv+1}) + b^v_{iv,jv} + a^v_{iv,jv}v^0_{iv,jv}\left(\frac{1}{\omega_v} - 1\right). \tag{9.2.24b}$$

Recall that $u' = u - u^*$, $v' = v - v^*$, and $p' = p - p'$. Subtracting Eq. (9.2.24) from Eq. (9.2.23), we obtain

$$\frac{a^u_{iu,ju}u'_{iu,ju}}{\omega_u} = \sum_{nb} a^u_{nb}u'_{nb} + A_u\left(p'_{iu,ju} - p'_{iu+1,ju}\right) \tag{9.2.25a}$$

$$\frac{a^v_{iv,jv}v'_{iv,jv}}{\omega_v} = \sum_{nb} a^v_{nb}v'_{nb} + A_v\left(p'_{iv,jv} - p'_{iv,jv+1}\right). \tag{9.2.25b}$$

As a part of consistent approximation $\sum_{nb} a^u_{nb}u'_{iu,ju}$ is subtracted from both sides of Eq. (9.2.25a)

$$\left(\frac{a^u_{iu,ju}}{\omega_u} - \sum_{nb} a_{nb}\right)u'_{iu,ju} = \underline{\sum_{nb} a^u_{nb}\left(u'_{nb} - u'_{iu,ju}\right)} + A_u\left(p'_{iu,ju} - p'_{iu+1,ju}\right). \tag{9.2.26}$$

In the SIMPLEC algorithm the underlined term in Eq. (9.2.26) is omitted, since this term is usually much smaller than $u'_{iu,ju}$. This consistent approximation leads to the following expression for u-correction term:

$$u'_{iu,ju} = \frac{A_u(p'_{iu,ju} - p'_{iu+1,ju})}{\left(\frac{a^u_{iu,ju}}{\omega_u} - \sum_{nb} a^u_{nb}\right)}. \tag{9.2.27}$$

Similarly, the v-correction term is:

$$v'_{iv,jv} = \frac{A_v(p'_{iv,jv} - p'_{iv,jv+1})}{\left(\frac{a^v_{iv,jv}}{\omega_v} - \sum_{nb} a^v_{nb}\right)}. \tag{9.2.28}$$

The SIMPLEC algorithmn is different from the SIMPLE algorithm in the following ways:

(1) Pressure is not under-relaxed, $\omega_p = 1$.

(2) The correction coefficients for u' and v' are:

$$d^u_{iu,ju} = \frac{A_u}{\left(\frac{a^u_{iu,ju}}{\omega_u} - \sum_{nb} a^u_{nb}\right)} \tag{9.2.29}$$

$$d^v_{iv,jv} = \frac{A_v}{\left(\frac{a^v_{iv,jv}}{\omega_v} - \sum_{nb} a^v_{nb}\right)}. \tag{9.2.30}$$

For SIMPLEC to work, the momentum equations have to be under-relaxed; that is, ω_u and $\omega_v < 1$.

SIMPLER Algorithm The Semi-Implicit Pressure-Linked Equations Revised (SIMPLER) algorithm was developed by Patankar [7]. The algorithm has more computational steps per iteration compared with the SIMPLE algorithm. However, the convergence time for the SIMPLER algorithm was less compared to that for the SIMPLE algorithm. As hardware capabilities improved through the 1990s, researchers and code developers deemed that, all things considered, it is more efficient to run the SIMPLE algorithm on the faster computers that became readily available in the 1990s. In other words, due to readily available superior hardware the extra effort in implementing the SIMPLER algorithm to improve convergence time became moot. Accordingly, details of the SIMPLER algorithm are omitted and readers interested in the specifics of this algorithm may refer to Patankar [7].

9.2.3 Residuals and Declaring Convergence

When iterative solution techniques are used to solve the discretized equations, the iterations are repeated until the change in variables are very small. At that point, we can declare that the solution has converged. Instead of monitoring the change in variables we can also monitor the change in residuals, which is equivalent to monitoring change in errors (see Chapter 3). Since the discretized equations in the FVM are nothing but flux conservation, the residuals indicates the amount of error or imbalance in the conservation of fluxes. The residuals are usually summed over all control volumes and scaled with the sum of local values of the variables. The relative residuals for momentum, pressure, and energy equations can be expressed in the following way:

$$R_u = \frac{\sum_{CVs}\left|a^u_P u_P - \sum_{nb} a^u_{nb} u_{nb} - b^u_P\right|}{\sum_{CVs}\left|a^u_P u_P\right|} \tag{9.2.31a}$$

$$R_v = \frac{\sum_{CVs}\left|a^v_P v_P - \sum_{nb} a^v_{nb} v_{nb} - b^v_P\right|}{\sum_{CVs}\left|a^v_P v_P\right|} \tag{9.2.31b}$$

$$R_p = \frac{\sum_{CVs}\left|(\rho_w u_w - \rho_e u_e)\Delta y + (\rho_s v_s - \rho_n v_n)\Delta x\right|}{\rho_{ref} u_{ref} L_{ref}} \tag{9.2.31c}$$

$$R_T = \frac{\sum_{CVs}\left|a_P^T T_P - \sum_{nb} a_{nb}^T T_{nb} - b_P^T\right|}{\sum_{CVs}\left|a_P^T T_P\right|}. \tag{9.2.31d}$$

For declaring convergence, usually the following tolerances are used: $R_u < 10^{-6}$, $R_u < 10^{-6}$, $R_p < 10^{-5}$, and $R_T < 10^{-5}$.

9.2.4 Boundary Conditions

In this section we will discuss the implementation of boundary conditions for velocity, pressure, and temperature fields in the staggered-grid approach. As discussed in Section 9.1 both velocity and pressure cannot be specified at the boundary at the same time.

9.2.4.1 Implementation of Boundary Condition for Momentum Equations when Velocities are Specified

As stated in Section 9.2.1, the discussion is confined to the staggered-grid approach only. First, we will consider when the u-velocity is specified at the boundary. As an example, consider that $u_{6,5}$ is specified in Fig. 9.2.4(b). At the outset, the corresponding velocity correction is set to zero, $u'_{6,5} = 0$, since actual velocity is already specified. Let $u_{6,5} = u_{given}$, then the momentum equation for the marked u-control volume is:

$$\begin{aligned} a^u(5,5)u_{5,5} = {} & A_u^u(p_{5,5} - p_{6,5}) + a_W^u(5,5)u_{4,5} + a_E^u(5,5)u_{given} \\ & + a_S^u(5,5)u_{5,4} + a_N^u(5,5)u_{5,6}. \end{aligned} \tag{9.2.32}$$

In Eq. (9.2.32) u_{given} is known so the term is kept on the right-hand side of the equation and regarded as the known or explicit term.

Similarly, consider v-control volume enclosing the unknown $v_{3,5}$. In this case $v_{3,6}$ is specified to be v_{given} in Fig. 9.2.4(c). The discretization equation for $v_{3,5}$ is written as :

$$\begin{aligned} a^v(3,5)v_{3,5} = {} & A_v^v(p_{3,5} - p_{3,6}) + a_W^v(3,5)v_{2,5} + a_E^v(3,5)v_{4,5} \\ & + a_S^v(3,5)v_{3,4} + a_N^v(3,5)V_{given}. \end{aligned} \tag{9.2.33}$$

In this case v_{given} is known and is kept on the right-hand side of the equation and regarded as a known or explicit term.

9.2.4.2 Implementation of the Boundary Condition for Momentum Equations when Derivatives of Velocities are Specified

As an example, consider the case in which the derivative of the velocity u is specified along the eastern boundary:

$$x = L, \left.\frac{\partial u}{\partial x}\right|_e = \alpha, \tag{9.2.34}$$

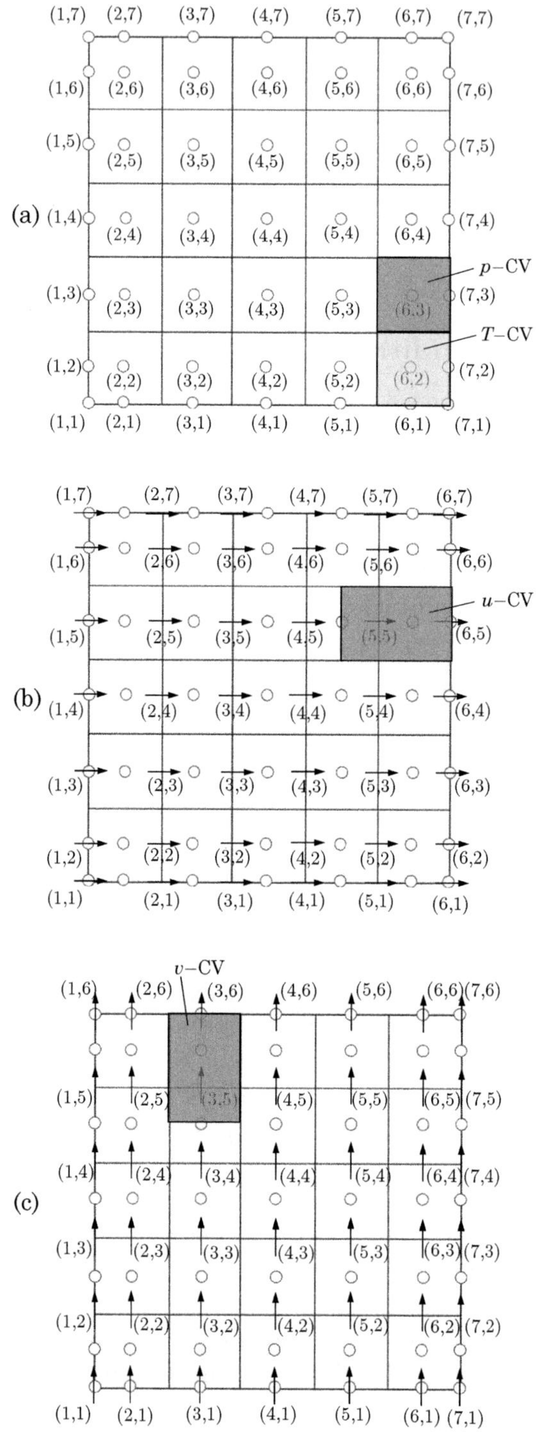

Fig. 9.2.4 Pressure, Temperature, u-velocity and v-velocity control volumes for the implementation of the boundary condition in a staggered-grid approach.

where α is a constant. If there is no flux normal to the east boundary, then α is equal to zero. In this case, we have a couple of options. One option is simply to represent the boundary condition using a finite difference expression. Thus, the discretization equation for the boundary node is

$$\frac{u_{6,5} - u_{5,5}}{\delta x} = \alpha \tag{9.2.35}$$

$$\left(-\frac{1}{\delta x}\right)u_{5,5} + \left(\frac{1}{\delta x}\right)u_{6,5} = \alpha. \tag{9.2.36}$$

When the derivative is zero, $\partial u/\partial x = 0$ or $u_{6,5} = u_{5,5}$. These types of boundary conditions are often referred to as natural conditions, which typically occur at the channel exit or far from a bluff body. In such cases the linking coefficient, that is, a_E^u, which links $u_{5,5}$ with $u_{6,5}$, is set to zero as u_E $(u_{6,5})$ will not appear in the discretization equation for $u_{5,5}$. Setting the linking a-coefficient to zero makes sense as it implies that information from the channel exit does not flow upstream, particularly for flows with high Péclet number.

9.2.4.3 Implementation of the Outflow Boundary condition

Outflow boundary conditions are used when the downstream condition has no impact on the upstream flow condition. Outflow boundary conditions can be applied to internal flows where the flow is fully developed, such as the channel flow exit, or the outlet of backward-facing step flow. Implementation of outflow boundary conditions for stream-wise velocity (u) and cross-stream velocity (v) are tied to convergence. At the channel exit both u and v velocities are subjected to a natural or zero-gradient boundary condition, $\frac{\partial u}{\partial x} = \frac{\partial v}{\partial x} = 0$. This ensures the continuation of flow at the channel exit. After discretization, the outflow boundary conditions for u- and v- velocities take the following form

For all ju

$$u_{iumax,ju} = u_{iumax-1,ju}. \tag{9.2.37}$$

For all jv

$$v_{ivmax,jv} = v_{ivmax-1,jv}. \tag{9.2.38}$$

Owing to the discretization and round-off errors, the mass fluxes may not be conserved between the channel inlet and exit, which may affect the convergence, of the numerical solution. To conserve the mass fluxes and to ensure convergence an overall mass-balance correction is enforced between exit and inlet at each iteration. At the kth iteration the ratio of mass fluxes is calculated as,

$$\dot{m}_r = \frac{\dot{m}_{out}}{\dot{m}_{in}} = \frac{\rho_{out} u_{out} A_{out}}{\rho_{in} u_{in} A_{in}}. \tag{9.2.39}$$

Then u-velocities at the channel exit are updated in the following way to ensure convergence:

For all ju

$$u_{iumax,ju} = \dot{m}_r * u_{iumax-1,ju}. \tag{9.2.40}$$

Note that care must be taken to ensure that the flow is fully developed to apply the outflow boundary condition. For cases where there is a flow reversal or recirculating flow, that is, part of the flow enters through the boundary, using an outflow boundary condition may produce physically unrealistic flow results. To avoid an ill-posed outflow boundary condition, the location of the boundary must be sufficiently far from the recirculating or reverse flow zones.

9.2.4.4 Boundary Conditions for Pressure

One may recall that in using the SIMPLE algorithm the pressure-correction (p') equation is solved rather than the equation for pressure (p) itself. So when pressure is specified at the boundary, no correction is needed, $p' = 0$. We end up having a Dirichlet-type boundary condition. On the other hand, implementation of a boundary condition when velocity is specified at a boundary requires some explanation. One may recall that the equation for pressure-correction is constructed by integrating the continuity equation over a p-control volume [see Fig. 9.2.4(a)]. Then each of the velocity terms is represented as the sum of the guessed value and the correction ($u = u^* + u'$), and u' is expressed as a multiplier of $\Delta p'$ ($u' = d_u \Delta p'$). Since velocity is specified at the boundary, $u' = 0$, and as a result p' at the boundary does not enter into the p' equation at all. Accordingly, the linking a-coefficient for the pressure-correction variable is set to zero.

So, in Fig. 9.2.4(a), if $u_{6,3}$ is specified then in the p' equation for the p-control volume ($p'_{6,3}$) the coefficient linking $p'_{6,3}$ with $p'_{7,3}$ is set equal to zero. If we write the discretization equation for $p'_{6,3}$ we get

$$a_p^{p'}(6,3)p'_{6,3} = a_W^{p'}(6,3)p'_{5,3} + a_S^{p'}(6,3)p'_{6,2} + a_N^{p'}(6,3)p'_{6,4}. \tag{9.2.41}$$

In Eq. (9.2.41) the linking coefficient $a_E^{p'}(6,3)$ is set to zero.

9.2.4.5 Boundary Conditions for Temperature

We will use the right bottom corner control volume (T-CV) in Fig. 9.2.4(a) to discuss the implementation of the temperature boundary conditions ($T(6,2)$).

(a) When temperature is specified at the boundary In this case let us say that $T(6,1)$ and $T(7,2)$ are specified so both $T(6,1)$ and $T(7,2)$ will be treated as explicit quantities and will be on the right-hand side of the discretization equation.

$$\begin{aligned} &- a_W^T(6,2)T_{5,2} + a_{P(6,2)}^T T_{6,2} - a_N^T(6,2)T_{6,3} \\ &= a_S^T(6,2)T_{6,1} + a_E^T(6,3)T_{7,2} + b^T. \end{aligned} \tag{9.2.42}$$

(b) When heat flux is specified at the boundary Let us consider a case where heat flux at $x = L$ (i.e., on the far right boundary) is specified:

$$x = L, \;\; q'' = -k\frac{\partial T}{\partial x}\bigg|_{x=L}. \tag{9.2.43}$$

Therefore, to develop a discretization equation for the control volume with temperature node $T_{6,2}$ we integrate the energy equation over the corner control volume:

$$\int_{CV}\left[\frac{\partial}{\partial x}\left(k\frac{\partial T}{\partial x}\right)+\frac{\partial}{\partial y}\left(k\frac{\partial T}{\partial y}\right)\right]dV=\int_{CV}S_T dV \tag{9.2.44}$$

$$k\frac{\partial T}{\partial x}\bigg|_w^e \Delta y + k\frac{\partial T}{\partial y}\bigg|_s^n \Delta x = S_T\Delta x\Delta y \tag{9.2.45}$$

$$\left[k\frac{\partial T}{\partial x}\bigg|_e - k\frac{\partial T}{\partial x}\bigg|_w\right]\Delta y + \left[k\frac{\partial T}{\partial y}\bigg|_n - k\frac{\partial T}{\partial y}\bigg|_s\right]\Delta x = S_T\Delta x\Delta y \tag{9.2.46}$$

$$-q''\Delta y - k_w\left(\frac{T_{6,2}-T_{5,2}}{\delta x_w}\right)\Delta y + k_n\left(\frac{T_{6,3}-T_{6,2}}{\delta y_n}\right)\Delta x$$
$$-k_s\left(\frac{T_{6,2}-T_{6,1}}{\delta y_s/2}\right)\Delta x = S_T\Delta x\Delta y \tag{9.2.47}$$

$$T_{6,2}\left(\frac{k_w\Delta y}{\delta x_w}+\frac{k_n\Delta x}{\delta y_n}+\frac{2k_s\Delta x}{\delta y_s}\right) = -q''\Delta y + \frac{k_w\Delta y}{\delta x_w}T_{5,2}$$
$$+\frac{k_n\Delta x}{\delta y_n}T_{6,3}+\frac{2k_s\Delta x}{\delta y_s}T_{6,1} - S_T\Delta x\Delta y \tag{9.2.48}$$

$$a_P^T(6,2)T_{6,2} = a_W^T(6,2)T_{5,2} + a_S^T(6,2)T_{6,1} + a_N^T(6,2)T_{6,3} + b_P^T(6,2), \tag{9.2.49}$$

where

$$a_W^T(6,2) = \frac{k_w\Delta y}{\delta x_w}$$

$$a_E^T(6,2) = 0$$

$$a_S^T(6,2) = \frac{2k_s\Delta x}{\delta y_s}$$

$$a_N^T(6,2) = \frac{k_n\Delta x}{\delta y_n}$$

$$b_P^T(6,2) = -q''\Delta y - S_T\Delta x\Delta y$$

$$a_P^T(6,2) = a_W^T(6,2) + a_E^T(6,2) + a_S^T(6,2) + a_N^T(6,2).$$

Equation 9.2.49 is incorporated into the system of equations once $T_{6,2}$ is obtained using a *line-by-line method* or the *strongly implicit procedure.* Then $T_{7,2}$ is calculated outside of the solution process. Once $T_{6,2}$ is solved then by using the given boundary condition we obtain:

$$-k_e\left(\frac{T_{7,2}-T_{6,2}}{\delta x_e/2}\right) = q'' \tag{9.2.50}$$

$$T_{7,2}-T_{6,2} = \frac{-q''\delta x_e}{2k_e} \tag{9.2.51}$$

$$T_{7,2} = T_{6,2} - \frac{q''\delta x_e}{2k_e}. \tag{9.2.52}$$

(c) Convective boundary conditions. Here we will consider a convective boundary condition:

$$x = L: \quad -k\frac{\partial T}{\partial x}\bigg|_{x=L} = h(T - T_\infty), \tag{9.2.53}$$

where h is the convective heat transfer coefficient and T_∞ is the ambient temperature. Again we will focus on obtaining a discretization equation for $T_{6,2}$ by integrating the two-dimensional steady diffusion equation over the bottom corner control volume enclosing nodal temperature $T_{6,2}$.

$$\int_{CV}\left[\frac{\partial}{\partial x}\left(k\frac{\partial T}{\partial x}\right) + \frac{\partial}{\partial y}\left(k\frac{\partial T}{\partial y}\right)\right]dV = \int_{CV} S_T dV \tag{9.2.54}$$

$$k\frac{\partial T}{\partial x}\bigg|_w^e \Delta y + k\frac{\partial T}{\partial y}\bigg|_s^n \Delta x = S_T \Delta x \Delta y \tag{9.2.55}$$

$$\left[k\frac{\partial T}{\partial x}\bigg|_e - k\frac{\partial T}{\partial x}\bigg|_w\right]\Delta y + \left[k\frac{\partial T}{\partial y}\bigg|_n - k\frac{\partial T}{\partial y}\bigg|_s\right]\Delta x = S_T \Delta x \Delta y \tag{9.2.56}$$

$$-h(T_{7,2} - T_\infty)\Delta y - k_w\left(\frac{T_{6,2} - T_{5,2}}{\delta x_w}\right)\Delta y + k_n\left(\frac{T_{6,3} - T_{6,2}}{\delta y_n}\right)\Delta x$$
$$-k_s\left(\frac{T_{6,2} - T_{6,1}}{\delta y_s/2}\right)\Delta x = S_T \Delta x \Delta y \tag{9.2.57}$$

$$T_{6,2}\left(\frac{k_w \Delta y}{\delta x_w} + \frac{k_n \Delta x}{\delta y_n} + \frac{2k_s \Delta x}{\delta y_s}\right) = -h(T_{7,2} - T_\infty)\Delta y$$
$$+\frac{k_w \Delta y}{\delta x_w}T_{5,2} + \frac{k_n \Delta x}{\delta y_n}T_{6,3} + \frac{2k_s \Delta x}{\delta y_s}T_{6,1} - S_T \Delta x \Delta y \tag{9.2.58}$$

$$a_P^T(6,2)T_{6,2} = a_W^T(6,2)T_{5,2} + a_E^T(6,2)T_{7,2} + a_S^T(6,2)T_{6,1}$$
$$+a_N^T(6,2)T_{6,3} + b_P^T(6,2), \tag{9.2.59}$$

where

$$a_W^T(6,2) = \frac{k_w \Delta y}{\delta x_w}, \quad a_E^T(6,2) = h\Delta y,$$

$$a_S^T(6,2) = \frac{2k_s \Delta x}{\delta y_s}, \quad a_N^T(6,2) = \frac{k_n \Delta x}{\delta y_n},$$

$$b_P^T(6,2) = hT_\infty \Delta y - S_T \Delta x \Delta y$$

$$a_P^T(6,2) = a_E^T(6,2) + a_W^T(6,2) + a_S^T(6,2) + a_N^T(6,2).$$

Equation 9.2.59 is incorporated into the system of discretization equations and then one can use either the *strongly implicit procedure* or the *line-by-line method* to solve. However, the problem is we need an additional equation for $T_{7,2}$. This is obtained by making an energy balance over a zero thickness control volume at $x = L$:

$$\int -k\frac{dT}{dx}\bigg|_{x=L} dy = \int h(T - T_\infty)dy \tag{9.2.60}$$

$$-k_w\left(\frac{T_{7,2}-T_{6,2}}{\delta x_w/2}\right)\Delta y = h(T_{7,2}-T_\infty)\Delta y \tag{9.2.61}$$

$$\left(\frac{2k_w}{\delta x_w}\right)T_{6,2} + \left(-\frac{2k_w}{\delta x_w} - h\right)T_{7,2} = -hT_\infty \tag{9.2.62}$$

$$a_P^T(7,2)T_{7,2} = a_W^T(7,2)T_{6,2} + b_P^T(7,2), \tag{9.2.63}$$

where

$$a_P^T(7,2) = \frac{2k_w}{\delta x_w} + h,\quad a_W^T(7,2) = \frac{2k_w}{\delta x_w},\quad b_P^T(7,2) = hT_\infty.$$

Therefore, the above relationship gives an additional equation for the unknown $T_{7,2}$ to have a closed system.

(d) Outflow Boundary Conditions for Temperature. For the purpose of discussion let us consider a flow of fluid in a two-dimensional rectangular channel subjected to heating at the wall. We have two cases: uniform wall temperature heating (UWT) and uniform wall heat flux (UHF) conditions.

To begin, we will consider the UWT case, as shown in Fig. 9.2.5. The variation of bulk temperature with axial distance X is shown in Fig. 9.2.6.

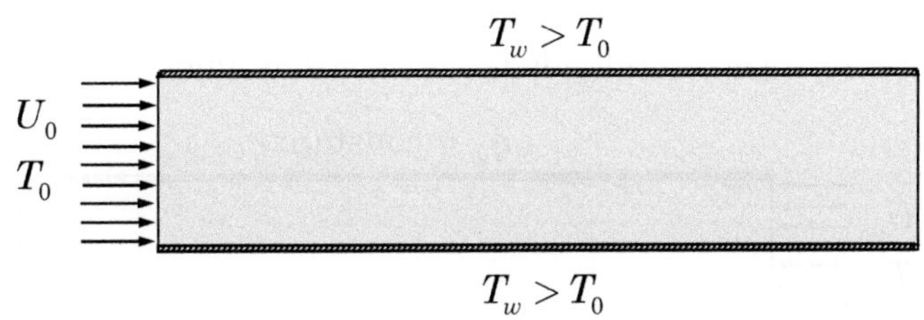

Fig. 9.2.5 The UWT condition.

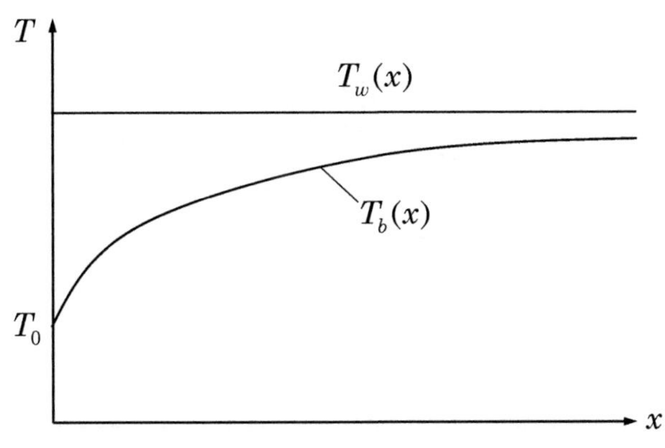

Fig. 9.2.6 Bulk-temperature and wall-temperature variation for the UWT condition.

Obviously, $\partial T_b/\partial x = 0$ at the channel exit, and we can extrapolate it to the temperature (T) as

$$x = L: \quad \frac{\partial T}{\partial x} = 0 \tag{9.2.64}$$

$$T_{imax,j} = T_{imax-1,j}. \tag{9.2.65}$$

Alternatively, we can set the east coefficient to zero, $a_E^T = 0$.

For the UHF case, we have (see Figs. 9.2.7 and 9.2.8)

$$x = L: \quad \frac{dT_b}{dx} = \text{constant} \neq 0. \tag{9.2.66}$$

So one can extrapolate

$$\left.\frac{\partial T}{dx}\right|_{imax,j} = \left.\frac{\partial T}{dx}\right|_{imax-1,j}. \tag{9.2.67}$$

That is, we can use a first order backward-difference scheme

$$\frac{T_{imax-1,j} - T_{imax-2,j}}{\delta x} = \frac{T_{imax,j} - T_{imax-1,j}}{\delta x} \tag{9.2.68}$$

$$-T_{imax-2,j} + 2T_{imax-1,j} = T_{imax,j}. \tag{9.2.69}$$

Equation (9.2.69) is the discretization equation for temperature at the exit. The recommendation is to treat $T_{imax,j}$ as a known quantity from a previous iteration and update $T_{imax,j}$ outside of the solution loop.

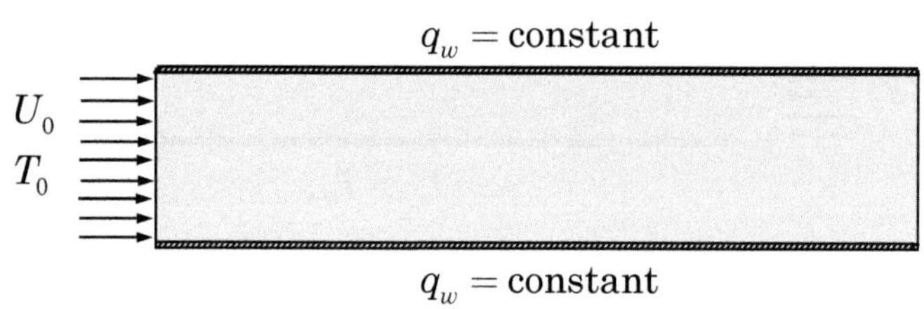

Fig. 9.2.7 The UHF condition.

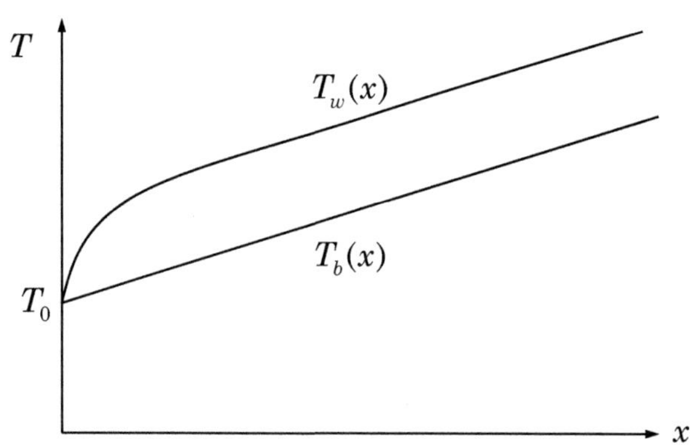

Fig. 9.2.8 Bulk-temperature and wall-temperature variation for the UHF condition.

9.2.5 Treatment of Source Terms

Body-force terms in momentum equations and heat-sink/source terms in the energy equation appear as source terms. Commonly encountered body forces are buoyancy, centrifugal, and Coriolis forces. Volumetric heating or cooling also appear as source terms. Source terms in x- and y-momentum equations are denoted by S_u and S_v in Eqs. (9.1.5) and (9.1.6), respectively. These source terms are to be integrated over appropriate control volumes in the staggered-grid approach. For example, when S_u is integrated over an internal u-control volume (Fig. 9.2.3), then the resulting source term $b^u_{iu,ju}$ will be:

$$b^u_{iu,ju} = \int_{CV} S_u dV = S_u \Delta x \Delta y. \tag{9.2.70}$$

For boundary control volumes in staggered-grid the integration is performed over 3/2 control volumes and the resulting source terms will be

$$b^u_{iu,ju} = \int_{CV} S_u dV = S_u 1.5 \Delta x \Delta y. \tag{9.2.71}$$

In the collocated grid approach, the issue of 3/2 control volume does not appear and source term integrated over the boundary control volume is same as Eq. (9.2.70). Treatment of source/sink terms has been described in Chapters 7 and 8.

9.3 Collocated-Grid Method

9.3.1 General Introduction

While discretizing the momentum equations, a straightforward application of the central difference scheme to the pressure-gradient term can result in a checkerboard-type pressure field, which might be the wrong depiction of the reality. Similarly, the discretization of the first derivative of the velocity terms in the continuity equation can lead us to a nonphysical wavy velocity field as a solution.

As discussed in Section 9.2, one remedy for this problem is to discretize the pressure and velocity variables in a staggered-grid, which was first proposed by Harlow and Welch [69]. However, the book-keeping of all the variables in displaced locations of a staggered-grid is often cumbersome and difficult, especially for non-orthogonal or curvilinear grids. The level of difficulty also increases if one tries to implement multi-grid techniques and parallel algorithms [70].

The main reason for the failure of early attempts to use the collocated-grid approach was the decoupling of the pressure from the velocities at the node of a given control volume when discretizing the pressure-gradient terms in the momentum equations. Rhie and Chow [66] developed a special interpolation technique to calculate velocities at the control volume faces of a collocated grid. The special interpolation technique is known as *momentum interpolation*. Instead of linear interpolation at the control volume interfaces, it uses discretized momentum equations of two adjacent control volumes to calculate the face velocities. The momentum interpolation technique has been widely used in pressure-based FVMs for unstructured grids [71, 72]. Here we will present a variation of this technique applied to structured collocated grid [73, 74].

A typical two-dimensional control volume for the collocated-grid approach is shown in Fig. 9.3.1. The two-dimensional velocity vector $\mathbf{V}$ contains the x-component u and the y-component v, that is, $\mathbf{V} = (u, v, 0)$. It can be seen from Fig. 9.3.1 that both pressure p and velocities u and v are computed at the same node. Integrating the momentum equations will produce formulations similar to Eqs. (9.2.1) and (9.2.3):

$$a^u_{i,j} u_{i,j} = \sum_{nb} a^u_{nb} u_{nb} + A_u (p_w - p_e)_{i,j} + b^u_{i,j} \tag{9.3.1}$$

$$a^v_{i,j} v_{i,j} = \sum_{nb} a^v_{nb} v_{nb} + A_v (p_s - p_n)_{i,j} + b^v_{i,j}. \tag{9.3.2}$$

Note that, unlike the staggered-grid approach, we are using the same location indices (i, j) for both u and v equations. To facilitate the interpolation, we first calculate the node velocities from the above equations:

$$u_{i,j} = H^u_{i,j} + d_u (p_w - p_e)_{i,j} + B^u_{i,j} \tag{9.3.3}$$
$$v_{i,j} = H^v_{i,j} + d_v (p_s - p_n)_{i,j} + B^v_{i,j}, \tag{9.3.4}$$

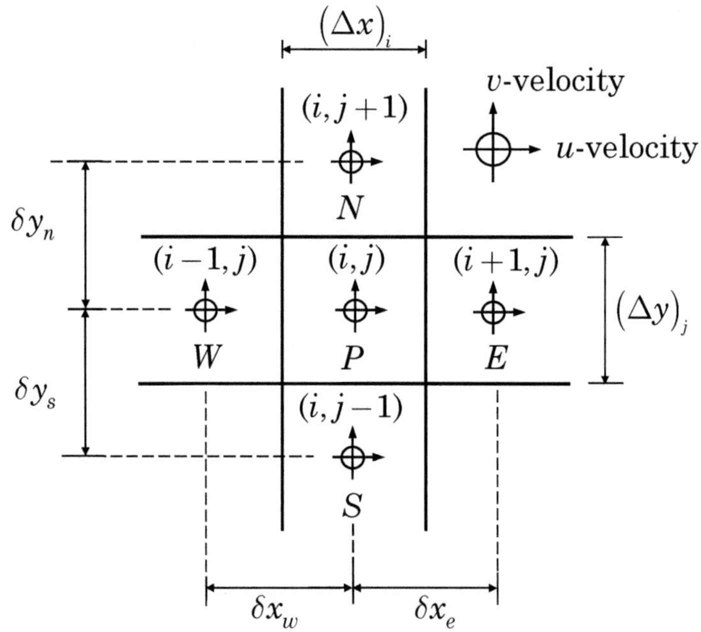

Fig. 9.3.1 A typical control volume in a collocated-grid approach.

where

$$H^u_{i,j} = \frac{\sum_{nb} a^u_{nb} u_{nb}}{a^u_{i,j}}, \ H^v_{i,j} = \frac{\sum_{nb} a^v_{nb} v_{nb}}{a^v_{i,j}} \tag{9.3.5}$$

$$d_u = \frac{A_u}{a^u_{i,j}}, \ d_v = \frac{A_v}{a^v_{i,j}} \tag{9.3.6}$$

$$B^u_{i,j} = \frac{b^u_{i,j}}{a^u_{i,j}} \ B^v_{i,j} = \frac{b^v_{i,j}}{a^v_{i,j}}. \tag{9.3.7}$$

9.3.2 Calculation of Control Volume Face Velocities

In the interpolation scheme of Rhie and Chow [66, 75], we calculate the face velocities directly from the momentum equations, not from an average of nodal velocities. Let us consider the control volume surrounding the node (i, j). The discretized momentum equation for east face can be written in a similar form as Eq. (9.3.3):

$$u_{e_{i,j}} = H^u_{e_{i,j}} + (d_u)_e(p_{i,j} - p_{i+1,j}) + B^u_{e_{i,j}}. \tag{9.3.8}$$

The terms with subscript e represent interpolated values at the east face of the control volume. These interpolated values can be calculated by the following linear interpolation formulas:

$$\frac{1}{a^u_{e_{i,j}}} = \frac{f_e}{a_{i,j}} + \frac{(1-f_e)}{a_{i+1,j}} \tag{9.3.9}$$

$$H^u_{e_{i,j}} = f_e H^u_{i,j} + (1-f_e) H^u_{i+1,j} \tag{9.3.10}$$

$$B^u_{e_{i,j}} = f_e B^u_{i,j} + (1-f_e) B^u_{i+1,j}, \tag{9.3.11}$$

where f_e is a geometric interpolation factor that depends on the distance between the nodes in the x-direction:

$$f_e = \frac{\Delta x_{i+1}}{\Delta x_i + \Delta x_{i+1}}. \tag{9.3.12}$$

The important point to note is instead of *interpolating* the pressure difference, we are explicitly *evaluating* it at the east face. To see how the face velocity is related to the node velocities, we use Eqs. (9.2.3) and (9.3.11) in Eq. (9.3.8):

$$\begin{aligned} u_{e_{i,j}} &= [f_e u_{i,j} + (1-f_e) u_{i+1,j}] + [(d_u)_e(p_{i,j} - p_{i+1,j}) \\ &\quad - f_e (d_u)_E (p_w - p_e)_{i,j} - (1-f_e)(d_u)_P (p_w - p_e)_{i+1,j}] \\ &\quad + \left[B^u_{e_{i,j}} - f_e B^u_{i,j} - (1-f_e) B^u_{i+1,j}\right]. \end{aligned} \tag{9.3.13}$$

For a uniform grid, the control volume nodes are equidistant, so $f_e = 0.5, (A_u)_P = (A_u)_E = \Delta y/\delta x$. Assuming a constant source term we get:

$$
\begin{aligned}
u_{e_{i,j}} &= [0.5u_{i,j} + 0.5u_{i+1,j}] + \Bigg[(d_u)_e(p_{i,j} - p_{i+1,j}) \\
&\quad - 0.5(d_u)_E\left(\frac{p_{i-1,j} + p_{i,j}}{2} - \frac{p_{i,j} + p_{i+1,j}}{2}\right) \\
&\quad + 0.5(d_u)_P\left(\frac{p_{i,j} + p_{i+1,j}}{2} - \frac{p_{i+1,j} + p_{i+2,j}}{2}\right)\Bigg] \\
&\quad + [B^u_{e_{i,j}} - 0.5B^u_{i,j} - 0.5B^u_{i+1,j}] \qquad (9.3.14) \\
&= 0.5[u_{i,j} + u_{i+1,j}] + \frac{\Delta y}{a^u_{i,j}4\delta x}[4p_{i,j} - 4p_{i+1,j} \\
&\quad - p_{i-1,j} + p_{i+1,j} - p_{i,j} + p_{i+2,j}] \\
&= 0.5[u_{i,j} + u_{i+1,j}] + \frac{\Delta y}{a^u_{i,j}}\left[\frac{p_{i+2,j} + 3p_{i,j} - 3p_{i+1,j} - p_{i-1,j}}{4\delta x}\right]. \qquad (9.3.15)
\end{aligned}
$$

In Eq. (9.3.15), the second term inside the bracket on the right-hand side is equivalent to the third derivative of pressure. Thus, the momentum-interpolation technique adds an extra term to the interpolated velocity proportional to the third derivative of pressure ($\frac{d^3p}{dx^3}$) which eliminates the velocity/pressure oscillations. If we put the expression of face velocities from Eq. (9.3.15) into the integral form of continuity equation, the dissipative term is proportional to the fourth derivative of pressure. In the original Rhie and Chow interpolation the source term was absent in the formulation. Later it was found that if the source term is neglected, the presence of large body forces can produce physically unrealistic solutions [76]. For the same reason, the solution can depend on the under-relaxation factor (see [77]) or time step for unsteady cases (see [73]). Thus it is important to include the appropriate source terms during the momentum interpolation.

9.3.3 Correction of Velocity and Pressure Fields by Enforcing the Incompressibility Condition

The calculated interface velocities from momentum interpolation will not, in general, satisfy the continuity equation. Thus, the velocity and pressure fields have to be corrected in each iteration. Similar to the staggered-grid algorithm, a pressure-correction equation is solved; that can be derived by expressing the velocities as a summation of a guessed value (u^*) and a correction term (u'), that is, $u = u^* + u'$, and substituting them into the continuity equation. After re-arrangement, the equation for the pressure-correction p' takes the following form:

$$a^{pc}_{i,j}p' = a^{pc}_{i+1,j}p'_{i+1,j} + a^{pc}_{i-1,j}p'_{i-1,j} + a^{pc}_{i,j+1}p'_{i,j+1} + a^{pc}_{i,j-1}p'_{i,j-1} + b^{pc}_{i,j}, \qquad (9.3.16)$$

where

$$a^{pc}_{i+1,j} = \rho \frac{(\Delta y)_j}{a_{e_{i,j}}} (\Delta y)_j \tag{9.3.17}$$

$$a^{pc}_{i-1,j} = \rho \frac{(\Delta y)_j}{a_{w_{i,j}}} (\Delta y)_j \tag{9.3.18}$$

$$a^{pc}_{i,j+1} = \rho \frac{(\Delta x)_i}{a_{n_{i,j}}} (\Delta x)_i \tag{9.3.19}$$

$$a^{pc}_{i,j-1} = \rho \frac{(\Delta x)_i}{a_{s_{i,j}}} (\Delta x)_i \tag{9.3.20}$$

$$b^{pc}_{i,j} = \rho(u^*_w - u^*_e)(\Delta y)_j + \rho(v^*_s - v^*_n)(\Delta x)_i. \tag{9.3.21}$$

The a-coefficients are calculated from Eq. (9.3.11). The source term for pressure [Eq. (9.3.21)] represents the mass imbalance of the control volume at node (i, j). Once the solution of the pressure-correction equation (9.3.16) is obtained, the interface velocities are corrected as follows:

$$u_{e_{i,j}} = u^*_{e_{i,j}} + \frac{(\Delta y)_j}{a_{e_{i,j}}} (p'_{i,j} - p'_{i+1,j}) \tag{9.3.22}$$

$$u_{w_{i,j}} = u^*_{w_{i,j}} + \frac{(\Delta y)_j}{a_{w_{i,j}}} (p'_{i-1,j} - p'_{i,j}) \tag{9.3.23}$$

$$v_{n_{i,j}} = v^*_{n_{i,j}} + \frac{(\Delta x)_i}{a_{n_{i,j}}} (p'_{i,j} - p'_{i,j+1}) \tag{9.3.24}$$

$$v_{s_{i,j}} = v^*_{s_{i,j}} + \frac{(\Delta x)_i}{a_{s_{i,j}}} (p'_{i,j-1} - p'_{i,j}). \tag{9.3.25}$$

Similarly, the nodal velocities are corrected:

$$u_{i,j} = u^*_{i,j} + \frac{(\Delta y)_j}{a_{i,j}} (p'_w - p'_e) \tag{9.3.26}$$

$$v_{i,j} = v^*_{i,j} + \frac{(\Delta x)_i}{a_{i,j}} (p'_s - p'_n). \tag{9.3.27}$$

The nodal pressures are corrected by the following equation:

$$p_{i,j} = p^*_{i,j} + \omega_p p'_{i,j}, \tag{9.3.28}$$

where ω_p is the under-relaxation factor for pressure. The interface pressures p'_w, p'_e, p'_s, and p'_n are calculated by an interpolation of the nodal values. The interpolation technique is similar to Eq. (9.3.11).

Programming Call Sequence

```
call geometry
call property
```

```
call initial_conditions
call boundary_conditions
call reset_pressure
call velocity_coefficients
call velocity_solve
call momentum_interpolation
call pccoefficient
call pcsolve
call velocity_correction
call pcorrection
call covergence_check
call print_results
```

9.4 Numerical Examples

Example 9.4.1

Consider a steady-state fully developed flow between two parallel plates (**Example 6.4.1**). The non dimensional distance between the plates is $b = 4$, and the length of the plate is $a = 1$. Determine the velocity field when (a) the both walls are stationary, and (b) bottom wall is stationary and the top wall is moving with a constant velocity V_0. Assume unit pressure difference ($\Delta P = P_L - P_R = 1$) along the length of the channel. Treat all variables and properties as nondimensional. Use an FVM with 8 control volumes.

Solution As stated in **Example 6.4.1**, the problem can be reduced to a one-dimensional problem where the velocity field v_x is a function of y only, that is, $v_x = v_x(y)$. The steady-state flow equations are given by

$$\frac{dP}{dy} = 0, \quad \frac{dP}{dx} = -\frac{d}{dy}\left(\mu\frac{du}{dy}\right). \tag{1}$$

Since the pressure gradient along x-direction is constant, let $\frac{dP}{dx} = \beta$. Therefore, the flow equation reduces to the following form:

$$\frac{d}{dy}\left(\mu\frac{du}{dy}\right) = -\beta. \tag{2}$$

Integrating Eq. (2) over the control volume:

$$\begin{aligned}
\int_{CV}\frac{d}{dy}\left(\mu\frac{du}{dy}\right)dy &= \int_{CV} -\beta dy \\
\left[\mu\frac{du}{dy}\right]_s^n &= -\beta\Big[y\Big]_s^n \\
\mu\frac{du}{dy}\bigg|_n - \mu\frac{du}{dy}\bigg|_s &= -\beta[y_n - y_s] \\
\mu_n\frac{u_N - u_P}{\delta y_n} - \mu_s\frac{u_P - u_S}{\delta y_s} &= -\beta\,\Delta y
\end{aligned}$$

$$\left[\frac{\mu_n}{\delta y_n} + \frac{\mu_s}{\delta y_s}\right]u_P = \frac{\mu_s}{\delta y_s}u_S + \frac{\mu_n}{\delta y_n}u_N + \beta\,\Delta y. \tag{3}$$

The final discretization equation has the following form:

$$a_P u_P = a_S u_S + a_N u_N + b_P, \tag{4}$$

where

$$a_S = \frac{\mu_s}{\delta y_s}, \quad a_N = \frac{\mu_n}{\delta y_n}, \quad a_P = a_S + a_N, \quad b_P = \beta \Delta y. \tag{5}$$

In order to compute the coefficient matrix, we need to calculate the diffusion coefficients. The number of control volumes is $NCV = 8$. The number of nodes in the y-direction are: $jmax = NCV + 2 = 10$. Then the control volume length is, $\Delta y = b/NCV = 4.0/8 = 0.5$.

For the internal control volumes, the diffusion lengths are:

$$\delta y_s = \delta y_n = \Delta y = 0.5. \tag{6}$$

For the boundary control volumes the diffusion lengths are:

$$\delta y_s(\text{bottom}) = \delta y_n(\text{top}) = \Delta y/2 = 0.25. \tag{7}$$

Let us assume that the viscosity $\mu = 1.0$. Thus the diffusion coefficients for the internal control volumes are:

$$D_s = \frac{\mu_s}{\delta y_s} = \frac{1.0}{0.5} = 2.0, \quad D_n = \frac{\mu_n}{\delta y_n} = \frac{1.0}{0.5} = 2.0. \tag{8}$$

Similarly, the diffusion coefficients for the boundary control volumes are:

$$\begin{aligned} D_s(\text{bottom}) &= \frac{\mu_s}{\delta y_s} = \frac{1.0}{0.25} = 4.0 \\ D_n(\text{top}) &= \frac{\mu_n}{\delta y_n} = \frac{1.0}{0.25} = 4.0. \end{aligned} \tag{9}$$

Now we can calculate the a-coefficients at each node. The velocities at the bottom and top boundaries are known. So, $a_S u_S$ for bottom control volume and $a_N u_N$ for the top control volume are known and can be moved to the right-hand side of the equation.

The resulting equations for all control volumes are as follows:

$$\begin{aligned} a_{P(2)}u_2 &= a_{S(2)}u_1 + a_{N(2)}u_3 + b_{P(2)} \\ a_{P(3)}u_3 &= a_{S(3)}u_2 + a_{N(3)}u_4 + b_{P(3)} \\ a_{P(4)}u_4 &= a_{S(4)}u_3 + a_{N(4)}u_5 + b_{P(4)} \\ a_{P(5)}u_5 &= a_{S(5)}u_4 + a_{N(5)}u_6 + b_{P(5)} \\ a_{P(6)}u_6 &= a_{S(6)}u_5 + a_{N(6)}u_7 + b_{P(6)} \\ a_{P(7)}u_7 &= a_{S(7)}u_6 + a_{N(7)}u_8 + b_{P(7)} \\ a_{P(8)}u_8 &= a_{S(8)}u_7 + a_{N(8)}u_9 + b_{P(8)} \\ a_{P(9)}u_9 &= a_{S(9)}u_8 + a_{N(9)}u_{10} + b_{P(9)}. \end{aligned} \tag{10}$$

Case (a) Here, $u_1 = u_{10} = 0$. Using the values of the a-coefficients and applying the boundary conditions, we get the following equations:

$$\begin{aligned} 6u_2 &= 4u_1 + 2u_3 + 0.5 = 2u_3 + 0.5 \\ 4u_3 &= 2u_2 + 2u_4 + 0.5 \\ 4u_4 &= 2u_3 + 2u_5 + 0.5 \\ 4u_5 &= 2u_4 + 2u_6 + 0.5 \\ 4u_6 &= 2u_5 + 2u_7 + 0.5 \\ 4u_7 &= 2u_6 + 2u_8 + 0.5 \\ 4u_8 &= 2u_7 + 2u_9 + 0.5 \\ 6u_9 &= 2u_8 + 4u_{10} + 0.5 = 2u_8 + 0.5. \end{aligned} \tag{11}$$

Equations (11) can be solved using the Jacobi iteration, the Gauss–Seidel iteration or a direct method, for example, the TDMA method. The solution for Eqs. (11) is

$$u = [0.0, 0.5, 1.25, 1.75, 2.0, 2.0, 1.75, 1.25, 0.5, 0.0]. \tag{12}$$

Case (b) Here, $u_1 = 0$, and $u_{10} = V_0 = 1$. This changes the equation associated with top control volume:

$$6u_9 = 2u_8 + 4u_{10} + 0.5 = 2u_8 + 4.5. \tag{13}$$

The solution for the set of equations is

$$u = [0.0, 0.5625, 1.4375, 2.0625, 2.4375, 2.5625, 2.4375, 2.0625, 1.4375, 1.0]. \tag{14}$$

Figure 9.4.1 shows the horizontal velocity profile along the y-coordinate for parallel plate flow ($v_x = V_0 = 0$) and Couette flow ($v_x = V_0 = 1$). The finite volume solution for both cases agree well with the exact solution.

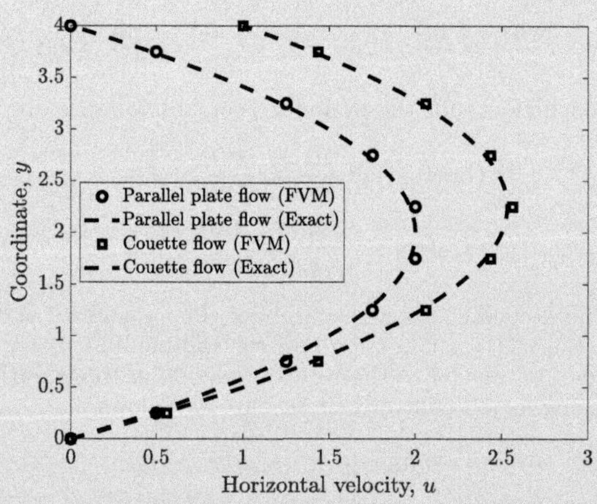

Fig. 9.4.1 Velocity profiles for flow between parallel plates ($V_0 = 0$) and Couette flow ($v_x = V_0 = 1$).

Example 9.4.2

Consider an incompressible viscous flow in a two-dimensional lid-driven cavity. The top lid of the unit square cavity (1 m×1 m) is moving at a constant velocity ($v_x = 1$ m/s; see Fig. 9.4.2). Assume viscosity $\mu = 1$ Pa-s and density $\rho = 1$ kg/m^3. Therefore, the Reynolds number is: $Re = \rho UL/\mu = 1$. Use an FVM to calculate the velocity field in the cavity. Use the SIMPLE algorithm to couple the velocity and pressure. Use the power-law scheme to represent the solution to the one-dimensional advection–diffusion equation. Plot the centerline velocities as a function of the distance along the horizontal and vertical directions. Use a staggered-grid approach with (a) 8×10, (b) 16×20 control volumes.

Solution The number of control volumes: $icv = 8, jcv = 10$. First we calculate the control volume lengths:

$$\begin{aligned} \Delta x &= \frac{L}{Nx} = \frac{1.0}{8} = 0.125 \text{ m} \\ \Delta y &= \frac{L}{Ny} = \frac{1.0}{10} = 0.1 \text{ m.} \end{aligned} \tag{1}$$

In staggered-grid approach, velocity and pressure control volumes are staggered. The diffusion lengths for the interior control volumes are the same as the control volume lengths. However,

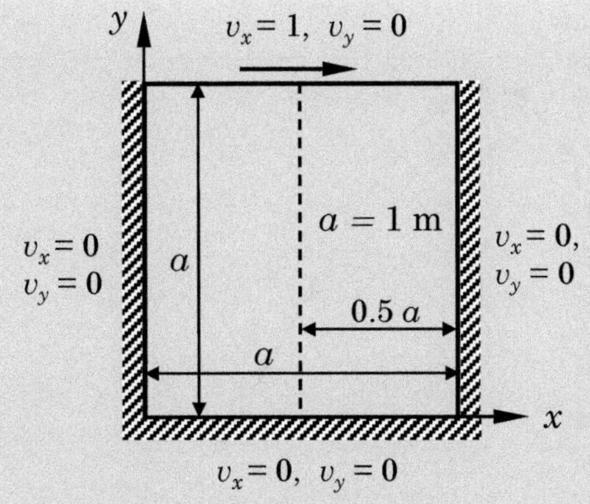

Fig. 9.4.2 Geometry of the square cavity and boundary conditions.

the diffusion lengths are different for boundary control volumes, as follows:

$$\begin{aligned}
\delta x_p(interior) &= 2\delta x_p(boundary) = \Delta x = 0.125 \text{ m}\\
\delta y_p(interior) &= 2\delta y_p(boundary) = \Delta y = 0.1 \text{ m}\\
\delta x_u(interior) &= \delta x_u(boundary) = \Delta x = 0.125 \text{ m}\\
\delta y_u(interior) &= 2\delta y_u(boundary) = \Delta y = 0.1 \text{ m}\\
\delta x_v(interior) &= 2\delta x_v(boundary) = \Delta x = 0.125 \text{ m}\\
\delta y_v(interior) &= \delta y_v(boundary) = \Delta y = 0.1 \text{ m}.
\end{aligned} \tag{2}$$

We have used the SIMPLE algorithm to solve the velocity and pressure variables. Figure 9.4.3 presents the steady state horizontal velocity profile at centerline for different mesh sizes. Figure 9.4.4 presents the steady state vertical velocity profile at center-line for different mesh sizes. From these two figures it is evident that the FEM and FVM solutions are in agreement.

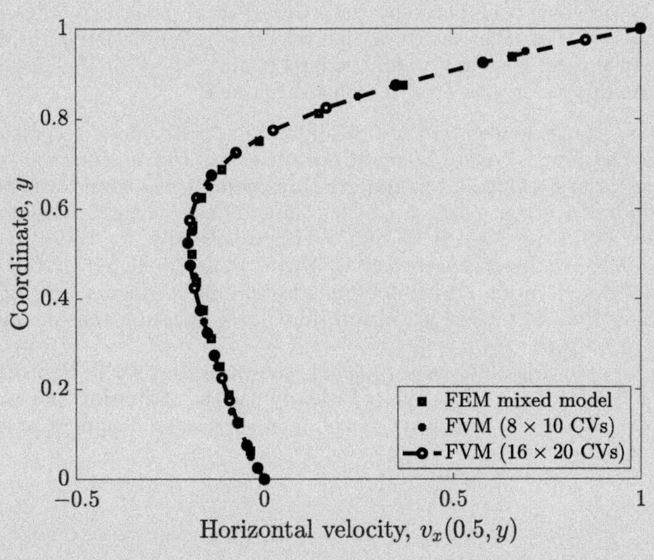

Fig. 9.4.3 Horizontal velocity $v_x(0.5, y)$ profile along y at $x = 0.5$.

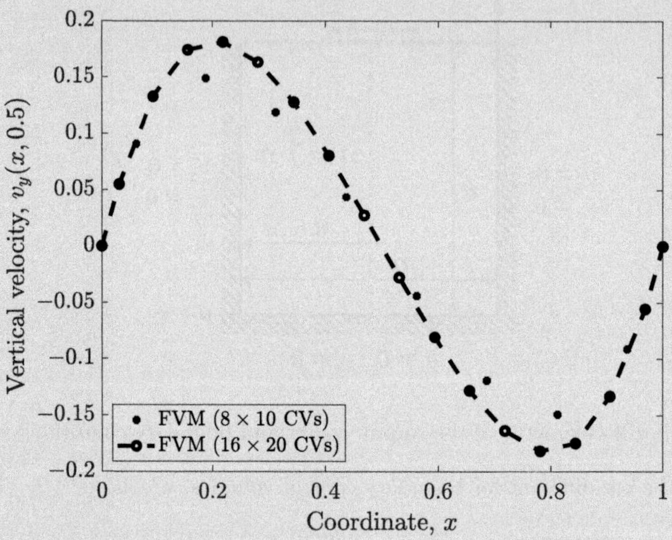

Fig. 9.4.4 Vertical velocity $v_y(x, 0.5)$ profile along x at $y = 0.5$.

Example 9.4.3

Consider a transient flow of an incompressible viscous fluid in a square cavity (1 m × 1 m). The flow is induced by the top surface having a specified velocity $v_x = 1$ and $v_y = 0$ (see Fig. 9.4.2). Assume that the Reynolds number is unity, $Re = \rho UL/\mu = 1$ (i.e., creeping flow or Stokes flow). Use the FVM with implicit Euler scheme to calculate the transient velocity field in the cavity. Link the pressure and velocity fields using the SIMPLE algorithm. Use the power-law to represent solution to the one-dimensional advection–diffusion equation. Use the same data as in the previous example, but study the effect of using two different time steps: $\Delta t = 0.001$ s and $\Delta t = 0.01$ s. Plot the centerline velocities as a function of the distance along horizontal and vertical directions at $t = 0.02$ s, 0.04 s, and 0.10 s for the case of $\Delta t = 0.01$. Use a staggered-grid approach with 16 × 20 control volumes.

Solution Here, the transient solution for the lid-driven cavity flow is computed using an implicit Euler time-stepping scheme. At any given time step, convergence was established for dependent variables prior to moving to the next time step. It was found that the number of iterations required for convergence decreased with time. The centerline horizontal and vertical velocity profiles are shown in Figs. 9.4.5 and 9.4.6, respectively, for times $t = 0.02$ s, 0.04 s, and 0.1 s. The FVM results compare very well with those predicted by the penalty FEM using a nonuniform, 8 × 10 mesh of nine-node quadratic finite elements (see Figs. 6.3.14 and 6.3.15). In obtaining the finite element results the Crank–Nicolson scheme with $\Delta t = 0.01$ was used.

The contour plot of *steady-state velocity field* is shown in Fig. 9.4.7. Figure 9.4.8 presents the velocity vector field in the cavity. We can clearly see the formation of a primary vortex near the top-middle region of the cavity. Secondary vortices will appear at corners if the Reynolds number is increased.

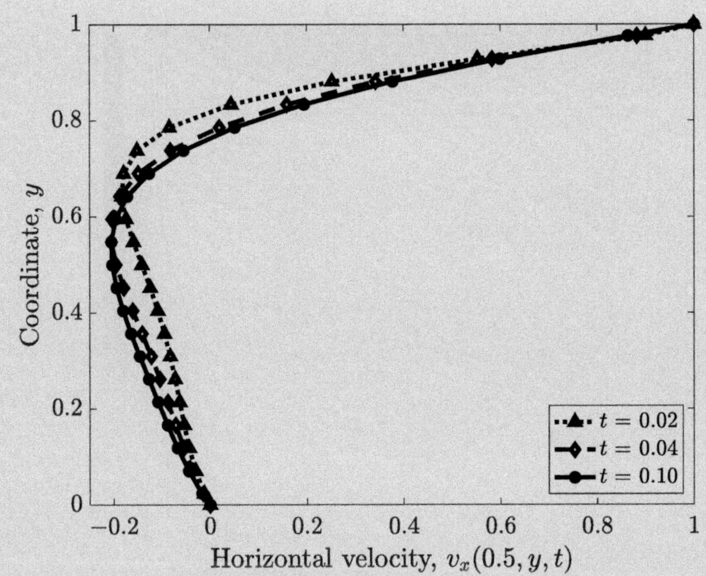

Fig. 9.4.5 Horizontal velocity profiles along y at $x = 0.5$ for various times.

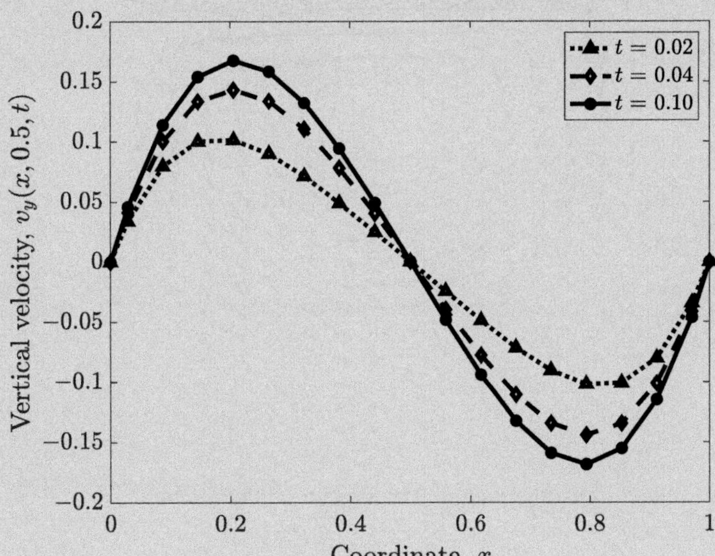

Fig. 9.4.6 Vertical velocity profiles along x at $y = 0.5$ for various times.

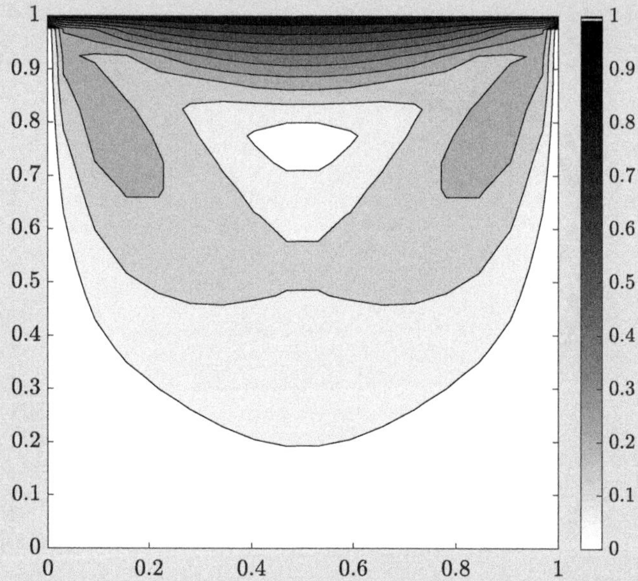

Fig. 9.4.7 Contour plot of steady-state velocity field for $Re = 1$.

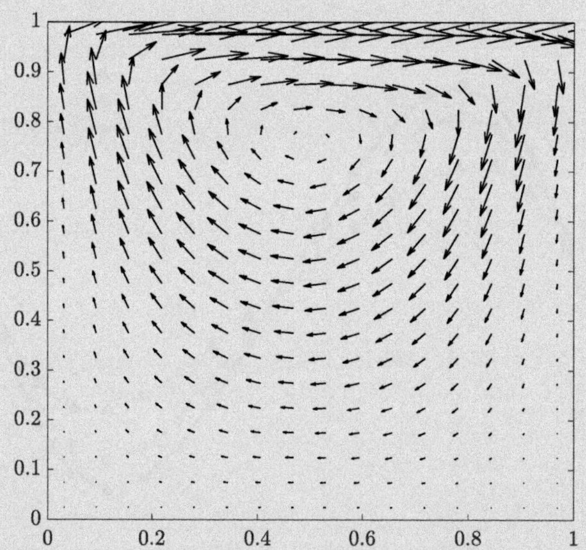

Fig. 9.4.8 Velocity vector plot of steady-state velocity field for $Re = 1$.

Example 9.4.4

Consider an incompressible viscous flow in a lid-driven square cavity (1 m×1 m). The Reynolds number is defined as $Re = \rho UL/\mu$, where ρ is the density, μ is the dynamic viscosity, U is the characteristic velocity, and L is the characteristic length. Solve the problem for $Re = 400$ and $Re = 1000$. Use an FVM to calculate the velocity field in the cavity. Use the SIMPLE algorithm to couple the velocity and pressure. Represent the solution to the one-dimensional advection–diffusion equation using the power law. Plot the centerline velocities as a function of the distance along horizontal and vertical directions. Use a collocated-grid approach with (a) 8×10, (b) 16×20, (c) 64×80 control volumes. Use the following under-relaxation factors: $\omega_u = \omega_v = 0.5$, and $\omega_p = 0.3$.

Solution We will solve the problem for the number of control volumes: $icv = 16, jcv = 20$. First we calculate the control volume lengths:

$$\Delta x = \frac{L}{Nx} = \frac{1.0}{16} = 0.0625 \text{ m}, \; \Delta y = \frac{L}{Ny} = \frac{1.0}{20} = 0.05 \text{ m}. \tag{1}$$

In a collocated-grid approach velocity and pressure control volumes are same. Therefore, the diffusion lengths for the interior and boundary control volumes are same for u, v, and p. The diffusion lengths can be calculated in the following way:

$$\begin{aligned} \delta x(interior) &= 2\delta x(boundary) = \Delta x = 0.0625 \text{ m}, \\ \delta y(interior) &= 2\delta y(boundary) = \Delta y = 0.05 \text{ m}. \end{aligned} \tag{2}$$

We used the SIMPLE algorithm to link the velocity and pressure variables. Figures 9.4.9 and 9.4.10 contain plots of the steady-state horizontal and vertical, respectively, velocity profiles at centerline for different mesh sizes for $Re = 400$. Similar plots for $Re = 1000$ are presented in Figs. 9.4.11 and 9.4.12. The nodal values of the horizontal velocity $v_x(0.5, y)$ are tabulated in Table 9.4.1 for $Re = 400$ and $Re = 1000$ while a comparison of the FVM and FEM results are included in Table 9.4.2 (also see Table 6.4.2).

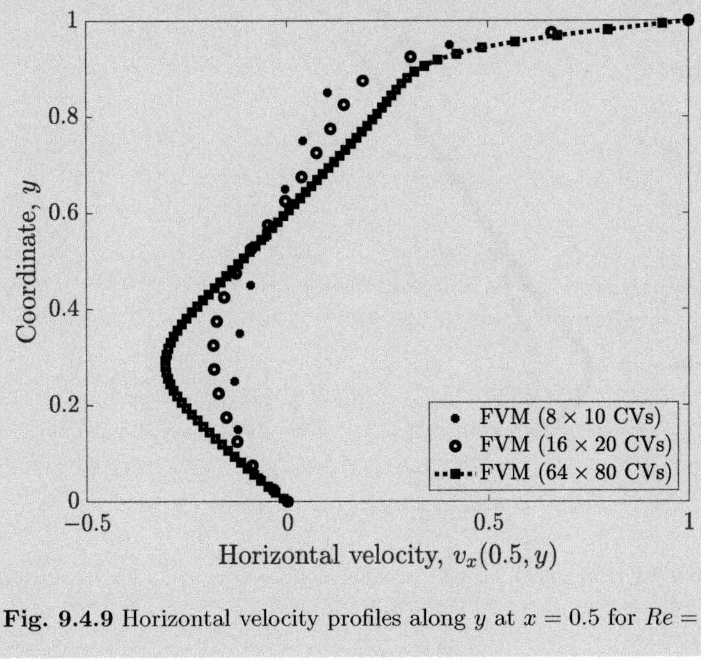

Fig. 9.4.9 Horizontal velocity profiles along y at $x = 0.5$ for $Re = 400$.

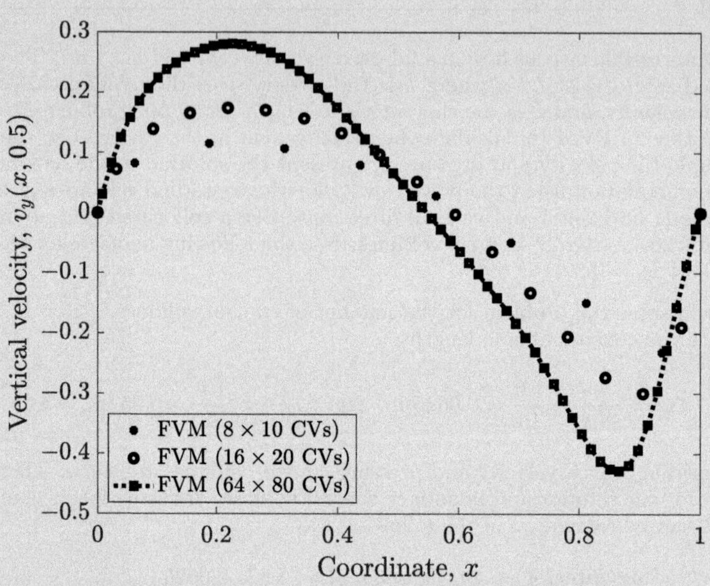

Fig. 9.4.10 Vertical velocity profiles along x at $y = 0.5$ for $Re = 400$; CV stands for control volume.

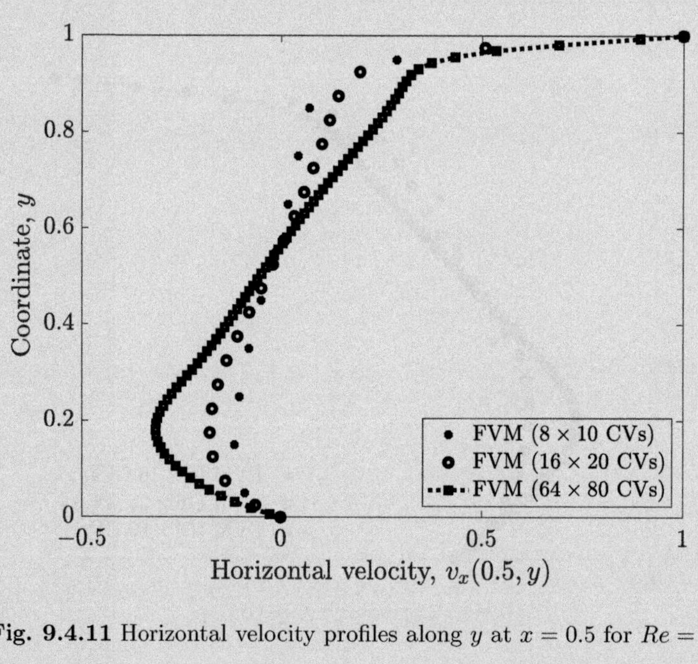

Fig. 9.4.11 Horizontal velocity profiles along y at $x = 0.5$ for $Re = 1000$

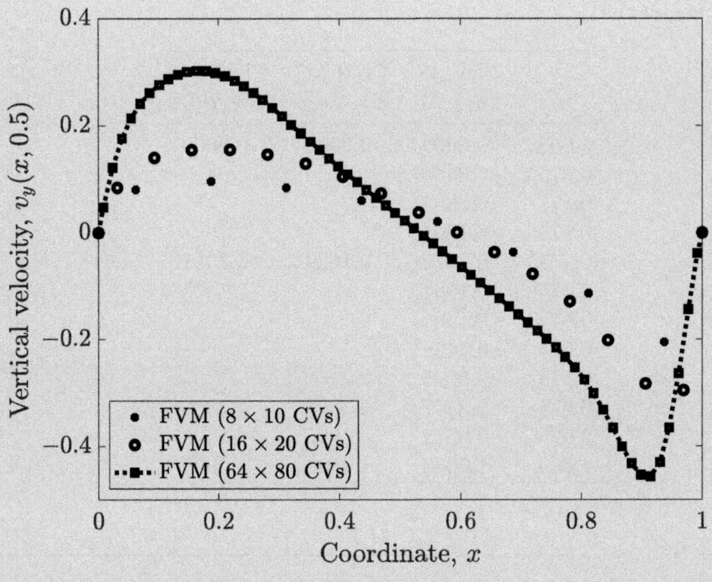

Fig. 9.4.12 Vertical velocity profiles along x at $y = 0.5$ for $Re = 1000$; CVs stand for control volumes.

Table 9.4.1 Values of $v_x(0.5, y)$ as a function of y for $Re = 400$ and $Re = 1000$ (mesh of 16 × 20 control volumes).

y	$Re = 400$	$Re = 1000$
0.0000	0.000000	0.000000
0.0250	−0.035500	−0.062350
0.0750	−0.087750	−0.137000
0.1250	−0.125000	−0.168000
0.1750	−0.152000	−0.176000
0.2250	−0.171000	−0.170000
0.2750	−0.182000	−0.155500
0.3250	−0.183500	−0.134000
0.3750	−0.175000	−0.107000
0.4250	−0.156500	−0.077750
0.4750	−0.127000	−0.048150
0.5250	−0.089650	−0.019550
0.5750	−0.047450	0.007685
0.6250	−0.004220	0.033550
0.6750	0.037000	0.057850
0.7250	0.074850	0.080500
0.7750	0.109500	0.101800
0.8250	0.143000	0.121000
0.8750	0.190000	0.142500
0.9250	0.310000	0.196500
0.9750	0.659500	0.507500
1.0000	1.000000	1.000000

Table 9.4.2 Velocity $v_x(0.5, y)$ obtained with the FVM ($Re = 1$) and FEM ($Re = 0$); (M = mixed and P = penalty).

y	FVM 16 × 20	FEM-M 8 × 10Q	FEM-P 16 × 20L
0.000	0.0000	0.0000	0.0000
0.075	−0.0449		
0.175	−0.0903		
0.275	−0.1304		
0.375	−0.1676	−0.1629	−0.1511
0.475	−0.1961		
0.575	−0.1988		
0.675	−0.1429		
0.775	0.0215		
0.875	0.3477	0.3694	0.3926
0.975	0.8512		
1.000	1.0000	1.0000	1.0000

Example 9.4.5

Consider the two-dimensional flow of a viscous incompressible fluid in a square cavity (1 m×1 m), where the flow is induced by the motion ($v_x = V_0$ and $v_y = 0$) of the top boundary (lid). Solve the problem for $Re = 100$ using the FVM, and calculate the velocity field in the cavity. Link the pressure and velocity and pressure fields using the SIMPLE algorithm. Use a staggered-grid approach with 8×8, 16×16, 32×32, 64×64, 128×128, and 256×256 control volumes. Use the following under-relaxation factors: $\omega_u = \omega_v = 0.5$, and $\omega_p = 0.5$. Use (a) power-law, and (b) hybrid schemes to represent solution for one-dimensional advection–diffusion equation.

(1) Find the minimum u and v velocities and tabulate their values for different grid sizes (see also Table 9.4.3). Compare the values with the solution of Ghia et al. [78].

(2) Calculate the orders of convergence and the GCI.

Solution The orders of convergence (i.e., U_{min} versus number of control volumes) for the lid-driven cavity are presented in Fig. 9.4.13.

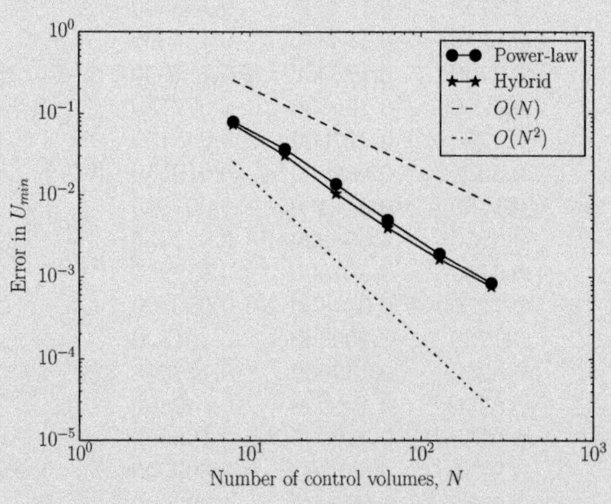

Fig. 9.4.13 Minimum u-velocity for different grid sizes.

Table 9.4.3 Values of U_{min}/U_{lid} and V_{min}/U_{lid} for $Re = 100$ as a function of the mesh size.

Grid #	Grid size	$\frac{U_{min}}{U_{lid}}$ (hybrid)	$\frac{V_{min}}{U_{lid}}$ (hybrid)	$\frac{U_{min}}{U_{lid}}$ (power law)	$\frac{V_{min}}{U_{lid}}$ (power law)
1	8×8	−0.14093	−0.16214	−0.13535	−0.15177
2	16×16	−0.18339	−0.20516	−0.17737	−0.20156
3	32×32	−0.20353	−0.23372	−0.20030	−0.23031
4	64×64	−0.21001	−0.24495	−0.20902	−0.24397
5	128×128	−0.21237	−0.24961	−0.21210	−0.24936
6	256×256	−0.21327	−0.25173	−0.21320	−0.25167
Ghia et al. [78]	129×129	−0.21090	−0.24533	−0.21090	−0.24533

The order of convergence p can be calculated from the slope of the curve of $\ln(error)$ versus $\ln(h)$, where h is the grid size. If the grid refinement ratio (r) is constant, the following formula can be used to calculate p:

$$p = \frac{\ln\left(\frac{f_3 - f_2}{f_2 - f_1}\right)}{\ln(r)}. \tag{1}$$

In this case, the refinement ratio is $r = 2$. Using formula (1) for U_{min}/U_{lid} and the three fine grid sizes, we obtain:

$$\text{hybrid: } p = \{1.394, 1.504, 1.488\}, \quad \text{power law: } p = \{1.634, 1.461, 1.380\}. \tag{2}$$

From the Taylor series analysis, it is not possible to assign the order of an FVM for either the power law or hybrid schemes. From these results we can see that the order of accuracy of both schemes is between first and second order. The same can be observed from Fig. 9.4.13.

The GCI gives an estimation of how far the solution is from the asymptotic numerical solution. The value of the GCI can be estimated using the formula

$$GCI_{fine} = \frac{F_s|\varepsilon|}{r^p - 1}. \tag{3}$$

where F_s is a factor of safety. The recommended value of F_s is 3.0 for two grids, and 1.25 for three or more grids. Since we have used more than two grids, we take $F_s = 1.25$. The ε is the relative error defined by

$$\varepsilon = \frac{f_1}{f_2} - 1, \tag{4}$$

where f_2 is the finer grid. Using the above formula, the GCI for the power-law and hybrid scheme can be obtained:

$$\text{hybrid: } GCI = \{5.876, 2.202, 0.865\}\%, \quad \text{power law: } p = \{8.793, 2.840, 1.004\}\%. \tag{5}$$

We can check the asymptotic range of convergence for three grids using the following formula:

$$\frac{GCI_{2,3}}{r^p GCI_{1,2}} \approx 1. \tag{6}$$

When the above expression is approximately 1, we are near the asymptotic range of convergence. For our example, we find that our solutions are within asymptotic convergence:

$$\text{hybrid: } \frac{GCI_{2,3}}{r^p GCI_{1,2}} = 1.0224, \quad \text{power law: } \frac{GCI_{2,3}}{r^p GCI_{1,2}} = 0.992. \tag{7}$$

Example 9.4.6

Consider a two-dimensional laminar flow of a viscous incompressible fluid inside a channel formed by two stationary parallel plates as shown in Figure 9.4.14. The fluid enters the channel with a uniform velocity U_{in} and temperature T_{in}. Channel walls are kept at a constant temperature T_w.

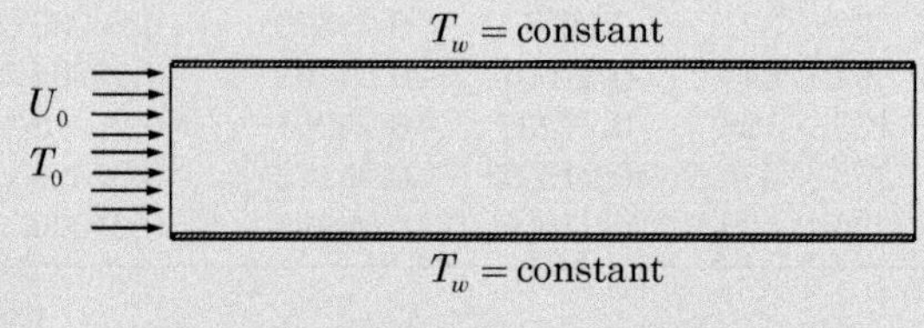

Fig. 9.4.14 Flow between parallel plates.

Write a finite volume code to predict flow and temperature fields for $Re = 200$. Represent the solution to one-dimensional advection–diffusion equation by the power law. Link the velocity and pressure fields using the SIMPLE algorithm. Declare convergence when R_U, R_V, R_P, and $R_T < 10^{-6}$. Use $\omega_u = \omega_v = 0.3$ and $\omega_p = 0.7$. Given: channel height ($H = 0.02$ m, channel length ($L = 2.0$ m, $T_{in} = 27\,°$C, and $T_w = 100°$ C, $\rho = 997.0$ kg/m^3, $k = 0.563$ W/m K, $\mu = 8.71 \times 10^{-4}$ N-s/m^2, $C_p = 4179.0$ J/kg K.

(1) Run your program with 10×5 control volumes. Print x-velocity (U), and temperature (T) fields to check for symmetry.

(2) Run your program using 20×10, 60×20, 120×40, 160×80 control volumes. Report the following cases:

(a) Plot U/U_{in} at the channel exit ($x = 2.0$ m) for the above grid sizes and show that the maximum normalized velocity (U_{max}/U_{in}) approaches 1.50 as the mesh is refined.

(b) Plot the nondimensional temperature ($\frac{T_w - T}{T_w - T_{bulk}}$) at $x = 2.0$ m as a function of y.

(c) Plot T_{bulk} and T_w as a function of x.

(d) Plot the local Nusselt number as a function of x. Compare the fully developed flow Nusselt number value (Nu_{fd}) with the exact analytical solution ($Nu_{exact} = 7.5407$) [79]. In addition, calculate the Fanning friction factor for fully developed flow and compare it with the analytical value ($f_{exact} = 24/Re$) [80].

Tips for accelerating convergence of iterative solver: (1) Correct the outlet velocity using a global mass-conservation technique. (2) At each cross section, make an additional correction for the longitudinal pressure gradient (dP/dx) and velocity (U) using the difference in the mass flow rate at the inlet and the corresponding cross section (see page 1795 of [67] for details).

Solution Table 9.4.4 contains the streamwise-velocity (U) for a grid size of 10×5 control volumes. It is important to note that the velocity values are symmetric about the channel axis up to seven decimal points. Checking for symmetric solutions for problems with symmetric geometry and subjected to symmetric boundary conditions is an important step in *verifying* a numerical method. While symmetric results will not guarantee *verification* of a numerical solution, the absence of symmetry in the solution most definitely indicates a shortcoming in either the problem formulation, implementation of a numerical technique, or the code. In other words, for symmetric problems symmetry in solution is a necessary condition for *verification*, but is not sufficient. To ensure convergence and symmetric results readers are encouraged to use a double precision format in the programming language.

Figure 9.4.15 shows the normalized stream-wise velocity (U/U_{in}) at the channel exit ($x = 2.0$ m) for various grid sizes. As expected, the fully developed x-velocity follows a parabolic profile. The results are almost indistinguishable for grid sizes finer than 60×20 control volumes.

Table 9.4.4 Values of $U(x,y)$ as a function of x and y for $Re = 200$ for the mesh size of 10×5 control volumes [$U(x,0) = U(x,0.02) = 0$].

	$U(x,y)$				
x (m)	$y = 0.002$ m	$y = 0.006$ m	$y = 0.010$ m	$y = 0.014$ m	$y = 0.018$ m
0.00	0.00436810	0.00436810	0.00436810	0.00436810	0.00436810
0.20	0.00260359	0.00526317	0.00610699	0.00526317	0.00260359
0.40	0.00244857	0.00533317	0.00627704	0.00533317	0.00244857
0.60	0.00242965	0.00533852	0.00630417	0.00533852	0.00242965
0.80	0.00242714	0.00533881	0.00630860	0.00533881	0.00242714
1.00	0.00242679	0.00533880	0.00630934	0.00533880	0.00242679
1.20	0.00242673	0.00533879	0.00630946	0.00533879	0.00242673
1.40	0.00242673	0.00533879	0.00630948	0.00533879	0.00242673
1.60	0.00242672	0.00533879	0.00630948	0.00533879	0.00242673
1.80	0.00242672	0.00533879	0.00630948	0.00533879	0.00242672
2.00	0.00242672	0.00533879	0.00630948	0.00533879	0.00242672

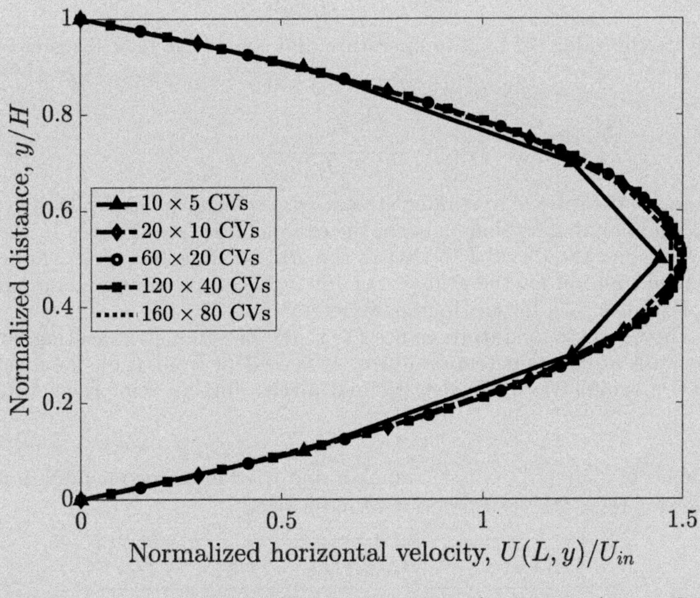

Fig. 9.4.15 Axial velocity profile at the channel exit ($x = 2.0$ m) as a function of y for different grid sizes; CV denotes control volume.

The *Fanning friction factor* is defined as the ratio of the wall shear stress and kinetic energy density:

$$f = \frac{\tau_w}{\frac{1}{2}\rho U_{avg}^2}, \tag{9.4.1}$$

where τ_w is the wall shear stress, ρ is the fluid density and U_{avg} is the average bulk velocity. We calculate the wall shear stress and average bulk velocity as a function of x using the following formulas:

$$\tau_{w,x} = \mu \frac{\partial U}{\partial y}\bigg|_{y=0}, \quad U_{avg,x} = \frac{\int \rho U dy}{\int \rho dy}. \tag{9.4.2}$$

Using the definition of the Reynolds number ($Re = \rho U_{avg} D_h/\mu$), we can calculate the Fanning friction factor as a function of x in the following way:

$$f_x = \frac{\frac{\partial U}{\partial y}\Big|_{y=0}}{2ReU_{avg,x}D_h}. \tag{9.4.3}$$

Here D_h is the hydraulic diameter and for channel flow $D_h = 2H$, where H is the height of the channel.

The Nusselt number is the ratio of the convective heat transfer to the conductive heat transfer, and it is defined by the following formula:

$$Nu = \frac{hD_h}{k}, \tag{9.4.4}$$

where h is the convective heat transfer coefficient and k is the thermal conductivity of the fluid. The convective heat transfer coefficient can be calculated from the wall heat flux in the following way:

$$q_w = -k\frac{\partial T}{\partial y}\bigg|_{y=0} = h(T_w - T_{bulk}), \tag{9.4.5}$$

where T_w is the wall temperature and T_{bulk} is the bulk temperature, which can be calculated as a function of x:

$$T_{bulk,x} = \frac{\int \rho UT dy}{\int \rho U dy}. \tag{9.4.6}$$

Once we substitute the values of bulk temperature and convective heat transfer coefficient in Eq. (9.4.4), we get:

$$Nu_x = \frac{D_h \frac{\partial T}{\partial y}\Big|_{y=0}}{(T_w - T_{bulk,x})}. \tag{9.4.7}$$

Table 9.4.5 shows the ratio of maximum stream-wise velocity to inlet velocity (U_{max}/U_{in}), Nusselt number, and Fanning friction factor at the channel exit ($x = 2.0$ m). It is evident from both the Table 9.4.5 and the Fig. 9.4.15 that as the grid is refined U_{max}/U_{in} approaches 1.50 [80]. The numerical solution for the grid size of 160 × 80 control volumes is almost identical to the analytical solution [79] for all three parameters.

Table 9.4.6 shows the temperature values ($T(x, y)$) for symmetric heating conditions by maintaining the walls at constant temperatures. It is evident from Table 9.4.6 that the temperature values are symmetric up to five decimal places, further reinforcing veracity of the solution.

Table 9.4.5 Values of U_{max}/U_{in} Nusselt number and friction factor for fully developed flow and heat transfer for $Re = 200$ as a function of mesh size.

Grid size	U_{max}/U_{in}	Nu_{fd}	f_{fd}
10 × 5	1.4444	7.5444	0.1111
20 × 10	1.4706	7.5416	0.1176
60 × 20	1.4925	7.5409	0.1194
120 × 40	1.4981	7.5408	0.1198
160 × 80	1.4995	7.5408	0.1200
Analytical [79, 80]	1.5000	7.5407	0.1200

Table 9.4.6 Values of $T(x, y)$ as a function of x and y for $Re = 200$ for the mesh size of 10×5 control volumes $[T(x, 0) = T(x, 0.02) = 100]$.

	$T(x,y)$				
x (m)	$y = 0.002$ m	$y = 0.006$ m	$y = 0.010$ m	$y = 0.014$ m	$y = 0.018$ m
0.00	27.00000000	27.00000000	27.00000000	27.00000000	27.00000000
0.10	56.44264403	37.39068674	32.17416080	37.39068872	56.44264498
0.30	73.29486084	44.56665491	36.65937199	44.56665828	73.29486363
0.50	80.31977606	50.97896541	41.68259745	50.97896832	80.31977817
0.70	83.83833266	56.55604424	46.87415428	56.55604578	83.83833333
0.90	86.06188512	61.37820468	51.93265619	61.37820476	86.06188457
1.10	87.72759158	65.57562145	56.69020235	65.57562044	87.72759033
1.30	89.09762554	69.26117654	61.07382771	69.26117499	89.09762412
1.50	90.27581055	72.52013612	65.06525554	72.52013457	90.27580935
1.70	91.31001579	75.41590026	68.67461338	75.41589910	91.31001504
1.90	92.22666669	77.99696330	71.92539548	77.99696247	92.22666625
2.00	92.68499214	79.28749483	73.55078653	79.28749416	92.68499185

Figure 9.4.16 shows the nondimensional temperature for various grid sizes at the channel exit. As the grid is refined the temperature profile becomes shaped towards a parabola. Another observation is that the fully developed temperature profile does not change significantly between 120×40 control volumes and 160×80 control volumes. So for all practical purposes one may declare grid independence with 120×40 control volumes.

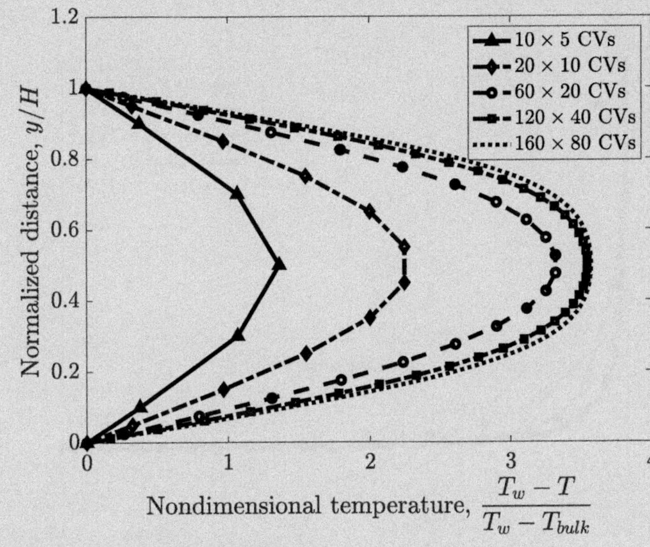

Fig. 9.4.16 Temperature profiles at channel exit ($x = 2.0$ m) as a function of y for different grid sizes; CV stands for control volume

Figure 9.4.17 shows the variation of bulk temperature as a function of stream-wise distance for various mesh sizes. Here again, the bulk temperature distribution points to 120×40 control volumes as a grid-independent solution. Figure 9.4.18 shows the variation of the local Nusselt number (Nu_x) as a function of stream-wise distance. Note that for all mesh sizes considered, the Nusselt numbers for fully developed flow approach the analytical value of 7.54 [79], further bolstering the veracity of the FVM solution.

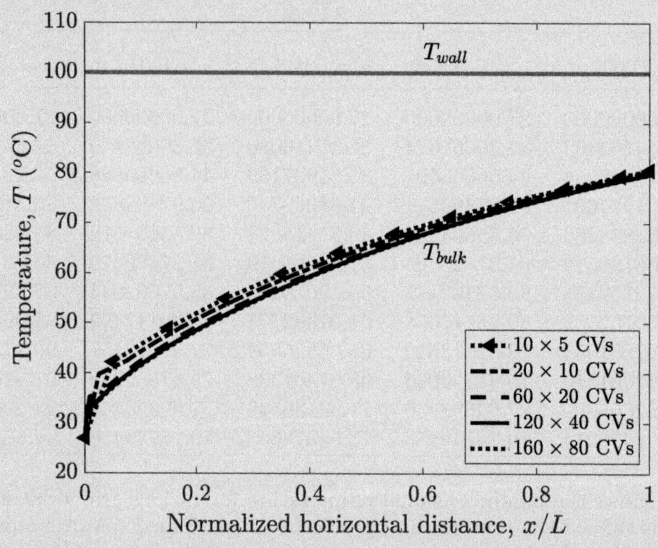

Fig. 9.4.17 Bulk temperature as a function of x for different grid sizes.

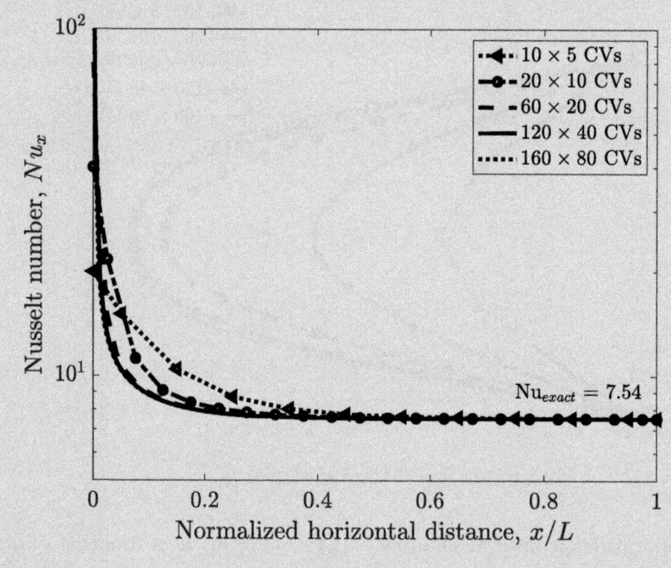

Fig. 9.4.18 Temperature profiles as a function of y for different grid sizes.

9.5 Treatment of Solid Obstacles in Flow Paths

9.5.1 Preliminary Comments

Discussion on the treatment of solid obstacles in flow paths using FVMs is presented in Patankar [7]. Here, we present the approach and illustrate it with an example.

In general, treatment of obstacles in flow paths is not an issue in the FEM as elements can be deployed to fit the domain. Similarly, treatment of obstacles in flow passages using a body-fitted or polyhedral grid in the FVM is a non-issue. However, for a rectangular mesh or non body-fitted coordinate system there are some challenges to be addressed.

To begin with, the obstacles can be classified into two categories:

(1) obstacles attached to one of the boundaries, and

(2) island-type boundaries that do not touch any boundaries.

These two types of obstacles are illustrated in Figs. 9.5.1 and 9.5.2.

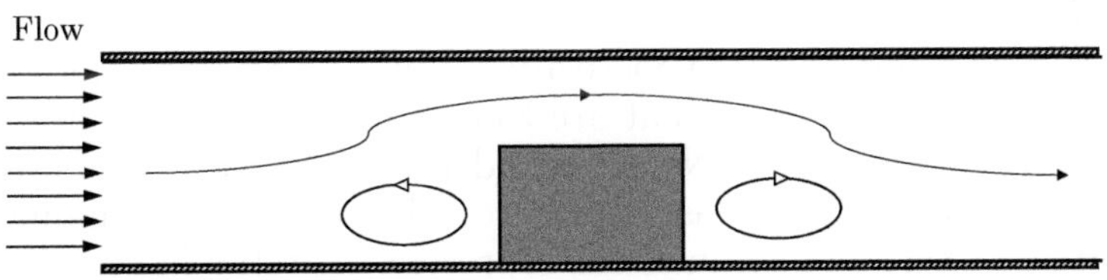

Fig. 9.5.1 Obstacle attached to the boundary.

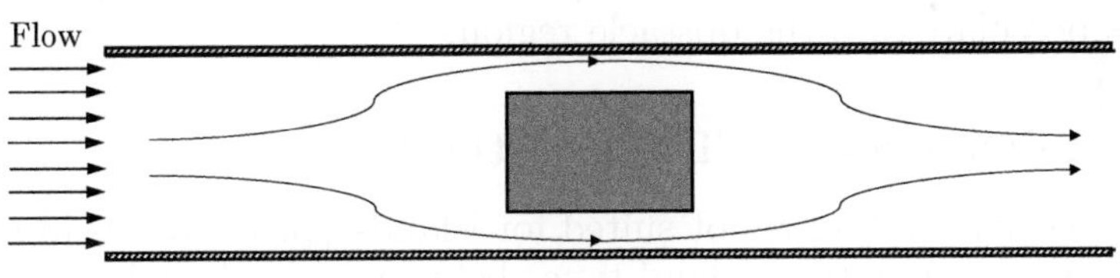

Fig. 9.5.2 Island type obstacle.

9.5.2 Domain Decomposition Method

For both types of obstacles a "domain decomposition" method can be applied. The idea is to decompose a large and complex domain into smaller, simpler subdomains and solve the governing equations in those subdomains. For the obstacle attached to the boundary, the domain is divided into three sub-domains Ω_1, Ω_2, and Ω_3, as shown in Fig. 9.5.3. For the case of island-type obstacles, the domain is divided into four subdomains Ω_1, Ω_2, Ω_3, and Ω_4 (see Fig. 9.5.4).

In the domain-decomposition method, discretization equations for each subdomain are solved and values are matched at the internal boundaries using the current values during the iterative process. This is a well-established method. For a review of the domain-decomposition method readers are referred to [81].

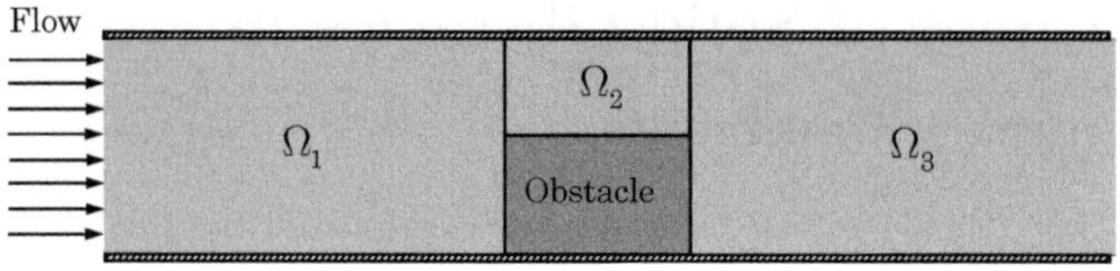

Fig. 9.5.3 Domain decomposition for obstacle attached to the boundary.

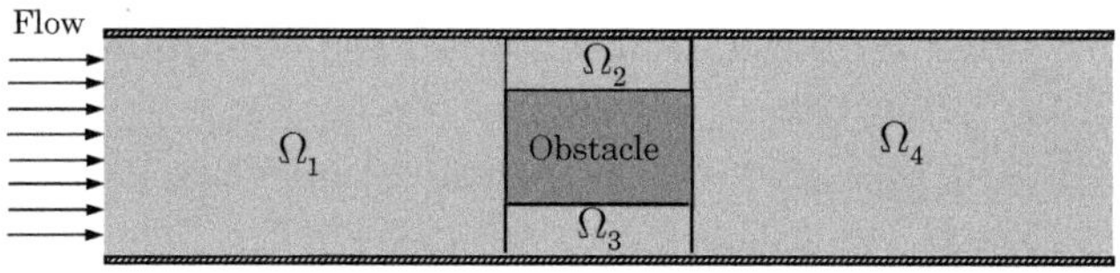

Fig. 9.5.4 Domain decomposition for island type obstacle

9.5.3 High-Viscosity Method

An alternative method is "high viscosity method." In this method, control volumes covering the obstacles that are touching the boundary are assigned a very high viscosity ($\mu \approx 10^{50}$ Ns/m^2), and thermophysical properties associated with flux transport such as thermal conductivity and dynamic viscosity at the interfaces of control volumes are represented by the harmonic mean and weighted harmonic mean for a nonuniform grid. Discretization equation structure is unaltered and the solution process is carried out as explained earlier. Such a process will result in near-zero velocities ($u = v \approx 10^{-50}$) in the region covering obstacles as no slip boundary condition ($u = v = 0$) values at the boundaries percolate into the obstacle region.

9.5.4 Dominant-Source-Term Method

The high-viscosity method is not suited for island-type obstacles in a flow passage. The best option to treat island-type obstacles is the domain-decomposition method. However, an alternative formulation was proposed by Patankar [7] that is referred to as *dominant-source-term method.* The objective is to modify the source term such that the resulting velocity field inside the obstacle is zero ($u = v = 0$). To illustrate this method let us revisit the discretization equations for x-momentum and y-momentum:

$$a^u_{i,j} u_{i,j} = \sum_{nb} a^u_{nb} u_{nb} + A_u(p_{iu,jv} - p_{iu+1,ju}) + b^u_{i,j} \tag{9.5.1}$$

$$a^v_{i,j} v_{i,j} = \sum_{nb} a^v_{nb} v_{nb} + A_v(p_{iv,jv} - p_{iv,jv+1}) + b^v_{i,j}. \tag{9.5.2}$$

First we linearize the source term:

$$b^u_{i,j} = b^u_1 - b^u_2 u_{i,j}, \quad b^v_{i,j} = b^v_1 - b^v_2 v_{i,j}. \tag{9.5.3}$$

Then we assume $b_1^u = \beta_u u_{desired}$ and $b_2^u = \beta_u$, $b_1^v = \beta_v v_{desired}$ and $b_2^v = \beta_v$. So, the above relations become:

$$b_{i,j}^u = \beta_u(u_{desired} - u_{i,j}), \quad b_{i,j}^v = \beta_v(v_{desired} - v_{i,j}). \tag{9.5.4}$$

To obtain the desired velocity in the solid-obstacle region, we set β_u and β_v to very high values ($\sim 10^{50}$) such that the other terms in Eqs. 9.5.1 and 9.5.2 become negligible and can be dropped. This way we can achieve the desired velocities in the obstacle region:

$$u_{i,j} = u_{desired}, \quad v_{i,j} = v_{desired}. \tag{9.5.5}$$

Here $u_{desired}$ and $v_{desired}$ can be any values (zero or nonzero). For stationary obstacles, one must use zero values. However, to simulate flow regions with constant velocities, nonzero values can be used.

Example 9.5.1

Consider an incompressible steady laminar flow over a backward-facing step as shown Fig. 9.5.5. The properties of the fluid are: $\rho = 1.177$ kg/m^3 and $\mu = 1.843 \times 10^{-5}$ Ns/m^2.

Write an FVM computer program to predict two-dimensional steady laminar flow field. Link the pressure and velocity fields using the SIMPLE algorithm. Represent the solution to one-dimensional advection–diffusion equation by the power law. Declare convergence when R_U, $R_V < 10^{-6}$, and $R_P < 10^{-5}$.

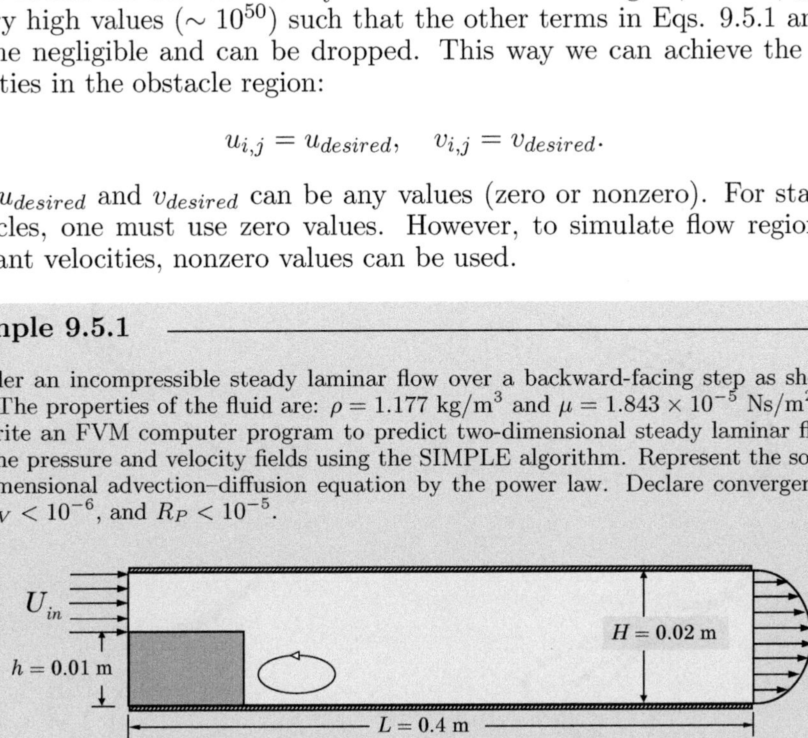

Fig. 9.5.5 Flow over a backward-facing step.

(1) Plot the u-velocity profiles (u/U_{in}) as a function of y at $x = 6.5h$, $8.5h$, $9.5h$, $10.5h$, $11.5h$, and $12.5h$. Use a mesh of 60×30 control volumes.

(2) Calculate (X_s/h) for $Re_h = 50$, 100, and 200, where X_s is the reattachment length. Verify your result with the empirical correlation of Goldstein [82]: $\frac{X_s}{h} = 2.13 + 0.021\,Re_h$. Present the comparison results in a tabular form.

(3) Carry out calculations for $Re_h = 200$ using 30×10, 60×20, 90×30, and 120×40 meshes of uniformly sized control volumes in the x- and y-directions, respectively. For each grid size, calculate the reattachment length (X_s), and present the dimensionless reattachment length (X_s/h) for each grid size in a tabular form.

Solution Flow over a backward-facing step is an important problem that is often used to benchmark and verify numerical codes. Flow over a backward-facing step is rich in physics as it includes both through-flow and flow recirculation. The primary recirculation zone appears downstream of the step. Typically for higher Reynolds number flows, a secondary recirculation zone may appear near the roof of the channel (i.e., at the upper wall). The secondary recirculation zone is difficult to capture with the mesh sizes used for classroom exercises.

The fluid enters the channel inlet, separates at the downstream top corner of the step, and then impinges on the floor of the channel, resulting in a recirculation bubble. The distance between the right (downstream) bottom corner of the step and the point where the flow impinges on the channel floor is termed the reattachment length (X_s). To find the point where the flow touches the channel floor, we can calculate the position where the velocity adjacent to the bottom wall changes sign, $u(iu, 2) * u(iu - 1, 2) \leq 0$. In this example we use the "dominant-source-term" method to simulate a zero-velocity flow field in the step region.

Figure 9.5.6 shows the normalized streamwise velocity u/U_{in} as a function of the normalized cross-stream distance (y/H) at various streamwise locations for $Re = 50, 100$, and 200, obtained using a 60×30 uniform mesh. For a given streamwise location, the maximum streamwise velocity increases with an increase in Re. The magnitude and the region of negative u-velocity decreases from left to right in the streamwise direction ($x = 6.5h$ to $12.5h$). These observations are consistent with the flow physics.

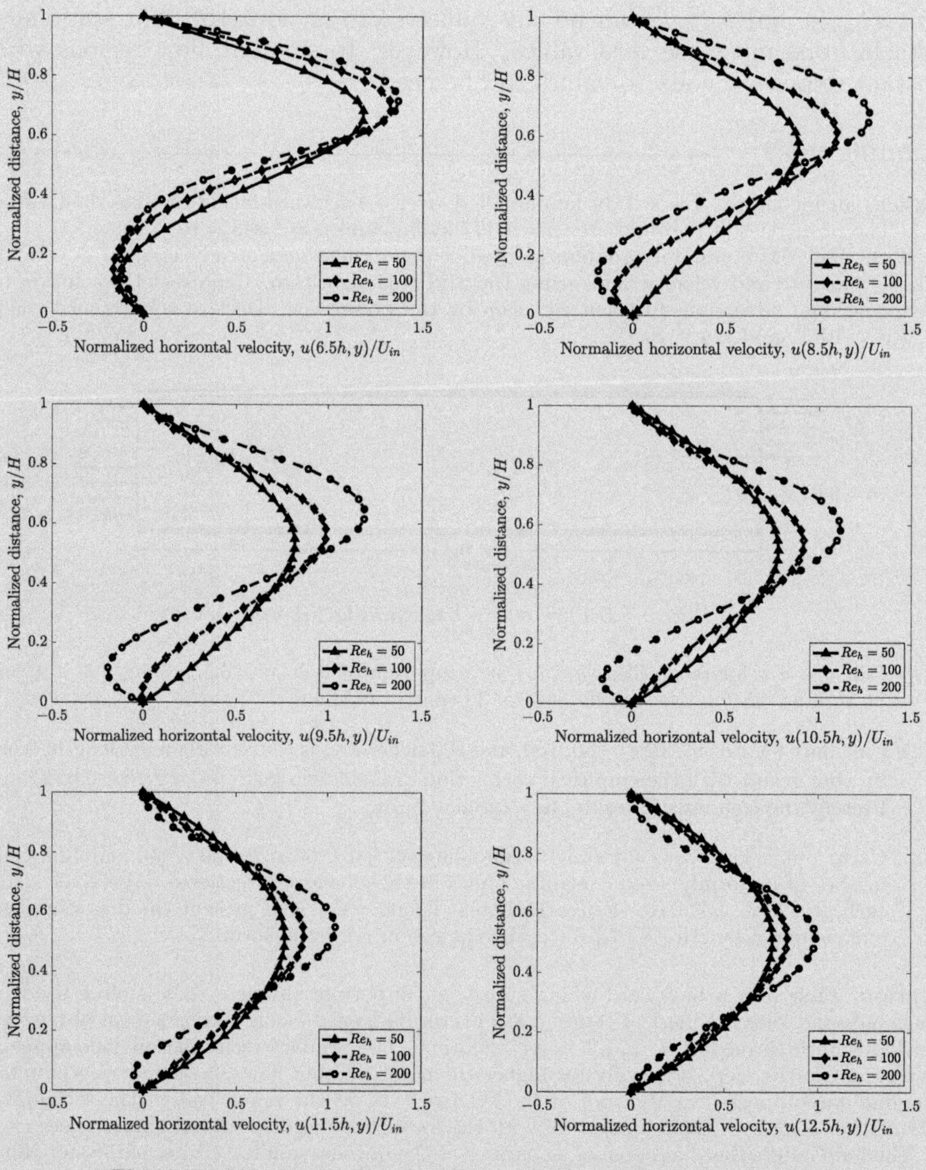

Fig. 9.5.6 Velocity profiles as a function of y at different x-locations.

The calculated values of the reattachment length are compared with the experimental data of Goldstein et al. [82] in Table 9.5.1. Results for a grid size of 60×30 control volumes with three different Reynolds numbers ($Re = 50, 100$, and 200) are presented. The maximum difference between the numerical prediction and the experimental data is approximately 19 percent. In Table 9.5.2 the numerically predicted dimensionless reattachment lengths are compared with the experimental data for $Re = 200$ for various mesh sizes. Oddly, the numerical prediction for 80×40 control volumes differs with the experimental data by about 19 percent. This difference can be attributed to the experimental uncertainties and not having established a grid-independent solution.

Table 9.5.1 Reattachment length X_s/H for different Reynolds numbers.

Re_h	X_s/H	X_s/H [82]
50	2.833	3.18
100	4.833	4.23
200	7.499	6.33

Table 9.5.2 Reattachment length X_s/H at $Re_h = 200$ for different grid sizes

Grid size	X_s/H	X_s/H [82]
20×10	6.00	6.33
40×20	7.00	6.33
60×30	7.33	6.33
80×40	7.50	6.33

The formulations used for fluid flows thus far were based on the Navier–Stokes equations in terms of the primitive variables, (u, v, p). In Section 9.6 we consider the vorticity–stream function formulation of the Navier–Stokes equations. As we will show, the vorticity–stream function formulation eliminates the pressure from the primary equations, and Poisson's equation for pressure is solved once the vorticity and stream function are determined.

9.6 Vorticity–Stream Function Equations

9.6.1 Governing Equations in Terms of Vorticity and Stream Function

In the vorticity–stream function approach, two mathematical variables, namely, vorticity (ω) and stream-function (ψ), are introduced for two-dimensional flows as follows:

$$\omega = \frac{\partial v}{\partial x} - \frac{\partial u}{\partial y} \tag{9.6.1}$$

$$u = \frac{\partial \psi}{\partial y}, \quad v = -\frac{\partial \psi}{\partial x}. \tag{9.6.2}$$

The elimination of the pressure field is carried out through cross-differentiation of the x- and y-momentum equations and subtracting one from the other, and invoking the continuity equation. Differentiating the x-momentum equation, Eq. (9.1.5), with respect to y, we obtain (where ρ and μ are treated as constants):

$$\rho\frac{\partial}{\partial y}\left(u\frac{\partial u}{\partial x}+v\frac{\partial u}{\partial y}\right)=-\frac{\partial^2 p}{\partial y\partial x}+\mu\frac{\partial}{\partial y}\left(\frac{\partial^2 u}{\partial x^2}+\frac{\partial^2 u}{\partial y^2}\right)$$

or

$$\rho\left(\frac{\partial u}{\partial y}\frac{\partial u}{\partial x}+u\frac{\partial^2 u}{\partial y\partial x}+\frac{\partial v}{\partial y}\frac{\partial u}{\partial y}+v\frac{\partial^2 u}{\partial y^2}\right)=-\frac{\partial^2 p}{\partial y\partial x}+\mu\left(\frac{\partial^3 u}{\partial y\partial x^2}+\frac{\partial^3 u}{\partial y^3}\right). \tag{9.6.3}$$

Differentiating the y-momentum equation, Eq. (9.1.6), with respect to x, we obtain

$$\rho\frac{\partial}{\partial x}\left(u\frac{\partial v}{\partial x}+v\frac{\partial v}{\partial y}\right)=-\frac{\partial^2 p}{\partial y\partial x}+\mu\frac{\partial}{\partial x}\left(\frac{\partial^2 v}{\partial x^2}+\frac{\partial^2 v}{\partial y^2}\right)$$

or

$$\rho\left(\frac{\partial u}{\partial x}\frac{\partial v}{\partial x}+u\frac{\partial^2 v}{\partial x^2}+\frac{\partial v}{\partial x}\frac{\partial v}{\partial y}+v\frac{\partial^2 v}{\partial x\partial y}\right)=-\frac{\partial^2 p}{\partial y\partial x}+\mu\left(\frac{\partial^3 v}{\partial x^3}+\frac{\partial^3 v}{\partial x\partial y^2}\right). \tag{9.6.4}$$

Now subtracting Eq. (9.6.4) from Eq. (9.6.3) we obtain:

$$\begin{aligned}\rho&\left[\frac{\partial u}{\partial x}\left(\frac{\partial u}{\partial y}-\frac{\partial v}{\partial x}\right)+\frac{\partial v}{\partial y}\left(\frac{\partial u}{\partial y}-\frac{\partial v}{\partial x}\right)+u\frac{\partial}{\partial x}\left(\frac{\partial u}{\partial y}-\frac{\partial v}{\partial x}\right)+v\frac{\partial}{\partial y}\left(\frac{\partial u}{\partial y}-\frac{\partial v}{\partial x}\right)\right]\\&=\frac{\partial^2 p}{\partial y\partial x}-\frac{\partial^2 P}{\partial y\partial x}+\mu\left[\frac{\partial^2}{\partial x^2}\left(\frac{\partial u}{\partial y}-\frac{\partial v}{\partial x}\right)+\frac{\partial^2}{\partial y^2}\left(\frac{\partial u}{\partial y}-\frac{\partial v}{\partial x}\right)\right]\end{aligned}$$

or

$$\begin{aligned}\rho&\left(\frac{\partial u}{\partial y}-\frac{\partial v}{\partial x}\right)\underbrace{\left(\frac{\partial u}{\partial x}+\frac{\partial v}{\partial y}\right)}_{0}+\rho u\frac{\partial}{\partial x}\underbrace{\left(\frac{\partial u}{\partial y}-\frac{\partial v}{\partial x}\right)}_{-\omega}+\rho v\frac{\partial}{\partial y}\underbrace{\left(\frac{\partial u}{\partial y}-\frac{\partial v}{\partial x}\right)}_{-\omega}\\&=\mu\frac{\partial^2}{\partial x^2}\underbrace{\left(\frac{\partial u}{\partial y}-\frac{\partial v}{\partial x}\right)}_{-\omega}+\mu\frac{\partial^2}{\partial y^2}\underbrace{\left(\frac{\partial u}{\partial y}-\frac{\partial v}{\partial x}\right)}_{-\omega}\end{aligned} \tag{9.6.5}$$

giving

$$\rho\left(u\frac{\partial \omega}{\partial x}+v\frac{\partial \omega}{\partial y}\right)=\mu\left(\frac{\partial^2 \omega}{\partial x^2}+\frac{\partial^2 \omega}{\partial y^2}\right). \tag{9.6.6}$$

Now, differentiating stream function definitions in Eq. (9.6.2) with respect to y and x, respectively, we obtain:

$$\frac{\partial u}{\partial y}=\frac{\partial^2 \psi}{\partial y^2},\quad \frac{\partial v}{\partial x}=-\frac{\partial^2 \psi}{\partial x^2}. \tag{9.6.7}$$

Subtracting the second equation from the first equation, we obtain

$$\frac{\partial^2 \psi}{\partial y^2}+\frac{\partial^2 \psi}{\partial x^2}=\frac{\partial u}{\partial y}-\frac{\partial v}{\partial x} \tag{9.6.8}$$

or

$$\frac{\partial^2 \psi}{\partial y^2} + \frac{\partial^2 \psi}{\partial x^2} = -\omega \quad \text{or} \quad -\nabla^2 \psi = \omega. \tag{9.6.9}$$

Thus, we have two scalar equations, Eq. (9.6.6) for vorticity (ω) and Eq. (9.6.9) for the stream function (ψ), and they are coupled. Once we solve for vorticity and the stream function the velocity fields can be calculated from Eq. (9.6.2). The gradient of the pressure can be calculated (and the pressure within a constant) from Eqs. (9.1.5) and (9.1.6). Next, we discuss an alternative approach to determine pressure.

9.6.2 Poisson's Equation for Pressure

Equations (9.1.5) and (9.1.6) involve higher derivatives and nonlinear terms of the velocity field if we were to calculate $\partial p/\partial x$ and $\partial p/\partial y$. Here we develop Poisson's differential equation for pressure p that can be solved using FVM formulation discussed in Chapter 7. In the following we assume that the density ρ is a constant.

Differentiating the x-momentum equation (9.1.5) with respect to x we obtain:

$$\frac{\partial}{\partial x}\left[\rho u \frac{\partial u}{\partial x} + \rho v \frac{\partial u}{\partial y}\right] = -\frac{\partial^2 p}{\partial x^2} + \frac{\partial}{\partial x}\left[\mu \frac{\partial^2 u}{\partial x^2} + \mu \frac{\partial^2 u}{\partial y^2}\right]$$

or

$$\rho u \frac{\partial^2 u}{\partial x^2} + \rho \left(\frac{\partial u}{\partial x}\right)^2 + \rho \frac{\partial v}{\partial x}\frac{\partial u}{\partial y} + \rho v \frac{\partial^2 u}{\partial x \partial y} = -\frac{\partial^2 p}{\partial x^2} + \mu \frac{\partial^3 u}{\partial x^3} + \mu \frac{\partial^3 u}{\partial x \partial y^2}. \tag{9.6.10}$$

Differentiating the y-momentum equation (9.1.6) with respect to y we obtain:

$$\frac{\partial}{\partial y}\left(\rho u \frac{\partial v}{\partial x} + \rho v \frac{\partial v}{\partial y}\right) = -\frac{\partial^2 p}{\partial y^2} + \frac{\partial}{\partial y}\left(\mu \frac{\partial^2 v}{\partial x^2} + \mu \frac{\partial^2 v}{\partial y^2}\right)$$

or

$$\rho \frac{\partial u}{\partial y}\frac{\partial v}{\partial x} + \rho u \frac{\partial^2 v}{\partial x \partial y} + \rho v \frac{\partial^2 v}{\partial y^2} + \rho \left(\frac{\partial v}{\partial y}\right)^2 = -\frac{\partial^2 p}{\partial y^2} + \mu \frac{\partial^3 v}{\partial x^2 \partial y} + \mu \frac{\partial^3 v}{\partial y^3}. \tag{9.6.11}$$

Adding Eqs. (9.6.10) and (9.6.11) we arrive at the result:

$$\rho u \frac{\partial}{\partial x}\underbrace{\left(\frac{\partial u}{\partial x} + \frac{\partial v}{\partial y}\right)}_{0} + \rho v \frac{\partial}{\partial y}\underbrace{\left(\frac{\partial u}{\partial x} + \frac{\partial v}{\partial y}\right)}_{0} + \rho \left(\frac{\partial u}{\partial x}\right)^2 + \rho \left(\frac{\partial v}{\partial y}\right)^2 + 2\rho \frac{\partial u}{\partial y}\frac{\partial v}{\partial x}$$

$$= -\frac{\partial^2 p}{\partial x^2} - \frac{\partial^2 p}{\partial y^2} + \mu \frac{\partial^2}{\partial x^2}\underbrace{\left(\frac{\partial u}{\partial x} + \frac{\partial v}{\partial y}\right)}_{0} + \mu \frac{\partial^2}{\partial y^2}\underbrace{\left(\frac{\partial u}{\partial x} + \frac{\partial v}{\partial y}\right)}_{0}$$

or

$$-\left(\frac{\partial^2 p}{\partial x^2} + \frac{\partial^2 p}{\partial y^2}\right) = \rho\left[\left(\frac{\partial u}{\partial x}\right)^2 + \left(\frac{\partial v}{\partial y}\right)^2 + 2\frac{\partial u}{\partial y}\frac{\partial v}{\partial x}\right]. \tag{9.6.12}$$

The right-hand side of Eq. (9.6.12) can be further simplified using the identity:

$$-\rho\left[\left(\frac{\partial u}{\partial x}\right)^2+\left(\frac{\partial v}{\partial y}\right)^2+2\frac{\partial u}{\partial y}\frac{\partial v}{\partial x}\right]=-\rho\left[\underbrace{\left(\frac{\partial u}{\partial x}+\frac{\partial v}{\partial y}\right)^2}_{0}-2\frac{\partial u}{\partial x}\frac{\partial v}{\partial y}+2\frac{\partial u}{\partial y}\frac{\partial v}{\partial x}\right]$$
$$=2\rho\left[\frac{\partial u}{\partial x}\frac{\partial v}{\partial y}-\frac{\partial u}{\partial y}\frac{\partial v}{\partial x}\right].$$

Then, the Poisson equation for pressure, Eq. (9.6.12), can be expressed as:

$$-\left(\frac{\partial^2 p}{\partial x^2}+\frac{\partial^2 p}{\partial y^2}\right)=2\rho\left[\frac{\partial u}{\partial y}\frac{\partial v}{\partial x}-\frac{\partial u}{\partial x}\frac{\partial v}{\partial y}\right]. \tag{9.6.13}$$

Equation (9.6.13) is valid in any formulation after the velocity field is obtained.

In the case of the vorticity–stream function formulation, Eq. (9.6.13) can be expressed in the alternative form by invoking the relationship between velocity components and the stream function as:

$$-\left(\frac{\partial^2 p}{\partial x^2}+\frac{\partial^2 p}{\partial y^2}\right)=2\rho\left[\left(\frac{\partial^2 \psi}{\partial x \partial y}\right)^2-\frac{\partial^2 \psi}{\partial x^2}\frac{\partial^2 \psi}{\partial y^2}\right]. \tag{9.6.14}$$

In the vorticity–stream function formulation, the pressure is calculated using Eq. (9.6.14) after obtaining the stream function. The solution procedure involves the following sequence of steps.

(1) Guess the velocity field (u and v).

(2) Solve the vorticity transport equation, Eq. (9.6.6), to obtain ω.

(3) Solve the Poisson equation for the stream function, Eq. (9.6.8), to obtain ψ.

(4) Obtain the velocity field (u and v) from Eq. (9.6.2) and return to step (2). Repeat the process until convergence.

(5) Using the converged velocity field (u and v) solve the Poisson equation, Eq. (9.6.13) by the FVM (or through an alternate approach) to determine p.

The vorticity–stream function formulation is very appealing, as calculations of the velocity and pressure fields are decoupled through the introduction of two new variables (ω and ψ). However, the application of the vorticity–stream function formulation for three-dimensional problems poses significant challenges as vorticity becomes a vector and the definition of the stream function becomes unclear. In addition, the boundary conditions on the vorticity and stream function have to be deduced from the physical boundary conditions on the primitive variables.

9.7 Summary

In this chapter, both staggered- and collocated-grid approaches to solve two-dimensional velocity and pressure fields using the FVM have been presented. The SIMPLE family of algorithms to solve two-dimensional pressure and velocity fields was discussed. Treatment of solid obstacles in flow paths in the context of finite volume formulation methods was discussed. Finally, the vorticity–stream function approach to solve two-dimensional velocity and pressure fields was presented.

Problems

9.1 Establish the relation in Eq. (9.6.14).

9.2 A viscous engine oil is trapped in a square cavity measuring 0.1 m $\times$ 0.1 m. Both top and bottom plates of the cavity are pulled to the right at a constant velocity. The Reynolds number is based on the velocity of the top plate (200 m/s) and the fluid Prandtl number ($Pr = 14.683$). The thermophysical properties of the engine oil at 40 °C are: density $\rho = 997.0$ kg/m^3, and viscosity $\mu = 0.000\,871$ Ns/m^2.
Write a computer code using a finite volume technique to predict the velocity and pressure fields. Represent a solution to the one-dimensional advection–diffusion equation using the power law. Link the velocity and pressure fields using the SIMPLE algorithm. Declare convergence when $R_u < 10^{-6}$, $R_v < 10^{-6}$, and $R_p < 10^{-5}$.

(a) Predict the velocity field using the 4 $\times$ 4, 8 $\times$ 8, 16 $\times$ 16, 32 $\times$ 32, 64 $\times$ 64, and 128 $\times$ 128 uniformly sized control volumes.

(b) Print the u-velocity matrix for 4$\times$4 control volume case to demonstrate symmetry of the velocity field.

(c) Plot centerline velocity as a function of distance for all six grid sizes on the same plot.

(d) Calculate the root mean square velocity of the centerline u-velocity.

(e) Calculate the GCI for 16 $\times$ 16, 32 $\times$ 32, 64 $\times$ 64, and 128 $\times$ 128 control volumes.

(f) Predict the order of the method using the GCI method.

(g) Plot the number of iterations required for convergence as a function of the grid size.

10
Advanced Topics

10.1 Introduction

10.1.1 General Remarks

There are several topics that are considered to be "advanced" for this book. We will briefly discuss some (but not all) of these topics to make the readers aware of the fact that the present coverage has precluded them, and then cover three topics in a greater detail.

When it comes to numerical simulations of engineering problems by any numerical method, two major topics of interest continue to receive attention: (1) inclusion of pertinent physics, especially from a constitutive behavior stand point, in the underlying mathematical models, and (2) novel and efficient use of solution methods (i.e., solution of resulting discretized equations). In fact, the constitutive models may dictate the coupling and require the solving of "multiphysics" problems (e.g., thermal–fluid, fluid–solid interactions, electromagnetics, chemically reacting flows, and a host of others). Each of these topics is so broad and there is so much to cover, a book can be devoted to each topic. Instead, this chapter includes a study of periodically fully developed flows, natural convection, and multigrid algorithms.

10.1.2 Periodic and Buoyancy-Driven Flows

Quite often, geometry in heat exchangers repeats itself from module to module. In the case of straight pipes or rectangular channels without obstructions, one can view the flow in a fully developed flow region as periodically fully developed with the flow repeating itself in a module length (l). The discussion of fully developed flow and periodically fully developed flow will be presented with examples in two-dimensional Cartesian coordinates (for the sake of simplicity). A comprehensive review of numerical treatment of periodically fully developed flow and heat transfer is provided by Lamourex and Baliga [83]. When flow attains a fully developed state, the stream-wise velocity (u) ceases to change with the stream-wise direction, the cross-stream velocity (v) will be zero, and the stream-wise pressure drop will be a nonzero constant.

Buoyancy-driven flows in enclosures are referred to as *natural convection* and buoyancy-driven flows over external surfaces are referred to as *free convection.* In Section 10.3, the finite volume formulation of natural convection in a rectangular cavity is illustrated using a contrived example (see Reddy and Gartling [3] for finite element analysis of natural convection problems).

10.1.3 Non-Newtonian Fluids

Concerning the inclusion of pertinent physics, the solution of an engineering problem is justified by the most dominating (or, the goal of the study may dictate considering only one, although the problem has several) physical phenomena. In the context of the present study, heat transfer and fluid flow were treated without the consideration of their two-way coupling. In addition, in Chapters 6 and 9, we have studied the FEM and FVM, respectively, of *Newtonian fluids* (i.e., fluids whose constitutive behavior is linear). Fluids that are not described by the Newtonian constitutive relations are not uncommon in a wide variety of industrial problems (i.e., motor oils, high-molecular-weight liquids such as polymers, slurries, pastes, and other complex mixtures). The processing and transport of such materials (treated as fluids) are central to the chemical, food, plastics, petroleum, and polymer industries. The key challenge in such problems is their material characterization (i.e., constitutive model development) and their incorporation into the balance equations.

Non-Newtonian behavior manifests itself in a number of different ways. Most non-Newtonian fluids exhibit a shear-rate-dependent viscosity, with "shear thinning" (i.e., decreasing viscosity with increasing shear rate) being the most prevalent behavior. Other phenomena associated with the elasticity and memory of the fluid, such as recoil, are also observed in many cases. Differences in the normal stress components occur in many flows and lead to such well-known effects as rod climbing or the Weissenberg effect, and the curvature of the free surface in an open channel flow. A comprehensive list and discussion of these and other non-Newtonian effects is given in the books by Bird et al. [32, 33].

Non-Newtonian fluids can be classified into two distinct categories: (1) inelastic fluids or fluids without memory, and (2) viscoelastic fluids or fluids which exhibit memory effects. Basically, inelastic fluids can be viewed as generalizations of the Newtonian fluid and are sometimes termed *generalized Newtonian fluids*. The generalized fluids are characterized by differential constitutive models in which the viscosity is assumed to be a function of the rate of deformation of the fluid, and thus allows "shear thinning" effects to be modeled. From FEM and FVM considerations, inelastic fluids can be treated using minor extensions to the conventional models developed for Newtonian fluids in Chapters 6 and 9. Viscoelastic fluids, on the other hand, represent a significant departure from the Newtonian limit in terms of both physical behavior and computational complexity. The primary difficulty here is the "memory" of the fluid; the motion of a material element depends not only on the present stress state, but also on the deformation history of the material element. These topics are outside the scope of this book.

10.1.4 Solution Methods

Throughout this book and at pertinent places (see, e.g., Chapter 3), we have discussed some of the numerical methods for the solution of resulting discretized equations. There exist many solution algorithms and strategies that make use of these numerical methods. Multigrid methods (see, e.g., Reddy and Reddy [38]) are one such class of methods, which will be discussed in Section 10.4.

10.2 Periodically Fully Developed Flows

10.2.1 Introduction

The fully developed conditions for two-dimensional steady flows can be stated as

$$\frac{\partial u}{\partial x} = v = 0 \tag{10.2.1}$$

$$\frac{\partial p}{\partial x} = \text{ constant } \neq 0. \tag{10.2.2}$$

The motivation to study periodically fully-developed flow and heat transfer is computational cost. Furthermore, heat transfer in periodically fully developed flow region forms the lower bound, which is valuable design information. Periodically fully developed flows set in when the flow is periodically interrupted by placing flow obstructions (e.g., baffles) and/or making the flow pass through periodic turns. Thermal boundary layers break and restart at each turn or flow interruption by a physical obstruction, thus creating a series of entry regions that are associated with a higher heat transfer coefficient. Consider two cases: one is a flow through a straight rectangular channel (Case A, Fig. 10.2.1) and the second is a channel with periodically positioned internal fins (Case B, Fig. 10.2.2). The qualitative Nusselt number distribution might look like that shown in Fig. 10.2.3.

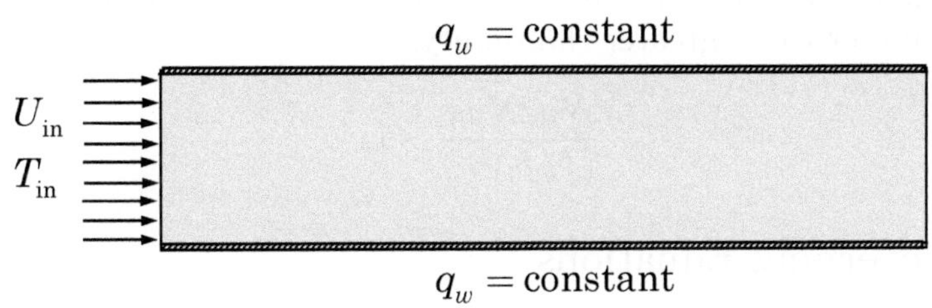

Fig. 10.2.1 Case A: Flow in a straight channel.

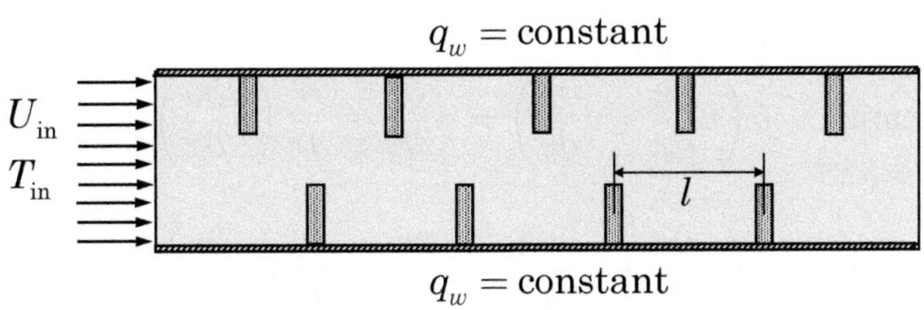

Fig. 10.2.2 Case B: Periodic flow in a channel.

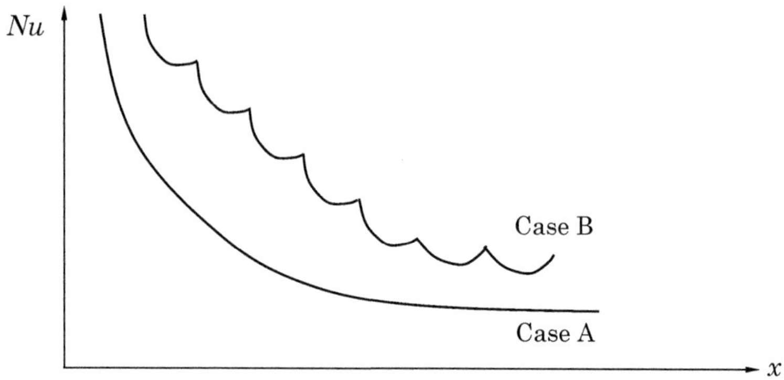

Fig. 10.2.3 Qualitative Nusselt number (Nu) variations for Case A and Case B.

The idea is to increase heat transfer by periodically interrupting the flow in a straight channel by placing internal fins that are equidistant. If, for a fully developed flow, the Nusselt number for Case A is Nu_0 and that for Case B is Nu, then the goal is to ensure that

$$\frac{Nu}{Nu_0} > 1. \tag{10.2.3}$$

In Case B, as result of the presence of internal fins, the flow experiences additional resistance, thus leading to increased pressure drop. But the pumping power is proportional to the one third of the friction factor ($f^{1/3}$). The goal is to achieve the ratio of increase in the heat transfer coefficient per unit increase in pumping power to be greater than unity,

$$\frac{Nu/Nu_0}{(f/f_0)^{1/3}} > 1. \tag{10.2.4}$$

10.2.2 Governing Equations

Consider a steady laminar, two-dimensional flow of a viscous fluid in a channel between two stationary parallel plates. The governing equations (after rewriting the original equations to make use of the divergence-free condition) are

$$\text{continuity:} \quad \frac{\partial u}{\partial x} + \frac{\partial v}{\partial y} = 0 \tag{10.2.5}$$

$$x\text{-momentum:} \quad \rho\left(u\frac{\partial u}{\partial x} + v\frac{\partial u}{\partial y}\right) = -\frac{\partial p}{\partial x} + \frac{\partial}{\partial x}\left(\mu\frac{\partial u}{\partial x}\right) + \frac{\partial}{\partial y}\left(\mu\frac{\partial u}{\partial y}\right) \tag{10.2.6}$$

$$y\text{-momentum:} \quad \rho\left(u\frac{\partial v}{\partial x} + v\frac{\partial v}{\partial y}\right) = -\frac{\partial p}{\partial y} + \frac{\partial}{\partial x}\left(\mu\frac{\partial v}{\partial x}\right) + \frac{\partial}{\partial y}\left(\mu\frac{\partial v}{\partial y}\right). \tag{10.2.7}$$

All variables appearing in Eqs. (10.2.6) and (10.2.8) have the same meaning as described previously.

Now consider a periodically fully developed flow module with internal fins as shown in Fig. 10.2.4. In periodically fully developed flows the velocity fields repeat themselves from module to module and the total pressure drop per module remains a constant in the periodically fully developed flow region. So the proper representation for pressure is

$$p(x, y) = -\beta x + \hat{p}(x, y), \tag{10.2.8}$$

where $\hat{p}(x, y)$ is the local pressure variation and the nonzero constant β is the global pressure drop per module.

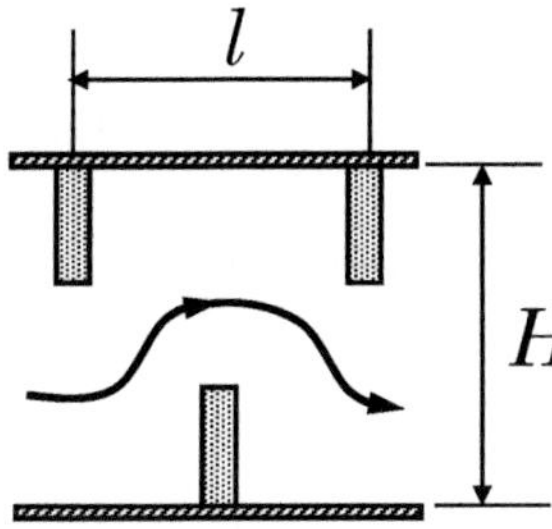

Fig. 10.2.4 Periodically fully developed flow module.

Substituting Eq. (10.2.8) for $p(x, y)$ in Eqs. (10.2.5)–(10.2.7), the governing equations for the periodically fully developed flow module take the following form:

$$\text{continuity:} \quad \frac{\partial u}{\partial x} + \frac{\partial v}{\partial y} = 0 \tag{10.2.9}$$

$$x\text{-momentum:} \quad \rho\left(u\frac{\partial u}{\partial x} + v\frac{\partial u}{\partial y}\right) = \beta - \frac{\partial \hat{p}}{\partial x} + \frac{\partial}{\partial x}\left(\mu\frac{\partial u}{\partial x}\right) + \frac{\partial}{\partial y}\left(\mu\frac{\partial u}{\partial y}\right) \tag{10.2.10}$$

$$y\text{-momentum:} \quad \rho\left(u\frac{\partial v}{\partial x} + v\frac{\partial v}{\partial y}\right) = -\frac{\partial \hat{p}}{\partial y} + \frac{\partial}{\partial x}\left(\mu\frac{\partial v}{\partial x}\right) + \frac{\partial}{\partial y}\left(\mu\frac{\partial v}{\partial y}\right). \tag{10.2.11}$$

The boundary conditions associated with these equations are

$$u(x, y) = u(x + l, y) \tag{10.2.12a}$$

$$v(x, y) = v(x + l, y) \tag{10.2.12b}$$

$$\left.\frac{\partial u}{\partial x}\right|_{(x,y)} = \left.\frac{\partial u}{\partial x}\right|_{(x+l,y)} \tag{10.2.12c}$$

$$\left.\frac{\partial v}{\partial x}\right|_{(x,y)} = \left.\frac{\partial v}{\partial x}\right|_{(x+l,y)} \tag{10.2.12d}$$

$$u = v = 0 \quad \text{at all solid surfaces.} \tag{10.2.12e}$$

In solving periodically fully developed flow fields, the flow rate or Reynolds number (Re) cannot be independently specified, but the global pressure drop per module (β) can be specified. Once a converged flow field is obtained, the flow rate or Reynolds number (Re) can be calculated. To achieve a specific flow rate, one may have to adopt an iterative procedure.

10.2.3 Thermally Fully Developed Flows

A temperature field is declared to be thermally fully developed if the nondimensional temperature field $\theta = \frac{T_w(x)-T(x,y)}{T_w(x)-T_b(x)}$ ceases to vary with the stream-wise direction, that is

$$\left.\frac{\partial \theta}{\partial x}\right|_{t,FD} = \frac{\partial}{\partial x}\left[\frac{T_w(x)-T(x,y)}{T_w(x)-T_b(x)}\right] = 0. \tag{10.2.13}$$

Since the temperature ratio is independent of x, the derivative of this ratio with respect to y must also be independent of x:

$$\frac{\partial}{\partial x}\left[\frac{T_w(x)-T(x,y)}{T_w(x)-T_b(x)}\right] = \frac{-1}{T_w(x)-T_b(x)}\left[\frac{\partial T}{\partial y}\right]_{y=0} \neq f(x). \tag{10.2.14}$$

However, the heat flux is

$$q_w = h(T_w(x)-T_b(x)) = -k\left[\frac{\partial T}{\partial y}\right]_{y=0}. \tag{10.2.15}$$

Therefore, we obtain

$$\frac{h}{k} = -\frac{1}{(T_w(x)-T_b(x))}\left[\frac{\partial T}{\partial y}\right]_{y=0} \neq f(x). \tag{10.2.16}$$

So, the average module heat transfer coefficient $\bar{h}$ is independent of x. Further analysis and prediction of the temperature field under thermally fully developed flow conditions depends on the boundary conditions at the wall, that is, uniform wall temperature (UWT) or uniform wall heat flux (UHF) conditions. The wall and bulk temperature distributions in a straight channel without obstruction for UWT and UHF cases are shown in Figs. 9.2.6 and 9.2.8, respectively.

Treatment of thermally fully developed flows for UHF and UWT cases are very different. Treatment of the UHF case is relatively easier, so we will consider this case first.

10.2.4 Uniform Heat Flux Condition

If the heat addition per module is σ, then temperature $T(x,y)$ can be represented as

$$T(x,y) = \hat{T}(x,y) + \sigma x, \tag{10.2.17}$$

where $\hat{T}(x, y)$ is the local variation in temperature that repeats itself from module to module. Substituting for $T(x, y)$ into the energy equation we obtain

$$\rho c u \frac{\partial \hat{T}}{\partial x} + \rho c v \frac{\partial \hat{T}}{\partial y} = \frac{\partial}{\partial x}\left(k^* \frac{\partial \hat{T}}{\partial x}\right) + \frac{\partial}{\partial y}\left(k^* \frac{\partial \hat{T}}{\partial y}\right) - \underbrace{\rho c u \sigma}_{\text{First source term}} + \underbrace{\frac{\partial}{\partial x}(k^* \sigma.)}_{\text{Second source term}} \tag{10.2.18}$$

Boundary conditions in the stream-wise direction are:

$$\hat{T}(x, y) = \hat{T}(x + l, y) \tag{10.2.19}$$

$$\left.\frac{\partial \hat{T}}{\partial x}\right|_{(x,y)} = \left.\frac{\partial \hat{T}}{\partial x}\right|_{(x+l,y)}. \tag{10.2.20}$$

At the walls, heat flux is specified. It should be noted that the second source term is important at the solid-fluid interface and in turbulent flows where thermal diffusivity varies throughout the flow domain [84].

10.2.5 Uniform Wall Temperature Condition

Here a nondimensional temperature (θ) is defined,

$$\theta = \frac{T_w(x) - T(x, y)}{T_w(x) - T_b(x)}, \tag{10.2.21}$$

where T_b is the bulk temperature given by

$$T_b = \frac{1}{\rho c \bar{u}} \int_0^H \rho c |u| T \, dy. \tag{10.2.22}$$

Substituting for θ into the energy equation and simplifying, we obtain:

$$\rho c u \frac{\partial \theta}{\partial x} + \rho c v \frac{\partial \theta}{\partial y} - \frac{\partial}{\partial y}\left(k^* \frac{\partial \theta}{\partial y}\right)$$
$$= \underbrace{\frac{1}{(T_b - T_w)} \frac{\partial \theta}{\partial x}\left[\frac{\partial}{\partial x}\left(k^* \frac{\partial T_b}{\partial x}\right)\right] + \frac{1}{(T_b - T_w)}\left[\frac{\partial}{\partial x}\left(k^* \theta \frac{\partial T_b}{\partial x}\right)\right]}_{\lambda(x,y)}. \tag{10.2.23}$$

On left-hand side of Eq. (10.2.23), u, v, and θ are all periodic, therefore the right-hand side of the equation, $\lambda(x, y)$, is also periodic. Thus Eq. (10.2.23) is homogeneous subject to homogeneous boundary conditions leading to an eigenvalue problem. So the process is to guess $\lambda(x, y)$ and solve for $\theta(x, y)$ and iterate until convergence (see Patankar et al. [85]). To overcome this mathematical complexity, Fullerton and Anand [86] proposed an alternative approach. In this approach the periodically fully developed flow field solution is used as an input to all modules of the developing flow domain and the temperature field is evaluated by monitoring the module average Nusselt number to declare thermally fully developed flow.

10.2.6 Cyclic Tri-Diagonal Matrix Algorithm

10.2.6.1 Solution Procedure for $N = 6$

In a periodically fully developed flow and heat transfer problem, the boundary conditions in the stream-wise direction repeat from module to module. In other words, the variables display cyclic behavior. Therefore, the Tri-Diagonal Matrix Algorithm (TDMA) needs to be modified to accommodate the cyclic elements of the matrix.

For the purpose of illustration, let us consider a one-dimensional diffusion problem with a cyclic condition,

$$\frac{d^2\phi}{dx^2} + \lambda = 0, \quad 0 < x < 1 \tag{10.2.24}$$

$$\phi(0) = \phi_0 \tag{10.2.25}$$

$$\phi(1) = \phi_0. \tag{10.2.26}$$

Consider a domain of unit length, with six nodes ($imax = 6$). There are $N = imax - 1$ unknowns as $\phi_1 = \phi_{imax}$, so we need to generate $(imax - 1)$ number of discretization equations. For illustration purpose, we will use a second-order finite difference scheme to generate the discretization equations, as follows:

$$\text{for } i = 1: \quad \frac{\phi_2 - 2\phi_1 + \phi_{-1}}{\Delta x^2} + \lambda = 0. \tag{10.2.27}$$

But ϕ_{-1} is outside the domain, so it is mapped to ϕ_5 or ϕ_{imax-1}. Thus, we have

$$\text{for } i = 1: \quad \phi_2 - 2\phi_1 + \phi_{imax-1} = -\lambda\Delta x^2. \tag{10.2.28a}$$

For $i = 2, 3, 4$, and 5, we have

$$\text{for } i = 2: \quad \phi_1 - 2\phi_2 + \phi_3 = -\lambda\Delta x^2 \tag{10.2.28b}$$

$$\text{for } i = 3: \quad \phi_2 - 2\phi_3 + \phi_4 = -\lambda\Delta x^2 \tag{10.2.28c}$$

$$\text{for } i = 4: \quad \phi_3 - 2\phi_4 + \phi_5 = -\lambda\Delta x^2 \tag{10.2.28d}$$

$$\text{for } i = 5: \quad \phi_4 - 2\phi_5 + \phi_1 = -\lambda\Delta x^2, \tag{10.2.28e}$$

where ϕ_6 is replaced with ϕ_1.

The five equations, Eqs. (10.2.28a)–(10.2.28e), can be assembled as shown in Table 10.2.1.

Table 10.2.1 Equations with the cyclic boundary condition; RHS stands for right-hand side.

Eq.	i	ϕ_1	ϕ_2	ϕ_3	ϕ_4	ϕ_5	RHS
1	1	−2	1			①	$-\lambda\Delta x^2$
2	2	1	−2	1			$-\lambda\Delta x^2$
3	3		1	−2	1		$-\lambda\Delta x^2$
4	4			1	−2	1	$-\lambda\Delta x^2$
5	5	①			1	−2	$-\lambda\Delta x^2$

There are two off-diagonal elements in the example. These two elements can be accommodated through a Gauss–Seidel iterative process. But the cyclic tri-diagonal matrix algorithm (CTDMA) provides a much more direct way to solve the set of discretization equations appearing in periodically fully developed flow and heat transfer problems.

Here we will discuss the solution procedure for the example ($N = 6$) considered above. The general solution procedure for a system of $(N-1)$ equations based on work of Patankar et al. [85] will be developed.

Following Eqs. (10.2.28a)–(10.2.28e) one can write

$$\phi_i = f(\phi_{i+1}, \phi_{i-1}). \tag{10.2.29}$$

Note that when $i = 1$, $\phi_{i-1} = \phi_{N-1}$. Therefore, Eq. (10.2.29) can be written as

$$\phi_1 = f(\phi_2, \phi_5) = f(\phi_2, \phi_{N-1}) \tag{10.2.30a}$$
$$\phi_2 = f(\phi_1, \phi_3) \tag{10.2.30b}$$
$$\phi_3 = f(\phi_2, \phi_4) \tag{10.2.30c}$$
$$\phi_4 = f(\phi_3, \phi_5) \tag{10.2.30d}$$
$$\phi_5 = f(\phi_4, \phi_1) \text{ because } \phi_6 = \phi_1. \tag{10.2.30e}$$

From Eq. (10.2.30a)

$$\phi_1 = f(\phi_2, \phi_5). \tag{10.2.31}$$

Then substituting for ϕ_1 from Eq. (10.2.30a) into Eq. (10.2.30b)

$$\phi_2 = g_2(\phi_3, \phi_5). \tag{10.2.32}$$

Substituting for ϕ_2 from Eq. (10.2.32) into Eq. (10.2.30c)

$$\phi_3 = g_3(\phi_4, \phi_5). \tag{10.2.33}$$

Again, substituting for ϕ_3 from Eq. (10.2.33) into Eq. (10.2.30d)

$$\phi_4 = g_4(\phi_5). \tag{10.2.34}$$

At this stage we have an equation Eq. (10.2.34), in terms of only two unknowns, ϕ_4 and ϕ_5 (ϕ_{N-1}). The next step is to generate another equation in terms of ϕ_4 and ϕ_5, so that both these equations can be used to solve for ϕ_4 and ϕ_5. Once ϕ_4 and ϕ_5 are obtained, then the rest of the unknowns (ϕ_3, ϕ_2, and ϕ_1) can be obtained by backward substitution. Because of the cyclic condition $\phi_6 = \phi_1$.

The second required equation is obtained by starting with Eq. (10.2.30e). Substituting for ϕ_1 from Eq. (10.2.30a) into Eq. (10.2.30e), one obtains,

$$\phi_5 = g_5(\phi_4, \phi_2). \tag{10.2.35}$$

Substituting for ϕ_2 from Eq. (10.2.32) into Eq. (10.2.35), one obtains,

$$\phi_5 = h_{51}(\phi_4, \phi_3). \tag{10.2.36}$$

Substituting for ϕ_3 from Eq. (10.2.33) into Eq. (10.2.36), one obtains,

$$\phi_5 = h_{52}(\phi_4). \tag{10.2.37}$$

Thus, we have two equations in terms of ϕ_4 (ϕ_{N-1}) and ϕ_5 (ϕ_N). Solving Eqs. (10.2.34) and (10.2.37), one can solve for ϕ_4 and ϕ_5. Once ϕ_4 and ϕ_5 are known, other unknowns can be determined by backward substitution. In the next section a Matlab function to solve a set of cyclic tri-diagonal equations is presented.

10.2.6.2 Matlab code

```
function [solution] = ctdma(a,b,N)
% Cyclic Tridiagonal Matrix Algorithm
% N-1, Number of equations
% a, coefficient matrix
% a(1,1) = right most corner element
% a(N-1,3) = left bottom corner element
% b , constant vector (input)
% b , solution vector (output)
e(1) = -a(1,3)/a(1,2);
f(1) = -a(1,1)/a(1,2);
g(1) = b(1)/a(1,2);
for i = 2:N-2
    denom = a(i,2)+a(i,1)*e(i-1);
    e(i) = -a(i,3)/denom;
    f(i) = -a(i,1)*f(i-1)/denom;
    g(i) = (b(i)-a(i,1)*g(i-1))/denom;
end
s(1) = a(N-1,2);
q(1) = -a(N-1,3);
r(1) = b(N-1);
for i=2:N-2
    s(i) = s(i-1)-(q(i-1)*f(i-1));
    q(i) = q(i-1)*e(i-1);
    r(i) = r(i-1)+q(i-1)*g(i-1);
end
b(N-1) = ((q(N-2)-a(N-1,1))*g(N-2) + r(N-2))/(s(N-2)
          -(q(N-2)-a(N-1,1))*(e(N-2)+f(N-2)));
for i = N-2:-1:1
    b(i) = e(i)*b(i+1) + f(i)*b(N-1) + g(i);
end
solution = b;
```

10.3 Natural Convection

10.3.1 Governing Equations

For nonisothermal, incompressible flows, the energy equation is decoupled from the momentum and continuity equations. However, for flows involving buoyancy forces, the equation of state is modified using the *Boussinesq approximation*, which allows the fluid properties to be expressed as functions of the thermodynamic state (e.g., temperature) and the density ρ to vary with temperature T according to the relation [see Section 2.4 and Eqs. (2.4.4) and (2.4.10)]

$$\rho = \rho_0[1 - \beta(T - T_{\text{ref}})], \tag{10.3.1}$$

where β is the coefficient of thermal expansion (1/°C) and the subscript zero indicates a reference condition; T_{ref} denotes a reference temperature. The variation of density as given in Eq. (10.3.1) is permitted only in the description of the body force; the density in all other situations is assumed to be that of the reference state, ρ_0. Other thermo-physical properties (e.g., viscosity and conductivity) of the fluid are assumed to remain constant while allowing the density to vary only in the buoyancy term.

For the purpose of illustration, we consider two-dimensional steady laminar flow in a rectangular cavity of width W and height H (see Fig. 10.3.1). In this example, the bottom and top walls are insulated. The left and right walls are maintained at uniform temperatures of T_h and T_c respectively, where $T_h > T_c$.

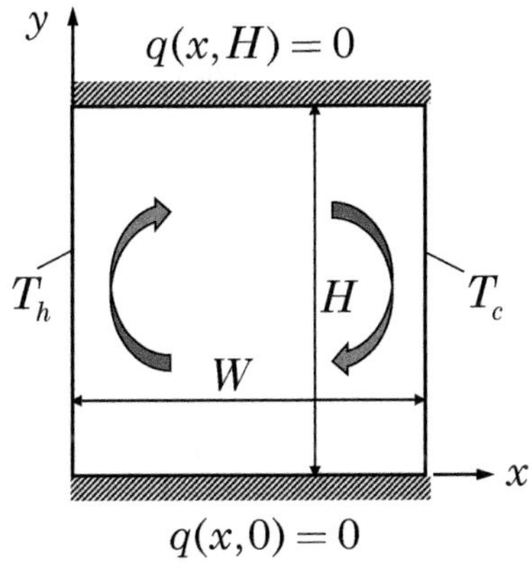

Fig. 10.3.1 Natural convection in a rectangular cavity

The fluid in the cavity is assumed to be air (taken here to be incompressible). Air becomes heated along the left wall and rises to the top of the cavity due to its decreased density. Along the right wall, the air cools down and moves downwards due to the gravity. Thus a recirculating flow is set up inside the cavity. In this case, due to the buoyancy force, the velocity and temperature fields are coupled. The buoyancy force term appears in the y-momentum equation (the assumed direction of gravity). Invoking the Boussinesq approximation, the governing equations and associated boundary conditions for steady flows can be written as:

$$\text{continuity:} \qquad \frac{\partial u}{\partial x} + \frac{\partial v}{\partial y} = 0 \tag{10.3.2}$$

$$x\text{-momentum:} \quad \rho\left(u\frac{\partial u}{\partial x} + v\frac{\partial u}{\partial y}\right) = -\frac{\partial p}{\partial x} + \frac{\partial}{\partial x}\left(\mu\frac{\partial u}{\partial x}\right) + \frac{\partial}{\partial y}\left(\mu\frac{\partial u}{\partial y}\right) \tag{10.3.3}$$

$$y\text{-momentum:} \quad \rho\left(u\frac{\partial v}{\partial x} + v\frac{\partial v}{\partial y}\right) = -\frac{\partial p}{\partial y} + \frac{\partial}{\partial x}\left(\mu\frac{\partial v}{\partial x}\right) + \frac{\partial}{\partial y}\left(\mu\frac{\partial v}{\partial y}\right) + \rho g_f\beta(T - T_{\text{ref}}) \tag{10.3.4}$$

$$\text{energy:} \quad \rho c\left(u\frac{\partial T}{\partial x} + v\frac{\partial T}{\partial y}\right) = \frac{\partial}{\partial x}\left(k\frac{\partial T}{\partial x}\right) + \frac{\partial}{\partial y}\left(k\frac{\partial T}{\partial y}\right). \tag{10.3.5}$$

Here T_{ref} is a reference temperature, which is usually computed as an arithmetic mean of the hot and cold wall temperatures, $T_{\text{ref}} = (T_h + T_c)/2$, and β is the thermal expansion coefficient. For an ideal gas, $\beta = 1/T_{\text{ref}}$, where T_{ref} is in absolute temperature scale (kelvin, k). For simplicity the subscript 0 on ρ is omitted.

The boundary conditions for this specific case are

$$u(0,y) = u(W,y) = u(x,0) = u(x,H) = 0 \tag{10.3.6a}$$
$$v(0,y) = v(W,y) = v(x,0) = v(x,H) = 0 \tag{10.3.6b}$$
$$T(0,y) = T_h, \quad T(W,y) = T_c \tag{10.3.6c}$$
$$\left.\frac{\partial T}{\partial y}\right|_{(x,0)} = 0, \quad \left.\frac{\partial T}{\partial y}\right|_{(x,H)} = 0. \tag{10.3.6d}$$

10.3.2 Discretized Equations

Integrating the x-momentum, y-momentum, and energy equations over the control volume on a collocated grid, we obtain the following discretization equations for u, v, and T:

$$a_{i,j}^u u_{i,j} = \sum_{nb} a_{nb}^u u_{nb} + A_u(p_{i,j} - p_{i+1,j}) + b_{i,j}^u \tag{10.3.7}$$
$$a_{i,j}^v v_{i,j} = \sum_{nb} a_{nb}^v v_{nb} + A_v(p_{i,j} - p_{i,j+1}) + b_{i,j}^v \tag{10.3.8}$$
$$a_{i,j}^T T_{i,j} = \sum_{nb} a_{nb}^T T_{nb} + b_{i,j}^T. \tag{10.3.9}$$

As we already know, the x-momentum and y-momentum equations are coupled. Additionally, the y-momentum equation is coupled with the energy equation since the buoyancy term appears as a source term in the y-momentum equation [Eq. (10.3.8)]. However, source terms for the x-momentum and energy equations are zero:

$$b_{i,j}^u = 0, \quad b_{i,j}^v = \rho g_f \beta (T_{i,j} - T_{\text{ref}}) \Delta x \Delta y, \quad b_{i,j}^T = 0. \tag{10.3.10}$$

The set of discretized equations along with the associated boundary conditions can be solved by using the SIMPLE algorithm, which has been discussed in Chapter 9. Note that, since the energy equation is coupled with the y-momentum equation, it should be solved inside the main iteration loop to ensure proper coupling.

The treatment of the source term in the y-momentum equation needs a brief discussion. In the case of the collocated grid, the term $b_{i,j}^v$ is integrated over a regular control volume with an area of $\Delta x\, \Delta y$, as shown in Eq. (10.3.10). In the case of the staggered-grid the internal control volumes measure $\Delta x\, \Delta y$, but the boundary control volumes are $1.5\Delta x\, \Delta y$. So, for the staggered-grid the source term has to be integrated over appropriate control volumes for the respective variable. Note that the problem in Example 10.3.1 will be solved using the collocated-grid approach.

Example 10.3.1

A viscous fluid (air) is trapped in a two-dimensional square cavity. The left wall is maintained at a uniform temperature of $T_h = 40\,^\circ$C and the right wall is maintained at $T_c = 20\,^\circ$C. The bottom and top walls are adiabatic (i.e., there is no heat flow across the boundary). Owing to the differential temperature of the side walls, a buoyancy-driven flow will be generated inside the cavity, as illustrated in Fig. 10.3.1. We wish to determine the flow characteristics and temperature distribution in the cavity.

Consider the air properties at a mean temperature of $T_{\text{ref}} = 30\,^\circ$C. The dimension of the cavity is $H = 0.01775$ m by $W = 0.01775$ m, which will produce a Rayleigh number ($Ra = c\rho\beta g_f(T_h - T_c)H^3/k\nu$) of 10^4 and Prandtl number ($Pr = c\mu/k = 0.71$) for the following choice of the data: density $\rho = 1.165$ kg/m^3, dynamic viscosity $\mu = 1.868 \times 10^{-5}$ kg/m.s, specific heat capacity $c = 1006.5$ J/kg.K, thermal conductivity $k = 0.02648$ W/m.K, $\beta = 1/\text{T}_{\text{ref}} = 1/303$ K^{-1}, and $g_f = 9.81$ m/s^2.

(a) Solve the governing equations for two-dimensional laminar steady flow with natural convection for this problem. Use the Boussinesq approximation to account for the variation in density. Discretize the governing equations with a finite volume technique using a collocated-grid approach. Solve the velocity and pressure fields by linking them through the SIMPLE algorithm. Represent the solution to the one-dimensional advection–diffusion problem using the power–law scheme. Run the program using 8×8 control volumes. Declare convergence when R_u, R_v, R_p, and R_T are less than 10^{-6}. Print your velocity fields up to five significant digits. Nondimensionalize the velocity fields with the reference velocity of $v_{ref} = \alpha/H$, where $\alpha = k/\rho c$ is the thermal diffusivity. So the nondimensional velocities are: $U = uH/\alpha$ and $V = vH/\alpha$.

(b) Make calculations for 8×8, 16×16, 32×32, 64×64, and 128×128 control volumes in each direction. Plot the centerline U and V velocities and centerline temperature T for each case (separate plots for U, V, and T). Report the U_{max}, V_{max}, and Nu_{max} for each case. Compare all the results with the benchmark solution of de Vahl Davis [87]. Note that, U_{max} is the maximum horizontal velocity on the vertical centerline of the cavity, and V_{max} is the maximum vertical velocity on the horizontal centerline of the cavity.

(c) Plot the velocity and temperature contours for 128×128 control volume case.

Solution The results are presented in Tables 10.3.1 and 10.3.2 and Figs. 10.3.2–10.3.8. The Nusselt number is defined as

$$Nu = \frac{hL}{k} = \frac{L}{T_h - T_c}\left(-\frac{\partial T}{\partial x}\bigg|_{y=0}\right). \tag{10.3.11}$$

Table 10.3.1 Results for natural convection in a square cavity for $Ra = 10^4$.

Grid size	U_{max}	V_{max}	Nu_{max}	Nu_{min}
8×8	15.5862	19.2431	4.3853	0.6079
16×16	15.8929	19.2636	3.8307	0.5855
32×32	16.1208	19.5319	3.6087	0.5854
64×64	16.1648	19.6138	3.5497	0.5851
128×128	16.1793	19.6203	3.5358	0.5850
de Vahl Davis [87]	16.1780	19.6217	3.5280	0.5860

The centerline x-velocity component (U), y-velocity component (V), and temperature (T) profiles for $Ra = 10^4$ are shown in Figs. 10.3.2, 10.3.3, and 10.3.4, respectively.

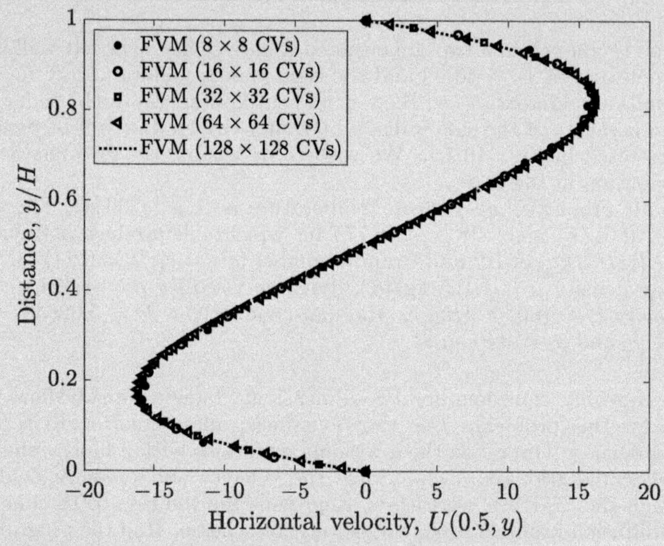

Fig. 10.3.2 Centerline x-velocity component (U) profile for $Ra = 10^4$ for a square cavity with left and right faces maintained at different temperatures and the bottom and top walls insulated; CVs stand for control volumes.

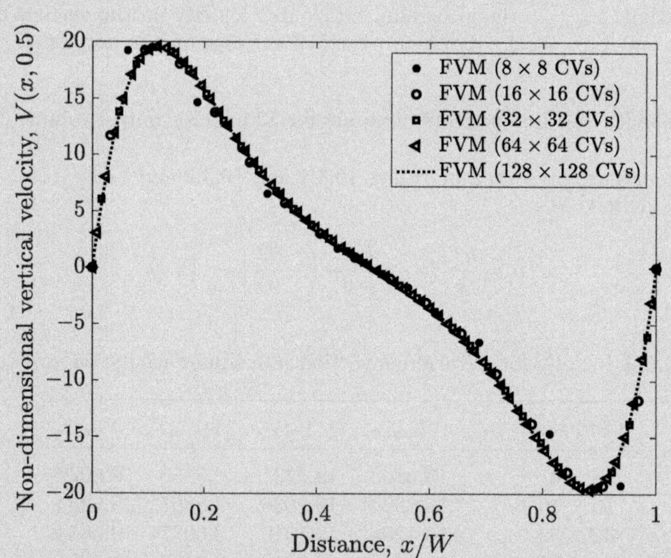

Fig. 10.3.3 Centerline y-velocity (V) profile for $Ra = 10^4$ for a square cavity with left and right faces maintained at different temperatures and the bottom and top walls insulated; CVs stand for control volumes.

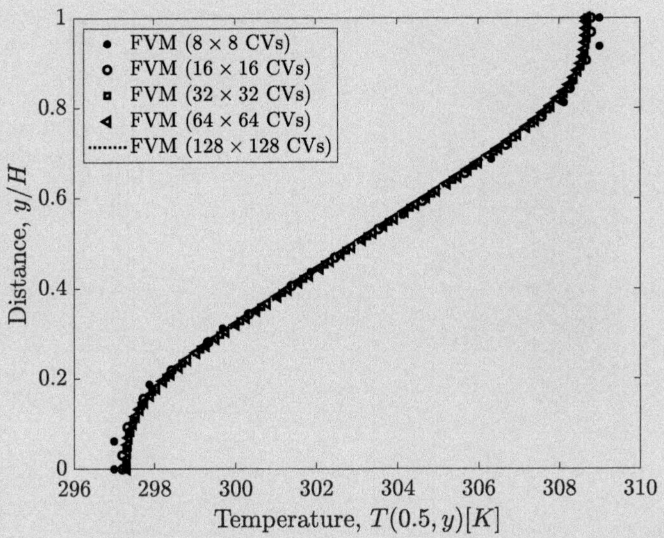

Fig. 10.3.4 Centerline temperature (T) profile for $Ra = 10^4$ for a square cavity with left and right faces maintained at different temperatures and the bottom and top walls insulated; CVs stand for control volumes.

The x-velocity component (U), y-velocity component, and temperature contours for $Ra = 10^4$ are shown in Figs. 10.3.5, 10.3.6, and 10.3.7, respectively. The velocity vectors for $Ra = 10^4$ with a mesh of 128×128 control volumes (CVs) are plotted in Fig. 10.3.8.

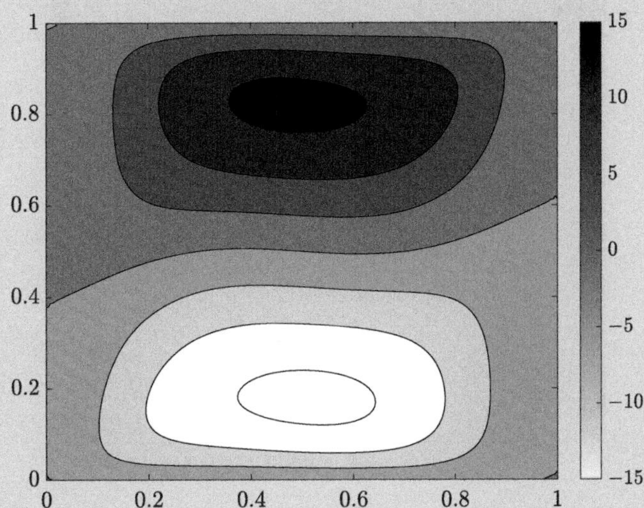

Fig. 10.3.5 x-velocity (U) contours for $Ra = 10^4$ of a square cavity with left and right faces maintained at different temperatures and the bottom and top walls insulated.

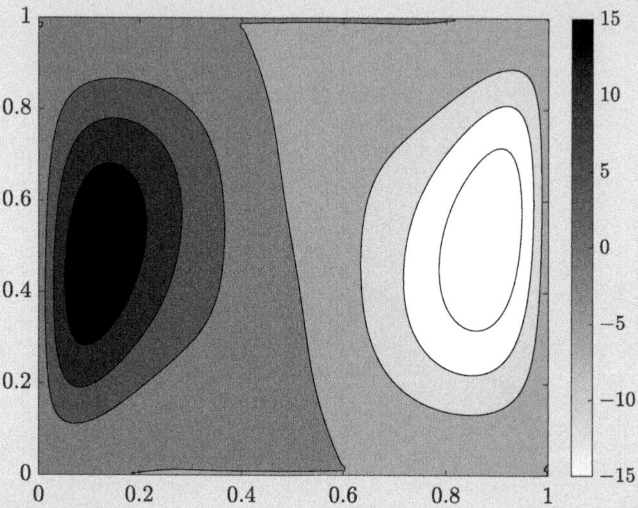

Fig. 10.3.6 y-velocity (V) contours for $Ra = 10^4$ of a square cavity with left and right faces maintained at different temperatures and the bottom and top walls insulated.

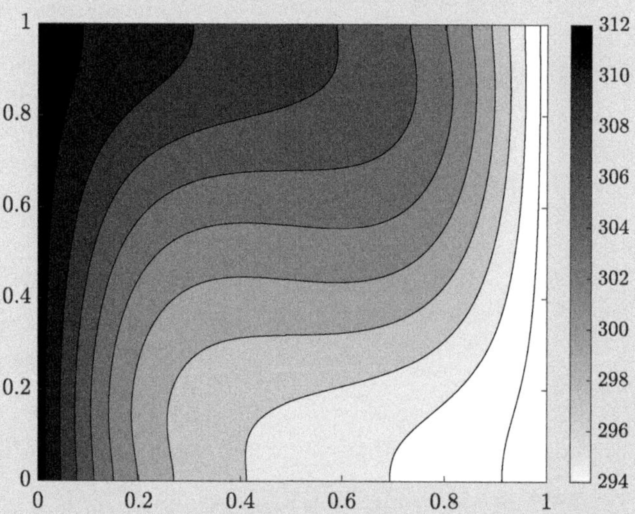

Fig. 10.3.7 Temperature (T) contours for $Ra = 10^4$ of a square cavity with left and right faces maintained at different temperatures and the bottom and top walls insulated.

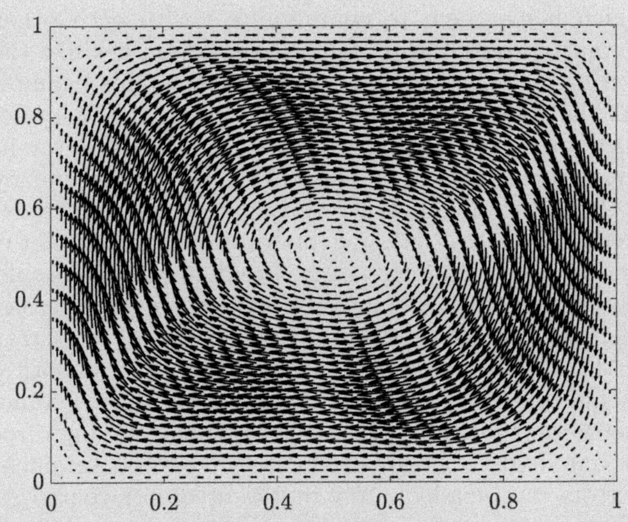

Fig. 10.3.8 Velocity vectors for $Ra = 10^4$ for a square cavity with left and right faces maintained at different temperatures and the bottom and top walls insulated.

Table 10.3.2 Velocity $u(x/W = 0.5, y/H)$ obtained with the FVM (8×8 mesh) and FEM (8×8 mesh).

y/H	FVM	y/H	FEM
0.0000	0.0000000	0.000	0.000000
0.0625	−0.0146455	0.125	−0.021202
0.1875	−0.0198299	0.250	−0.020574
0.3125	−0.0142276	0.375	−0.011224
0.4375	−0.0050020	0.500	0.000001
0.5625	0.0050020	0.625	0.011217
0.6875	0.0142276	0.750	0.020577
0.8125	0.0198299	0.875	0.021204
0.9375	0.0146455	—	—
1.0000	0.0000000	1.000	0.000000

10.4 Multigrid Algorithms

10.4.1 Preliminary Comments

Multigrid methods are increasingly being used as an acceleration technique for the solution of Naiver–Stokes equations. The basic idea of a multigrid technique relies on the principle of error smoothing of iterative solvers. A simple analogy can be the propagation of sound in air. The high-frequency component of sound, commonly known as "treble," dissipates quickly due to the friction of sound waves with the air. However, the low-frequency component, known as "bass," is not easily damped by the air medium and can propagate for a long distance. A few animals for example elephants, can detect these low-frequency components which helps them to communicate with each other over a long distance. The computational grid of iterative solvers act like the air medium and the solution error resembles the sound wave in this example. Similar to the sound wave, the solution error of iterative solvers is composed of high frequency components and low frequency components. During the initial phase of iterations, the iterative solvers tend to smooth the high frequency components (treble) of the solution errors quickly, resulting in a high convergence rate. However, the low-frequency parts (bass) of the solution error are not easily damped by the solver and propagate a long way before dying down completely. For this reason, the convergence rate of most iterative solvers decreases significantly after exhibiting fast convergence during the initial stage of iterations.

As most of the standard iterative methods tend to smooth out the high-frequency or oscillatory components of the error, they are commonly known as *smoothers.* After a few iterations of these smoothers, the high frequency errors are quickly eliminated while the low-frequency errors remain unchanged. These low-frequency errors can be adequately represented on a coarse grid (fewer grid points compared with a fine grid) due to the inherent smoothness. However, the low-frequency errors in a fine grid are treated as high-frequency errors in the coarse grid, and hence iterations in a coarse grid can remove the low-frequency errors quickly. The basic idea of multigrid methods is to accelerate the convergence of iterative solvers by eliminating high and low-frequency errors through repeated application of fine and coarse-grid iterations.

As depicted by Achi Brandt [88], the golden rule of multigrid methods is: *"The amount of computational work should be proportional to the amount of real physical changes in the computed system."* In a single grid, the computational work increases as $O(N^d)$, where N is the number of grid points and d is the dimension of the system. Thus, for a two-dimensional system, the computational work increases quadratically whereas for a three-dimensional system, the increase is cubic in nature. The golden rule of Brandt predicts that an efficient multigrid method should be linearly dependent on the number of grid points N; that is, it should exhibit an $O(N)$ increase in the computational work.

The basic principle of the multigrid algorithm is to eliminate the approximate solution error efficiently by alternately iterating on fine and coarse grids. The first step of this algorithm is to derive a set of coarse-grid equations from the existing fine-grid discretized equations. It is important to note that, in the coarse grid, the original discretized equations for SIMPLE algorithm are not solved, rather correction equations are solved. Owing to the nonlinear nature of the governing equations, the correction equations are derived from the discretized equations for the SIMPLE algorithm. This process of obtaining the correction equations is known as the full approximation scheme (FAS). After deriving the coarse grid correction equations with FAS, suitable grid transfer operators are chosen to interchange the coarse- and fine-grid values without losing essential information. Transferring the fine grid information to the coarse grid is known as *restriction.* On the other hand, transferring the coarse-grid information into the fine grid is called *prolongation.* In Sections 10.4.2–10.4.4, these essential multigrid elements will be described. Interested readers are referred to [70] for further details.

10.4.2 Coarse-Grid Equations

The system of linear equations can be expressed in the following form:

$$A\phi = b. \tag{10.4.1}$$

Here, ϕ is the exact solution of the problem, A is the coefficient matrix and b is the source term. A is composed of a linear part, $L(\phi)$ (the pressure difference term) and a nonlinear part, $N(\phi)$ (advection–diffusion term). Thus, we can write:

$$N\phi + L\phi = b. \tag{10.4.2}$$

After a few outer iterations (ν_1), the approximate solution ϕ^* satisfies the following equation:

$$N^*\phi^* + L\phi^* = b^* + R^*, \tag{10.4.3}$$

where R^* is the residual. By subtracting (10.4.3) from (10.4.2), we get:

$$N\phi - N^*\phi^* + L\phi' = b - b^* - R^*. \tag{10.4.4}$$

Here, ϕ' is the difference between the exact and approximate solution, $\phi' = \phi - \phi^*$. The coarse grid equation is assembled from Eq. (10.4.4) by restricting the fine grid variables and using them as a fixed source term on the coarse grid:

$$\widehat{N}\widehat{\phi} = \widehat{b} - \widehat{L}(\phi') + \underline{\widetilde{N}\widetilde{\phi} - \widetilde{b} - \widetilde{R}}. \tag{10.4.5}$$

The coarse grid variables are denoted by caret ˆ, and the restricted quantities are denoted by tilde ˜. The underlined terms are kept frozen during the coarse-grid iterations. Thus, Eq. (10.4.5) estimates the difference ($\hat{\phi} - \tilde{\phi}$) that is driven by residual R. This difference ($\hat{\phi} - \tilde{\phi}$) is known as the coarse grid correction.

In view of Eq. (10.4.5), the coarse grid correction equations for x-velocity and y-velocity in collocated-grid configuration are as follows:

$$\widehat{a}_P\widehat{u}_P = \sum_{nb}\widehat{a}_{nb}\widehat{u}_{nb} + \Delta y(p'_w - p'_e) + \widehat{S}_u + \underline{\widetilde{a}_P\widetilde{u}_P - \sum_{nb}\widetilde{a}_{nb}\widetilde{u}_{nb} - \widetilde{S}_u - \widetilde{R}_u} \tag{10.4.6}$$

$$\widehat{a}_P\widehat{v}_P = \sum_{nb}\widehat{a}_{nb}\widehat{v}_{nb} + \Delta x(p'_s - p'_n) + \widehat{S}_v + \underline{\widetilde{a}_P\widetilde{v}_P - \sum_{nb}\widetilde{a}_{nb}\widetilde{v}_{nb} - \widetilde{S}_v - \widetilde{R}_v}. \tag{10.4.7}$$

Denoting the underlined terms with $\widetilde{f}_u$ and $\widetilde{f}_v$, we can compute the face velocities similarly:

$$(\widehat{a}_P)_e\widehat{u}_e = \left(\sum_{nb}\widehat{a}_{nb}\widehat{u}_{nb}\right)_e + (\Delta y)_e\,(p'_P - p'_E) + (\widehat{S}_u)_e + (\widetilde{f}_u)_e \tag{10.4.8}$$

$$(\widehat{a}_P)_n\widehat{v}_n = \left(\sum_{nb}\widehat{a}_{nb}\widehat{v}_{nb}\right)_n + (\Delta x)_n\,(p'_P - p'_N) + (\widehat{S}_v)_n + (\widetilde{f}_v)_n. \tag{10.4.9}$$

The coarse-grid steps for the SIMPLE algorithm can be summarized as follows:

(1) Start with the guess:

$$\widehat{u} = \widetilde{u}, \quad \widehat{v} = \widetilde{v}, \quad p' = 0. \tag{10.4.10}$$

(2) A number of relaxation sweeps on coarse-grid momentum equations are performed to get approximate solutions $\widehat{u}^*$ and $\widehat{v}^*$.

(3) Face velocities are calculated with the modified momentum interpolation method:

$$\widehat{u}_e^* = (1-\alpha_u)\widehat{u}_e^l + \frac{\alpha_u}{(\widehat{a}_P)_e}\left[\left(\sum_{nb}\widehat{a}_{nb}\widehat{u}_{nb}\right)_e + (\Delta y)_e (p'_P - p'_E) + (\widehat{S}_u)_e + (\widetilde{f}_u)_e\right] \tag{10.4.11}$$

$$\widehat{v}_n^* = (1-\alpha_v)\widehat{v}_n^l + \frac{\alpha_v}{(\widehat{a}_P)_n}\left[\left(\sum_{nb}\widehat{a}_{nb}\widehat{v}_{nb}\right)_n + (\Delta x)_n (p'_P - p'_N) + (\widehat{S}_v)_n + (\widetilde{f}_v)_n\right]. \tag{10.4.12}$$

(4) An equation for correction to the pressure-correction is solved:

$$a_P p''_P = \sum_{nb} a_{nb} p''_{nb} + b''_p \tag{10.4.13}$$

$$b''_p = \rho[(\widehat{u}_w^* - \widetilde{u}_w) - (\widehat{u}_e^* - \widetilde{u})](\Delta y) + \rho[(\widehat{v}_s^* - \widetilde{v}_s) - (\widehat{v}_n^* - \widetilde{v}_n)](\Delta x). \tag{10.4.14}$$

(5) Nodal and face velocities are corrected by using p'':

$$\widehat{u}_e = \widehat{u}_e^* + \frac{(\Delta y)}{(\widehat{a}_P)_e}(p''_P - p''_E) \tag{10.4.15}$$

$$\widehat{v}_n = \widehat{v}_n^* + \frac{(\Delta x)}{(\widehat{a}_P)_n}(p''_P - p''_N) \tag{10.4.16}$$

$$\widehat{u}_P = \widehat{u}_P^* + \frac{(\Delta y)}{(\widehat{a}_P)}(p''_w - p''_e) \tag{10.4.17}$$

$$\widehat{v}_P = \widehat{v}_P^* + \frac{(\Delta x)}{(\widehat{a}_P)}(p''_s - p''_n). \tag{10.4.18}$$

Similarly, the pressure-correction variables are updated through:

$$p'_P = p'_P + \alpha_p p''_P. \tag{10.4.19}$$

(6) Return to step (2) to complete one outer relaxation sweep.

The coarse-grid sequence may look similar to those described by Lien and Leschziner [89]. However, there is a subtle difference which lies in the expression of coarse-grid-face velocities. In their coarse-grid formulations, the cell-face velocities become equal to the average of the node velocities when the pressure-correction terms are zero. This is not the case in the present coarse-grid formulation.

10.4.3 Grid-Transfer Operators

In order to achieve the favorable $O(N)$ convergence behavior, as predicted by Brandt [88], a multigrid method should satisfy the *smoothing property* and the *approximation property* [90]. The smoothing property is responsible for the rapid elimination of the oscillatory or high-frequency modes of the error leaving the smooth modes [91]. Most iterative solvers, including the one used here (line-by-line method), exhibit the smoothing property. It should be noted that, although in single-grid methods, this smoothing property can be seen as a serious drawback, for multigrid methods this smoothing property is desirable. The approximation property requires that $m_P + m_R > M$, where m_P and m_R are the order of accuracy plus one for the prolongation and restriction schemes, respectively, and M is the highest derivative of the variable under consideration in the governing equations. In the incompressible Navier–Stokes equations, $M = 1$ for pressure, and $M = 2$ for velocity components. Thus, first order accurate interpolation schemes (e.g., bilinear interpolation for two dimensions and trilinear interpolation for three dimensions) should be sufficient to satisfy the approximation property.

In the collocated grid arrangements, the variables are located at the center of the control volumes. In this type of cell-centered discretization, the coarse grid points do not coincide with the fine-grid points, as shown in Fig. 10.4.1. As a result, special care is needed for grid-transfer operations for restriction and prolongation. A four-point average was used for restriction and a bilinear interpolation was used for prolongation [89, 92]. During the restriction, the coarse-grid variables are calculated as an area-weighted average of the surrounding fine grid variables. Mathematically, the fine to coarse-grid operation can be expressed as:

$$\phi^{2h} = I_h^{2h} \begin{Bmatrix} \phi_1 \\ \phi_2 \\ \phi_3 \\ \phi_4 \end{Bmatrix}^h . \tag{10.4.20}$$

For inner region (excluding the boundary cells), the restriction operator is

$$I_h^{2h} = \tfrac{1}{4}\{1 \ \ 1 \ \ 1 \ \ 1\}. \tag{10.4.21}$$

For west boundary cells we have

$$I_h^{2h} = \tfrac{1}{4}\{2 \ \ 2 \ \ 0 \ \ 0\}. \tag{10.4.22}$$

For north-east corner cells,

$$I_h^{2h} = \tfrac{1}{4}\{4 \ \ 0 \ \ 0 \ \ 0\}. \tag{10.4.23}$$

The residuals are restricted differently from the variables. In the finite volume formulations, the residuals represent flux imbalances through the control volume faces. Therefore, the coarse-grid residual must be computed as the sum of the surrounding fine-grid residuals. The resulting prolongation operator for residuals is:

$$I_h^{2h} = \{1 \ \ 1 \ \ 1 \ \ 1\}. \tag{10.4.24}$$

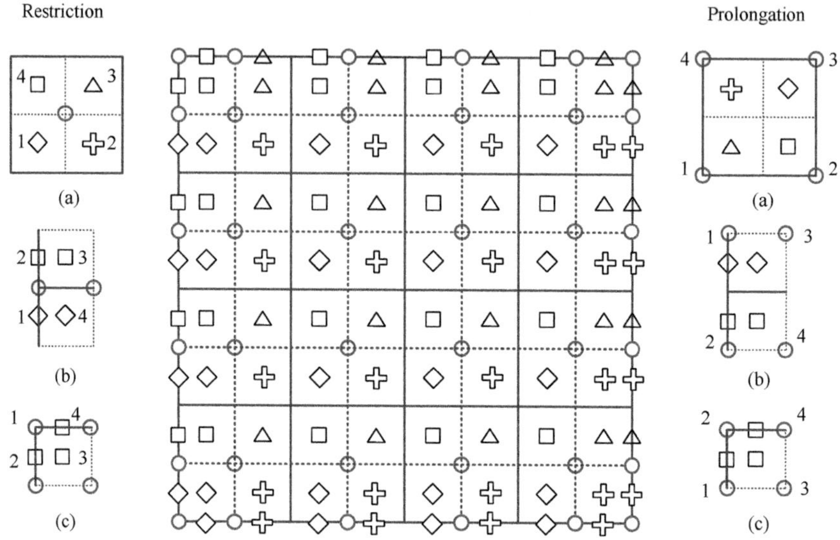

Fig. 10.4.1 Finite volume collocated grid with symbols explaining the grid transfer operators [70]. Circles denote the coarse-grid points and all other symbols denote the fine-grid points. For restriction and prolongation: (a) interior cells, (b) west boundary cells, and (c) north-west cells.

The fine-grid variables are corrected in the following way:

$$\phi_h^{new} = \phi_h^{old} + \alpha_{MG}\delta\phi_h = \phi_h^{old} + \alpha_{MG} I_{2h}^h(\hat{\phi}_{2h} - \widetilde{\phi}_{2h}), \tag{10.4.25}$$

where, α_{MG} is an underrelaxation factor. The coarse-grid to fine-grid operation for prolongation can be expressed as:

$$\begin{Bmatrix} \delta\phi_1 \\ \delta\phi_2 \\ \delta\phi_3 \\ \delta\phi_4 \end{Bmatrix}^h = I_{2h}^h \begin{Bmatrix} \delta\phi_1 \\ \delta\phi_2 \\ \delta\phi_3 \\ \delta\phi_4 \end{Bmatrix}^{2h} . \tag{10.4.26}$$

The prolongation operators are as follows. For the interior region:

$$I_{2h}^h = \frac{1}{16}\begin{bmatrix} 9 & 3 & 3 & 1 \\ 3 & 9 & 1 & 3 \\ 3 & 1 & 9 & 3 \\ 1 & 3 & 3 & 9 \end{bmatrix} . \tag{10.4.27}$$

For west boundary cells and north-west corner cells, we have

$$I_{2h}^h = \frac{1}{8}\begin{bmatrix} 0 & 0 & 6 & 2 \\ 0 & 0 & 2 & 6 \\ 3 & 1 & 3 & 1 \\ 1 & 3 & 1 & 3 \end{bmatrix} , \tag{10.4.28}$$

$$I_{2h}^h = \frac{1}{4}\begin{bmatrix} 2 & 2 & 0 & 0 \\ 0 & 0 & 0 & 4 \\ 1 & 1 & 1 & 1 \\ 0 & 2 & 0 & 2 \end{bmatrix} . \tag{10.4.29}$$

10.4.4 Multigrid Cycles

Once the coarse-grid correction equations are derived and suitable grid transfer operators are chosen, the multigrid algorithm can be applied by a recursive operation of smoothing and inter-grid transfer. This recursive process is commonly known as a multigrid cycle. The most basic cycle is called a V-cycle. In a V-cycle, the fine-grid errors are restricted to the coarse-grid. After a few iterations, namely pre-smoothing operations, the coarse-grid errors are restricted to the next coarser level. This process is continued until the coarsest grid level is reached, where the computational work for solving the residual equation is so cheap that a converged solution to the discretized equation can easily be calculated, either by using a direct solver or an iterative solver. Once the converged solution is found in the coarsest grid the error is prolongated into the next fine grid and the old error on the fine grid is corrected using this prolongated value. After a few post-smoothing operations on the corrected error, it is further prolongated to the next finer level, and the process is continued until the finest grid level is reached. A schematic of the V-cycle is shown in Fig. 10.4.2. The down arrows indicate a restriction, whereas the up arrows indicate a prolongation. The circles at each level imply smoothing operations to get an approximate solution. The square box at the coarsest level signifies a converged solution of the residual equation.

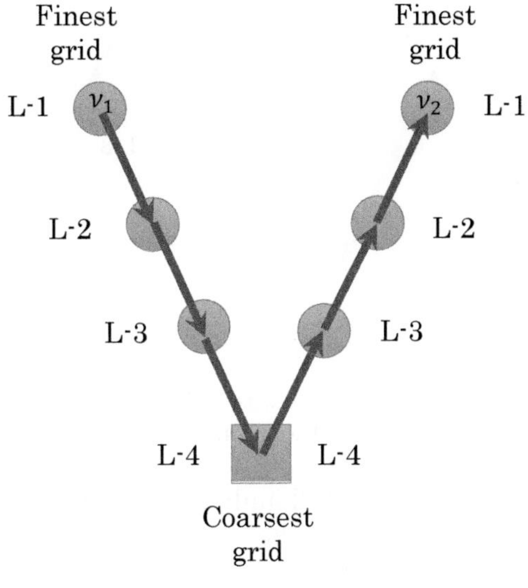

Fig. 10.4.2 Multigrid V cycle with ν_1 pre-smoothing and ν_2 post-smoothing iterations. The circles denote approximate solutions and the squares denote converged solutions; L-1 through L-4 denote multigrid levels.

There are a number of variations to the V-cycle, of which the full multigrid V-cycle is the most popular one [89, 93]. Instead of starting the multigrid V-cycle on a given fine grid, we can first apply it on a very coarse grid, where a few cycles would produce a converged solution. This converged solution on the coarse grid then can be prolongated on the next fine grid to provide a good initial guess to the solution of the next level. If this process is applied consecutively, then we will already have a very good initial guess when the finest grid level is reached. Thus, a faster convergence rate than a V-cycle can be achieved by using the full multigrid V-cycle.

Figure 10.4.3 presents a schematic of the full multi-grid V-cycle. In terms of computational complexity, the multi-grid V-cycle has a cost of $O(N)$, where N is the number of unknown variables in the fine grid system, as described earlier. However, the initial guess for the pressure/velocity variables introduce an error of $O(1)$, and it takes $O(NlogN)$ V-cycle operations to reduce the error below the convergence criteria [94]. Although this is a large reduction in the computational work from $O(N^d)$ operations in a single grid, it does not exactly satisfy the "golden rule" of multigrid methods [88, 95] . Fortunately, this limitation is overcome by the full multigrid V-cycle, which can effectively reduce the number of operations to $O(N)$ by successively introducing good initial guesses on each fine grid.

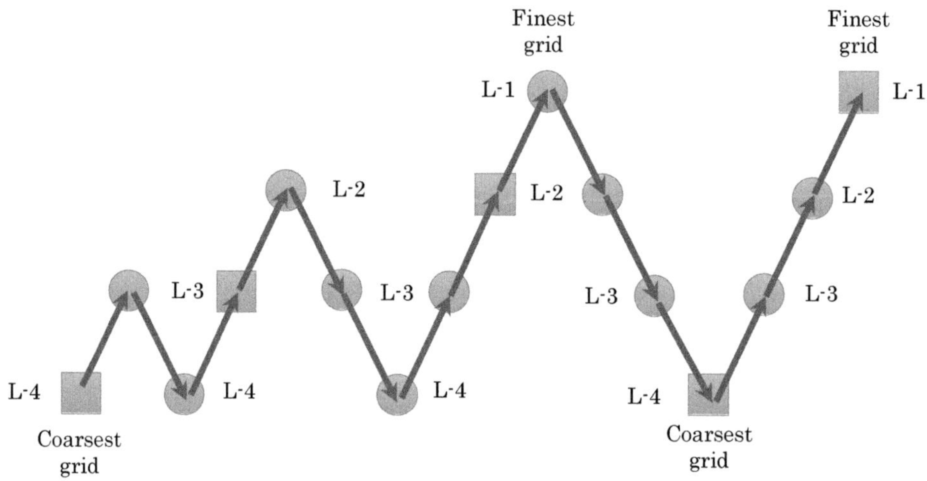

Fig. 10.4.3 Full multigrid (FMG) V-Cycle with 4 levels. The circles denote approximate solutions and the squares denote converged solutions; L-1 through L-4 denote multigrid levels.

As discussed before, during the full multigrid V-cycle, the converged solution in the coarse grid is prolongated to the next fine-grid. The prolongation can be performed in several ways, of which the bilinear interpolation described in Section 10.4.3 and the upwind based interpolation are very common. In our previous work [70], a bilinear interpolation was used for node velocities and pressures. A momentum interpolation was applied to obtain the face velocities. It was observed that the momentum interpolation method results in better initial guesses for the fine-grid iterations than the upwind-based interpolation scheme.

10.5 Summary

This chapter has been dedicated to a brief discussion of some advanced topics. Three topics that were considered in detail are: (1) fully-developed periodic flows, (2) natural convection, which couples heat transfer to fluid flow through the buoyancy effect, and (3) multigrid solution strategies. Mathematical formulation of periodically developed heat transfer under constant wall temperature and constant heat flux boundary conditions were presented. The CTDMA algorithm to solve a system of discretization equations resulting from cyclic boundary conditions was illustrated with a contrived example. The presence of the second source term resulting from variable thermo-physical properties of the fluid was identified. In the case of natural convection, the coupled Navier–Stokes equations and the energy equation were solved to determine the flow fields in the presence of temperature difference. Finally, the ideas behind multigrid solution strategies were discussed.

Owing to the limited scope of the present book, many other advanced topics (e.g., flows of generalized and viscoelastic fluids, electro-magnetics, chemically-reacting flows with heat transfer, and so on) have not been considered here.

References

[1] J. N. Reddy, *An Introduction to the Finite Element Method*, 4th ed. New York, NY: McGraw Hill, 2019.

[2] B. Szabó and I. Babuška, *Introduction to Finite Element Analysis: Formulation, Verification and Validation.* New York, NY: John Wiley & Sons, 2011.

[3] J. N. Reddy and D. K. Gartling, *The Finite Element Method in Heat Transfer and Fluid Dynamics*, 3rd ed. Boca Raton, FL: CRC Press, 2010.

[4] P. M. Gresho and R. L. Sani, *Incompressible Flow and the Finite Element Method.* London, UK: John Wiley & Sons, 1998.

[5] K. S. Surana and J. N. Reddy, *The Finite Element Method for Boundary Value Problems, Mathematics and Computations.* Boca Raton, FL: CRC Press, 2017.

[6] K. S. Surana and J. N. Reddy, *The Finite Element Method for Initial Value Problems, Mathematics and Computations.* Boca Raton, FL: CRC Press, 2018.

[7] S. V. Patankar, *Numerical Heat Transfer and Fluid Flow.* Boca Raton, FL: CRC Press, 1980.

[8] H. K. Versteeg and W. Malalasekera, *Computational Fluid Dynamics, The Finite Volume Method*, 2nd ed. Harlow, England, UK: Pearson Education (Prentice Hall), 2007.

[9] J. H. Ferziger and M. Perić, *Computational Methods for Fluid Dynamics*, 3rd ed. New York, NY: Springer Verlag, 2002.

[10] F. Moukalled, L. Mangani, and M. Darwish, *The Finite Volume Method in Computational Fluid Dynamics.* New York, NY: Springer, 2016.

[11] S. Mazumder, *Numerical Methods for Partial Differential Equations. Finite Difference and Finite Volume Methods.* New York, NY: Elsevier, 2016.

[12] T. J. Chung, *Computational Fluid Dynamics.* New York, NY: Cambridge University Press, 2002.

[13] J. N. Reddy, *Energy Principles and Variational Methods in Applied Mechanics*, 3rd ed. New York, NY: John Wiley & Sons, 2017.

[14] J. N. Reddy, *The Principles of Continuum Mechanics*, 2nd ed. New York, NY: Cambridge University Press, 2018.

[15] J. N. Reddy, *An Introduction to Continuum Mechanics*, 2nd ed. New York, NY: Cambridge University Press, 2013.

[16] B. Carnahan, H. A. Luther, and J. O. Wilkes, *Applied Numerical Methods.* New York, NY, 1969; reprinted by Krieger Publishing, Malabar, FL: John Wiley & Sons, 1990.

[17] S. C. Chapra and R. P. Canale, *Numerical Methods for Engineers*, 8th ed. New York, NY: McGraw Hill, 2021.

[18] E. Isaacson and H. B. Keller, *Analysis of Numerical Methods.* New York: John Wiley & Sons, 1966.

[19] N. Kim, N. K. Anand, and D. L. Rhode, "A study on convergence criteria for a simple-based finite-volume algorithm," *Numerical Heat Transfer, Part B: Fundamentals: An International Journal of Computation and Methodology*, vol. 34, pp. 401–417, 2007.

[20] P. J. Roach, "Perspective: a method for uniform reporting of grid refinement studies," *Journal of Fluids Engineering*, vol. 116, pp. 405–413, 2002.

[21] M. P. McHale, J. R. Friedman, and J. Karian, "Standard for verification and validation in computational fluid dynamics and heat transfer," *The American Society of Mechanical Engineers, ASME V&V*, vol. 20, 2009.

[22] P. J. Roache, *Verification and Validation in Computational Science and Engineering.* Albuquerque, NM: Hermosa Publishers, 1998.

[23] I. Babuška and J. T. Oden, "Verification and validation in computational engineering and science: basic concepts," *Computer Methods in Applied Mechanics and Engineering*, vol. 193, pp. 4057–4066, 2004.

[24] W. L. Oberkampf and C. J. Roy, *Verification and Validation in Scientific Computing.* New York, NY: Cambridge University Press, 2010.

[25] A. Bejan, *Convection Heat Transfer*, 4th ed. New York, NY: John Wiley & Sons, 2013.

[26] A. Bejan, *Advanced Engineering Thermodynamics*, 3rd ed. New York, NY: John Wiley & Sons, 2006.

[27] H. S. Carslaw and J. C. Jaeger, *Conduction of Heat in Solids*, 2nd ed. New York, NY: Oxford University Press, 1959.

[28] J. P. Holman, *Heat Transfer*, 10th ed. New York, NY: McGraw Hill, 2010.

[29] F. Kreith, R. M. Manglik, and M. S. Bohn, *Principles of Heat Transfer*, 7th ed. Stamford, CT: Cengage Learning, 2011.

[30] G. E. Myers, *Analytical Methods in Conduction Heat Transfer.* New York, NY: McGraw Hill, 1971.

[31] M. N. Özisik, *Heat Conduction*, 2nd ed. New York, NY: John Wiley & Sons, 1993.

[32] R. B. Bird, R. C. Armstrong, and O. Hassager, *Dynamics of Polymeric Liquids, Vol. 1: Fluid Mechanics*, 2nd ed. New York, NY: John Wiley & Sons, 1987.

[33] R. B. Bird, W. E. Stewart, and E. N. Lightfoot, *Transport Phenomena*, 2nd ed. New York, NY: John Wiley & Sons, 2007.

[34] B. R. Munson, D. F. Young, T. H. Okiishi, and W. W. Huebsch, *Fundamentals of Fluid Mechanics*, 6th ed. New York, NY: John Wiley & Sons, 2009.

[35] H. Schlichting and K. Gersten, *Boundary Layer Theory*, 9th ed. Berlin: Springer Verlag, 2017.

[36] J. B. Scarborough, *Numerical Mathematical Analysis.* Baltimore, MD: Johns Hopkins Press, 1955.

[37] W. J. Minkowycz, *Handbook of Numerical Heat Transfer.* New York, NY: John Wiley, 1988.

[38] M. P. Reddy and J. N. Reddy, "Multigrid methods to accelerate convergence of element-by-element solution algorithms for viscous incompressible flows," *Computer Methods in Applied Mechanics and Engineering*, vol. 132, no. 3-4, pp. 179–193, 1996.

[39] J. Crank and P. Nicolson, "A practical method for numerical evaluation of solution of partial differential equations of the heat conduction type," *Proceedings of Cambridge Philosophical Society*, vol. 43, pp. 50–67, 1947.

[40] J. N. Reddy, *An Introduction to Nonlinear Finite Element Analysis*, 2nd ed. Oxford, U.K: Oxford University Press, 2015.

[41] J. N. Reddy, "On the finite element method with penalty for incompressible fluid flow problems," in *The Mathematics of Finite Elements and Applications III*, J. R. Whiteman, Ed. New York, NY: Academic Press, 1979, pp. 227–285.

[42] J. N. Reddy, "On penalty function methods in the finite element analysis of flow problems," *International Journal of Numerical Methods in Fluids*, vol. 2, pp. 151–171, 1982.

[43] M. P. Reddy and J. N. Reddy, "Finite-element analysis of flows of non-newtonian fluids in three-dimensional enclosures," *International Journal of Non-Linear Mechanics*, vol. 27, pp. 9–26, 1992.

[44] R. S. Marshall, J. C. Heinrich, and N. O. C. Zienkiewicz, "Convection in a square enclosure by a finite element, penalty function method, using primitive fluid variables," *Numerical Heat Transfer*, vol. 1, pp. 315–330, 1978.

[45] T. J. Hughes, W. K. Liu, and A. Brooks, "Review of finite element analysis of incompressible viscous flows by penalty function formulation," *Journal of Computational Physics*, vol. 30, pp. 1–60, 1979.

[46] J. T. Oden, "RIP-methods for Stokesian flows," *Finite Elements in Fluids*, vol. 4, pp. 305–318, 1982.

[47] B. N. Jiang, *The Least-Squares Finite Element Method. Theory and Applications in Computational Fluid Dynamics and Electromagnetics.* New York, NY: Springer Verlag, 1998.

[48] P. B. Bochev and M. D. Gunzburger, *Least-Squares Finite Element Methods.* New York, NY: Springer Verlag, 2009.

[49] K. S. Surana, A. R. Ahmadi, and J. N. Reddy, "The k-version of finite element method for non-self-adjoint operators in BVP," *International Journal of Computational Engineering Science*, vol. 4, pp. 737–812, 2003.

[50] K. S. Surana, A. R. Ahmadi, and J. N. Reddy, "The k-version of finite element method for nonlinear operators in BVP," *International Journal of Computational Engineering Science*, vol. 5, pp. 133–207, 2004.

[51] K. S. Surana, S. Allu, P. W. TenPas, and J. N. Reddy, "Finite element method in gas dynamics: higher order global differentiability numerical solutions," *International Journal for Numerical Methods in Engineering*, vol. 69, pp. 1109–1157, 2007.

[52] J. P. Pontaza and J. N. Reddy, "Spectral/hp least-squares finite element formulation for the Navier–Stokes equations," *Journal of Computational Physics*, vol. 190, pp. 523–549, 2003.

[53] J. P. Pontaza and J. N. Reddy, "Least-squares finite element formulations for viscous compressible and incompressible fluid flows," *Computer Methods in Applied Mechanics and Engineering*, vol. 195, pp. 2454–2494, 2006.

[54] V. Prabhakar and J. N. Reddy, "Spectral/hp penalty least-squares finite element formulation for the steady incompressible Navier–Stokes equations," *Journal of Computational Physics*, vol. 215, pp. 274–297, 2006.

[55] G. S. Payette and J. N. Reddy, "On the roles of minimization and linearization in least squares finite element models of nonlinear boundary-value problems," *Journal of Computational Physics*, vol. 230, pp. 3589–3613, 2011.

[56] A. Nadai, *Theory of Flow and Fracture of Solids.* New York, NY: McGraw Hill, 1963.

[57] J. F. Thompson, "Grid generation techniques in computational fluid dynamics," *AIAA Journal*, vol. 22, No.11, p. 1505–1523, 1984.

[58] J. F. Thompson, Z. Warsi, and C. W. Mastin, *Numerical Grid Generation: Foundations and Applications.* Amsterdam, The Netherlands: North Holland, 1985, vol. 45.

[59] J. F. Thompson, "Grid generation," in *Handbook of Numerical Heat Transfer*, W. J. Minkowycz, E. M. Sparrow, G. E. Schneider, and R. H. Pletcher, Eds. New York, NY: John Wiley & Sons, 1988.

[60] J. N. Reddy, "A dual mesh finite domain method for the numerical solution of differential equations," *Int. J. Comput. Methods Eng. Sci. Mech.*, vol. 20, no. 3, pp. 212–228, 2019.

[61] J. N. Reddy, N. Kim, and M. Martinez, "A dual mesh control domain method for the solution of nonlinear Poisson's equation and the navier-stokes equations for incompressible fluids," *Physics of Fluids*, vol. 32, no. 9, p. 093608, 2020.

[62] J. N. Reddy, M. Martinez, and P. Nampally, "A novel numerical method for the solution of nonlinear equations with applications to heat transfer," *International Journal of Numerical Methods in Heat Transfer and Fluid Flow*, 2021.

[63] B. D. Spalding, "A novel finite difference formulation for differential expressions involving both first and second derivatives," *International Journal for Numerical Methods in Engineering*, vol. 4, pp. 551–559, 1972.

[64] B. P. Leonard, "A stable and accurate convective modelling procedure based on quadratic upstream interpolation," *Computer Methods in Applied Mechanics and Engineering*, vol. 19, pp. 59–98, 1979.

[65] J. A. Ekaterinaris, "High-order accurate, low numerical diffusion methods for aerodynamics," *Progress in Aerospace Sciences*, vol. 41, no. 3-4, pp. 192–300, 2005.

[66] C. Rhie and W. Chow, "Numerical study of the turbulent flow past an airfoil with trailing edge separation," *AIAA Journal*, vol. 21, pp. 1525–1532, 1983.

[67] S. V. Patankar and D. B. Spalding, "A calculation procedure for heat, mass and momentum transfer in three-dimensional parabolic flows," *International Journal of Heat and Mass Transfer*, vol. 15, pp. 1787–1806, 1972.

[68] J. P. Van Doormaal and G. D. Raithby, "Enhancements of the simple method for predicting incompressible fluid flows," *Numerical Heat Transfer*, vol. 7, no. 2, pp. 147–163, 1984.

[69] F. H. Harlow and J. E. Welch, "Numerical calculation of time-dependent viscous incompressible flow of fluid with free surface," *Physics of Fluids*, vol. 8, No. 12, 2182, 1965.

[70] P. Roy, N. K. Anand, and D. Donzis, "A parallel multigrid finite-volume solver on a collocated grid for incompressible Navier–Stokes equations," *Numerical Heat Transfer, Part B: Fundamentals*, vol. 67, pp. 376–409, 2015.

[71] I. Demirdžić, Ž. Lilek, and M. Perić, "A collocated finite volume method for predicting flows at all speeds," *International Journal for Numerical Methods in Fluids*, vol. 16, no. 12, pp. 1029–1050, 1993.

[72] F.-S. Lien, "A pressure-based unstructured grid method for all-speed flows," *International Journal for Numerical Methods in Fluids*, vol. 33, pp. 355–374, 2000.

[73] S. K. Choi, "Note on the use of momentum interpolation method for unsteady flows," *Numerical Heat Transfer: Part A: Applications*, vol. 36, pp. 545–550, 1999.

[74] M. Ijaz and N. K. Anand, "Co-located variables approach using implicit runge-kutta methods for unsteady incompressible flow simulation," *Numerical Heat Transfer, Part B: Fundamentals*, vol. 54, pp. 291–313, 2008.

[75] C. M. Rhie and W. L. Chow, "Numerical study of the turbulent flow past an airfoil with trailing edge separation," *AIAA Journal*, vol. 21, pp. 1525–1532, 1983.

[76] C. Y. Gu, "Computation of flows with large body forces," *Numerical Methods in Laminar and Turbulent Flow*, vol. 7, no. part 2, pp. 1568–1578, 1991.

[77] S. Majumdar, "Role of underrelaxation in momentum interpolation for calculation of flow with nonstaggered grids," *Numerical Heat Transfer*, vol. 13, pp. 125–132, 1988.

[78] U. Ghia, K. N. Ghia, and C. T. Shin, "High-resolution for incompressible flow using the Navier–Stokes equations and the multigrid method," *Journal of Computational Physics*, vol. 48, pp. 387–411, 1982.

[79] R. Shah and A. London, *Laminar Flow Forced Convection in Ducts: a Source Book for Compact Heat Exchanger Analytical Data.* New York, NY: Academic Press, 2014.

[80] W. Rohsenow, J. Hartnett, Y. Cho, and et al., *Handbook of Heat Transfer.* New York, NY: McGraw Hill, 1998.

[81] H. S. Tang, R. D. Haynes, and G. Houzeaux, "A review of domain decomposition methods for simulation of fluid flows: Concepts, algorithms, and applications," *Archives of Computational Methods in Engineering*, vol. 28, pp. 841–873, 2021.

[82] R. J. Goldstein, V. L. Eriksen, R. M. Olson, and E. R. Eckert, "Laminar separation, reattachment, and transition of the flow over a downstream-facing step," *Journal of Fluids Engineering*, vol. 92, pp. 732–741, 1970.

[83] A. Runchal, *50 Years of CFD in Engineering Sciences: A Commemorative Volume in Memory of D. Brian Spalding.* Nature: Springer, 2020.

[84] S. H. Kim and N. K. Anand, "Periodically fully developed flow in channels with conducting blockages," *Journal of Thermophysics and Heat Transfer*, vol. 6, pp. 91–97, 1992.

[85] S. V. Patankar, C. H. Liu, and E. M. Sparrow, "Fully developed flow and heat transfer in ducts having streamwise-periodic variations of cross-sectional area," *Journal of Heat Transfer*, vol. 99, no. 2, pp. 180–186, 1977.

[86] T. L. Fullerton and N. K. Anand, "An alternative approach to study periodically fully-developed flow and heat transfer problems subject to isothermal heating conditions," *International Journal of Engineering Science*, vol. 48, pp. 1253–1262, 2010.

[87] G. de Vahl Davis, "Natural convection of air in a square cavity: a bench mark numerical solution," *International Journal for Numerical Methods in Fluids*, vol. 3, no. 3, pp. 249–264, 1983.

[88] A. Brandt, "Multi-level adaptive solutions to boundary-value problems," *Mathematics of Computation*, vol. 31, No. 138, pp. 333–390, 1977.

[89] F. S. Lien and M. A. Leschziner, "Multigrid acceleration for recirculating laminar and turbulent flows computed with a non-orthogonal, collocated finite-volume scheme," *Computer Methods in Applied Mechanics and Engineering*, vol. 118, pp. 351–371, 1994.

[90] W. Hackbusch, *Multi-grid Methods and Applications, vol. 4.* Berlin: Springer Verlag, 1985.

[91] W. L. Briggs, V. E. Henson, and S. F. McCormick, *A Multigrid Tutorial.* SIAM, 2000.

[92] U. Trottenberg, C. W. Oosterlee, and A. Schuller, *Multigrid.* New York, NY:Academic Press, 2000.

[93] M. Hortmann, M. Perić, and G. Scheuerer, "Finite volume multigrid prediction of laminar natural convection: bench-mark solutions," *International Journal for Numerical Methods in Fluids*, vol. 11, pp. 189–207, 1990.

[94] J. E. Jones and S. F. McCormick, "Parallel multigrid methods," in *Parallel Numerical Algorithms, vol. 4.* Vol 4. Springer, Dordrecht: ICASE/LaRC Interdisciplinary Series in Science and Engineering, 1997, pp. 203–224.

[95] A. Brandt and O. E. Livne, *Multigrid Techniques: 1984 Guide with Applications to Fluid Dynamics.* SIAM, Philadelphia, PA, 2011, vol. 67.

Index